W9-ABL-359

PROCEEDINGS OF SYMPOSIA
IN PURE MATHEMATICS
Volume XXX, Part 1

SEVERAL COMPLEX VARIABLES

AMERICAN MATHEMATICAL SOCIETY
PROVIDENCE, RHODE ISLAND
1977

PROCEEDINGS OF THE SYMPOSIUM IN PURE MATHEMATICS
OF THE AMERICAN MATHEMATICAL SOCIETY

HELD AT WILLIAMS COLLEGE
WILLIAMSTOWN, MASSACHUSETTS
JULY 28 – AUGUST 15, 1975

EDITED BY
R. O. WELLS, JR.

Prepared by the American Mathematical Society
with partial support from National Science Foundation grant MPS 75–01436

Library of Congress Cataloging in Publication Data

Symposium in Pure Mathematics, Williams College, 1975.
Several complex variables.

(Proceedings of symposia in pure mathematics ; v. 30, pt. 1-2)
"Twenty-third Summer Research Institute."
Includes bibliographies and index.
1. Functions of several complex variables--Congresses.
I. Wells, Raymond O'Neil, 1940- II. American Mathematical Society. III. Title. IV. Series.
QA331.S937 1975 515'.94 77-23168
ISBN 0-8218-0249-6 (v. 1)
ISBN 0-8218-0250-X (v. 2)

AMS (MOS) subject classifications (1970). Primary 32–XX.
Copyright © 1977 by the American Mathematical Society
Printed by the United States of America
All rights reserved except those granted to the United States Government.
This book may not be reproduced in any form without the permission of the publishers.

1568769

Math Sci Sep

CONTENTS

Principal Lecture 1

Function Theory and Real Analysis

Principal Lecture 2

Compact Complex Manifolds

Principal Lecture 3

These proceedings are dedicated to

Lillian Casey

who was the guiding force
behind six years of AMS meetings

PREFACE

The American Mathematical Society held its twenty-third Summer Research Institute at Williams College, Williamstown, Massachusetts, from July 28 to August 15, 1975. Several Complex Variables was selected as the topic for the institute. Members of the Committee on Summer Institutes at the time were Louis Auslander (chairman), Richard E. Bellman, S. S. Chern, Richard K. Lashof, Walter Rudin, and John T. Tate. The institute was supported by a grant from the National Science Foundation.

ORGANIZATION. The Organizing Committee for the institute consisted of Ian Craw, Hans Grauert, Robert C. Gunning (cochairman), David Lieberman, James Morrow, R. Narasimhan, Hugo Rossi (cochairman), Yum-Tong Siu, and R. O. Wells, Jr. (Editor of these PROCEEDINGS).

PROGRAM. The topic of the 1975 summer institute was the theory of functions of several complex variables. The emphasis in arranging the program was on the more analytical aspects of that subject, with particular attention to the relations between complex analysis and partial differential equations, to the properties of pseudoconvexity and of Stein manifolds, and the relations between currents and analytic varieties. However, there were also lectures and seminars on other aspects of that broad and active field of investigation, such as deformation theory, singularities of analytic spaces, value distribution theory, compact complex manifolds, and approximation theory.

There were six series of invited expository lectures, as well as twenty-two hour lectures of a general or survey nature; there were also eight series of seminars on current developments in the subject, six of which were planned and partially arranged in advance.

PRINCIPAL LECTURE SERIES.
Chern-Moser invariants by Daniel M. Burns, Jr. and Steven Shnider (2 lectures);
Power series methods in deformation theorems by Otto Forster (5 lectures);
Holomorphic chains and their boundaries by Reese Harvey (4 lectures);

Methods of PDE in complex analysis by Joseph J. Kohn (5 lectures);
Tangential Cauchy-Riemann equations by Masatake Kuranishi (3 lectures);
Analysis on noncompact Kähler manifolds by Hung-Hsi Wu and Robert E. Greene (4 lectures).

HOUR LECTURES.

Theta functions with characteristic and distinguished subspaces of the Heisenberg manifold by Louis Auslander;
Hardy spaces and local estimates on boundaries of strongly pseudo-convex domains. I by Ronald Coifman;
Some recent developments in the theory of the Bergman kernel function. A survey by Klas Diederich;
Fibering of residual currents by Miguel Herrera;
The Monge-Ampere equation and complex analysis by Norbert Kerzman;
Stable homology and positive currents on abelian varieties by H. Blaine Lawson, Jr.;
A method of inverse functions for plurisubharmonic functions; applications to positive and closed currents by Pierre Lelong;
Holomorphic vector fields on projective varieties by David Lieberman;
Diffusion estimates in complex analysis by Paul Malliavin;
Rational singularities, Moisheson spaces, and compactifications of $\boldsymbol{C}^3$ by James A. Morrow;
Rankin-Selberg method in the theory of automorphic forms by Mark Novodvorsky;
Hardy spaces and local estimates on boundaries of strongly pseudo-convex domains. II by Richard Rochberg;
Analysis of the boundary Laplacian and related hypoelliptic differential operators by Linda Rothschild;
Classification of quasi-symmetric domains by Ichiro Satake;
The Levi problem by Yum-Tong Siu;
Boundary values of the solutions of the $\bar{\partial}$ *equation and characterization of zeros of functions in the Nevanlinna class* by Henri Skoda;
Value distribution on parabolic spaces by Wilhelm Stoll;
Extending functions from submanifolds of the boundary by Edgar L. Stout;
Equisingularity and local analytic geometry by J. Stutz;
Constructability by Jean-Louis Verdier;
Boundary values of holomorphic functions on a Siegel domain and Cauchy-Riemann tangential equations by Michelle Vergne;
Deformations of strongly pseudoconvex domains in $\boldsymbol{C}^2$ by R. O. Wells, Jr.

SEMINAR SERIES.

(1) Singularities of analytic spaces (Egbert Brieskorn, chairman);
(2) Function theory and real analysis (Stephen Greenfield, chairman);
(3) q-convexity and noncompact manifolds (Yum-Tong Siu, chairman);
(4) Value distribution theory in several variables (Wilhelm Stoll, chairman);
(5) Compact complex manifolds (A. Van de Ven, chairman);
(6) Problems in approximation (John Wermer, chairman);
(7) Group representations and harmonic analysis (Kenneth I. Gross, chairman);
(8) Differential geometry and complex variables (Peter Gilkey, chairman).

SUMMARY. The broad areas emphasized by this summer institute can be summarized as follows:

The $\bar{\partial}$-equation. The relation between complex analysis and partial differential equations centers about the Cauchy-Riemann or $\bar{\partial}$-equations. Kohn gave a series of lectures which provided a survey of the role of the $\bar{\partial}$-equations. Several lectures and seminar talks were devoted to the regularity up to the boundary of solutions of the $\bar{\partial}$-equation, with applications to complex analysis. The structure imposed by the $\bar{\partial}$-equation on the smooth boundary of a region in $\boldsymbol{C}^n$ was discussed in several seminars as well. Most lectures on these topics were in the seminars led by S. Greenfield and Y.-T. Siu.

Holomorphic chains. Reese Harvey gave a survey of the characterization of holomorphic chains and their boundaries among other currents, with a number of applications to classical problems in complex analysis; other lectures discussed recent research on holomorphic chains, which includes, for instance, value distribution theory.

Differential geometry. R. E. Greene and H. Wu gave a survey of the characterization of Stein manifolds by curvature conditions on an underlying Kähler metric. A seminar was organized by P. Gilkey on differential geometry and complex analysis in an extension of this. The survey lectures by D. M. Burns, and S. Shnider discussed recent developments in the theory of higher order invariants associated to real hypersurfaces in $\boldsymbol{C}^n$.

Singularities. A seminar led by E. Brieskorn was devoted to the singularities of analytic spaces, with an emphasis on the complex analytic properties rather than the algebraic properties treated in the preceding summer institute on algebraic geometry.

Value distribution theory. Wilhelm Stoll lectured on general value distribution theory and led a seminar in which a number of recent developments in this area were discussed.

Compact complex manifolds. A number of analytic results on compact manifolds were discussed in a seminar led by A. Van de Ven.

Approximation theory. While no attempt was made to cover recent developments in complex analysis and function algebras, there was some discussion of approximation theory in a seminar led by J. Wermer.

Harmonic analysis. Symmetric spaces and boundaries of homogeneous domains have been much investigated in complex analysis and lead quite naturally to recent work on group representation and harmonic analysis, some of which was discussed in a seminar led by K. Gross.

PROCEEDINGS. The proceedings of the 1975 summer institute are published here in two volumes. The hour lectures and seminar papers accepted for publication appear in the seminar series most appropriate to the subject matter of the given paper. These are principally research reports describing current research of the authors, while some are of a general expository nature in a given area. The principal lecture series are represented by six survey articles which have been interlaced in these volumes with the seminar series, with an attempt being made for some relationship between the seminar series and the survey articles they juxtapose.

Volume 1: *Seminar Series*: Singularities of Analytic Spaces
Principal Lecture 1: M. Kuranishi

Seminar Series: Function Theory and Real Analysis
Principal Lecture 2: J. J. Kohn
Seminar Series: Compact Complex Manifolds
Principal Lecture 3: Reese Harvey

Volume 2: *Seminar Series*: Noncompact Complex Manifolds
Principal Lecture 4: R. Greene and H. Wu
Seminar Series: Differential Geometry and Complex Analysis
Principal Lecture 5: D. Burns, Jr., and S. Shnider
Seminar Series: Problems in Approximation
Principal Lecture 6: O. Forster
Seminar Series: Value Distribution Theory
Seminar Series: Group Representation and Harmonic Analysis

The detailed list of authors and titles of papers is given in the table of contents for each volume. At the conclusion of each volume is an author index for authors of articles, as well as authors of papers cited in the bibliographies for each particular part.

R. O. WELLS, JR.
Houston, Texas
May 1976

SEMINAR SERIES

SINGULARITIES OF ANALYTIC SPACES

Proceedings of Symposia in Pure Mathematics
Volume 30, 1977

HIGHER DERIVATIONS AND THE ZARISKI-LIPMAN CONJECTURE

JOSEPH BECKER

1. A well-known conjecture in algebraic geometry is that if R is the affine ring of an algebraic variety V over a field k of characteristic zero, and $\mathrm{Der}_k^1(R)$ the module of derivations of R over k is free, then R is a regular ring, e.g., the embedding dimension equals the Krull dimension. In this note we show how this conjecture is related to some questions about higher derivations that have been studied recently by several authors.

First we review the literature on the above conjecture. Lipman [**14**] has proven that $\mathrm{Der}_k^1(R)$ free implies that R is normal, e.g., integrally closed in its quotient field. Scheja and Storch [**23**] have proven R is regular under the additional hypothesis that R has embedding codim $\leqq 1$, e.g., the variety is a hypersurface. Moen [**16**] and Hochster [**10**] have proven the same under the additional hypothesis that R is a homogenous complete intersection. Recently [**11**] Hochster has improved this to require only that R be a quasi-homogenous domain.

Assuming V is normal, so that everything extends across the supposed singular locus, the conjecture has a nice geometric statement for analytic varieties: If the tangent bundle to the regular points of V is holomorphically trivial, then V is actually nonsingular. It is not enough just to require that the tangent bundle be topologically trivial, because Brieskorn's example [**8**] of a locally contractible hypersurface singularity has topologically trivial tangent bundle, but its module of tangent vector fields must be nonfree by the above result of Scheja and Storch. It is interesting to note that the above discussion does not contradict the Grauert equivalence of holomorphic and continuous vector bundles because, for a singular variety, Reg V is not Stein. Unfortunately, all progress on this problem to date has been by algebraic methods.

It is elementary (in terms of well-established machinery in commutative algebra) to prove the following [**3**], [**20**]:

AMS (MOS) subject classifications (1970). Primary 14B05, 13B25, 13D99.

© 1977, American Mathematical Society

(a) There is an R-module $\Omega_k^1(R)$ of differentials of R over k such that $\mathrm{Hom}_R(\Omega_R^1(R), R) = \mathrm{Der}_k^1(R)$; just let $\Omega_k^1(R)$ be the free R-module on the symbols dr, $r \in R$, modulo the submodule generated by the relations $d(r+s) = dr + ds$, $d(rs) = rds + sdr$, and $dl = 0$ for $l \in k$, $r, s \in R$. The embedding dim of R is the rank of $\Omega_k^1(R)$ as an R-module.

(b) For prime ideal p in R, $\Omega_k^1(R)_p = \Omega_k^1(R) \otimes_R R_p$ and R_p is flat over R since localization is an exact functor. By Noetherianness of R, $\Omega_k^1(R)$ has a finite presentation; so by tensoring and Homing in the opposite orders and comparing the resulting exact sequences we see that $\mathrm{Der}_k^1(R_p) = \mathrm{Der}_k^1(R)_p$. Hence it suffices to consider only local rings, to prove the conjecture.

(c) The algebraic closure $\bar{k}$ of k is faithfully flat over k, so $\mathrm{Der}_k(R) \otimes_k \bar{k} = \mathrm{Der}_{\bar{k}}(R \otimes_k \bar{k})$. So it suffices to assume k is algebraically closed.

(d) The analogous conjecture for commutative local rings R over an algebraically closed field of characteristic zero can easily be seen to be a first-order theorem in logic (this means the hypothesis and conclusion can be stated in finitely many symbols) so by Tarski's theorem [**26**] on the elimination of quantifiers, it is true in all algebraically closed fields if it is true in one algebraically closed field. Hence we may take the field to be the complex numbers $\boldsymbol{C}$.

(e) If m is the maximal ideal of R and $\hat{R}$ the completion of R in its m-adic topology then $\Omega_k^1(R)^\wedge = \Omega_k^1(\hat{R})/\bigcap_{i=1}^{\infty} m^i\Omega_k^1(R)$, $\mathrm{Der}_k^1(\hat{R}) = \mathrm{Der}_k^1(R)^\wedge$ and $\hat{R}$ is faithfully flat over R so it suffices to consider only complete local rings, that is formal power series rings modulo an ideal (by Cohen's theorem).

(f) If $n =$ Krull dim of $R \equiv$ geometric dim of V, then $\Omega_C^1(R)$ is free if and only if it is generated over R by n elements.

In the following we will work in the category of complex analytic varieties although, as indicated above, we might just as well work over complete local rings. We let $\mathcal{O}$ denote the ring of germs at the origin of convergent power series in n variables. If V is a complex analytic variety with local ring at a point $R = \mathcal{O}/I$, then a derivation of R is a holomorphic tangent vector field

$$D = \sum_{i=1}^{n} a_i(z) \frac{\partial}{\partial z_i}, \qquad a_i \in \mathcal{O}/I,$$

such that $D(I) \subset I$. By analogy we introduce the higher order differential operators

$$\mathrm{Der}^k(V) = \left\{D = \sum_{|\alpha| \leqq k} a_\alpha(z) \frac{\partial}{\partial z^\alpha} : D(I) \subset I, a_\alpha \in \mathcal{O}/I\right\},$$

$$\mathrm{Der}\,(V) = \bigcup_{k=1}^{\infty} \mathrm{Der}^k(V).$$

If V is a manifold, it is easily seen that this is just the usual definition of differential operator. Also we introduce the constant coefficient differential operators

$$C(V) = \left\{D = \sum^{\text{finite}} c_\alpha \frac{\partial}{\partial z^\alpha} : D(f)(0) = 0 \text{ for all } f \in I\right\}.$$

Clearly we have a map $\rho : \mathrm{Der}\,(V) \to C(V)$ given by $\rho(\sum a_\alpha(z)\partial^\alpha) = \sum a_\alpha(0)\partial^\alpha$. It is trivial to check all of the following:

(i) $\mathrm{Der}\,(V)$ is a noncommutative ring under composition and a module over $\mathcal{O}(V) = \mathcal{O}/I$ (note $\partial_t(a(t)\partial_t) - (a(t)\partial_t)\,\partial_t = a'(t)\partial_t$), and the associated graded ring

Gr Der(V) $= \bigoplus_{k=0}^{\infty} \mathrm{Der}^k(V)/\mathrm{Der}^{k-1}(V)$ is a commutative ring and a module over $\mathcal{O}(V)$.

(ii) Not every constant coefficient differential operator (ccdo) is a variable coefficient differential operator (vcdo): Consider the example $\partial/\partial x$ on $x^2 - y^3$ in $\boldsymbol{C}^2$.

(iii) $D = \sum_{|\alpha| \leqq k} a_\alpha \partial^\alpha \in \mathrm{Der}^k(V)$ if and only if $D(z^\beta f) \in I$ for all $f \in I$ and all $|\beta| \leqq k - 1$. As a consequence the differential operators of degree $\leqq k$ form a coherent sheaf of $\mathcal{O}$-modules.

(iv) If V is irreducible, then Der(V) is a domain: This is not at all clear from the ring point of view, but is clearly true at a regular point by looking at the symbol. So this claim just follows from (iii) above, and the obvious fact that the above sheafs have no embedded components in their primary decomposition.

(v) If $V = \bigcup_{i=1}^{n} V_i$ is the decomposition of V into irreducible components and $D \in$ Der (V), then D restricts to be an element **[22]** of each Der (V_i).

(vi) If $\phi : V \to W$ is a holomorphic mapping, it induces a pull-back of differential operators; $\phi^*(\mathrm{Der}\,(W))$ is just the meromorphic differential operators (means the coefficients are meromorphic) on V which preserve the subring $\phi^*(\mathcal{O}(W))$. If codim Sg $V \geqq 2$, these operators have holomorphic coefficients. For additional details of this construction, see **[24]**.

We now put **[5]** a topological structure on $\mathcal{O}(V)$ which shows why the ccdo are important. Let $\hat{\mathcal{O}}(V)$ be the completion of $\mathcal{O}$ in its m-adic topology. The simple topology on $\hat{\mathcal{O}}$ as a Frèchet space given by the seminorms $\rho_\beta(\sum a_\alpha z^\alpha) = a_\beta$ is not as fine as the Krull topology determined by the metric $\|f - g\| = e^{-\mathrm{ord}(f-g)}$ and so has fewer continuous functions (but does have the same linear continuous functions). It is well known **[27]** that the vector space continuous dual to the formal power series is the polynomials: Every polynomial $\sum_{|\alpha| \leqq k} c_\alpha y^\alpha$ induces a map: $\hat{\mathcal{O}} \to \boldsymbol{C}$ by $f \to \sum_{|\alpha| \leqq k} c_\alpha(\partial f/\partial x^\alpha)(0)$. On the other hand for any continuous linear L: $\hat{\mathcal{O}} \to \boldsymbol{C}$, $L = \sum (L(x^\alpha)/\alpha!)\,\partial/\partial x^\alpha$, and the sum is finite or else $L(\sum x^\alpha/L(x^\alpha)) = 1 + 1 + \cdots = \infty$. ($L$ commutes with infinite sums because it is continuous.) We will denote $C(V) = \hat{\mathcal{O}}(V)^*$.

Two questions frequently occurring in the literature are:

(I) Is the ring of constant coefficient operators representable? That is, is ρ onto? For example, is every constant coefficient operator at p the specialization at p of a section of the sheaf of variable coefficient operators?

(II) Is the ring of differential operators at a point finitely generated? That is, does there exist $k > 0$ so that the set of kth order operators $\mathrm{Der}^k(V)$ generates Der (V) as an algebra over $\mathcal{O}(V)$? Is the associated graded ring finitely generated?

Neither is true in general, but both have been shown to hold for irreducible curves. Clearly Der (V) finitely generated implies Gr Der (V) finitely generated but not conversely.

Remark. The set of $\boldsymbol{C}$ linear continuous (Krull topology) functions from $\hat{\mathcal{O}}$ to $\hat{\mathcal{O}}$ is precisely the set of infinite length differential operators $\sum a_\alpha(z)\,\partial/\partial z^\alpha$ such that ord $a_\alpha \to \infty$ as $|\alpha| \to \infty$. However there exists **[9]** an $\mathcal{O}(V)$-module M such that $\mathrm{Hom}_{\mathcal{O}}(M, \mathcal{O}) = \mathrm{Der}(V)$ and $\mathrm{Hom}_{\boldsymbol{C}}(M, \boldsymbol{C}) = C(V)$. There is a commutative diagram

$$\begin{array}{ccc} & \nearrow & \mathcal{O} \\ M & & \downarrow \\ & \searrow & \boldsymbol{C} \end{array}$$

Hence the name representable stems from the fact that the functor $C(V)$ is represented as the specialization of the functor $\mathrm{Der}(V)$.

EXAMPLE. The curve $x^2 = y^3$ has normalization $t \to (t^3, t^2)$ so the local ring is isomorphic to $\boldsymbol{C}\{t^3, t^2\}$. Then ∂_t is a ccdo but not a vcdo. It is represented by the vcdo $\partial_t - t\partial_t^2$.

The history and present state of the art on these questions is:

(A) If V is reducible at p, then V is not representable at p.

PROOF. Let $V = V_1 \cup V_2$, $F \in I(V_1) - I(V_2)$. Since ideals in a Noetherian ring are closed in the Krull topology, by the Hahn-Banach theorem for Frèchet spaces there exists $D \in C(V)_p$ so that $D(F)_p = 1$, $DI(V_2) = 0$. If D has a representative $\tilde{D} \in \mathrm{Der}(V)$, $1 = D(F)_p = \tilde{D}(F)_p = \lim_{q \in V_2} \tilde{D}(F)_q = \lim_{q \in V_2} 0 = 0$.

(B) Bloom [**5**] has proven that if V is an irreducible curve, then V is representable. In fact every ccdo of degree d can be represented by a vcdo of degree $= \max(d, e)$, where e is the exponent of the conductor. ($\mathcal{O}(V)$ is a subring of $\boldsymbol{C}\{t\}$ and e is the smallest integer such that $t^n \in \mathcal{O}(V)$ for all $n \geqq e$.) He also gives an example ($x^2 = yz^2$) of a variety which is not representable at an irreducible point which is the limit of reducible points. It seems likely that this should always be the case.

(C) Stutz [**24**] extended Bloom's result to varieties having a Puiseux series normalization, a condition [**25**] which holds on an arbitrary variety in the complement of a subvariety of codim 2.

(D) M. Jaffee [**12**] has completely determined all the differential operators on the curve $x^p = y^q$ by studying the double grading on Der given by degree and strength, where

$$\mathrm{strength}\left(\sum_{\alpha, \beta} a_{\alpha\beta} x^\alpha \frac{\partial}{\partial y^\beta}\right) = \inf(|\alpha| - |\beta|) \quad \text{over all nonzero } a_{\beta\alpha}.$$

Note str $\delta = s$ implies $\delta(m^k) \subset m^{k+s}$, where m is the maximal ideal of the local ring.

(E) Using Jaffee's ideas, Bloom [**6**] and Vigué [**28**] independently proved that, for irreducible curves, Gr Der (V) is finitely generated by operators of degree $\leqq 2e + 1$. Vigué proved and Bloom showed by example that, if codim Sg $V \geqq 2$ and V has nonsingular normalization, then Gr Der (V) is not finitely generated. In addition Vigué gave such an example with $C(V)$ representable and Der (V) finitely generated.

(F) Kantor [**13**] has shown that if V is the quotient of $\boldsymbol{C}^n$ by a finite group of automorphisms of $\boldsymbol{C}^n$ then Gr Der (V) is finitely generated. It is interesting to note that, for $n = 2$, Brieskorn [**7**] has shown that all such quotients have only curves of germs zero in the exceptional sets of their minimal resolutions.

(G) Bernstein, Gel'fand, and Gel'fand [**4**] have shown that, for the cubic cone $x^3 + y^3 + z^3 = 0$ in $\boldsymbol{C}^3$, Der (V) is not finitely generated and that every $D \in$ Der (V) has nonnegative strength. This also gives an everywhere locally irreducible example with $C(V)$ not representable because $\partial/\partial x \in C(V)$ and has strength $+1$. This paper is published in Russian; see [**29**] for a French translation. The basic technique is the following: Since $\mathcal{O}(V)$ is a homogeneous ring, the differential operators can be decomposed into sums of operators of homogenous degree and strength. Such operators may be considered to be acting on the dehomogenization of the cone to the projective curve $x^3 + y^3 + 1 = 0$, which has genus one. The Riemann-Roch theorem gives us information about the differentials on the curve which can then be used to prove the result.

This suggests lifting the general problem to the resolution of the variety, in particular, for 2 dim normal singularities. The exceptional set is closely related to the projectivized tangent cone since the fiber over a point blow-up is just the tangent cone.

(H) A natural idea is to study the structure of the set of points where (I) and (II) are false. An easy exercise in elimination theory shows **[2]** that the set of nonrepresentable points and the set of points where Der (V) and Gr Der (V) are not finitely generated are each the countable union of analytic subvarieties of V.

(I) I conjecture all of the following: Gr Der V finitely generated implies V is irreducible; Gr Der V finitely generated implies V is representable; V representable implies Der V is finitely generated. If V is a normal 2 dim singularity with the exceptional set of its minimal resolution containing a curve of genus one, then Der (V) is not finitely generated. The above sets in (H) should be analytic or at worst constructible.

(J) The Nakai conjecture **[19]** is that if $\mathrm{Der}^1(R)$ generates $\mathrm{Der}^n(R)$ over R for all $n \geqq 0$, then R is a regular ring. Mount and Villamayor **[18]** have proven this when R is the affine ring of an irreducible algebraic curve over a field of char 0.

(K) If $\mathrm{Der}^1(R)$ is a free R-module, then it follows that $\mathrm{Der}^1(R)$ generates Gr Der (R) over R. (A sketch of the proof is given at the end of this section.) Hence the validity of the Nakai conjecture would imply the Zariski-Lipman conjecture.

We now show how the Nakai conjecture would follow from the second conjecture in part I above.

PROPOSITION. [Gr Der V fin gen *over* $\mathcal{O}(V) \Rightarrow C(V)$ rep] $\Rightarrow$ [$\mathrm{Der}^1(V)$ gen Der (V) *over* $\mathcal{O}(V) \Rightarrow V$ *nonsingular*].

PROOF. Recall the Zariski fundamental lemma on derivations which states that if δ is a derivation of R, $x \in m$, and $\delta(x) = 1$, then R is an analytic product; there is a subring R' of $\hat{R}$ and indeterminant t over R' in R such that $\hat{R} = R'[[t]]$. This is known to analytic geometers as Rossi's lemma and is stated as: If there is a nonvanishing holomorphic tangent vector field on V, then there is a variety V' so that V is biholomorphically equivalent to $V' \times \boldsymbol{C}$. An elementary computation shows Der $(V' \times \boldsymbol{C}) =$ Der $(V')\,[z_n, \partial/\partial z_n]$ and hence all of the conditions in the statement of the proposition hold for V if and only if they hold for V'. Hence one may assume that V is singular and $\delta \in \mathrm{Der}^1(V)$, $x \in m \Rightarrow \delta(x) \in m$, that is str $\delta \geqq 0$. Since strength is a valuation and $\mathrm{Der}^1(V)$ generates Der (V), it follows that str $D \geqq 0$ for all $D \in$ Der (V). Hence ρ: Der $(V) \rightarrow C(V)$ is the zero map. But ρ is surjective and $C(V)$ is nonzero by the Hahn-Banach theorem so we have a contradiction with the assumption that V is singular.

PROOF OF (K). We will just do the case $n = 2$, as the general case is quite similar. We use well-known facts in homological algebra.

(1) There exists a module of nth order differentials $\Omega^n(R)$ such that $\mathrm{Hom}_R(\Omega^n(R), R) = \mathrm{Der}^n(R)$. Just take the free R-module on the symbols dr, $r \in R$ modulo the relation obtained from the Leibnitz rule: For every $a_0, a_1, \cdots, a_n \in R$,

$$d(a_0a_1\cdots a_n) = \sum_{s=1}^{n} (-1)^{s-1} \sum_{i_1<\cdots<i_s} a_{i_1} \cdots a_{i_s}\, d(a_0 \cdots \hat{a}_{i_1} \cdots \hat{a}_{i_s} \cdots a_n).$$

Alternately [17], one can define $\Omega^n(R)$ to be I/I^{n+1} where I is the kernel of the multiplication map $R \otimes_C R \to R$.

(2) The dual of a finitely generated R-module is reflexive, e.g.,

$$\mathrm{Hom}_R(\mathrm{Hom}_R(\mathrm{Hom}_R(M, R), R), R) = \mathrm{Hom}_R(M, R).$$

(3) Hence $\mathrm{Der}^n(R)$ is reflexive by (1) and (2).

(4) By the Serre criteria for normality, grad $R \geqq 2$.

(5) A reflexive module over a ring of grad $\geqq 2$ has grad $\geqq 2$.

(6) Hence grad $\mathrm{Der}^n(R) \geqq 2$, by (3), (4), and (5).

(7) For $D_1, \cdots, D_n \in \mathrm{Der}^1(V)$, by the canonical products generated by the D_i we mean the set $D_1, D_2, \cdots, D_n, D_1D_1, D_1D_2, \cdots, D_1D_n, D_2D_2, D_2D_3, \cdots, D_2D_n, D_3D_3, \cdots, D_3D_n, \cdots, D_{n-1}D_{n-1}, D_{n-1}D_n, D_nD_n$. If V is nonsingular at a point q and $D_1, \cdots, D_n$ generate $\mathrm{Der}^1(V)_q$, then the canonical products generate $\mathrm{Der}^2(V)_q$; if $D_1, \cdots, D_n$ are independent over $\mathcal{O}_q(V)$, then the canonical products are independent over $\mathcal{O}_q(V)$.

(8) By coherence of $\mathrm{Der}^1(V)$, if $D_1, \cdots, D_n$ are a basis of $\mathrm{Der}^1(V)$ at p, the canonical products are a basis of $\mathrm{Der}^2(V)$ at all nearly regular points. Let $\mathcal{S}$ be the $\mathcal{O}(V)$ subsheaf of $\mathrm{Der}^2(V)$ generated by the canonical products. $\mathcal{S}$ is a free $\mathcal{O}(V)$ sheaf by construction and $\mathcal{S} \mid \mathrm{Reg}\, V = \mathrm{Der}^2(V) \mid \mathrm{Reg}\, V$.

(9) If $\mathcal{F}_1 \subset \mathcal{F}_2$ are both reflexive sheafs of grad $\geqq 2$, and $\mathcal{F}_1 = \mathcal{F}_2$ off a set of codim $\geqq 2$, then $\mathcal{F}_1 = \mathcal{F}_2$.

(10) Hence $\mathcal{S} = \mathrm{Der}^2(V)$, by (6), (8), and (9).

In view of these results, one wants to know if Gr Der V finitely generated implies $C(V)$ representable. This approach has the following advantage: Since the Zariski-Lipman conjecture is likely to be true, it is hard to find any examples that give insight to the problem, but since there are many examples with Gr Der (V) not finitely generated and $C(V)$ not representable, one can learn from examples.

2. Additional topics. Seidenberg [21] has shown that a derivation of a Noetherian domain R containing the rationals extends to $\bar{R}$, the integral closure of the ring in its field of quotients. This is not necessarily the case for higher derivations. In fact [1] every vcdo on the variety extends to the normalization if and only if codim Nat Sg $V \geqq 2$, where Nat Sg $V = \{p \in V\colon p$ is a singular point of some irreducible component of $V\}$. If every vcdo on the normalization is induced from V, then codim Sg $V \geqq 2$. (The converse is false and it is not known what the correct criteria should be.) One might hope that if all higher derivations extend and contract, to and from the integral closure, then the variety is normal. However the example

$$R = \boldsymbol{C}[\text{all monomials of degree 2 or 3 in } x, y, \text{ and } z],$$
$$R = \boldsymbol{C}[x, y, z]/(x^3 + y^3 + z^3),$$

shows this is false. In fact, this example shows that one cannot distinguish a variety from its normalization by its ring of higher derivations!

Let $\pi\colon W \to V$ be the normalization of V, q be the finite set $\pi^{-1}(p)$ and let $\mathcal{O}(W, q) = \bigoplus_{p' \in q} \mathcal{O}(W, p')$; we will identify $\mathcal{O}(V, p)$ with its image $\pi^*\mathcal{O}(V, p)$ in $\mathcal{O}(W, q)$. Since $\mathcal{O}(W, q)$ is a finite integral extension of $\mathcal{O}(V, p)$, the natural and induced topologies agree on the closed subspace $\mathcal{O}(V, p)$ so π^* extends to an injection $\hat{\mathcal{O}}(V, p) \to \hat{\mathcal{O}}(W, q)$. Also it is not hard to see that the natural and induced Frèchet topologies on $\hat{\mathcal{O}}(V, p)$ agree because this is a finite extension. Let Ann $(\mathcal{O}(V, p)) = \{D \in C(W)_q : DF(p) = 0$ for all $D \in \hat{\mathcal{O}}(V, p)\} = (\hat{\mathcal{O}}(W, q)/\hat{\mathcal{O}}(V, p))^*$. If $F \in \mathcal{O}(W, q)$ and $DF = 0$ for each $D \in$ Ann, then $F \in \mathcal{O}(V, p)$. (By the Hahn-Banach theorem, the common zero of the D's is just the zero element of $\hat{\mathcal{O}}(W, q)/\hat{\mathcal{O}}(V, p)$ so $F \in \hat{\mathcal{O}}(V, p) \cap \mathcal{O}(W, q) = \hat{\mathcal{O}}(V, p)$.) Furthermore $C(V, p) \subset C(W, q)$ since each $D: \hat{\mathcal{O}}(V, p) \to \boldsymbol{C}$ extends (by Hahn-Banach) to a map $\hat{\mathcal{O}}(W, q) \to \boldsymbol{C}$.

So the constant coefficient operators can always detect subrings of $\mathcal{O}(W, q)$, but the variable coefficient operators sometimes do not. This again illustrates the usefulness of $C(V)$.

REFERENCES

1. J. Becker, *Higher derivations and integral closure* (preprint).

2. ———, *Differential operators on analytic spaces* (preprint).

3. R. Berger, R. Kiehl, E. Kunz and H.-J. Nastold, *Differential rechnung in der analytischen Geometrie*, Lecture Notes in Math., vol. 38, Springer-Verlag, Berlin and New York, 1967. MR **37** #469.

4. I. N. Bernšteĭn, I. M. Gel'fand and S. I. Gel'fand, *Differential operators on a cubic cone*, Uspehi Mat. Nauk **27** (1972), no. 1 (163), 185–190 = Russian Math. Surveys **27** (1972), no. 1, 169–174.

5. T. Bloom, *Opérateurs différentials sur les espaces analytiques*, Séminaire Lelong, Exposé 1, Lecture Notes in Math., vol. 71, Springer-Verlag, New York, 1968.

6. ———, *Differential operators on curves*, Rice Univ. Studies **59** (1973), no. 2, 13–19. MR **48** #4341.

7. E. Brieskorn, *Rationale Singularitäten komplexer Flächen*, Invent. Math. **4** (1967/68), 336-358. MR **36** #5136.

8. ———, *Examples of singular normal complex spaces which are topological manifolds*, Proc. Nat. Acad. Sci. U.S.A. **55** (1966), 1395–1397. MR **33** #6652.

9. W. C. Brown, *The algebra of differentials of infinite rank*, Canad. J. Math. **25** (1973), 141–155. MR **47** #3364.

10. M. Hochster, *The Zariski Lipman conjecture for homogenous complete intersections*, Proc. Amer. Math. Soc. **49** (1975), 261–262.

11. ———, *The Zariski Lipman conjecture for the graded case* (to appear).

12. M. Jaffee, *Differential operators on the curve* $x^a - y^b = 0$, Dissertation, Brandeis University, 1972.

13. J. M. Kantor, *Opérateurs différentiels sur les singularitiés-quotients*, C.R. Acad. Sci. Paris Sér. A-B **273** (1971), A897-A899. MR **44** #5510.

14. J. Lipman, *Free derivation modules on algebraic varieties*, Amer. J. Math. **87** (1965), 874–898, MR **32** #4130.

15. B. Malgrange, *Analytic spaces*, Ensignement Math. (2) **14** (1968), 1–28. MR **38** #6105.

16. S. Moen, *Free derivation modules and a criterion for regularity*, Thesis, Univ. of Minnesota, 1971; Proc. Amer. Math. Soc. **39** (1973) 221–227. MR **47** #1794.

17. K. Mount and O. E. Villamayor, *Taylor series and higher derivations*, Departments de Mathematicas Facultad de Ciencias Exactas y Naturales Universidade Buenos Aires, Serie no. 18, Buenos Aires, 1969.

18. ———, *On a conjecture of Y. Nakai*, Osaka J. Math. **10** (1973), 325–327.

19. Y. Nakai, *On the theory of differentials in commutative rings*, J. Math. Soc. Japan **13** (1961), 63–84. MR **23** #A2437.

20. ———, *Higher order derivations*. I, Osaka J. Math. **7** (1970), 1–27. MR **41** #8404.

21. A. Seidenberg, *Derivations and integral closure*, Pacific J. Math. **16** (1966), 167–173. MR **32** #5686.

22. ———, *Differential ideals in rings of finitely generated type*, Amer. J. Math. **89** (1967), 22–42. MR **35** #2902.

23. G. Scheja and U. Storch, *Differentielle Eigenschaften der Lokalisierungen analytischer Algebren*, Math. Ann. **197** (1972), 137–170. MR **46** #5299.

24. J. Stutz, *The representation problem for differential operators on analytic sets*, Math. Ann. **189** (1970), 121–133. MR **44** #2943.

25. ———, *Analytic sets as branched coverings*, Trans. Amer. Math. Soc. **166** (1972), 241–259. MR **48** #2420.

26. A. Tarski, *A decision method for elementary algebra and geometry*, The Rand Corporation, Santa Monica, Calif., 1948; 2nd ed., Univ. of California Press, Berkeley, Calif., 1951. MR **10**, 499; **13**, 423.

27. F. Treves, *Topological vector spaces, distributions and kernels*, Academic Press, New York and London, 1967. MR **37** #726.

28. J.-P. Vigué, *Opérateurs différentiels sur les espaces analytiques*, Invent. Math. **20** (1973), 313–336.

29. ———, *Opérateurs différentials sur les cônes normaux de dimension* 2, C.R. Acad. Sci. Paris Sér. A-B **278** (1974), A1047–A1050. MR **50** #4999.

30. O. Zariski and P. Samuel, *Commutative algebra*, Vols. 1,2, Univ. Ser. in Higher Math., Van Nostrand, Princeton, N. J., 1958, 1960. MR **19**, 833; **22** #11006.

PURDUE UNIVERSITY

Proceedings of Symposia in Pure Mathematics
Volume 30, 1977

FIBERING OF RESIDUAL CURRENTS

N. COLEFF AND M. HERRERA

Introduction. The purpose of this paper is to present a fibering theory of residual currents [12], [13] which will be given in full detail in a forthcoming paper. These currents are naturally associated with the semimeromorphic differential forms on a complex space, giving high-codimensional generalizations of the one-codimensional currents introduced in [5] and [14].

Residual currents have proved themselves useful to the duality theory on complex spaces [19], [20]. As another kind of application, we give (cf. Theorem 2.1) an extension of P. Griffiths' recent result on the sum of the punctual residues of a meromorphic n-form on a compact complex manifold [22].

The Fibering Theorem 3.1 is needed to prove the "Purity Property" 1.7.2 of residual currents. The fibered residue $\pi_* \operatorname{res}_{X;\pi}(u)$ associated in n·4 to any fibering $\pi: X \to T$ of complex spaces generalizes Grothendieck's residue symbol, which is essentially defined when the fibers $\pi^{-1}(t)$, $t \in T$, are regular [4], [6], [7], [11]. In this case the residue symbol determines the fibered residue [7].

Fibering theorems for geometric (0-continuous) currents have been obtained by Andreotti and Norguet [1], Stoll [21], Federer [8], King [15], Hardt [10], Poly [18] and Dubson (unpublished).

1. Definitions and general properties. $\boldsymbol{N}$, $\boldsymbol{R}$ and $\boldsymbol{C}$ denote the set of natural, real and complex numbers, respectively. $\boldsymbol{R}_>$ denotes the set of positive real numbers.

1.1. A complex n-cycle $[X, c]$ is a reduced and paracompact complex space X of pure dimension n, together with a given class $c \in H_{2n}(X; Z)$ of the (Borel-Moore) $2n$-homology of X with closed supports [3]. The notions of semianalytic chain or cycle are defined similarly [2].

An ordered family $\mathscr{F} = \{Y_1, \cdots, Y_{p+1}\}$ of $p + 1$ hypersurfaces in X $(0 \leqq p \leqq n)$

AMS (MOS) subject classifications (1970). Primary 32A25.

Key words and phrases. Residues, principal values, complex spaces, Grothendieck's residue symbol, meromorphic forms, currents, resolution of singularties.

© 1977, American Mathematical Society

is *admissible* if each Y_i is locally defined by one equation. An equation of Y_i on the open subspace $W \subset X$ is a function $\phi_i \in \mathcal{O}(W)$ such that $Y_i \cap W = V(\phi_i)$. We set

$$\begin{aligned} \bigcup \mathcal{F} &= \bigcup (Y_i \colon 1 \leqq i \leqq p+1), \\ \bigcap \mathcal{F} &= \bigcap (Y_i \colon 1 \leqq i \leqq p+1), \\ \mathcal{F}' &= \{Y_1, \cdots, Y_p\}, \\ Y &= \bigcap \mathcal{F}', \end{aligned} \tag{1}$$

and $i \colon Y \to X$ the inclusion. Unless specifically mentioned, no restrictions are made on the dimensions of Y or $\bigcap \mathcal{F}$.

1.2. We shall make use of the following additional notations:

(1) $\mathcal{E}^{\cdot}_X$ and $\Omega^{\cdot}_X$ are the sheaves of smooth and holomorphic differential forms on X, respectively. On a local model $X' \to W$ of X (W open in $\boldsymbol{C}$), they are induced by the usual forms on W, modulo those which are zero when restricted to the regular points of X' **[2]**. The differential and bigrading of forms on W are carried over to forms on X.

(2) $\mathcal{D}^{\cdot}(X) = \Gamma_c(X, \mathcal{E}^{\cdot}_X)$, with its usual locally convex inductive topology **[2]**. $'\mathcal{D}_{\cdot}(X)$, the space of currents on X, is the topological dual of $\mathcal{D}^{\cdot}(X)$, with the differential $b.T(\alpha) = (-1)^{q+1}\, T(d\alpha)$ for $T \in {'\mathcal{D}_q}(X)$ and $\alpha \in \mathcal{D}^q(X)$.

(3) $\Omega^{\cdot}_X(*\hat{Y})$ and $\mathcal{E}^{\cdot}_X(*\hat{Y})$ are the sheaves of meromorphic and semimeromorphic differential forms on X, respectively, with poles on the hypersurface $\hat{Y}$ of X; for instance if $U = X - \hat{Y}$ and $j \colon U \to X$ is the immersion, $\mathcal{E}^{\cdot}_X(*\hat{Y})$ is the subsheaf of $j_*\mathcal{E}^{\cdot}_U$ formed by those germs $\tilde{\omega} \in j_*\mathcal{E}^{\cdot}_U$ such that $\phi\tilde{\omega}$ has a smooth extension on X for some local equation ϕ of $\hat{Y}$.

(4) Set $\mathcal{E}^{\cdot}_X(*) = \bigoplus(\mathcal{E}^{\cdot}_X(*\hat{Y}) \colon \hat{Y}$ is an admissible hypersurface in $X)$; we define the sheaf $\mathcal{E}^{\cdot}_X(**)$ of *locally semimeromorphic forms on* X as the sheaf generated by the presheaf $W \to \mathcal{E}^{\cdot}_W(*)$ (W open in X).

1.3. *Orientation of tubes.* Let $\gamma = [X, c]$ be a complex n-cycle and $\mathcal{F} = \{Y_1, \cdots, Y_{p+1}\}$ an admissible family in X. Consider equations ϕ_i of Y_i, $1 \leqq i \leqq p+1$, on the open subspace $W \subset X$, and the mappings:

$$\begin{aligned} |\phi| &= (|\phi_1|, \cdots, |\phi_{p+1}|) \colon W - \bigcup \mathcal{F} \to \boldsymbol{R}^{p+1}_>, \\ |\phi'| &= (|\phi_1|, \cdots, |\phi_p|) \colon W - \bigcup \mathcal{F}' \to \boldsymbol{R}^p_>; \end{aligned}$$

for each $\boldsymbol{\delta} = (\delta_1, \cdots, \delta_{p+1}) \in \boldsymbol{R}^{p+1}_>$, denote $\boldsymbol{\delta}' = (\delta_1, \cdots, \delta_p)$ and consider the local cohomology class $\Delta\boldsymbol{\delta}' \in H^p_{\{\boldsymbol{\delta}'\}}(\boldsymbol{R}^p; \boldsymbol{Z})$ defined as the Poincaré dual of the canonical 0-homology class $[\boldsymbol{\delta}'] \in H_0^{\{\boldsymbol{\delta}'\}}(\boldsymbol{R}^p; \boldsymbol{Z})$ of $\boldsymbol{R}^p$ with support on the point $\boldsymbol{\delta}'$. Then the cap product

$$c_{\boldsymbol{\delta}'} = c \cap |\phi|^*(\Delta\boldsymbol{\delta}') \in H_{2n-p}^{\{|\phi'| = \boldsymbol{\delta}'\}}(X; \boldsymbol{Z}) \tag{2}$$

of c with the inverse image $|\phi|^*(\Delta\boldsymbol{\delta}') \in H^p_{\{|\phi'| = \boldsymbol{\delta}'\}}(X; Z)$ belongs to the $(2n-p)$-homology of X with support on the set $\{|\phi'| = \boldsymbol{\delta}'\}$. If

$$\dim_{\boldsymbol{R}}\{|\phi'| = \boldsymbol{\delta}'\} \leqq 2n - p, \tag{3}$$

we define the real analytic $(2n-p)$-cycle

$$T^p_{\boldsymbol{\delta}}(\phi) = (-1)^{p(p-1)/2}[\{|\phi'| = \boldsymbol{\delta}'\}, c_{\boldsymbol{\delta}'}] \tag{4}$$

and the semianalytic $(2n-p)$-chain

$$D_{\delta}^{p+1}(\phi) = T_{\delta}^{p}(\phi) \cap [|\phi_{p+1}| > \delta_{p+1}] \tag{5}$$

which we shall use for the definition of the residual currents on γ. They are tubes around Y, oriented according to the class c, and connected by the formula $\partial D_{\delta}^{p+1}(\phi) = T_{\delta}^{p}(\phi)$, where ∂ denotes the (Borel-Moore) topological boundary. The tube

$$T_{\delta}^{p+1}(\phi) = (-1)^{(p+1)p/2}[\{|\phi| = \delta\}, c_{\delta}]$$

is defined similarly.

The integration currents on these tubes are denoted by $I[T_{\delta}^{p}(\phi)]$ and $I[D_{\delta}^{p+1}(\phi)]$, respectively.

1.4. *Admissible trajectories.* Condition (3) will not be satisfied in general for any sufficiently small δ' unless $\dim Y \leqq n - p$. Nevertheless, one can prove that (3) is satisfied on any relatively compact open set $W \subset X$, for all sufficiently small δ', regardless of the dimension of Y, if δ' is taken along an *admissible trajectory* in $\boldsymbol{R}^{p}$. We so call any continuous mapping $\boldsymbol{\delta} = (\delta_1(\delta), \cdots, \delta_p(\delta))\colon (0, 1) \to \boldsymbol{R}_{>}^{p}$ of the unit interval into $\boldsymbol{R}_{>}^{p}$ such that

$$\lim_{\delta \to 0} \delta_i/\delta_{i+1}^{q}(\delta) = 0, \quad 1 \leqq i \leqq p - 1,$$

for all $q \in \boldsymbol{N}$.

1.5. *Local definitions.* Using resolution of singularities to replace $\bigcup \mathscr{F}$ by a hypersurface with only normal crossings, one can prove the existence, for any admissible trajectory $\boldsymbol{\delta}$ in $\boldsymbol{R}^{p+1}$, of the limits

$$RP_{\gamma;\mathscr{F}}(\tilde{\omega}) = \lim_{\delta \to 0} I[D_{\delta}^{p+1}(\phi)](\tilde{\omega}), \qquad \tilde{\omega} \in \Gamma_c(W, \mathscr{E}_X^{2n-p}(* \cup \mathscr{F})) \tag{6}$$

and

$$R_{\gamma;\mathscr{F}}(\tilde{\nu}) = \lim_{\delta \to 0} I[T_{\delta}^{p+1}(\phi)](\tilde{\nu}), \qquad \tilde{\nu} \in \Gamma_c(W, \mathscr{E}_X^{2n-p-1}(* \cup \mathscr{F})), \tag{7}$$

and their independence of the particularly chosen trajectory $\boldsymbol{\delta}$ and equations ϕ_i of $Y_i \cap W$, $1 \leqq i \leqq p + 1$.

This allows one to define the *multiple residue-principal value* $RP_{\gamma;\mathscr{F}}(\tilde{\omega})$ and the *multiple residue* $R_{\gamma;\mathscr{F}}(\tilde{\nu})$ of any forms $\tilde{\omega} \in \Gamma_c(X, \mathscr{E}_X^{2n-p}(* \cup \mathscr{F}))$, and $\tilde{\nu} \in \Gamma_c(X, \mathscr{E}_X^{2n-p-1}(* \cup \mathscr{F}))$, by patching together the local definitions (6) and (7).

In the case where $p = 0$ and c is the fundamental class of X, $RP_{\gamma;\mathscr{F}}$ reduces to the principal value PV introduced in [**14**] (or [**5**]), which will be denoted here as P.

We want to mention here the following facts about these operators (cf. [**12**] and [**13**] for additional properties):

1.5.1. The operator $RP_{\gamma;\mathscr{F}}$ has bigrade $(n, n - p)$: It has value zero on all forms of bigrade $(i, j) \neq (n, n - p)$.

1.5.2. Suppose $\dim Y = n - p$ and $\dim \bigcap \mathscr{F} < n - p$. Then

$$R_{\gamma;\mathscr{F}'}(\tilde{\nu}) = RP_{\gamma;\mathscr{F}}(\tilde{\nu})$$

for all $\tilde{\nu} \in \Gamma_c(X, \mathscr{E}_X^{2n-p}(* \cup \mathscr{F}'))$.

1.5.3. $\mathrm{R}_{\gamma;\mathscr{F}}$ depends essentially on the order of $\mathscr{F}$, as shown by the example $\phi_1 = z_1 \cdot z_2$, $\phi_2 = z_2$ in $\boldsymbol{C}^2$, and

$$\tilde{\omega} = \frac{a dz_1 \wedge dz_2}{\phi_1\phi_2}, \qquad a \in \mathscr{D}^0(C^2);$$

in this case, if $\mathscr{F} = \{V(\phi_1), V(\phi_2)\}$ and $\hat{\mathscr{F}} = \{V(\phi_2), V(\phi_1)\}$, one has $R_{C^2;\mathscr{F}}(\tilde{\omega}) = (\partial a/\partial z_2)(0, 0)$, $R_{C^2;\hat{\mathscr{F}}}(\tilde{\omega}) = 0$.

In the case where Y is a complete intersection ($\dim Y = n - p$), however, the equality

$$R_{\gamma;\mathscr{F}} = (-1)^{\mathrm{sg}(\sigma)} R_{\gamma;\sigma\mathscr{F}}$$

holds, for any permutation $\sigma\mathscr{F}$ of $\mathscr{F}$.

1.6. *Residual currents.* In the conditions of 1.3 (no restrictions on dim Y), choose $\tilde{\mu} \in \Gamma(X, \mathscr{E}^q(* \cup \mathscr{F}))$. Then the operators

$$RP_{\gamma;\mathscr{F}}[\tilde{\mu}](\alpha) = RP_{\gamma;\mathscr{F}}(\tilde{\mu} \wedge \alpha), \qquad \alpha \in \mathscr{D}^{2n-p-q}(X),$$
$$R_{\gamma;\mathscr{F}}[\tilde{\mu}](\beta) = R_{\gamma;\mathscr{F}}(\tilde{\mu} \wedge \beta), \qquad \beta \in \mathscr{D}^{2n-p-q-1}(X),$$

are continuous. They are the *residual currents* associated to $\tilde{\mu}$ and $\mathscr{F}$, and define sheaf homomorphisms

$$RP_{\gamma;\mathscr{F}}: \mathscr{E}^q(* \cup \mathscr{F}) \longrightarrow {}'\mathscr{D}_{2n-p-q},$$
$$R_{\gamma;\mathscr{F}}: \mathscr{E}^q(* \cup \mathscr{F}) \longrightarrow {}'\mathscr{D}_{2n-p-q-1},$$

into the sheaf of currents on X.

A local application of Stokes' theorem gives the relation:

$$(8) \qquad b.RP_{\gamma;\mathscr{F}}[\tilde{\mu}] + (-1)^{p+1} RP_{\gamma;\mathscr{F}}[d\tilde{\mu}] = R_{\gamma;\mathscr{F}}[\tilde{\mu}],$$

for each $\tilde{\mu} \in \Gamma(X, \mathscr{E}^{\cdot}_X(* \cup \mathscr{F}))$.

To describe the support of the residual currents, it is convenient to introduce the notion of *essential intersection* of the family $\mathscr{F}$.

1.7. *The essential intersection.* Consider our admissible family $\mathscr{F} = \{Y_1, \cdots, Y_{p+1}\}$ in X, and define inductively the following sequence of subvarieties of X:

$Y'_0 = X$, and for each j, $1 \leq j \leq p$,

Y'_j = union of the irreducible components of $Y'_{j-1} \cap Y_j$ which are not included in Y_{j+1}.

The subvarieties of X, $\tilde{V}_e(\mathscr{F}) = Y'_p$ and $V_e(\mathscr{F}) = Y'_p \cap Y_{p+1}$, are called the *essential intersections* of $\mathscr{F}$. They have the following properties:

(a) $\tilde{V}_e(\mathscr{F}) \subset \bigcap \mathscr{F}'$ and, if it is not empty, it has pure dimension $n - p$; $V_e(\mathscr{F}) \subset \bigcap \mathscr{F}$ and, if it is not empty, it has pure dimension $n - p - 1$.

(b) If $\dim \bigcap \mathscr{F}' = n - p$, $\tilde{V}_e(\mathscr{F})$ is the union of the irreducible components of $\bigcap \mathscr{F}'$ not included in Y_{p+1}.

(c) If $\dim \bigcap \mathscr{F}' = n - p$ and $\dim \bigcap \mathscr{F} = n - p - 1$ (the case of a complete intersection), one has

$$\tilde{V}_e(\mathscr{F}) = \bigcap \mathscr{F}' \quad \text{and} \quad V_e(\mathscr{F}) = \bigcap \mathscr{F}.$$

Moreover:

1.7.1. For each $\tilde{\mu} \in \Gamma(X, \mathscr{E}^{\cdot}_X(* \cup \mathscr{F}))$, the supports of $RP_{\gamma;\mathscr{F}}[\tilde{\mu}]$ and $R_{\gamma;\mathscr{F}}[\tilde{\mu}]$ are contained in $\tilde{V}_e(\mathscr{F})$ and $V_e(\mathscr{F})$, respectively.

1.7.2. Suppose that X is regular and that $\bigcap \mathscr{F}'$ and $\bigcap \mathscr{F}$ are complete intersections. Then, if an s-dimensional complex variety exists which contains the support of $RP_{\gamma:\mathscr{F}}[\tilde{\mu}]$, and $s < n - p$, one has $RP_{\gamma:\mathscr{F}}[\tilde{\mu}] = 0$.

For instance, in the example of 1.5.3 we have $V_e(\mathscr{F}) = \{0\}$, $V_e(\hat{\mathscr{F}}) = \emptyset$.

2. The punctual residue. Let $\gamma = [X, c]$ be a complex p-cycle and $\mathscr{F} = \{Y_1, \cdots, Y_p\}$ an admissible family in X. For each $\tilde{\lambda} \in \Gamma(X, \mathscr{E}^p_X(* \cup \mathscr{F}))$, the current $R_{\gamma:\mathscr{F}}[\tilde{\lambda}] \in {}'\mathscr{D}_0(X)$ has support contained in the 0-dimensional variety $V_e(\mathscr{F})$ (1.7.1) of X; therefore, we can define *the punctual residue* $\mathrm{res}_{\gamma:\mathscr{F}:y}(\tilde{\lambda})$ of $\mathrm{Res}_{\gamma:\mathscr{F}}[\tilde{\lambda}]$ at any point $y \in V_e(\mathscr{F})$ by:

$$\text{(1)} \qquad \mathrm{res}_{\gamma:\mathscr{F}:y}(\tilde{\lambda}) = \mathrm{Res}_{\gamma:\mathscr{F}}[\tilde{\lambda}](\alpha_y), \qquad \alpha_y \in \mathscr{D}^0(W),$$

where W is a neighborhood of y such that $\bar{W} \cap (V_e(\mathscr{F})) = \{y\}$ and $\alpha_y \equiv 1$ on some neighborhood of y.

If $\phi_1, \cdots, \phi_p \in \mathscr{O}(W)$ are equations of $Y_1, \cdots, Y_p$ on W, our definitions give

$$\text{(2)} \qquad \mathrm{res}_{\gamma:\mathscr{F}:y}(\tilde{\lambda}) = \lim_{\delta \to 0} I[T^p_\delta(\phi)](\tilde{\lambda}),$$

for any admissible trajectory $\boldsymbol{\delta}$; if $\tilde{\lambda}$ is meromorphic, the right-hand side integral is constant for δ sufficiently small, by Stokes' theorem; if, moreover, $\mathscr{F}$ is a complete intersection at y, $(\bigcap \mathscr{F}) \cap W = \{y\}$, (2) reduces to Grothendieck's residue

$$\mathrm{Res}_y[\phi_1, \overset{\lambda}{\ldots}, \phi_p]$$

of $\tilde{\lambda} = \lambda/\phi_1 \cdots \phi_p$ at y **[4]**, **[6]**, **[22]**.

We can now prove the following generalization of a result of P. Griffiths **[22]**:

2.1. THEOREM. *Suppose that X is compact and $\tilde{\lambda} \in \Gamma(X, \Omega^p_X(* \cup \mathscr{F}))$. Then the sum of the punctual residues of $\tilde{\lambda}$ at the points of the essential intersection of $\mathscr{F}$ is zero:*

$$\sum_{y \in V_e(\mathscr{F})} \mathrm{res}_{\gamma:\mathscr{F}:y}(\tilde{\lambda}) = 0.$$

(d) It is clear from the definitions that this sum is equal to the evaluation $R_{\gamma:\mathscr{F}}[\tilde{\lambda}]$ (1) on the unit function on X. By 1.6(1) this evaluation is zero, since $d\tilde{\mu} = 0$ and $b.RP_{\gamma:\mathscr{F}}[\tilde{\mu}]\,(1) = RP_{\gamma;\mathscr{F}}[\tilde{\mu}]\,(d1) = 0$.

We remark that Griffiths' result is stated for the case where X is a complex manifold and $\dim \bigcap \mathscr{F} = 0$; his proof does not use resolution of singularities, but it is not immediately extendible to the present situation.

3. The fibered residue. Consider now a complex n-cycle $\gamma = [X, c]$ and an admissible family $\mathscr{F} = \{Y_1, \cdots, Y_p\}$ in X $(0 \leqq p \leqq n)$ such that $\dim Y = n - p$, $Y = \bigcap \mathscr{F}$; suppose to simplify that c is the fundamental class of X and denote by $i\colon Y \to X$ the immersion. Let $\pi\colon X \to T$ be a morphism of X into a reduced complex space T of pure dimension $n - p$, such that

(1) $\dim \pi^{-1}(t) = p$,

(2) $\dim(\pi \circ i)^{-1}(t) = 0$ $(t \in T)$.

We suppose that T is oriented by its fundamental class.

Let sT denote the singular set of T and $T_* = T - sT$. For each $t \in T_*$, we define the complex p-cycle $\pi^{-1}[t]$, *the inverse image of the* 0-*cycle* $[t]$ *by* π, as:

$$\pi^{-1}[t] = [\pi^{-1}(t), c_t], \qquad c_t = c \cap \pi^*(\Delta[t]) \in H_{2p}(\pi^{-1}(t); \boldsymbol{Z});$$

as before, $\Delta[t] \in H^{2(n-p)}_{\{t\}}(T_*; Z)$ denotes the Poincaré dual of the canonical 0-

homology class $[t] \in H_0^{\{t\}}(T_*; \boldsymbol{Z})$ of T_* with support on $\{t\}$, and

$$\pi^*(\Delta[t]) \in H^{2(n-p)}_{\pi^{-1}(t)}(X - \pi^{-1}(sT); \boldsymbol{Z}).$$

For each $y \in Y - \pi^{-1}(sT)$, denote by X_y the reduced subspace $\pi^{-1}[\pi(y)]$ of X and $[X_y]$ the complex p-cycle $\pi^{-1}[\pi(y)]$. Let $i_y : X_y \to X$ be the immersion and observe that, by (1), $\mathscr{F}_y = \{Y_1 \cap X_y, \cdots, Y_p \cap X_y\}$ is an admissible family in X_y with dim $\bigcap \mathscr{F}_y = 0$.

If $\tilde{\mu} \in \Gamma(X, \mathscr{E}^p_X(* \cup \mathscr{F}))$ and $y \in Y - \pi^{-1}(sT)$, we have

$$i_y^*(\tilde{\mu}) \in \Gamma(X_y; \mathscr{E}^p_{X_y}(* \cup \mathscr{F}_y)),$$

where i_y^* is the homomorphism induced by i_y on differential forms. Therefore, we can define on $Y - \pi^{-1}(sT)$ the function $\mathrm{res}_{Y;\mathscr{F}}(\tilde{\mu})$:

$$\mathrm{res}_{Y;\mathscr{F}}(\tilde{\mu})(y) = \mathrm{res}_{[X_y];\mathscr{F}_y;y}(i_y^*(\tilde{\mu})),$$

$\tilde{\mu} \in \Gamma(X, \mathscr{E}^p_X(* \cup \mathscr{F}))$, $y \in Y - \pi^{-1}(sT)$, which is called *the fibered residue of* $\tilde{\mu}$ *by* π.

3.1. THEOREM. (i) *For each* $\tilde{\mu} \in \Gamma(X, \mathscr{E}^p_X(* \cup \mathscr{F}))$, $\mathrm{res}_{\mathscr{F};\pi}(\tilde{\mu})$ *is a locally semi-meromorphic function on* Y.

(ii) *Suppose that* $\tilde{\mu} \in \Gamma_c(X; \mathscr{E}^p_X(* \cup \mathscr{F}))$ *and* $\xi \in \mathscr{D}^{2(n-p)}(T)$. *Then*

$$R_{Y;\mathscr{F}}(\tilde{\mu} \wedge \pi^*\xi) = P_Y(\mathrm{res}_{\mathscr{F};\pi}(\tilde{\mu}) \wedge \pi^*(\xi)).$$

SKETCH OF THE PROOF. (i) It suffices to consider the case where T is a closed subspace of the open set U in $\boldsymbol{C}^l$, X is a closed subspace of the open set W in $\boldsymbol{C}^l \times \boldsymbol{C}^m$, and π is induced by the projection

$$\boldsymbol{C}^l \times \boldsymbol{C}^m \to \boldsymbol{C}^l, \qquad (t^{(l)}, t^{(m)}) \to t^{(l)},$$

where $t^{(l)} = (t_1, \cdots, t_l)$, $t^{(m)} = (t_{l+1}, \cdots, t_{l+m})$.

Choose a point $y_0 \in Y$ and an open neighborhood $W' \subset W$ of y_0 in $\boldsymbol{C}^l \times \boldsymbol{C}^m$ such that functions $\phi_i \in \mathscr{O}(W')$ exist with the property:

$$X \cap V(\phi_i) = W' \cap Y_i, \qquad 1 \leqq i \leqq p.$$

Let $a \in \mathscr{D}^0(W')$, $A \subset \{l+1, \cdots, l+m\}$ be a set of p elements and

$$\tilde{\mu} = a\,dt_A/\phi_1 \cdots \phi_p \,|\, X \in \Gamma_c(X \cap W', \mathscr{E}^p_X(* \cup \mathscr{F})),$$
$$dt_A = dt_{i_1} \wedge \cdots \wedge dt_{i_p}, \quad A = \{i_1, \cdots, i_p\}.$$

We define now a family $k[r, A]$ of meromorphic functions on $Y \cap W'$ that will allow us to represent $\mathrm{res}_{\mathscr{F};\pi}(\tilde{\mu})$ conveniently. For each $y \in Y - \pi^{-1}(sT)$, $y = (y^{(l)}, y^{(m)})$, and $r \in \boldsymbol{N}^m$, let $\beta[r, y]$ denote the following meromorphic function on $(y^{(l)} \times \boldsymbol{C}^m) \cap W'$:

$$\beta[r, y](t^{(m)}) = \left(\prod_{i=1}^{p} \phi_i(y^{(l)}, t^{(m)})\right)^{-1} \cdot \prod_{l<j<l+m} (t_j - y_j)^{r_j}, \quad t^{(m)} \in \boldsymbol{C}^m;$$

then

$$k[r, y](y) = \mathrm{res}_{[,X];\mathscr{F}_y;y}(\beta[r, y]dt_A).$$

In these conditions one can prove that a polydisk $D = D^l \times D^m$ centered at y_0 exists, together with a hypersurface S of $Y \cap D$ and $n_1 \in \boldsymbol{N}$, such that

$$k[r, A] \in \Gamma(Y \cap D, \mathscr{O}_Y(*S)), \qquad r \in \boldsymbol{N}^m, \tag{3}$$

and

(4) $$\operatorname{res}_{\mathscr{F};\pi}(\tilde{\mu})(y) = \sum_{|r|\leq n_1} \frac{1}{r!}\partial^r a(y)\cdot k[r, A](y), \qquad y \in (Y - S) \cap D,$$

where $r! = r_{l-1}! \cdots r_{l+m}!$, $|r| = r_{l+1} + \cdots + r_{l+m}$ and

$$\partial^r = \partial^{|r|}/\partial t_{l+1}^{r_{l+1}} \cdots \partial t_{l+m}^{r_{l+m}}.$$

The fact that $\operatorname{res}_{\mathscr{F};\pi}(\tilde{\mu}) \in \Gamma(D \cap Y; \mathscr{E}_Y^0(*S))$ follows immediately from (3) and (4).

(ii) Let $\psi \in \mathcal{O}(D')$ be such that $V(\psi) \supset sT$ and $\hat{S} = X \cap D \cap \pi^{-1}(V(\psi))$ verifies $S \subset \hat{S} \cap Y$. The admissible family in $X \cap D$

$$\mathscr{G} = \{Y_1 \cap D, \cdots, Y_p \cap D, \hat{S}\}$$

satisfies then the conditions of 1.5.2, so that

(5) $$R_{\gamma;\mathscr{F}}(\tilde{\nu}) = RP_{\gamma;\mathscr{G}}(\tilde{\nu}), \qquad \tilde{\nu} \in \Gamma_c(X \cap D, \mathscr{E}_X^{2n-p}(* \cup \mathscr{F})).$$

The right-hand side of (5) can also be calculated as the iterated limit:

(6) $$RP_{\gamma;\mathscr{G}}(\tilde{\nu}) = \lim_{\delta_{p+1}\to 0} \lim_{\delta\to 0} I[D^p_{\delta;\delta_{p+1}}(\hat{\phi})](\tilde{\nu}),$$

where $\boldsymbol{\delta}$ is any admissible trajectory in $\boldsymbol{R}^p$ and $D^p_{\delta;\delta_{p+1}}(\hat{\phi})$ is the tube associated in 1.3 to $\hat{\phi} = (\phi_1, \cdots, \phi_p, \psi)$ and to $(\boldsymbol{\delta}; \delta_{p+1}) \in \boldsymbol{R}_>^{p+1}$.

Using resolution of singularities one can prove that, if $\tilde{\nu} = \tilde{\mu} \wedge \pi^*\xi$, then

$$\lim_{\delta\to 0} I[D^p_{\delta;\delta_{p+1}}(\hat{\phi})](\tilde{\nu}) = I[Y \cap (|\psi| > \delta_{p+1})](\operatorname{res}_{\mathscr{F};\pi}(\tilde{\mu}) \wedge \pi^*\xi).$$

The limit as $\delta_{p+1} \to 0$ of the expression on the right is exactly $P_Y(\operatorname{res}_{\mathscr{F};\pi}(\tilde{\mu}) \wedge \pi^*\xi)$, so that by (5) and (6) we obtain the equality of the theorem.

3.2. REMARKS. (3) There is the following more general version of Theorem 3.1:

$$RP_{\gamma;\mathscr{F}}(\tilde{\mu} \wedge \pi^*\xi) = P_Y(\operatorname{res}_{\mathscr{F};\pi}(\tilde{\mu}) \wedge \pi^*\xi), \qquad \tilde{\mu} \in \Gamma_c(X, \mathscr{E}_X^p(* \cup \hat{\mathscr{F}})),$$

in the case $\hat{\mathscr{F}} = \{\mathscr{F}, Y_{p+1}\}$ and $\dim (Y \cap Y_{p+1}) < n - p$.

(b) The proof of (3) uses a kind of Lojasiewicz's fibered inequality and Hardt and Poly's fibering of geometric currents **[10]**, **[18]**.

4. Relation with Grothendieck's residue symbol. Some particular cases of Theorem 4.1 are worth mentioning here:

4.1. If $\tilde{\mu} \in \Gamma(X, \Omega_X^p(* \cup \mathscr{F}))$ one can prove directly, without using resolution of singularities, that in the case $\pi|Y: Y \to T$ is finite, the "trace" $\pi_* \operatorname{res}_{\mathscr{F};\pi}(\tilde{\mu})$:

$$\pi_* \operatorname{res}_{\mathscr{F};\pi}(\tilde{\mu})(t) = \sum(\operatorname{res}_{\mathscr{F};\pi}(\tilde{\mu})(y): y \in \pi^{-1}(t))$$

is a weakly holomorphic function on T (i.e., it is holomorphic on the simple points of T and locally bounded on T).

4.2. Suppose moreover that T is a subvariety of the open subspace U of $\boldsymbol{C}^l$, that $X = T \times F$, where F is a pure p-dimensional complex space, and that $\pi: X \to T$ is induced by the projection $\hat{\pi}: U \times F \to U$ (or suppose that π is locally isomorphic to such a morphism). Then $\pi_* \operatorname{res}_{\mathscr{F};\pi}(\tilde{\mu})$ is *holomorphic* on T.

In fact, it can be deduced in this case that the germ $\tilde{\mu}_x$ of $\tilde{\mu}$ at $x \in Y$ is the restriction on X of a germ $\hat{\mu}_x \in \Omega^p_{U\times F,x}(* \cup \mathscr{G})$, where $\mathscr{G}$ is an admissible family in $U \times F$ such that $\mathscr{G}|X = \mathscr{F}$ and $\hat{\pi}|\bigcap\mathscr{G}: \bigcap\mathscr{G} \to U$ is still finite at x.

We have then that $\hat{\pi}_*$ $\text{res}_{\mathscr{G};\hat{\pi}}$ $\hat{\pi}(\hat{\mu}_x)$ is defined and weakly holomorphic at $\hat{\pi}(x)$ and therefore is holomorphic at $\hat{\pi}(x)$. Since in our case $\text{res}_{\mathscr{F};\pi}(\tilde{\mu})(x) = \text{res}_{\mathscr{G};\hat{\pi}}(\hat{\mu})(x)$, one concludes that the germ of π_* $\text{res}_{\mathscr{F};\pi}(\tilde{\mu})$ at $\pi(x)$ is the restriction on T of a sum of holomorphic germs in U at $\pi(x)$, one for each $y \in \pi^{-1}\pi(x)$. Hence π_* $\text{res}_{\mathscr{F};\pi}(\tilde{\mu})$ is holomorphic at $\pi(x)$.

4.3. Suppose that π is smooth (i.e., locally isomorphic to the morphism described in 4.2, with F regular), that $\phi_i \in \mathcal{O}(X)$ are global equations of Y_i on X ($1 \leqq i \leqq p$), and that $\tilde{\mu} = \mu/\phi_1 \cdots \phi_p$, $\mu \in \Gamma(X, \Omega_X^p)$. Then **[6]**

$$\pi_* \ \text{res}_{\mathscr{F};\pi}(\tilde{\mu}) = \text{Res}\begin{bmatrix} \mu \\ \phi_1 \cdots \phi_p \end{bmatrix},$$

where the expression on the right denotes Grothendieck's residue symbol. In this way, π_* $\text{res}_{\mathscr{F};\pi}(\tilde{\mu})$ is the generalization of this symbol to the case where $\pi\colon X \to T$ is any morphism (satisfying 3(1) and 3(2)) and $\tilde{\mu}$ is semimeromorphic.

References

1. A. Andreotti and F. Norguet, *La convexité holomorphe dans l'espace analitique des cycles d'une variété algébrique*, Ann. Scuola Norm. Sup. Pisa (3) **21** (1967), 31–82. MR **39** #477.

2. T. Bloom and M. Herrera, *de Rham cohomology of an analytic space*, Invent. Math. **7** (1969), 275–296. MR **40** #1601.

3. A. Borel, and A. Haefliger, *La classe d'homologie fondamentale d'un espace analytique*, Bull. Soc. Math. France **89** (1961), 461–513. MR **26** #6990.

4. A. Beauville, *Une notion de résidus en géométrie analytique*, Séminaire P. Lelong 1969/70, Lecture Notes in Math., vol. 205, Springer-Verlag, Berlin and New York, 1970.

5. P. Dolbeault, *Valeurs principales sur les espaces complexes*, Séminaire P. Lelong 1970/71, Exposé du 13 janvier 1971, Springer Lecture Notes.

6. F. Elzein, *Résidus en géométrie algébrique*, Compositio Math. **23** (1971), 379–405. MR **46** #178.

7. ———, *Comparaison des résidus de Grothendieck et Herrera*, C.R. Acad. Sci. Paris Sér. A **278** (1974), 863–866.

8. H. Federer, *Geometric measure theory*, Die Grundlehren der math. Wissenschaften, Band 153, Springer-Verlag, New York, 1969. MR **41** #1976.

9. ———, *Some theorems on integral currents*, Trans. Amer. Math. Soc. **117** (1965), 43–67. MR **29** #5984.

10. R. M. Hardt, *Slicing and intersection theory for chains associated with real analytic varieties*, Acta Math. **129** (1972), 75–136. MR **47** #4110.

11. R. Hartshorne, *Residues and duality*, Lecture Notes in Math., no. 20, Springer-Verlag, Berlin and New York, 1966. MR **36** #5145.

12. M. Herrera, *Résidus multiples sur les espaces complexes*, Exposé aux Journées Complexes de Metz, (1972), I.R.M.A., Université L. Pasteur, Strasbourg.

13. ———, *Les courants résidus multiples*, Journées Géom. Analyt. (1972) Poitiers, Bull. Soc. Math. France, Mém. **38** (1974), 27–30.

14. M. Herrera and D. Lieberman, *Residues and principal values on complex spaces*, Math. Ann. **194** (1971), 259–294. MR **45** #5413.

15. J.R. King, *The currents defined by analytic varieties*, Acta Math. **127** (1971), 185–220.

16. ———, *Global residues and intersections on a complex manifold* (preprint).

17. R. Narasimhan, *Introduction to the theory of analytic spaces*, Lecture Notes in Math., no. 25, Springer-Verlag, Berlin and New York, 1966. MR **36** #428.

18. J.B. Poly, *Formule des résidus et intersection des chaines sousanalytiques*, Thesis, Université de Poitiers, 1974.

19. G. Ruget, *Complexe dualisant et résidus*, Journées Géom. Analytique (1972) Poitiers, Bull. Soc. Math. France, Mém. **38** (1974), 31–38.

20. J.P. Ramis and G. Ruget, *Résidus et dualité*, Invent. Math. **26** (1974), 89–131. MR **50** #5009.

21. W. Stoll, *The fiber integral in constant*, Math. Z. **104** (1968), 65–73. MR **37** #467.

22. P. Griffiths, *Lectures on the Abel's theorem* (preprint).

UNIVERSIDAD NACIONAL DE LA PLATA

Proceedings of Symposia in Pure Mathematics
Volume 30, 1977

CARTESIAN PRODUCT STRUCTURE OF SINGULARITIES

ROBERT EPHRAIM

In this note I announce some structure theorems for germs of analytic sets over k ($k = \boldsymbol{R}$ or $k = \boldsymbol{C}$), in which the structure of such a germ as a cartesian product of lower dimensional germs is considered. An application of the main structure theorem is stated in which the germs of complex analytic sets which are real analytically equivalent to a fixed complex analytic germ are classified. Proofs of these results will appear elsewhere.

Definitions. Throughout this section V will denote a germ of an analytic set over k.

DEFINITION 1. By a decomposition of V of length n, I mean an n-tuple $(V_1, \cdots, V_n)$ of germs of analytic sets over k, such that V is k-analytically equivalent to the cartesian product $V_1 \times \cdots \times V_n$, and such that $\dim V_j \neq 0$, $j = 1, \cdots, n$.

DEFINITION 2. V is said to be indecomposable if it has no decomposition of length greater than one.

DEFINITION 3. V will be said to be uniquely k-decomposable if, given any two decompositions of V, $(V_1, \cdots, V_n)$ and $(W_1, \cdots, W_m)$ for which all V_i and W_i are indecomposables, one has $n = m$, and, after a permutation of the W_i, one has V_j k-analytically isomorphic to W_j, $j = 1, \cdots, n$.

Complex germs. The main structure theorem I have proven is:

THEOREM 1. *Let V be an irreducible germ of a complex analytic set. Then V is uniquely* $\boldsymbol{C}$*-decomposable.*

This theorem is applied to prove:

THEOREM 2. *Let V and W be irreducible germs of complex analytic sets which are*

AMS (MOS) subject classifications (1970). Primary 32B10, 32C05; Secondary 32C40.

Key words and phrases. Germs of analytic sets (real or complex), real structure of complex germs, complexification, cartesian product decomposition, real analytic isomorphism, complex analytic isomorphism.

© 1977, American Mathematical Society

real analytically isomorphic to each other. Then there exist decompositions (V_1, V_2) *and* (W_1, W_2) *of* V *and* W *respectively, such that* V_1 *is complex analytically isomorphic to* W_1, *and* V_2 *is complex analytically isomorphic to* $\bar{W}_2$ *(the complex conjugate of* W_2*).*

As an easy application of Theorems 1 and 2 one gets the fact that, up to complex analytic isomorphism, there are only finitely many germs W real analytically isomorphic to any given V.

Real germs. It is reasonable to ask for a real analytic analogue of Theorem 1. In fact, the proof of Theorem 1 will, in the real analytic category, prove:

THEOREM 3. *Let* V *be an irreducible germ of a real analytic set such that* $\mathrm{Sg}(V)$, $\mathrm{Sg}(\mathrm{Sg}(V))$, *and all higher iterates* $\mathrm{Sg}^n(V)$ *are real analytic or empty. Then* V *is uniquely R-decomposable.*

In particular, since a complex analytic germ is automatically also a real analtyic germ, one has:

COROLLARY 4. *Let* V *be an irreducible germ of a complex analytic set. Then* V *is uniquely R-decomposable.*

Actually, the condition on $\mathrm{Sg}^n(V)$ is unnecessary. I wish to thank Professor Joseph Becker for pointing out that only a minor change need be made in the proof of Theorem 1 to establish:

THEOREM (BECKER). *Let* V *be an irreducible germ of a real analytic set. Then* V *is uniquely* $\boldsymbol{R}$*-decomposable.*

The real structure of complex germs. Because of Theorem 1 and Corollary 4, any irreducible germ of a complex analytic set is both uniquely $\boldsymbol{C}$-decomposable and uniquely $\boldsymbol{R}$-decomposable. It is natural to ask how the real and the complex decompositions into indecomposables relate to each other. The answer is given by the following theorem due independently to myself and to Professor Becker.

THEOREM 5. *Let* V *be an irreducible germ of a complex analytic set which is not a regular germ. If* V *is* $\boldsymbol{C}$*-indecomposable, then* V *is also* $\boldsymbol{R}$*-indecomposable.*

The germ $\boldsymbol{C}$ of course has a $\boldsymbol{R}$-decomposition $(\boldsymbol{R}, \boldsymbol{R})$.

As a consequence of Theorem 5 one gets that there is no essential difference between the $\boldsymbol{C}$- and the $\boldsymbol{R}$-decompositions into indecomposables of an irreducible germ of a complex analytic set. If one has a $\boldsymbol{C}$-decomposition, and one further decomposes any regular germ which appears in this decomposition as $(\boldsymbol{R}, \boldsymbol{R})$, then one obtains the $\boldsymbol{R}$-decomposition of this germ.

As a special case, suppose V is an irreducible complex germ with $C_1(V, 0) = 0$ (the first Whitney tangent cone). Then the two decompositions are in fact identical.

Idea of proofs. There are two ideas that feed into the proofs of the structure theorems. The first is that one can get at the terms in a decomposition for a germ V by examining $\mathrm{Sg}(V)$. A simple example illustrates this idea. Let $V = W \times W'$ with both W and W' isolated singularities, then $\mathrm{Sg}(V) = W \cup W'$.

The second idea is given by a cancellation theorem:

LEMMA 6. *Suppose* V *and* W *are germs of k-analytic sets such that* $V \times k = W$

$\times\ k$; *then* $V = W$. ($=$ *means k-analytically isomorphic.*)

The proof of this lemma is very technical, and it depends on a theorem about a certain functional equation.

These two elements combine to prove the structure theorems by induction on dim V.

The proof of Theorem 2 is a straightforward application of the structure Theorem 1, and of the geometric fact:

LEMMA 7. *Let V be an irreducible germ of a complex analytic set. Regard V as a real analytic germ. Then the complexification of V is just $V \times \bar{V}$.*

REFERENCES

1. R. Ephraim, *C^∞ and analytic equivalence of singularities*, Rice Univ. Studies **59** (1973), no. 1, 11–32. MR **48** #8834.

2. ———, *The cartesian product structure and C^∞ equivalences of singularities* (preprint).

HERBERT A. LEHMAN COLLEGE (CUNY)

Proceedings of Symposia in Pure Mathematics
Volume 30, 1977

A GEOMETRIC STUDY OF THE MONODROMY OF COMPLEX ANALYTIC SURFACES

GERALD LEONARD GORDON*

1. Statement of problem.

1.1. Suppose we have $\pi: W \to D$ where D is the unit disk in the t-plane, W is a complex manifold and π a holomorphic map. We set $\pi^{-1}(t) = V_t$ and we assume that for $t \neq 0$, V_t is a connected complex manifold of complex dimension 2 such that $\pi: W - V_0 \to D - \{0\}$ is a holomorphic fibre bundle. V_0 can acquire arbitrary singularities under the analytic deformation. Then if T_* is the map induced from going once around the unit circle, i.e., the monodromy, and I_* the map induced from the identity, we wish to study $T_*^N - I_*$ on $H_*(V_1)$, where we take coefficients in a fixed field, usually thought of as the rationals.

We note here that the techniques are valid for V_t any dimension whatsoever. Furthermore, we believe that these techniques are also applicable to the study of the monodromy of surfaces over arbitrary fields. Some of these results have been found independently by Todorov [**10**]. He has communicated to the author that he has some generalizations to higher dimensions.

1.2. We now make some topological reductions. We can first of all resolve singularities so that V_0 is locally given by $\prod_{i=1}^k z_i^{a_i} = 0$. Then by the semistable reduction theorem (cf. Kempf et al. [**6**]) we can assume that V_0 is locally given by $\prod_{i=1}^k z_i = 0$ and that T_* is unipotent, i.e., $(T_* - I_*)^3 = 0$.

We let $V_0 = \bigcup_{i \in I} X_i$, where the X_i are connected, nonsingular complex surfaces in general position, and let $I_0 \subset I$ be those X_i which belong to the proper transform. Moreover, we can also assume that $X_{ij} = X_i \cap X_j$ and $X_{ijk} = X_i \cap X_j \cap X_k$ are connected. We let M_3 be the triple points of V_0, $M_2 = \bigcup(X_{ij} - M_3)$, the double points, and M_1 the nonsingular part of V_0.

AMS (MOS) subject classifications (1970). Primary 32G10, 32L05, 14J99.

Key words and phrases. Monodromy, complex analytic surfaces, Kähler manifolds, deformations.

*This research was partially supported by NSF grant GP 38964A # 1.

© 1977, American Mathematical Society

1.3. We now take a regular tubular neighborhood of V_0 in W, call it $\tau(V_0)$, such that $\tau(V_0) = \partial W$ and $\tau(V_0) \cap V_1 = C_1$, is a "bundle" over V_0, i.e., if $P \in M_k$, the fibre over M_k is a $(k - 1)$-torus, where a 0-torus is a point; see Clemens **[1]** for details of this construction. Locally, what we have is that if $P \in M_k$ with $\prod_{i=1}^{k} z_i = 0$ a locally defining equation of V_0 at P, then $\prod_{i=1}^{k} z_i = 1$ defines V_1 near P; and the cartesian product $(|z_1| = 1) \times \cdots \times (|z_k| = 1)$ defines $\tau(V_0)$ near P, so that $(\prod_{i=1}^{k} z_i = 1) \cap \times_{i=1}^{k} (|z_i| = 1)$ is a k-torus.

We let $g_{\#}^{-1} : C_*(V_0) \to C_*(V_1)$ be the chain map which associates to each p-chain $\Gamma \subset M_k$ the $(k + p - 1)$-chain over Γ in C_1. Then if $\tau_1 \cdots \tau_k$ corresponds to the tubular map, i.e., locally it is the product with $\times_{i=1}^{k} (|z_i| = 1)$, then $g_{\#}^{-1}\Gamma = \tau_1 \cdots \tau_k \Gamma$.

Then the action of T_* follows τ, i.e.,

1.3.1. If $\Gamma \subset M_k$, then $(T_{\#} - I_{\#})g_{\#}^{-1}(\Gamma) = (\partial_W)_{\#}\tau_1 \cdots \tau_k \Gamma = (-1)^k \tau_1 \cdots \tau_k (\partial_{V_0})_{\#}\Gamma$.

Hence the study of $T_* - I_*$ will be essentially the study of boundary operators on relative homology groups.

1.4. We wish to study the image of $T_* - I_*$. To do this, we consider the Wang sequence

$$
\begin{array}{ccccccc}
\longrightarrow H_p(V_1) & \xrightarrow{T_* - I_*} & H_p(V_1) & \xrightarrow{i_*} & H_p(\partial W) & \longrightarrow & H_{p-1}(V_1) \\
 & & & \searrow & \downarrow \wr & & \\
 & & & g_* & H_p(W - V_0) & & \\
 & & & & \downarrow j_* & & \\
 & & & & H_p(W) & & \\
 & & & & \downarrow \wr & & \\
 & & & & H_p(V_0) & &
\end{array}
$$

where i, j are the inclusions, the isomorphisms are induced from deformation retractions, and g is the composition map which makes the diagram commute.

1.4.1. *Locally invariant cycle problem* (cf. Griffiths **[5**, p. 249**]**). Is $j_* | \operatorname{im} i_*$ one-to-one?

In cohomology, this question becomes: If $T^*\phi = \phi$, does this imply $\phi = j^*\phi$, i.e., can we extend ϕ across the singular fibre? Since we are working with coefficients in a field, in homology this question becomes: If $g_*\gamma = 0$, does this imply $\gamma \in \text{Image}(T_* - I_*)$? Hence we need to study the kernel of g_*.

1.4.2. To study the index of nilpotency we know that $(T_* - I_*)^3 = 0$, but when is $(T_* - I_*)^2 = 0$? If so, when is $T_* - I_* = 0$, i.e., when is the original monodromy of finite order?

Information about (1.4.1) and (1.4.2) will come from the topological study of the X_i and the ker g_*. The statements of the results we obtain are stated in §3 below. §2 is a discussion of some of the techniques.

2. Study of the ker g_*.

2.1. In what follows, g_*^{-1} will correspond to the homology map induced from the chain map defined in (1.3) and will not mean the inverse of g_*. Since we never use the inverse of g_*, there is no ambiguity.

2.1.1. THEOREM. *$\ker g_* \cap H_2(V_1)$ is generated by 3 subspaces G_1, G_2, G_3 where*

(i) *G_1 is the subspace generated by $g_*^{-1}(\gamma)$ for $0 \neq \gamma \in H_0(X_{ijk})$,*

(ii) *G_2 is the subspace generated by $g_*^{-1}(\gamma)$ for $\gamma = \sum_J \gamma_{ij}$ where $\gamma_{ij} \in$*

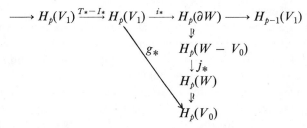

$H_1(X_{ij}, \bigcup_k X_{ijk})$ *with* $\partial_* \gamma_{1,ij} \neq 0$, *and*

(iii) G_3 *is the subspace generated by* $g_*^{-1}(\gamma)$ *for* $\gamma \in H_1(X_{ij})$.

2.1.2. DEFINITION. A polygon Δ_2 is *simple* if at each vertex there are either three or two edges. It is *closed* if $\partial \Delta_2 = 0$.

Hence, topologically, G_1 is generated by tori, G_3 by circles over 1-cycles of the curves X_{ij} and G_2 by cylinders which are plumbed together three at an end: Namely, at X_{ijk}, we have three curves X_{ij}, X_{ik}, and X_{jk}, coming together; and on each curve we have a real line with endpoint X_{ijk}. Over each point on the line we have a circle, which forms the cylinder. If $\alpha \times \beta$ represents the "torus fibre" over X_{ijk}, then the boundaries of the cylinders are α, β and $-(\alpha + \beta)$, which are fitted together to form a cycle.

2.1.3. CONJECTURE. $\ker g_* \cap H_2(V_1) = G_1 \oplus G_2 \oplus G_3$.

2.2. We now discuss G_1.

We let $I(i) = \{j \mid X_{ij} \neq \varnothing\}$, $I(i,j) = \{k \mid X_{ijk} \neq \varnothing\}$ and $I^c \subset I$ (respectively $I^c(i) \subset I(i)$) be those j with X_j (respectively X_{ij}) compact. Recall that X_{ij} and X_{ijk} are always assumed to be connected.

Then the X_{ij} will give relations among the $g_*^{-1}(X_{ijk}) \in G_j$.

2.2.1. DEFINITION. Let G^F (respectively G^c) be the abelian group generated by $g_*^{-1}(X_{ijk})$ subject to the relations that for all $i \in I$, $j \in I(i)$ (respectively, for all $i \in I$, $j \in I^c(i)$) (where ε_{ijk} is -1 if $i < k < j$ and $+1$ otherwise)

$$\sum_{k \in I(i,j); i<j} \varepsilon_{ijk} g_*^{-1} X_{ijk} = 0.$$

We let G_1^F or G_1^c mean we consider closed or compact support in $H_*(V_1)$, which we denote by $H_*^F(V_1)$ or $H_*^c(V_1)$ respectively.

2.2.2. CONJECTURE. rank $G^* = \dim G_1^*$ *for* $* = c$ *or* F.

2.3. Associated in V_0, we now construct a 2-dimensional polygon Γ. The construction is done in four steps, but we will only give the first two of them, which will be sufficient if one studies closed support or compact support if V_0 dominates a surface having only isolated singularities (which implies $|I_0| = 1$, i.e., there is only one proper transform). The complete details for the construction of the polyhedron in any dimension can be found in [3].

First step. To each $i \in I$, we correspond a 2-dimensional plane Y_i. If $j \in I(i)$, then the two 2-dimensional planes Y_i and Y_j will intersect in a line Y_{ij}. Finally, if $k \in I(i,j)$, then the three lines Y_{ij}, Y_{jk} and Y_{ik} will intersect in a vertex Y_{ijk}.

Second step. Suppose $i \in I^c$. Then Y_i is subdivided into a collection of polygonal regions by the lines Y_{ij}. Then let W_i be the simplicial subcomplex of Y_i obtained by removing all the open unbounded regions of Y_i, i.e., W_i is the one-skeleton of Y_i plus any finite region that the one-skeleton bounds.

Next, suppose $j \in I^c(i)$. If Y_{ij} has no vertices, then suppress the line Y_{ij}, i.e., separate Y_i (or W_i) and Y_j (or W_j). If Y_{ij} has vertices, throw away the open unbounded part, and call the remaining simplex W_{ij}.

Finally, if $i \notin I^c$ is such that Y_i has its 1-skeleton compact after the above excisions, excise the open unbounded part of Y_i and call it W_i.

Let $W_{ijk} = Y_{ijk}$.

The last two steps, which we omit, are analogous to looking at the three real planes $xyz = 0$ in R^3 and excising all but the part in the first octant.

2.3.1. PROPOSITION. *If Γ is the above polygon associated to V_0, then $G^F \simeq H^2_c(\Gamma; Z)$, $G^c \simeq H^F_2(\Gamma; Z)$, and Γ is generated by simple, closed 2-dimensional polygons.*

2.3.2. Let us consider a specific example in constructing Γ.

Let $\tilde{V} = \{(x, y, z) \in C^3 \mid x^2y^2z^2 + x^8 + y^8 + z^8 = 0\}$, the example studied by Malgrange [7]. Then we get the following polyhedron.

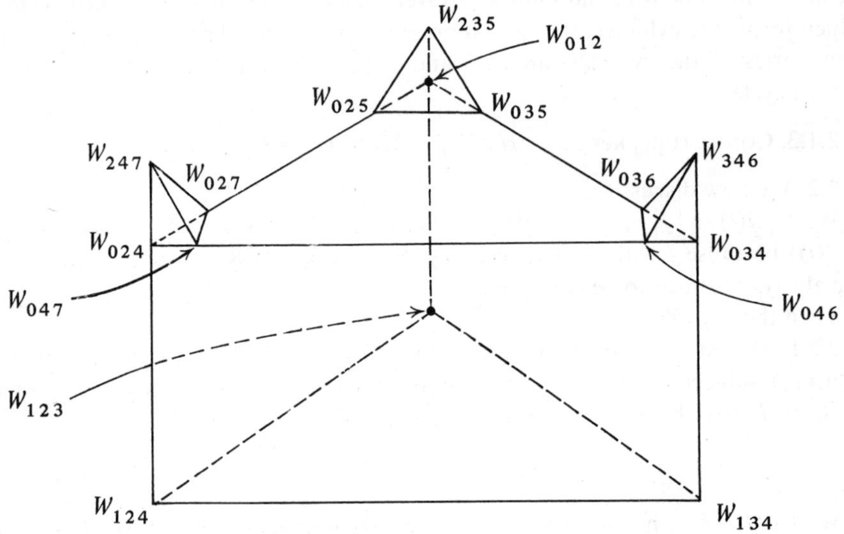

and $H_2(\Gamma; Z) = Z \oplus Z \oplus Z \oplus Z$.

2.4. Similar analyses are made for G_2 and G_3.

For G_2, the polygon Γ used in the study of G_1 is also used here as well, but nothing else must be introduced.

For G_3, the $\gamma_3 \in H_3(X_i)$ give relations among the $g_*^{-1}(\gamma)$ for $\gamma \in H_1(X_{ij})$ analogous to the X_{ij} for $g_*^{-1}(X_{ijk})$. To study G_3, we form a one-dimensional polygon whose lines correspond to $\gamma_3 \in H_3(X_i)$ and vertices to $\gamma \in H_1(X_{ij})$.

3. Some applications.

3.1. The details of the results of §3 can be found in [**2**].

We first consider the local invariant cycle problem (1.4.1).

3.1.1. DEFINITION. We say X_i is *quasi-Kähler* if X_i is bianalytically homeomorphic to a compact Kähler manifold minus, at most, a proper subvariety.

3.1.2. THEOREM. *Suppose $\{V_t\}_{t \in D}$ is a family of complex varieties of complex dimension two such that the generic fibre is nonsingular. Moreover, suppose that all the components of the proper transform are quasi-Kähler. Then* (1.4.1) *is true for $H^F_p(V_1)$ for all p.*

We need quasi-Kähler for two reasons. One is (2.3.1) cited above and the other is the *principle of the two types* which tells us that if we have a family of curves S_i in general position in a nonsingular Kähler manifold and we form a 1-cycle in this family of curves by joining the double points S_{ij} by real lines in S_i which do not wrap around any handles, then the 1-cycle bounds in the ambient space.

An important tool we use is the following theorem, cf. [**4**].

3.1.3. THEOREM. *If V is a nonsingular compact curve in W, a nonsingular compact Kähler surface, then $H_1(V) = \ker i_* \oplus I(H_3(W))$, where $i\colon V \subset W$ is the inclusion and $I\colon H_3(W) \to H_1(V)$ is transverse intersection. Furthermore the decomposition is nondegenerate with respect to the intersecting pairing.*

This is important in studying G_1 (cf. (2.4)). The corresponding result is true for surfaces as submanifolds of compact Kähler manifolds, but false in higher dimensions. These examples in higher dimension let us construct the examples of (3.4.1), below.

3.2. Then, using the results of (2.3), we have

3.2.1. THEOREM. *$(T_* - I_*)^2 H_2^F(V_1) \subset G_1^F$, and if each of the components of V_0 is quasi-Kähler, we have equality. If V_0 dominates a globally irreducible subvariety, then $(T_* - I_*)^2 H_2^c(V_1) \subset G_1^c$, and equality holds if the proper transform is quasi-Kähler.*

Hence, from (2.3.1), we have that $H_2^c(\Gamma; Z) = 0$ implies $(T_* - I_*)^2 H_2^F(V_1) = 0$. If V_0 dominates an irreducible subvariety, then the result is still true when we interchange c and F. The converse is true if each of the components of V_0 is quasi-Kähler.

We need the hypothesis of irreducibility for compact support, for if one considers $0 = xyz$ in $\boldsymbol{C}^3$, then $0 \neq G_1$, but $T_* = I_*$ on $H_2^c(V_1)$. We also note that we need quasi-Kähler for the converse. For we consider $\{(x^2 + y^3)(y^3 + x^2) + z^2 = t\} = V_t$ and let $\tilde{V}_t = V_t \cap D_\varepsilon$ where D_ε is a small ball about the origin. Then $\tilde{V}_t$ is not quasi-Kähler, and when one resolves the singularity of $0 \in \tilde{V}_0$, one has $H_2^F(\Gamma; Z) \neq 0$, but $(T_*^{20} - I_*)^2 = 0$ by the results of Sebastian and Thom [**9**] which states the monodromy in question is minus that of $((x^2 + y^3)(y^3 + x^2) = t) \cap D_\varepsilon$.

3.3. Suppose $(T_* - I_*)^2 = 0$. Then when is T_* of finite order?

3.3.1. PROPOSITION. *Suppose $H_2(\Gamma; Z) = 0$. Then $(T_* - I_*)H_2(V_1) \subset G_3$. Conversely, if all the X_i are quasi-Kähler, then $G_3 \subset (T_* - I_*)H_2(V_1)$.*

To determine when $G_3 = 0$, one should look for $X_{ij} \subset X_i$ such that the inclusion map is injective on $H_1(X_{ij})$. One special case is the proposition

3.3.2. PROPOSITION. *Let $\tilde{V} \subset \boldsymbol{C}^3$ be a surface given by a polynomial having only isolated singularities. Let $V_0 = \bigcup_{i \in I} X_i$ be a semistable reduction (cf. (1.2)) and let X_0 denote the proper transform. Suppose that $H_2(\Gamma; Z) = 0$. Then the following are equivalent:*

(i) *T_* is of finite order.*
(ii) $0 \longrightarrow \bigoplus_i H_1(X_{i0}) \longrightarrow H_1(X_0)$.

3.3.3. COROLLARY. *Let $0 \in \boldsymbol{C}^3$ be an isolated singularity of a hypersurface $\tilde{V}$ and suppose $\tilde{V}$ is contractible. Then let $V_0 = \bigcup_{i \in I} X_i$ be any resolution of $\tilde{V}$ and let Γ_0 be the graph associated to the X_{i_0} in X_0, the proper transform. Then $H_1(\Gamma_0) = 0$ and T_* is of finite order.*

(3.3.3) contains as a special case the weighted homogeneous polynomials studied by Orlik and Wagreich [**8**].

3.4. Finally, we give some applications to construct some projective algebraic varieties which are not deformations of nonsingular Kähler manifolds.

3.4.1. PROPOSITION. *Let* $i_1 : X_1 \subset Y_1$ *where* Y_1 *is a nonsingular projective variety of complex dimension n and let* X_1 *be a compact submanifold of complex codimension* $k \geqq 2$. *Let* $0 < p < 2n - 2k$ *and suppose* $H_p(X_1) \neq 0$. *Then suppose either* (i) (a) $p + 2i = n$ *and* (b) $i \neq k - 1$, *or if* $i = k - 1$, *then* $n < 2k - 1$, *or* (ii) $H_{p+2k}(Y_1) = 0$. *Finally let* $\pi : Y \to Y_1$ *be the monoidal transform with center* X_1 *and let* $X = \pi^{-1}(X_1)$. *Then the projective variety* $Y \cup_X Y$ *cannot be embedded in a one-dimensional analytic deformation of Kähler varieties whose generic fibre is nonsingular and compact.*

BIBLIOGRAPHY

1. C. H. Clemens, *Picard-Lefschetz theorem for families of nonsingular algebraic varieties acquiring ordinary singularities,* Trans. Amer. Math. Soc. **136** (1969), 93–108. MR **38** #2135.
2. G. L. Gordon, *A geometric study of the monodromy of complex analytic surfaces* (to appear).
3. ———, *On a polyhedron associated to normal crossings* (to appear).
4. ———, *On the homology of submanifolds of compact Kähler manifolds,* J. Differential Geometry (to appear).
5. P. A. Griffiths, *Periods of integrals on algebraic manifolds: Summary of main results and discussion of open problems,* Bull. Amer. Math. Soc. **76** (1970), 228–296. MR **41** #3470.
6. G. Kempf, et al., *Toroidal embeddings.* I, Lecture Notes in Math., vol. 339, Springer-Verlag, Berlin and New York, 1973. MR **49** #299.
7. B. Malgrange, *Letters to the editors,* Invent. Math. **20** (1973), 171–172. MR **48** #8839.
8. P. Orlik and P. Wagreich, *Isolated singularities of algebraic surfaces with C* action,* Ann. of Math. (2) **93** (1971), 205–228. MR **44** #1662.
9. M. Sebastiani and R. Thom, *Un résultat sur la monodromie,* Invent. Math. **13** (1971), 90–96. MR **45** #2201.
10. A. H. Todorov, Izv Acad. Nauk USSR (to appear).

UNIVERSITY OF ILLINOIS AT CHICAGO CIRCLE

Proceedings of Symposia in Pure Mathematics
Volume 30, 1977

REMARKS ON ASYMPTOTIC INTEGRALS, THE POLYNOMIAL OF I. N. BERNSTEIN AND THE PICARD-LEFSCHETZ MONODROMY

HELMUT A. HAMM*

B. Malgrange has pointed out in [7] that there is an intimate connexion between asymptotic integrals and the local Picard-Lefschetz monodromy. Starting with the case of asymptotic integrals in the real domain, consider the asymptotic behaviour of the integral

$$I(\tau) = \int_{\boldsymbol{R}^n} e^{i\tau f(x)} g(x)\, dx$$

for $\tau \to \infty$, where f and g are real-valued C^∞ functions on $\boldsymbol{R}^n$ and g has compact support. We can assume that the support of g is contained in a small neighbourhood of the origin and that $f(0) = 0$. If f is real analytic there and the origin is an isolated critical point of f, there is a method of reduction to asymptotic integrals in the complex domain which is classically well known, e.g., for the case $n = 1$, $f(x) = x^2$, as "method of steepest descent". These complex integrals have an asymptotic development of the form

$$\sum_{\mu; 0 \leqq q < n} e_{\mu, q} \tau^{-\mu} (\log \tau)^q,$$

where μ runs through a countable set of positive rational numbers. The connexion with the local Picard-Lefschetz monodromy of f at 0 is now established by the fact that the numbers $e^{2\pi i \mu}$ are the eigenvalues of the monodromy.

Of course, the monodromy can give only partial information about the numbers μ; but one can get more from another invariant: the local "Bernstein" polynomial whole roots are related to the numbers μ. If f is a polynomial, there is a corresponding global analogue which has been studied by I. N. Bernstein. B. Malgrange

AMS (MOS) subject classifications (1970). Primary 14B05, 41A60; Secondary 14D05.

*Supported in part by Deutsche Forschungsgemeinschaft and the National Science Foundation.

© 1977, American Mathematical Society

[9] has shown the connexion between this global Bernstein polynomial and the global Picard-Lefschetz monodromy of f. For the case of an isolated singularity, he has complete results about the local Bernstein polynomial in [8].

But one of the main purposes of this paper is to include the case of nonisolated singularities, too. The methods are adopted from [7] and [9]. We need auxiliary results—e.g., the regularity theorem for nonisolated singularities—from [6]. Unfortunately, the results about the Bernstein polynomial are by no means complete.

I. **Asymptotic integrals in the complex domain.** Let $f: D \to \mathbf{C}$ be holomorphic, where $D = \{z \in \mathbf{C}^n \mid \| z \| < \varepsilon\}$, $f(0) = 0$. If $S' = \{t \in \mathbf{C} \mid 0 < |t| < \alpha\}$ and $X' = f^{-1}(S')$, the mapping $f' = f \mid X' : X' \to S'$ defines a C^∞ fibre bundle provided that ε was chosen small enough and $0 < \alpha \ll \varepsilon$; cf. [5, 1.6]. Possibly changing f, we can assume that -1 belongs to S'. For each i, the canonical generator of $\pi_1(S', -1)$ induces an automorphism $h = h_f^i$ of $\tilde{H}^i(X_{-1}; \mathbf{C})$, where X_t denotes the fibre of f over t; h is the (reduced local complex) Picard-Lefschetz monodromy (in dimension i).

There is a basis of $\tilde{H}_i(X_{-1})$—homology always being taken with complex coefficients—which consists of elements $[\gamma(-1)]$ such that

$$(1) \qquad (h - \lambda \, \mathrm{id})^p [(\gamma(-1)] = 0, \qquad (h - \lambda \, \mathrm{id})^{p-1} [\gamma(-1)] \neq 0,$$

where λ is an eigenvalue of h. We fix such a $[\gamma(-1)]$ and therefore also λ and p; let $[\Gamma] \in H_{i+1}(X, X_{-1})$ correspond to $[\gamma(-1)]$ under $H_{i+1}(X, X_{-1}) \cong \tilde{H}_i(X_{-1})$, where $X = f^{-1}(S)$, $S = \{t \in \mathbf{C} \mid |t| < \alpha\}$.

Let $\Omega_X^\cdot$ be the de Rham complex of germs of holomorphic differential forms on X, $\mathscr{S}^\cdot = \ker(df: \{\Omega_X^\cdot, d\} \to \{\Omega_X^{\cdot+1}, -d\})$, $[\omega] \in H^{i+1}(\Gamma(X, \mathscr{S}^\cdot))$. We want to study the asymptotic behaviour of $\int_\Gamma e^{\tau f} \omega$ for $\tau \to \infty$. We refer to [7] for the manner in which asymptotic integrals in the real domain are related to asymptotic integrals of this type with $i = n - 1$.

THEOREM 1. *For* $\tau \to \infty$, $\int_\Gamma e^{\tau f} \omega$ *has the following asymptotic expansion*:

$$\sum_{e^{2\pi i \mu} = \lambda; 0 \leqq q \leqq p-1} e_{\mu,q}(\omega) \tau^{-\mu} (\log \tau)^q$$

where $e_{\mu,q}(\omega) \in \mathbf{C}$ *vanishes for* $\mu \leqq 0$. *The expansion is independent of the choice of representatives for* $[\Gamma]$ *and* $[\omega]$.

This means that

$$\frac{d^k}{d\tau^k} \left[\int_\Gamma e^{\tau f} \omega - \sum_{\substack{e^{2\pi i \mu} = \lambda; \mu \leq \nu; \\ 0 \leqq q \leqq p-1}} e_{\mu,q}(\omega) \tau^{-\mu} (\log \tau)^q \right] = O(\tau^{-\nu-k})$$

for $\tau \to \infty$ and all k. The case of an isolated singularity, $i = n - 1$, has been proved by B. Malgrange; cf. [7, 6.3].

The essential tool for the proof is the "regularity theorem" for the local singular Gauss-Manin connexion. It can be stated as follows: If $[\gamma(t)]$ is a horizontal family of homology classes in $\tilde{H}_i(X_t; \mathbf{C})$, $t \in S'$, and if $\varphi \in \Gamma(X, \Omega_X^i)$ satisfies $df \wedge d\varphi = 0$, there is a number N and a constant C such that for all $t \in S'$:

$$(2) \qquad \left| \int_{\gamma(t)} \varphi \right| \leqq C \cdot |t|^{-N}.$$

The case of an isolated singularity has been proved by E. Brieskorn [**3**]; for the general case cf. [**6**, II, 3.5].

Now we choose $\gamma(t)$ such that $[\gamma(-1)]$ is the homology class fixed before. From (1) and (2) one can deduce that

$$(3) \qquad \int_{\gamma(t)} \varphi = \sum_{e^{2\pi i\mu}=\lambda; 0\leq q\leq p-1} c_{\mu,q}(\varphi) t^{\mu}(\log t)^q$$

where $c_{\mu,q} = 0$ for $\mu \ll 0$. But one also has

$$(4) \qquad \lim_{t>0, t\to 0} \int_{\gamma(t)} \varphi = 0,$$

cf. [**7**, Lemme 4.5]; therefore $c_{\mu,q} = 0$ for $\mu \leqq 0$.

If $[\omega] \in H^{i+1}(\Gamma(X, \mathcal{S}^\cdot))$, ω can be written in the form $d\varphi$, where $\varphi \in \Gamma(X, \Omega_X^i)$ with $df \wedge d\varphi = 0$. From

$$(5) \qquad \frac{d}{dt}\int_{\gamma(t)} \varphi = \int_{\gamma(t)} \frac{d\varphi}{df},$$

we get

$$(6) \qquad \int_{\gamma(t)} \frac{\omega}{df} = \sum_{e^{2\pi i\mu}=\lambda; 0\leq q\leq p-1} d_{\mu,q}(\omega) t^{\mu}(\log t)^q$$

with $d_{\mu,q} = 0$ for $\mu \leqq -1$. Recall the definition of ω/df: Since $df \wedge \omega = 0$, there is for large ν a form $\psi \in \Gamma(X, \Omega_X^i)$ such that $f^\nu \omega = df \wedge \psi$. Then $\omega/df = f^{-\nu}\psi$, and this form is, modulo $df \wedge \Omega_{X}^{i-1}$, independent of the choice of ψ.

PROOF OF THEOREM 1. For $[\tilde{\omega}] \in H^{i+1}(\Gamma(X, \mathcal{S}^\cdot))$, we have because of (4) and (5):

$$\int_\Gamma \tilde{\omega} = \int_0^{-1} dt \int_{\gamma(t)} \frac{\tilde{\omega}}{df}.$$

Therefore

$$\int_\Gamma e^{\tau f}\omega = \int_0^{-1} e^{\tau t}\, dt \int_{\gamma(t)} \frac{\omega}{df}.$$

Now the theorem follows from (6).

It is obvious that $H^{i+1}(\Gamma(X, \mathcal{S}^\cdot))$ is a $\Gamma(S, \mathcal{O}_S)$-module; let $H^{i+1}(\Gamma(X, \mathcal{S}^\cdot))^\wedge$ be the completion with respect to $0 \in S$. By continuity, the asymptotic expansion can be defined even for $[\omega] \in H^{i+1}(\Gamma(X, \mathcal{S}^\cdot))^\wedge$. Furthermore, the maximal ideals of $\mathcal{O}_{S,0}$ and $\mathcal{O}_{X,0}$ induce the same topologies on $H^{i+1}(\Gamma(X, \mathcal{S}^\cdot))^\wedge$; this can be shown using the methods of [**6**, I, 3]. Therefore the reduction of asymptotic integrals in the real domain to those in the complex domain is established in the same way as in the proof of [**7**, 7.6]; of course, one has to assume then that the origin is an isolated critical point of $f|\mathbf{R}^n: \mathbf{R}^n \to \mathbf{R}$.

Let us return to the complex case and define $M^{i+1} = H^{i+1}(\Gamma(X, \mathcal{S}^\cdot)) \otimes_{\mathbf{C}[[t]]} K$, where K is the quotient field of $\mathbf{C}[[t]]$. The usual local singular Gauss-Manin connexion defines a K-derivation on the finite-dimensional K-vector space M^{i+1} (cf. [**6**, II, 4])

$$\nabla_t : M^{i+1} \to M^{i+1}.$$

But we can consider $H^{i+1}(\Gamma(X, \mathcal{S}^\cdot))^\wedge$ as a $\mathbf{C}[[\tau^{-1}]]$-module, too; the multiplication by τ^{-1} comes from the following C-endomorphism of $H^{i+1}(\Gamma(X, \mathscr{S}^{\cdot}))$: $[\omega] \mapsto [df \wedge \varphi]$, where $d\varphi = \omega$. Let K' be the quotient field of $C[[\tau^{-1}]]$; then M^{i+1} is a K'-module, for we have $\nabla_t[\omega] = \tau \cdot [\omega]$. On M^{i+1} we can introduce a K'-derivation

$$\nabla_\tau : M^{i+1} \to M^{i+1}$$

by $\nabla_\tau[\omega] = t \cdot [\omega]$. There is a nice duality theorem.

THEOREM 2. *∇_τ is regular singular, and its monodromy is inverse to that of ∇_t.*

The proof is the same as that of [7, 6.6] for the case of an isolated singularity. The theorem can be applied to prove the following theorem which in the case of an isolated singularity goes back to R. Thom and M. Sebastiani [**10**].

THEOREM 3. *If $f \in C\{x_1, \cdots, x_n\}$, $g \in C\{y_1, \cdots, y_m\}$, $f(0) = 0$, $g(0) = 0$, then*

$$h^{i-1}_{f+g} = \bigoplus_{\mu+\nu=i} h^{\mu-1}_f \otimes h^{\nu-1}_g.$$

The proof is the same as in [7, 6.8]; one uses ∇_τ instead of ∇_t.

II. The local Bernstein polynomial. We use the same notations as in § I. According to Björk [**2**], there is a polynomial $B(s) \in C[s] - \{0\}$ and a linear partial differential operator $P(s, z, D_z)$ with coefficients in $(C\{z_1, \cdots, z_n\})[s]$ such that $Pf^s = B(s)f^{s-1}$. These $B(s)$, together with 0, form an ideal; the generator with leading coefficient 1 is called the local Bernstein polynomial $b(s)$ of f at 0. It is divisible by s; $\tilde{b}(s) = b(s)/s$ is the reduced local Bernstein polynomial. We fix $\gamma(-1)$, λ and p as in § I and introduce the numbers

$$\mu_k = \inf \{\mu \mid e^{2\pi i \mu} = \lambda, \exists\, q \geqq k - 1 \;\; \exists\, \omega \in \Gamma(X, \mathscr{S}^{i+1}) \text{ such that } d_{\mu, q}(\omega) \neq 0\}, \qquad 1 \leqq k \leqq p.$$

We assume that $i = n - 1$.

THEOREM 4. *The polynomial $(s + \mu_1), \cdots, (s + \mu_p)$ divides $\tilde{b}$.*

Note that $\mu_1, \cdots, \mu_p > -1$ according to (6). The proof is the same as that of [**9**, 3.3]. It is not clear whether the theorem holds also if $i < n - 1$.

Since the numbers $e_{\mu, q}(\omega)$ can be computed from the $d_{\mu, q}(\omega)$, the μ_k give information about the lowest power of τ^{-1} which can occur in the asymptotic expansion of $\int_\Gamma e^{\tau f} \omega$.

III. The global Bernstein polynomial. Now let $f: C^n \to C$ be a polynomial; the other notations are the same as in §I, so $S' = \{t \mid 0 < |t| < \alpha\}$, $X' = f^{-1}(S')$ and so on. According to P. Deligne [**4**], f' defines a C^∞ fibre bundle, and the regularity theorem for the Gauss-Manin connexion holds in this case. All differential forms are now supposed to have polynomial coefficients.

According to I. N. Bernstein [**1**], there is a linear partial differential operator $P(s, z, D_z)$ with polynomial coefficients and $B(s) \in C[s] - \{0\}$ such that $Pf^s = B(s)f^{s-1}$; so we get a "global" Bernstein polynomial $b(s)$ and a corresponding $\tilde{b}(s)$.

Then Theorem 4 remains true. It has been proved by B. Malgrange in [**9**] in a weaker form, where $\gamma(-1)$ is supposed to be a "vanishing cycle", cf. [**9**, 3.3]; but the proof works also in general. In the case where $\gamma(-1)$ is a vanishing cycle, (4) holds and therefore $\mu_1, \cdots, \mu_p > -1$. The same is true for the larger class of "bounded cycles". The definition is given below.

There is another global fibre bundle associated to f besides that defined by f'. Let us consider $X'_R = f^{-1}(S') \cap D_R, f'_R = f' \mid X'_R$, where $D_R = \{z \in \mathbf{C}^n \mid \|z\| < R\}$. If R is large enough and $0 < \alpha \ll 1/R, f'_R$ defines a C^∞ fibre bundle; cf. [**6**, II, 1.1].

It is a natural question whether this bundle is always equivalent to that defined by f'. The answer is "no", though it is difficult to find examples.

EXAMPLE FOR $n = 2$. $f(x, y) = x^3 + x^2y - x$. Then $f_R'^{-1}(t)$ is diffeomorphic to $f^{-1}(0)$ and therefore to the disjoint union of $\mathbf{C}$ and $\mathbf{C}^*$, whereas $f'^{-1}(t)$ is diffeomorphic to $\mathbf{C}^*$.

Now $\gamma(-1)$ is called a "bounded" cycle if it is the image of some cycle in $f_R'^{-1}(-1)$. Instead of (4) one knows then at least that $|\int_{\gamma(t)} \varphi|$ is bounded, so $\mu_1, \cdots, \mu_p > -1$. For unbounded cycles I know no restriction of this kind on the numbers $\mu_1, \cdots, \mu_p$.

ADDED IN PROOF. M. Kashiwara showed that the roots of the local Bernstein polynomial are rational and <1 (unpublished). As B. Malgrange pointed out to me, a coherence argument implies the same for the global Bernstein polynomial. If we replace the roots α of the local Bernstein polynomial by $e^{-2\pi i\alpha}$ we get a multiple of the minimal polynomial of the local Picard-Lefschetz monodromy in every dimension (B. Malgrange, unpublished).

REFERENCES

1. I. N. Bernšteĭn, *Analytic continuation of generalized functions with respect to a parameter*, Funkcional. Anal. i Priložen. **6** (1972), no. 4, 26–40 = Functional Anal. Appl. **6** (1972), 273–285 (1973), MR **47** #9269.

2. J. E. Björk, *Dimensions over algebras of differential operators* (to appear).

3. E. Brieskorn, *Die Monodromie der isolierten Singularitäten von Hyperflächen*, Manuscripta Math. **2** (1970), 103–161. MR **42** #2509.

4. P. Deligne, *Equations différentielles à points singuliers réguliers*, Springer-Verlag, Berlin and New York, 1970.

5. H. A. Hamm, *Lokale topologische Eigenschaften komplexer Räume*, Math. Ann. **191** (1971), 235–252. MR **44** #3357.

6. ———, *Zur analytischen und algebraischen Beschreibung der Picard-Lefschetz-Monodromie*, Habilitationsschrift Göttingen, 1974 (mimeographed).

7. B. Malgrange, *Intégrales asymptotique et monodromie*, Ann. Sci. Ecole. Norm. Sup. (4) **7** (1974), 405–430.

8. ———, *Le polynôme de Bernšteĭn d'une singularité isolée*, Fourier Integral Operators and Partial Differential Equations, Springer-Verlag, Berlin and New York, 1975.

9. ———, *Sur les polynômes de I. N. Bernstein*, Séminaire Goulaouic-Schwartz 1973–1974, Exposé XX.

10. M. Sebastiani and R. Thom, *Un résultat sur la monodromie*, Invent. Math. **13** (1971), 90–96. MR **45** #2201.

UNIVERSITÄT GOTTINGEN

Proceedings of Symposia in Pure Mathematics
Volume 30, 1977

DEFORMATIONS OF CUSP SINGULARITIES

ULRICH KARRAS

Introduction. Let Γ be the Hilbert modular group of a totally real field of degree n over the rationals. Γ operates effectively on H^n where H is the upper half-plane. The orbit space is a normal complex space with quotient singularities. Schlessinger [**21**] proved that quotient singularities of dimension > 2 are rigid. In the case $n = 2$ the orbit space has only cyclic quotient singularities. Their interesting deformation theory was worked out by Brieskorn [**6**], Arnold [**1**] and Riemenschneider [**18**]. By passing from H^n/Γ to the compactification $\overline{H^n/\Gamma}$ normal isolated singularities arise which are not quotient singularities. In this note we deal with their deformation theory. We first want to give a more precise definition of these singularities.

Let K be a totally real field of degree n over the rationals and M an additive subgroup of K which is a free abelian group of rank n. Let U_M^+ be the group of those units ε of K which are totally positive and satisfy $\varepsilon M = M$. For a given pair (M, V) with $V \subset U_M^+$ (where V has rank $n - 1$) one defines

$$G(M, V) : \left\{ \begin{pmatrix} \varepsilon & \mu \\ 0 & 1 \end{pmatrix} \Big|, \varepsilon \in V, \mu \in M \right\}.$$

The group $G(M, V)$ operates freely and properly discontinuously on H^n by $z_j \mapsto \varepsilon^{(j)} z_j + \mu^{(j)}$, where $x \mapsto x^{(j)}$, $1 \leqq j \leqq n$, denote the n different embeddings of K into the reals. Then $H^n/G(M, V)$ defines a complex manifold which acquires a normal singularity when an additional point ∞ is added with neighborhoods $|\operatorname{Im}(z_1) \cdots \operatorname{Im}(z_n)| >$ const. The singularity at ∞ will be called a *cusp singularity* of type (M, V).

THEOREM 1 (FREITAG AND KIEHL [7]). *Cusp singularities of dimension* >2 *are rigid.*

We shall show that the deformation theory of two-dimensional cusp singularities is nontrivial. This result is not surprising because at present no example of a rigid two-dimensional singularity is known.

AMS (MOS) subject classifications (1970). Primary 13D10, 14H20, 32C40, 32G05.

© 1977, American Mathematical Society

THEOREM 2. *Let (S, p) be a two-dimensional cusp singularity and let Z be its fundamental divisor. Then the simple elliptic singularities of degree $d = -Z \cdot Z$ are specializations of (S, p).*

Recall the simple elliptic singularities of degree d, $d \geqq 3$, are exactly the affine cones over smooth elliptic curves of degree d in $\boldsymbol{P}^{d-1}$. For cusp singularities being complete intersections we get a complete description of all possible specializations; see Theorem 9.

In this paper we announce the results and give outlines of their proofs. A more detailed treatment of the subject will appear elsewhere.

I would like to thank Professor H. Laufer for helpful correspondence and many stimulating conversations. I am also grateful to Professor E. Brieskorn for many useful references and for inviting me to talk about this work in his seminar at the Summer Institute on Several Complex Variables of the American Mathematical Society (Williams College, Williamstown, Massachusetts, 1975). Finally I wish to acknowledge the support given by University Dortmund for the trip to Williamstown.

1. Deformations of cusp singularities being complete intersections. In the following we will deal only with two-dimensional cusp singularities. They can be characterized as follows. Let (S, p) be a normal complex surface singularity and let $\pi : \tilde{S} \to S$ be (a representative of) the minimal resolution of (S, p). Let $A := \pi^{-1}(p)$ be the exceptional set. Then (S, p) is a cusp singularity if and only if A is an irreducible rational curve with a node singularity or A is a "cycle" of nonsingular rational curves A_i. The configuration is illustrated by the diagram:

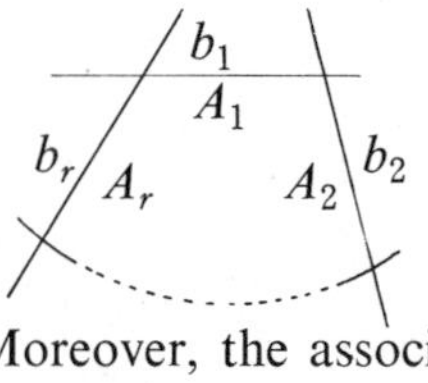

where $b_i = A_i A_i$, $1 \leqq i \leqq r$. Moreover, the associated cycle $((b_1, \cdots, b_r))$ of self-intersection numbers determines the singularity (S, p) up to complex-analytic equivalence; see **[9]**, **[14]** and **[11]**.

In this section we consider only cusp singularities which are complete intersections.

THEOREM 3. *A cusp singularity (S, p) is a complete intersection if and only if* ebdim $(S, p) \leqq 4$. *Defining equations are given as follows.*

(a) ebdim $(S, p) = 3$: $x^k + y^l + z^m = xyz$ *with* $1/k + 1/l + 1/m < 1$.

(b) ebdim $(S, p) = 4$: $x^p + w^q = yz$; $y^r + z^s = xw$ *with* $p, q, r, s \geqq 2$ *and at least one exponent* $\geqq 3$.

The first statement is a special case of a much more general result of Laufer **[13**, Theorem 3.13**]**. In **[12]** we proved that "ebdim $(S, p) \leqq 4$" is a necessary condition. Let Z be the fundamental divisor of (S, p). By **[11]** we have ebdim $(S, p) = \max(3, -Z \cdot Z)$; see also **[13]**. It is easy to write down all cusp singularities with embedding dimension $\leqq 4$ in terms of their associated cycles. Comparing them with the minimal resolutions of singularities given by the equations in Theorem 3 completes the proof.

In **[10]** Kas and Schlessinger gave an explicit description of semi-universal deformations of isolated complete intersections in terms of their defining equations; see also [**24**]. Hence we get by direct computations:

LEMMA 4. (a) *The semi-universal deformation of the hypersurface singularity* $\{f(x, y, z) := xyz - x^k - y^l - z^m = 0\}$ *is given by the natural projection* $C^3 \times C^n \supset V: \to C^n$, $n = k + l + m - 2$, where

$$V := \Big\{(x, y, z, \lambda) \mid \lambda_0 = f(x, y, z) + \sum_1^{k-1} \lambda_i x^i + \sum_k^{k+l-2} \lambda_i y^{i-k+1} + \sum_{k+l-1}^{n-1} \lambda_i z^{i-k-l+2}\Big\}.$$

(b) *The semi-universal deformation of* $\{f_1 := yz - x^p - w^q = 0;\ f_2 := xw - y^r - z^s = 0\} \subset C^4$ *is given by the natural projection* $C^4 \times C^n \supset V: \to C^n$, $n = p + q + r + s - 2$, *where*

$$V := \{(x, y, z, w, \lambda) \mid f_1 + \lambda_2 x + \cdots + \lambda_p x^{p-1} + \lambda_{p+1} w + \cdots + \lambda_{p+q-1} w^{q-1} = \lambda_0;\ f_2 + \lambda_{p+q} y + \cdots + \lambda_{n-s} y^{r-1} + \lambda_{n-s+1} z + \cdots + \lambda_{n-2} z^{s-1} = \lambda_1\}.$$

As an easy application we get a formula for the number of moduli. Let (X, x) be an isolated singularity. Then the set $T^1_{X,x}$ of isomorphism classes of first order infinitesimal deformations of (X, x) is a complex vector space of finite dimension. If (X, x) is a complete intersection, then $\dim_C T^1_{X,x}$ equals the dimension of the parameter space of the semi-universal deformation of (X, x).

COROLLARY 5. *Let* (S, p) *be a cusp singularity with* ebdim $(S, p) \leqq 4$ *and Milnor number* μ. *Let* $\pi: \tilde{S} \to S$ *be the minimal resolution and* $((b_1, \cdots, b_r))$ *the associated cycle. Then we have*

$$\dim_C T^1_{S,p} = \mu - 1 = 10 + \sum_{i=1}^r 3 + b_i.$$

The formula for the Milnor number also follows from a much more general result.

PROPOSITION 6. *Let* (X, x) *be an isolated two-dimensional complete intersection with minimal resolution* $\pi : \tilde{X} \to X$. *Let* $\chi(A)$ *be the (topological) Euler characteristic of the exceptional set* $A := \pi^{-1}(x)$. *Then we have*

$$1 + \mu = \chi(A) + K \cdot K + 12 \dim_C H^1(\tilde{X}; \mathcal{O}_{\tilde{X}})$$

where K *is the canonical divisor on* $\tilde{X}$.

Laufer [**15**] proved this for hypersurface singularities. His proof also works in the general case.

DEFINITION. Let (X_0, x_0) be an isolated singularity and let $(X, x) \to (T, t)$ be a deformation of (X_0, x_0). An isolated singularity (Y, y) is called a *specialization* of (X_0, x_0) (notation: $(Y, y) \leqq (X_0, x_0)$) if a suitable representative of the map germ$(X, x) \to (T, t)$ has the following property. There are fibres X_s over points s in any neighborhood of t in T having x_s as their only singularities so that (X_s, x_s) and (Y, y) are isomorphic.

THEOREM 7. *Let* (S, p) *be a cusp singularity which is a complete intersection. Then any specialization of (S, p) is a complete intersection with finite or solvable local fundamental group. Conversely every two-dimensional isolated complete intersection with this property is a specialization of a cusp singularity.*

To describe all possible specializations of a cusp singularity we associate a "Dynkin diagram" to every two-dimensional isolated complete intersection with finite or solvable local fundamental group as follows:

(1) The "Dynkin diagram" of rational double points equals the usual Dynkin diagram of A_k, D_k, E_k in the theory of semisimple Lie algebras.

(2) The "Dynkin diagram" of the simple elliptic singularities of degree d, $1 \leqq d \leqq 4$, equals $\tilde{E}_8, \tilde{E}_7, \tilde{E}_6, \tilde{D}_5$ where the index is $9 - d$. The diagrams $\tilde{E}_1$ are the so-called extended Dynkin diagrams for E_1 in the theory of semisimple Lie groups (see [**3**, p. 199]) and

$\tilde{D}_5 :=$ 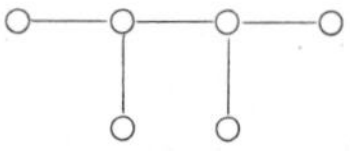

(3) By analogy the "Dynkin diagram" of a hypersurface cusp singularity $\{x^k + y^l + z^m = xyz\}$ is defined by

$E_{k,l,m} :=$ 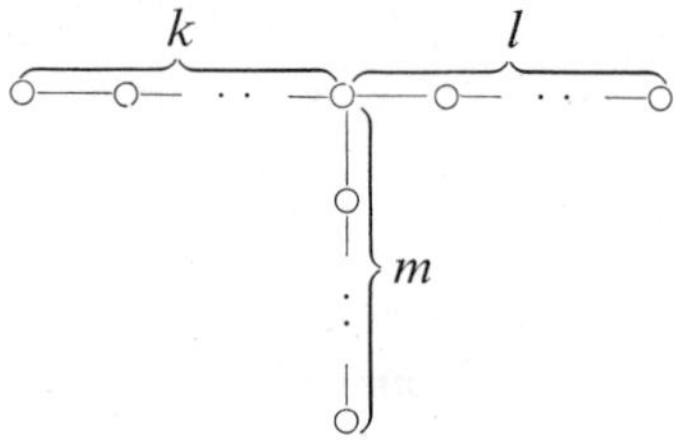

(4) The "Dynkin diagram" of $\{yz = x^p + w^q\,; xw = y^r + z^s\}$ is defined by

$D_{p,q,r,s} :=$ 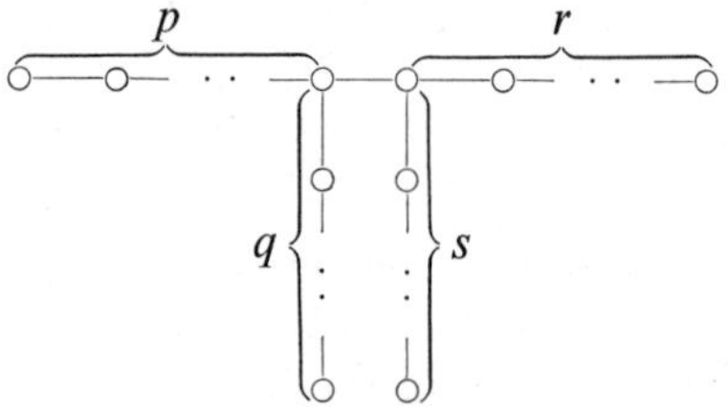

From Theorem 3 and the classification theorems ([**4**] and [**11**]) it follows that any two-dimensional complete intersection with finite or solvable local fundamental group is isomorphic to one of the singularities above. As is well known there is a one-to-one correspondence between Dynkin diagrams of rational double points and intersection forms of their Milnor fibres.

Question. Does a similar relation hold in the other cases?

Recently Gabrielov [**8**] gave an affirmative answer to this question in the hypersurface case. Actually he computed intersection matrices of the so-called unimodal singularities. As an application he proved:

PROPOSITION 8 (GABRIELOV [**8**]). *Let (S, p) be a hypersurface singularity with Milnor number μ. Assume (S, p) is simple elliptic (or a cusp singularity). Then the monodromy group of (S, p) contains as a normal subgroup the free abelian group* $\boldsymbol{Z}^r$, *by which the factor group is isomorphic to the Coxeter group with Dynkin diagram* (2) ((3) *respectively*). *The number r is equal to* $\mu - 2$ *if* (S, p) *is simple elliptic and* $\mu - 1$ *otherwise.*

We next formulate a slightly stronger version of Theorem 7.

THEOREM 9. *Assume* (X, x) *and* (Y, y) *are isolated two-dimensional complete intersections with finite or solvable local fundamental groups. Then we have* $(Y, y) \leqq (X, x) \Leftrightarrow$ *Dynkin diagram* $(Y, y) \subset$ *Dynkin diagram* (X, x).

OUTLINE OF PROOF. The result is known for rational double points and simple elliptic hypersurface singularities; see **[1]**, **[20]** and **[23]**. Recently I learned that Pinkham (correspondence with Brieskorn) can describe all possible specializations of any simple elliptic singularity. Hence it remains to study cusp singularities. First we consider only deformations with constant multiplicity. Let (S, p) be a cusp singularity with multiplicity > 2. We may assume that (S, p) is defined by an equation given in Theorem 3. Then the restriction of the semi-universal deformation of (S, p) (see Lemma 4) to $V' := \{(x, y, z, w, \lambda) \mid \lambda_0 = \lambda_1 = \lambda_2 = \lambda_{p+q} = \lambda_{p+1} = \lambda_{n-s+1} = 0\} \subset V$ ($V' := \{(x, y, z, \lambda) \mid \lambda_0 = \lambda_1 = \lambda_k = \lambda_{k+l-1} = 0\} \subset V$ respectively) is a deformation of (S, p) which is semi-universal for deformations of (S, p) with constant multiplicity. The fibres of this deformation only have absolutely isolated singularities placed at the origin. By blowing up we can now easily verify Theorem 9 on the understanding that the multiplicities are > 2 and coincide. The next step reduces our problem to the case of double points.

(5) Let (S, p) be a cusp singularity of multiplicity > 2. Then there exists a specialization (S', p') of (S, p) satisfying mult $(S', p') <$ mult (S, p) and the following property: $(Y, y) \leqq (S, p)$ and mult $(Y, y) <$ mult $(S, p) \Rightarrow (Y, y) \leqq (S', p')$. It is trivial to find a candidate. But we have to do rather tedious computations to check the wanted properties. Hence it remains to describe all possible specializations of double point cusp singularities. A double point always has defining equation of the form $z^2 = f(x, y)$. For cusp singularities we have $f(x, y) = x^k + x^2y^2 + y^l$, $k \leqq l$; see **[12]**. $f(x, y)$ defines a plane curve singularity at (0, 0). It suffices to study its semi-universal deformation. It is easy to see that any specialization of $\{f(x, y) = 0\}$ has defining equation of the following general form:

$$F_\lambda(x, y) = x^k + x^2y^2 + y^l + \lambda_1 xy + \lambda_2 x^2 + \lambda_3 y^2 + \lambda_4 x^2 y + \lambda_5 xy^2 + \lambda_6 x^3 + \cdots + \lambda_{k+2} x^{k-1} + \lambda_{k+3} y^3 + \cdots + \lambda_{k+l-1} y^{l-1}.$$

The new plane curve B arising after the blow-up at (0, 0) looks like

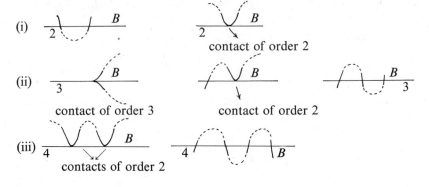

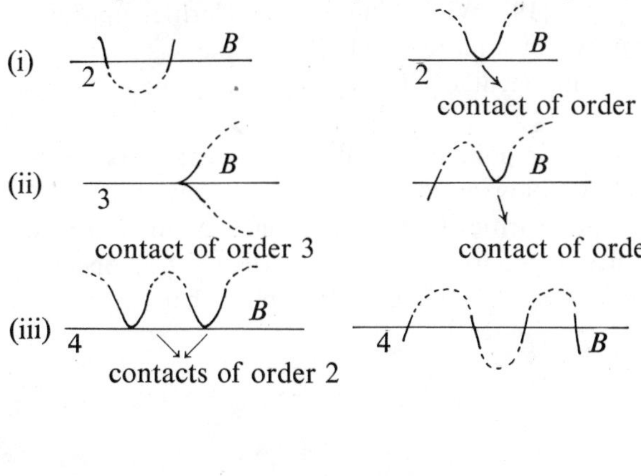

It is easy to determine the types of the corresponding two-dimensional singularities. If the plane curve B still has singularities at the intersection points of higher order, we need more quadratic transformations to resolve the singularity. But quadratic transformations define relations between the parameters λ_i. Now a careful analysis of them completes the proof.

2. Proof of Theorem 2. Let (S, p) be a cusp singularity with minimal resolution M. In this section we shall study deformations of the complex structure of M "smoothing" the exceptional set $A \subset M$. The basic idea is to deform the complex structure only near the singular points of A. We wish to describe in detail this local construction.

Choose two copies D_1 and D_2 of the polydisc $\{|z_i| < 1, i = 1, 2\} \subset C^2$. In D_i, $i = 1, 2$, $U_1 := \{|x_1| < 2\varepsilon, |y_1| < 2\varepsilon\}$ and $U_2 := \{|x_2| < \varepsilon, \varepsilon < |y_2| < 1\} \cup \{\varepsilon < |x_2| < 1, |y_2| < \varepsilon/2\}$ define open submanifolds. We next define maps $\varphi_t: U_2 \to U_1$ depending on a parameter t by:

$$\begin{aligned} \varphi_t(x_2, y_2) &:= (x_2 + t/y_2, y_2) \quad \text{if } |x_2| < \varepsilon,\ \varepsilon < |y_2| < 2\varepsilon, \\ &:= (x_2, y_2 + t/x_2) \quad \text{if } \varepsilon < |x_2| < 2\varepsilon,\ |y_2| < \varepsilon/2. \end{aligned}$$

We may choose $|t|$ sufficiently small so that the inverse function φ_t^{-1} is well defined on the image of φ_t. It is clear that φ_t is biholomorphic. By pasting U_1 and U_2 together via φ_t we obtain a regular one-parameter family (U_t) of smooth complex surfaces $U_t := U_1 \cup_{\varphi_t} U_2$. In U_t, $t \neq 0$, we have nonsingularly embedded curves $C_t := \{(x_2 y_2 = 0) \cup (x_1 y_1 = t)\}$. In other words the deformation (U_t) defines a "smoothing" of the singular point $(0, 0)$ of C_0.

Let now (S, p) be a cusp singularity with minimal resolution $\tilde{S}$ and exceptional set $A_0 \subset \tilde{S}$. Then it is always possible to choose a covering of a suitable strictly pseudoconvex neighborhood $M_0 \subset \tilde{S}$ of A_0 such that we may simultaneously construct a deformation as above near the singular points of A_0 without changing the complex structure of M_0 outside. Thus we get a one-parameter family (M_t) of strictly pseudoconvex manifolds M_t having the following property. The exceptional set A_t of M_t, $t \neq 0$, has one irreducible component which is biholomorphically equivalent to a smooth elliptic curve E_t. One can easily compute the self-intersection number $b = E_t \cdot E_t$ in terms of the self-intersection numbers b_i of the irreducible components of A_0, namely $b = \sum b_i + 2$. As $E_t \subset A_t$ and the function $t \mapsto \dim_C H^1(M_t, \mathcal{O}_{M_t})$ is upper semicontinous (see **[17]** and **[22]**) we have $\dim_C H^1(M_t, \mathcal{O}_{M_t}) = 1$. Hence after blowing down the family (U_t) one gets a deformation $(X, x) \to (T, 0)$ of (S, p); see **[19]**. What kinds of singularities appear in the fibres of $X \to T$? To see this we must more closely look at the sets A_t.

(6) Let $\tilde{E}_t$ be the connected component of A_t with $E_t \subset \tilde{E}_t$. Then we have $\tilde{E}_t = E_t$.

Let $x_1, \cdots, x_k$, $k \geqq 1$, be the singular points of the fibre X_t, t near 0. Assume x_1 is the singular point one gets by blowing down the exceptional set $\tilde{E}_t$. As $\dim_C H^1(M_t, \mathcal{O}_{M_t}) = 1$ the other singularities in X_t are rational. Artin **[2]** proved that every rational singularity has a smooth deformation. Hence the "openness property" of versal deformations (see, e.g., **[5**, p. 35]) implies that (X_t, x_1) is a specialization of (S, p).

LEMMA 10. *Let* (V, p) *be an isolated singularity and let* $\mathcal{O}_{V,p}$ *be Gorenstein. Assume* (Y, y) *is a specialization of* (V, p). *Then* $\mathcal{O}_{Y,y}$ *is Gorenstein.*

From this and from Laufer's characterization of minimally elliptic singularities (see [**13**, Theorem 3.10]), it follows the (X_t, x_1) is a minimally elliptic singularity. Hence $\tilde{E}_t = E_t$. In particular this completes the proof of Theorem 2.

Actually, we can show that $A_t = E_t$, i.e. the following stronger result is true.

PROPOSITION 11. *Let* (S, p) *be a two-dimensional cusp singularity with fundamental divisor* Z. *Then there exists a regular one-parameter family* (M_t) *of strictly pseudo-convex nonsingular complex surfaces* M_t *with* $\dim_{\mathbf{C}} H^1(M_t; \mathcal{O}_{M_t})$ *constant and the following property.* M_0 *is a minimal resolution of* (S, p) *and the exceptional sets* E_t *of* M_t *are smooth elliptic curves with* $E_t \cdot E_t = Z \cdot Z$.

REFERENCES

1. V. I. Arnol'd, *Normal forms for functions near degenerate critical points. The Weyl groups of* A_k, D_k, E_k, *and Lagrangian singularities*, Funkcional. Anal. i Priložen. **6** (1972), no. 4, 3–25 = Functional Anal. Appl. **6** (1972), 254–272. MR **50** #8595.

2. M. Artin, *Algebraic construction of Brieskorn's resolutions*, J. Algebra **29** (1974), 330–348. MR **50** #7143.

3. N. Bourbaki, *Groupes et algèbres de Lie*. Chaps. 4, 5, 6, Actualités Sci. Indust., nos. 1292, 1308, Hermann, Paris, 1961, 1964. MR **30** #2027; **33** #2660.

4. E. Brieskorn, *Rationale Singularitäten komplexer Flächen*, Invent. Math. **4** (1967/68), 336–358. MR **36** #5136.

5. ———, *Special singularities—resolution, deformation and monodromy*, Bonn, 1975 (preprint).

6. ———, *Singular elements of semi-simple algebra groups*, Proc. Internat. Congress Math. (Nice, 1970), vol. 2, Gauthier-Villars, Paris, 1971, pp. 279–284.

7. E. Freitag and R. Kiehl, *Algebraische Eigenschaften der lokalen Ringe in den Spitzen der Hilbertschen Modulgruppen*, Invent. Math. **24** (1974), 121–148. MR **50** #324.

8. A. M. Gabrielov, *Dynkin diagrams for unimodal singularities*, Functional Anal. Appl. **8** (1974), 1–6.

9. F. Hirzebruch, *Hilbert modular surfaces*, Enseignement Math. **19** (1973), 183–281.

10. A. Kas and M. Schlessinger, *On the versal deformation of a complex space with an isolated singularity*, Math. Ann. **196** (1972), 23–29. MR **45** #3769.

11. U. Karras, *Klassifikation zweiaimensionaler Singularitaten mit auflösbaren lokalen Fundamentalgruppen*, Math. Ann. **213** (1975), 231–255.

12. ———, *Eigenschaften der lokalen Ringe in zweiaimensionalen Spitzen*, Math. Ann. **215** (1975), 119–129.

13. H. B. Laufer, *On minimally elliptic singularities*, 1975 (preprint).

14. ———, *Taut two-dimensional singularities*, Math. Ann. **205** (1973), 131–164. MR **48** #11563.

15. ———, *On* μ *for surface singularities*, 1975 (preprint).

16. H. Pinkham, *Deformation of alegbraic varieties with* G_m *actions*, Asterisque no. 22, Soc. Math. France, 1975.

17. O. Riemenschneider, *Halbstetigkeitssätze für* 1*-konvexe holomorphe Abbilaungen*, Math. Ann. **192** (1971), 216–226. MR **45** #2209.

18. ———, *Deformationen von Quotientensingularitäten (nach zyklischen Gruppen)*, Math. Ann. **209** (1974), 211–248.

19. ———, *Familien komplexer Räume mit streng pseudokonvexer spezieller Faser*, 1975 (preprint).

20. K. Saito, *Einfach-elliptische Singularitäten*, Invent. Math. **23** (1974), 289–325. MR **50** #7147.

21. M. Schlessinger, *Rigidity of quotient singularities*, Invent. Math. **14** (1971), 17–26. MR **45** #1912.

22. Y.-T. Siu, *Dimension of sheaf cohomology groups in holomorphic defoarmtions*, Math. Ann. **192** (1971), 203–215. MR **45** #8878.

23. D. Siersma, *Classification and deformation of singularities*, Thesis, Amsterdam, 1974.

24. G. N. Tjurina, *Locally semiuniversal flat deformations of isolated singularities of complex spaces*, Izv. Akad. Nauk SSSR Ser. Mat. **33** (1969), 1026–1058 = Math. USSR Izv. **3** (1969), 967–1000. MR **40** #5903.

ABTEILUNG MATHEMATIK DER UNIVERSITÄT DORTMUND

Proceedings of Symposia in Pure Mathematics
Volume 30, 1977

ON μ FOR SURFACE SINGULARITIES

HENRY B. LAUFER

I. Introduction. Let $f(z_0, \cdots, z_n)$ be a holomorphic function defined near $0 = (0, \cdots, 0)$ such that $V = \{(z_0, \cdots, z_n) | f(z_0, \cdots, z_n) = 0\}$ has an isolated singularity at 0. We choose V to be Stein. Let μ be the Milnor number of the singularity at 0. The Milnor number, originally defined for polynomial f [**13**, §§6–7, pp. 45–64] is also defined for holomorphic f [**16**], [**10**], [**4**] and is a topological invariant of the local embedding near 0 of V in C^{n+1} [**11**].

For $n = 1$, the plane curve case, Milnor [**13**, Theorem 10.5, p. 85] showed that

$$\mu = 2\delta - r + 1 \tag{1}$$

where r is the number of irreducible components of V and δ is the "number" of nodes and cusps at 0.

In this paper, we shall prove an analogous formula to (1) for the case $n = 2$. This formula, equation (3) of Theorem 1, expresses μ in terms of a resolution $\pi : M \to V$ of the singularity. In (3), $\chi_T(A)$ and $K \cdot K$ are topological invariants of π. $H^1(M, \mathcal{O})$, which is isomorphic to $R^1\pi_*\mathcal{O}_{M,0}$ [**9**, Lemma 3.1, p. 599], [**2**], is an analytic invariant of M. Since μ and the topology of M are often readily computable from the equation for f, Theorem 1 thus provides a practical means for computing $\dim_C H^1(M, \mathcal{O})$ in many cases.

In the last section of this paper, we give some applications, examples and conjectures relating to Theorem 1.

The author would like to thank Joseph Lipman for several helpful suggestions in connection with this work.

II. A formula for μ. Let $\{(x, y, z) | \ f(x, y, z) = 0\}$ define V as above. Then

$$\omega = \frac{dx \wedge dy}{f_z} = \frac{dy \wedge dz}{f_x} = \frac{dz \wedge dx}{f_y} \tag{2}$$

AMS (MOS) subject classifications (1970). Primary 32C40, 32C45, 14E15; Secondary 32G05.
Key words and phrases. Milnor number, Riemann-Roch, Gorenstein, direct image.

© 1977, American Mathematical Society

is a nonzero holomorphic 2-form on $V - \{0\}$. $\pi^*(\omega)$ extends to a meromorphic 2-form on M with a pole set contained in $A = \pi^{-1}(0)$. K, the divisor of ω and also called the canonical divisor, may be characterized topologically by the adjunction formula [**21**, Proposition IV.5, p. 75]

$$A_i \cdot K = -A_i \cdot A_i + 2g_i - 2 + 2\delta_i$$

where A_i is an irreducible component of A, g_i is the genus of A_i, and δ_i is the "number" of nodes and cusps on A_i.

THEOREM 1. *Let $f(x, y, z)$ be holomorphic in N, a Stein neighborhood of* $(0, 0, 0)$ *with* $f(0, 0, 0) = 0$. *Let* $V = \{(x, y, z) \in N \mid f(x, y, z) = 0\}$ *have* $(0, 0, 0)$ *as its only singular point. Let μ be the Milnor number of* $(0, 0, 0)$. *Let $\pi : M \to V$ be a resolution of V. Let $A = \pi^{-1}(0, 0, 0)$. Let $\chi_T(A)$ be the topological Euler characteristic of A. Let K be the canonical divisor on M. Then*

$$1 + \mu = \chi_T(A) + K \cdot K + 12 \dim_C H^1(M, \mathcal{O}). \tag{3}$$

PROOF. Any holomorphic function which agrees with f to sufficiently high order defines a holomorphically equivalent singularity at (0, 0, 0) [**20**], [**5**], [**8**, Theorem 7.1, p. 134]. So we may take f to be a polynomial. Compactify $\boldsymbol{C}^3$ to $\boldsymbol{P}^3$. Let $\bar{V}_t$ be the closure in $\boldsymbol{P}^3$ of $V_t = \{(x, y, z) \in \boldsymbol{C}^3 \mid f(x, y, z) = t\}$. By adding a suitably general high order homogeneous term of degree e to the polynomial f, we may additionally assume that $\bar{V}_0$ has $(0, 0, 0) \in \boldsymbol{C}^3$ as its only singularity and that $\bar{V}_t$ is nonsingular for small $t \neq 0$. We may also assume that the highest order terms of f define, in homogeneous coordinates, a nonsingular curve of order e in $\boldsymbol{P}^2 = \boldsymbol{P}^3 - \boldsymbol{C}^3$. $\bar{V}_t$ is then necessarily irreducible for all small t. Without loss of generality, we take $N = \boldsymbol{C}^3$. Then $V = V_0$.

For any analytic space S with structure sheaf $\mathcal{O}$, the analytic Euler characteristic $\chi_A(S)$ is defined by

$$\chi_A(S) = \sum (-1)^i \dim_C H^i(S, \mathcal{O}), \qquad 0 \leqq i < \infty.$$

Let $\bar{M}$ be the resolution of $\bar{V}$ which has M as an open subset. We prove that

$$\chi_A(\bar{M}) = \chi_A(\bar{V}) - \dim_C H^1(M, \mathcal{O}) \tag{4}$$

as follows. By the Mayer-Vietoris sequence [**1**, p. 236], the rows of the following commutative diagram are exact.

$$\begin{array}{ccccccccc}
0 \to & H^0(\bar{M}, \mathcal{O}) & \to & H^0(M, \mathcal{O}) \oplus H^0(\bar{M} - A, \mathcal{O}) & \to & H^0(M - A, \mathcal{O}) & & & \\
& \uparrow \pi_1 & & \uparrow \pi_2 \qquad \uparrow \pi_3 & & \uparrow \pi_4 & & & \\
0 \to & H^0(\bar{V}, \mathcal{O}) & \to & H^0(V, \mathcal{O}) \oplus H^0(\bar{V} - \{0\}, \mathcal{O}) & \to & H^0(V - \{0\}, \mathcal{O}) & & & \\
\to & H^1(\bar{M}, \mathcal{O}) & \to & H^1(M, \mathcal{O}) \oplus H^1(\bar{M} - A, \mathcal{O}) & \to & H^1(M - A, \mathcal{O}) & \to & H^2(\bar{M}, \mathcal{O}) & \to 0 \\
& \uparrow \pi_5 & & \uparrow \pi_6 \qquad \uparrow \pi_7 & & \uparrow \pi_8 & & \uparrow \pi_9 & \\
\to & H^1(\bar{V}, \mathcal{O}) & \to & H^1(V, \mathcal{O}) \oplus H^1(\bar{V} - \{0\}, \mathcal{O}) & \to & H^1(V - \{0\}, \mathcal{O}) & \to & H^2(\bar{V}, \mathcal{O}) & \to 0
\end{array} \tag{5}$$

The higher terms in (5) are 0 by [**22**]. In (5), π_1, π_2, π_3, π_4, π_7, and π_8 are isomorphisms. $H^1(V, \mathcal{O}) = 0$. The kernel and cokernel of the map $H^1(\bar{V} - \{0\}, \mathcal{O}) \to H^1(V - \{0\}, \mathcal{O})$ are finite dimensional. (4) now follows easily from exactness in (5).

In a similar manner, using a tubular neighborhood of A rather than M in (5), one sees that

(6) $$\chi_T(\bar{M}) = \chi_T(\bar{V}) + \chi_T(A) - 1.$$

ω, defined above in (2), is a nonzero holomorphic 2-form on V_t, $t \neq 0$, and on $V - \{0\}$. Let $K_{\infty,t}$ be the part of the divisor of ω on $\bar{V}_t$ which is supported on $\bar{V}_t - V_t$, for t small. ω has, in fact, a zero of order $e - 4$ on $\bar{V}_t - V_t$. $K_{\infty,t} \cdot K_{\infty,t}$ is independent of t since the family $\{\bar{V}_t\}$ is differentiably trivial away from (0, 0, 0) $\in \boldsymbol{C}^3$. Let $K_\infty \cdot K_\infty$ denote this constant value for $K_{\infty,t} \cdot K_{\infty,t}$. Noether's formula or the Riemann-Roch theorem [**25**, especially p. 74 and p. 89], [**6**, especially Theorem 4.10.1, p. 70] says

(7) $$\chi_A(\bar{M}) = (1/12)(K \cdot K + K_\infty \cdot K_\infty + \chi_T(\bar{M})),$$

(8) $$\chi_A(\bar{V}_t) = (1/12)(K_\infty \cdot K_\infty + \chi_T(\bar{V}_t)), \qquad t \neq 0.$$

By Milnor's theorem [**13**, Theorem 5.11, pp. 53, 64] and the differentiable triviality of the family $\{V_t\}$ away from $(0, 0, 0) \in \boldsymbol{C}^3$,

(9) $$\chi_T(\bar{V}_t) = \chi_T(\bar{V}) + \mu, \qquad t \neq 0.$$

Since $\chi_A(\bar{V}_t) = \chi_A(\bar{V})$ for all small t [**18**], (3) follows from (4), (6), (7), (8) and (9).

III. Applications and examples. Let M be a given holomorphically convex two-dimensional manifold with exceptional set A. Assume that A is connected. Then the Remmert reduction $\pi : M \to V$ blows down A to a normal singular point in V. Given a holomorphic family $\{M_t\}$ of such M, with $M_0 = M$, the family simultaneously blows down if $\dim_{\boldsymbol{C}} H^1(M, \mathcal{O})$ is independent of t [**24**], [**19**]. If V_0 is a hypersurface, then so are all nearby V_t.

COROLLARY 2. *Let $\{M_t\}$ be a holomorphic family of holomorphically convex two-manifolds such that the exceptional sets $\{A_t\}$ form a topologically trivial holomorphic family. Suppose that* $\dim_{\boldsymbol{C}} H^1(M, \mathcal{O})$ *is independent of t and that each component of A_0 blows down to a hypersurface singularity. Then the Milnor fiberings of V_t are fiber homotopy equivalent for all t.*

PROOF. Consider each component of A_0 separately. By Theorem 1, μ is independent of t. The conclusion now follows immediately from [**12**].

A double point singularity at (0, 0, 0) can be written in the form $z^2 = f(x, y)$. The plane curve singularity $f(x, y) = 0$ at (0, 0) is uniquely determined [**7**, Theorem 2, p. 601]. Because of the explicit form for their versal deformations [**23**], double points necessarily deform into only double or simple points and there is an induced deformation of the corresponding plane curve singularity.

COROLLARY 3. *Let $\{M_t\}$ be a holomorphic family of holomorphically convex two-manifolds such that the exceptional sets $\{A_t\}$ form a topologically trivial holomorphic family. Suppose that* $\dim_{\boldsymbol{C}} H^1(M, \mathcal{O})$ *is independent of t, that A_0 is connected and that V_0 has a double point at $\pi(A_0)$. Then the plane curve singularities determined by the double points of $\{V_t\}$ form an equisingular family.*

PROOF. μ for the singularity at (0, 0, 0) for $z^2 - f(x, y) = 0$ equals μ for the singularity at (0, 0) for $f(x, y) = 0$. Hence the family of plane curves has constant μ and so is equisingular by [**12**], [**26**] and [**27**].

Observe that in Brenton's [**3**] Riemann-Roch theorem for two-dimensional surfaces with only isolated hypersurface singularities, the correction term $R(X)$ equals μ, as given by Theorem 1.

It is not true that the topology of A in M determines μ (and hence $\dim_C H^1(M, \mathcal{O})$) for hypersurface singularities. Consider the singularities given at (0, 0, 0) by $V = \{(x, y, z) | x^2 + y^7 + z^{14} = 0\}$ and $V' = \{(x, y, z) | x^3 + y^4 + z^{12} = 0\}$. By **[15]** or **[14]** both V and V' resolve to have exceptional sets A and A' which are non-singular Riemann surfaces of genus 3 and with $A \cdot A = A' \cdot A' = -1$. But $\mu = 78$ for V and $\dim_C H^1(M, \mathcal{O}) = 9$ while $\mu = 66$ for V' and $\dim_C H^1(M', \mathcal{O}) = 8$.

Given a resolution $\pi : M \to V$ of a normal, Stein two-dimensional analytic space with only one singularity $p \in V$, the right side of (3) makes sense so long as $K \cdot K$ is defined. In particular $K \cdot K$ is defined if p is Gorenstein.

CONJECTURE 1. *Let p be a Gorenstein singularity of the normal two-dimensional analytic space V. Suppose that V is smoothable to V′ near p. Then*

$$\chi_T(V') = \chi_T(A) + K \cdot K + 12 \dim_C H^1(M, \mathcal{O}).$$

The following conjecture agrees with, but is weaker than, examples of Pinkham **[17]**.

CONJECTURE 2. *Let p be a Gorenstein singularity of the normal two-dimensional analytic space V. If*

$$\chi_T(A) + K \cdot K + 12 \dim_C H^1(M, \mathcal{O}) \leqq 0,$$

then V is not smoothable.

REFERENCES

1. A. Andreotti and H. Grauert, *Théorèmes de finitude pour la cohomologie des éspaces complexes*, Bull. Soc. Math. France **90** (1962), 193–259 .MR **27** #343.
2. M. Artin, *On isolated rational singularities of surfaces*, Amer. J. Math. **88** (1966), 129–136. MR **33** #7340.
3. L. Brenton, *Complex singularities: A Riemann-Roch theorem for singular surfaces and extensions of holomorphic differential forms*, Dissertation, University of Washington, 1974.
4. E. Brieskorn, *Die Monodromie der isolierten Singularitäten von Hyperflächen*, Manuscripta Math. **2** (1970), 103–161. MR **42** #2509.
5. H. Hironaka and H. Rossi, *On the equivalence of imbeddings of exceptional complex spaces*, Math. Ann. **156** (1964), 313–333. MR **30** #2011.
6. F. Hirzebruch, *Topological methods in alegbraic geometry*, Die Grundlehren der math. Wissenschaften, Band 131, Springer-Verlag, New York, 1966. MR **34** #2573.
7. D. Kirby, *The structure of an isolated multiple point of a surface*. I, Proc. London Math. Soc. (3) **6** (1956), 597–609. MR **19**, 319.
8. H. B. Laufer, *Normal two-dimensional singularities*, Ann. of Math. Studies, no. 71, Princeton Univ. Press, Princeton, N. J.; Univ. of Tokyo Press, Tokyo, 1971. MR **47** #8094.
9. ———, *On rational singularities*, Amer. J. Math. **94** (1972), 597–608. MR **48** #8837.
10. Lê Dũng Tráng, *Singularités isolées des hypersurfaces complexes*, Ecole Polytechnique, May 1969 (preprint).
11. ———, *Topologie des singularités des hypersurfaces complexes*, Astérisque, nos. 7/8, Soc. Math. France, 1973.
12. Lê Dũng Tráng and C. P. Ramanujan, *The invariance of Milnor's number implies the invariance of the topological type* (to appear).
13. J. Milnor, *Singular points of complex hypersurfaces*, Ann. of Math. Studies, no. 61, Princeton Univ. Press, Princeton, N. J.; Univ. of Tokyo Press, Tokyo, 1968. MR **39** #969.

14. I. Ono and K. Watanabe, *On the singularity of* $z^p + y^q + x^{pq} = 0$, Sci. Rep. Tokyo Kyoika Daigaku Sect. A **12** (1974), 123–128.

15. P. Orlik and P. Wagreich, *Isolated singularities of alegbraic surfaces with* C^* *action*, Ann. of Math. (2) **93** (1971), 205–228. MR **44** #1662.

16. V. P. Palamodov, *The multiplicity of a holomorphic transformation*, Funkcional. Anal. i Priložen. **1** (1967), no. 3, 54–65. (Russian) MR **38** #4720.

17. H. Pinkham, *Deformations of algebraic varieties with* G_m *action*, Astérisque **20** (1974).

18. O. Riemenschneider, *Über die Anwendung algebraischer Methoden in der Deformationstheorie komplexer Räume*, Math. Ann. **187** (1970), 40–55. MR **41** #5659.

19. ———, *Bermerkungen zur Deformationstheorie nicht-rationaler Singularitäten*, Manuscripta Math. **14** (1974), 91–100.

20. P. Samuel, *Algébricité de certains points singuliers algébroids*, J. Math. Pures Appl (9) **35** (1956), 1–6. MR **17**, 788.

21. J.-P. Serre, *Groupes algébriques et corps de classe*, Actualitiés Sci. Indust., 1264, Hermann, Paris, 1959. MR **21** #1973; erratum, **30**, p. 1200.

22. Y.-T. Siu, *Analytic sheaf cohomology groups of dimension n of n-dimensional complex spaces*, Trans. Amer. Math. Soc. **143** (1969), 77–94. MR **40** #5902.

23. G. Tjurina, *Locally semiuniversal flat deformations of isolated singularities of complex spaces*, Izv. Akad. Nauk SSSR Ser. Mat. **33** (1969), 1026–1058 = Math. USSR Izv. **3** (1969), 967–1000. MR **40** #5903.

24. J. Wahl, *Equisingular deformations of normal complex singularities*. I (to appear).

25. O. Zariski, *Algebraic surfaces*, 2nd ed. Springer-Verlag, New York, 1971.

26. ———, *Studies in equisingularity*. II. *Equisingularity in codimension* 1 (*and characteristic* 0), Amer. J. Math. **87** (1965), 972–1006. MR **33** #125.

27. ———, *Contributions to the problem of equisingularity*, Questions on Algebraic Varieties (C.I.M.E., III Ciclo, Varenna, 1969), Edizioni Cremonese, Rome, 1970, pp. 261–343. MR **43** #1987.

STATE UNIVERSITY OF NEW YORK AT STONY BROOK

Proceedings of Symposia in Pure Mathematics
Volume 30, 1977

SMOOTHING PERFECT VARIETIES

R. MANDELBAUM* AND M. SCHAPS

0. Introduction. In this research report we discuss the deformation theory of intersections of germs of perfect analytic varieties. It is well known that hypersurface singularities are always smoothable and that the parameter space S of the versal deformation space of a hypersurface singularity is isomorphic to the parameter space of the space of infinitesimal first-order deformations of the given hypersuface. As noted in [5] the same results are true for complete intersections of hypersurfaces. If we move on from hypersurfaces to pure codimension two analytic objects and in addition add the hypothesis of perfectness we find similar phenomena occurring. In particular in [9], [11] it is shown that a germ X of a perfect analytic variety of codimension two in $\boldsymbol{C}^n$ ($n \leqq 5$) will always be smoothable and if $n > 5$ then even though X is not generically smoothable it nevertheless has a well-understood generic form X' whose singular locus $\mathscr{S}(X')$ has codimension 4 in X'. In Theorem 1 we show that a proper intersection $X = \bigcap X_i$ of perfect germs of analytic varieties has smoothness properties at least as good as those of the individual germs X_i. Thus if the X_i all have codimension at most two then X will always be deformable to a germ X' with $\operatorname{codim}(\mathscr{S}(X'), X') \geqq 4$. In particular if $\dim X \leqq 3$ it will always be smoothable.

In [9], [11] it is also shown that all first-order deformations of germs of perfect analytic varieties of codimension two in $\boldsymbol{C}^n$ can be lifted unobstructedly to flat analytic deformations of the germs. In Theorem 2 we show that the same is true for proper intersections of such germs.

We deal throughout with germs of analytic subvarieties at the origin in some $\boldsymbol{C}^n$ as defined, for example, in [3], [4]. $\mathcal{O}_X$ will denote the structure sheaf of the subvariety X, $\mathscr{I}(X)$ its defining ideal and $\mathscr{S}(X)$ its singular locus. All other definitions and notation will be as in [3], [4].

AMS (MOS) subject classifications (1970). Primary 32G05.

*The first author was supported in part by the European Research Office under contract DAERO-12474-G0069 during the period when this research was conducted.

© 1977, American Mathematical Society

1. Definitions. We recall that $\mathscr{V} = (V, \pi, T)$ is a flat deformation of X in Y if $\pi : V \to T$ is a flat map of germs of analytic varieties, X, Y are germs of analytic varieties, X a subvariety of Y, V a subvariety of $Y \times T$, and $X \rightrightarrows V_0$. We can assume without loss of generality that $V_0 = X$ is defined in some open neighborhood of the origin in $V = \boldsymbol{C}^n$ by holomorphic equations $f_i(x) = 0$, $i = 1, \cdots, m$, and that V has equations $f_i(x, t) = 0$, $i = 1, \cdots, m$, in $\boldsymbol{C}^n \times T$ with $f_i(x, 0) = f_i(x)$. As a working definition of flatness we take $\pi : V \to T$ is flat if every relation $r(x) = (r_1(x), \cdots, r_m(x))$ between the $f_1(x), \ldots, f_m(x)$ (i.e., $\sum r_i(x) f_i(x) = 0$) can be lifted to a relation $r(x, t) = (r_1(x, t), \cdots, r_m(x, t))$ between the $f_i(x, t)$.

If $\mathscr{V}$ is a flat deformation of X in $\boldsymbol{C}^n$ we shall say $\mathscr{V}$ is a smoothing of X to order k if the generic fiber V_t of $\mathscr{V}$ has singular locus Σ_t with codim $(\Sigma_t, V_t) \geqq k$. If V_t is nonsingular then $\mathscr{V}$ is a smoothing of X. We say X is smoothable to order k if it has a smoothing of this order ($k = \infty$ if and only if X is smoothable).

We call X rigid if all flat deformations of X are locally trivial. In particular a germ of a rigid singular variety X is not smoothable. Even nonrigid X may not be smoothable as shown by examples of Mumford and Schlessinger **[10]**, **[12]**. In particular there exist curves in $\boldsymbol{P}^n$ which are not smoothable. On the other hand all analytic curves in $\boldsymbol{C}^3$ are smoothable. [The question for reduced irreducible curves in $\boldsymbol{P}^3$ is still open.][1]

We recall that given any germ V of a k-dimensional variety at the origin, there exists a finite-analytic mapping f: $V \to \boldsymbol{C}^k$ exhibiting ${}_V\mathcal{O}$ as a finitely generated ${}_k\mathcal{O}$-module. We say V is perfect if ${}_V\mathcal{O}$ is free as a ${}_k\mathcal{O}$-module.

This is of course equivalent to the Cohen-Macaulay condition that depth ${}_{n\mathcal{O}}({}_V\mathcal{O}) = \dim {}_V\mathcal{O} = \dim V$ where V is a subvariety of $\boldsymbol{C}^n$. Now by **[9]**, **[11]** if $V \subset \backslash \boldsymbol{C}^n$ is a perfect germ of codimension 2 and $n \leqq 5$ then V is smoothable. Since all pure 1-dimensional varieties are perfect we find that all curves in $\boldsymbol{C}^3$ are smoothable. The above results are in a sense best possible. If $n = 6$ the familiar example of the cone of the Segre embedding X of $\boldsymbol{P}^1 \times \boldsymbol{P}^2$ in $\boldsymbol{P}^6$ is perfect of codimension 2 but of course not smoothable.

The key aspect of the proof of the above results of [Schaps, Loday] is showing that a germ of a perfect subvariety of codimension 2 is necessarily determinantal. (We recall that a germ of a variety V is determinantal of type (m, n, l) if $\mathscr{I}(V) \subset {}_N\mathcal{O}_0$ is generated by the $l \times l$ minors of some $m \times n$ matrix R with coefficients in ${}_N\mathcal{O}_0$ and ht $\mathscr{I} = \operatorname{codim} V = (m - l + 1)(n - l + 1)$.) In particular if V is perfect of codimension 2 then $\mathscr{I}(V)$ is generated by the maximal minors of an $n \times (n - 1)$ matrix. Now it can be shown that if V is determinantal of type (m, n, l) then generically its singular locus will have codimension $(m - l + 2)(n - l + 2)$ and thus $\operatorname{codim}(\mathscr{S}(V), V) = m + n - 2l + 3$.

Perfect subvarieties of codim 2 are determinantal of type $(n, n - 1, n - 1)$ with codimension $(\mathscr{S}(V), V) = 4$, thus giving us the [Schaps, Loday] result. This also furnishes us with examples of perfect codim 2 varieties which are smoothable to order k, but not $k + 1$. The variety given by the 2×2 minors of

$$\begin{bmatrix} x_1 & x_4 \\ x_1x_2 & x_1x_5 \\ x_3 & x_6 \end{bmatrix}$$

[1]ADDED IN PROOF. Tannenbaum has recently shown [Thesis, Harvard, 1976] that such curves are not necessarily smoothable, although they will be weakly smoothable.

will have singular locus of codimension one, will be smoothable to a variety with one isolated singular point, but not smoothable.

2. Smoothing intersections. To determine to what extent smoothability is preserved under intersection we first need some preliminary lemmas.

LEMMA 2.1 (CF. [7]). *Let P be a Noetherian local ring and suppose J is an ideal in P such that $B = P / J$ has projective dimension m as a P-module. Let N be a finite P-module.*

Then for all $i > m -$ depth $_J N$, $\mathrm{Tor}_i(B, N) = 0$.

PROOF. Induction on depth $_J N$.

LEMMA 2.2. *Suppose X_1, X_2 are perfect germs of analytic subvarieties at the origin in $\mathbf{C}^n$. Let $\mathscr{J}_1 = \mathscr{J}(X_1)$, $\mathscr{J}_2 = \mathscr{J}(X_2)$, $\mathcal{O}_1 = {}_n\mathcal{O}_0/\mathscr{J}_1$ and $\mathcal{O}_2 = {}_n\mathcal{O}_0/\mathscr{J}_2$. Then, if* $\mathrm{ht}\,(\mathscr{J}_1 + \mathscr{J}_2) = \mathrm{ht}\,\mathscr{J}_1 + \mathrm{ht}\,\mathscr{J}_2$,

(1) $\mathrm{Tor}_i^{{}_n\mathcal{O}}(\mathcal{O}_1, \mathcal{O}_2) = 0$ *for* $i > 0$,

(2) $\mathscr{J}_1\mathscr{J}_2 = \mathscr{J}_1 \cap \mathscr{J}_2$,

(3) $X_1 \cap X_2$ *is a perfect germ.*

PROOF. (1) $\mathrm{ht}\,(\mathscr{J}_1 + \mathscr{J}_2) = \mathrm{ht}\,\mathscr{J}_1 + \mathrm{ht}\,\mathscr{J}_2$ implies depth $_{\mathscr{J}_1}\,{}_n\mathcal{O}_0 =$ depth $_{\mathscr{J}_1}\mathcal{O}_2$. Then, by Lemma 1, $\mathrm{Tor}_i(\mathcal{O}_1, \mathcal{O}_2) = 0$ for $i <$ proj dim $_n\mathcal{O}_1 -$ depth $_{\mathscr{J}_1}\mathcal{O}_2$ $=$ depth $_{\mathscr{J}_1}{}_n\mathcal{O}_0 -$ depth $_{\mathscr{J}_1}\mathcal{O}_2 = 0$ since X_1 is perfect.

(2) Since $\mathscr{J}_1\mathscr{J}_2 \subset \mathscr{J}_1 \cap \mathscr{J}_2$ it suffices to show $\mathscr{J}_1 \cap \mathscr{J}_2 \subset \mathscr{J}_1\mathscr{J}_2$. Let

$${}_n\mathcal{O}^m \xrightarrow{d_2} {}_n\mathcal{O}^r \xrightarrow{d_1} {}_n\mathcal{O} \xrightarrow{\pi} \mathcal{O}_1 \to 0$$

be a segment of a free resolution of $\mathcal{O}_1$ obtained by setting $d_1(a) = a \cdot f$ where $f = (f_1, \cdots, f_r)$ and $\{f_1, \cdots, f_r\}$ generate $\mathscr{J}_1$. Then tensoring by $\mathcal{O}_2$ and using $\mathrm{Tor}_1(\mathcal{O}_1, \mathcal{O}_2)$ gives the desired result.

(3) is a straightforward calculation showing

$$\mathrm{codim}\,(X_1 \cap X_2, \mathbf{C}^n) < \mathrm{proj\,dim}_n\, X_1.$$

LEMMA 2.3. *Suppose X_1, X_2 are germs of analytic subvarieties at $0 \in \mathbf{C}^n$ with $\mathscr{J}_1 = \mathscr{J}(X_1)$ and $\mathscr{J}_2 = \mathscr{J}(X_2)$ and suppose $\mathscr{J}_1\mathscr{J}_2 = \mathscr{J}_1 \cap \mathscr{J}_2$. Suppose $\mathscr{V}_1 = (V_1, \pi_1, T)$, $\mathscr{V}_2 = (V_2, \pi_2, T)$ are flat deformations of X_1, X_2 in $\mathbf{C}^n$. Let $\tilde{V} = V_1 \cap V_2$, considered a subvariety in $\mathbf{C}^n \times T$ with projection $\pi : \tilde{V} \to T$, and set $\tilde{\mathscr{V}} = (V, \pi, T)$. Then $\tilde{\mathscr{V}}$ is a flat deformation of $X_1 \cap X_2$ in $\mathbf{C}^n$.*

PROOF. To show $\pi : \tilde{V} \to T$ is flat we must show that all the relations on $\mathscr{J}_1 + \mathscr{J}_2$ lift. We can demonstrate that since $\mathscr{J}_1 \cap \mathscr{J}_2 = \mathscr{J}_1\mathscr{J}_2$ all such relations are generated by relations on $\mathscr{J}_1$ and $\mathscr{J}_2$ and by trivial relations. But all such relations lift, so π is flat.

THEOREM 1 (CF. [8]). *Let X_1, X_2 be germs of perfect analytic subvarieties of $\mathbf{C}^n$ smoothable to order k. Suppose* $\mathrm{codim}\,(X_1 \cap X_2, \mathbf{C}^n) = \mathrm{codim}\,(X_1, \mathbf{C}^n) + \mathrm{codim}\,(X_2, \mathbf{C}^n)$. *Then $X = X_1 \cap X_2$ is a germ of a perfect analytic subvariety of $\mathbf{C}^n$ smoothable to order k.*

PROOF. Let $\mathscr{V}_i = (V_i, \pi_i, T_i)$ be the hypothesized smoothing of X_i. Let $G = GA(n, \mathbf{C})$ be the affine transformations of $\mathbf{C}^n$ and let $T = G \times T_1 \times T_2$. Let $\tilde{\mathscr{V}}_1$ be the deformation of X_1 over T given by $\tilde{V}_{1,(g,t_1,t_2)} = g(V_{1,t_1})$ and $\tilde{\mathscr{V}}_2$ the deformation of X_2 given by $\tilde{V}_{2,(g,t_1,t_2)} = V_{2,t_2}$. Let $\tilde{V} = \tilde{V}_1 \cap \tilde{V}_2$ in $\mathbf{C}^n \times T$ and $\tilde{\mathscr{V}}$ be the corresponding deformation of X. Then $\mathscr{V}$ is a flat deformation by Lemmas 2, 3 and it thus remains to show that the generic fiber V_t of V has singular locus Σ_t with codim $(\Sigma_t, V) \geqq k$. Let $c_1 = \operatorname{codim}(X_1, \boldsymbol{C}^n)$, $c_2 = \operatorname{codim}(X_2, \boldsymbol{C}^n)$, $\Sigma_{1,s} = \mathscr{S}(V_{1,s})$, and $\Sigma_{2,t} = \mathscr{S}(V_2, t)$. Let $P: \boldsymbol{C}^n \times T \to T_1 \times T_2$ be the canonical projection, $Z_{s,t} = P^{-1}(s, t)$ for $(s, t) \in T_1 \times T_2$, $\tilde{V}_{s,t} = \tilde{V} \cap Z_{s,t}$ and $p: \tilde{V}_{s,t} \to G$ the obvious projection. Define $F: Z_{s,t} \to \boldsymbol{C}^n \times \boldsymbol{C}^n$ by $F(z, g) = (g^{-1}(z), z)$ so that $\tilde{V}_{s,t} = F^{-1}(V_{1,s} \times V_{2,t})$. Let $\Sigma_{s,t} = \mathscr{S}(\tilde{V}_{s,t})$ and $\tilde{\Sigma}_{s,t} = F^{-1}(\Sigma_{1,s} \times V_{2,t}) \cup F^{-1}(V_{1,s} \times \Sigma)$.

Then $\tilde{\Sigma}_{s,t} \subset \Sigma_{s,t}$ and it can be shown that, for generic g, $\Sigma_{s,t} \cap p^{-1}(g) = \tilde{\Sigma}_{s,t} \cap p^{-1}(g)$. Now since F is flat [**1**], we obtain

$$\operatorname{codim}(\tilde{\Sigma}_{s,t}; \tilde{V}_{s,t}) \geqq \min(\operatorname{codim}(\Sigma_{1,s}; V_{1,s}), \operatorname{codim}(\Sigma_{2,t}; V_{2,t})) \geqq k,$$

for generic s, t. However codim $(\tilde{\Sigma}_{s,t} \cap p^{-1}(g); V_{(g,s,t)}) = \operatorname{codim}(\Sigma_{s,t}, \tilde{V}_{s,t})$ for generic g, s, t. Thus for generic $\tau \in T$ we find codim $(\mathscr{S}(\tilde{V}_\tau), \tilde{V}_\tau) \geqq k$, as desired.

Clearly our theorem can be inductively extended to any sequence $X_1, \cdots, X_r$ of germs of perfect analytic subvarieties of $\boldsymbol{C}^n$ satisfying

$$\operatorname{codim}\left(\bigcap_{i=1}^{t} X_i, \boldsymbol{C}^n\right) = \sum_{i=1}^{t} \operatorname{codim}(X_i, \boldsymbol{C}^n), \quad \text{for all } t \leqq r.$$

We call such a sequence a proper sequence.

If the sequence satisfies the stonger condition

$$\operatorname{codim}\left(\bigcap_{j=1}^{t} X_{i_j}, \boldsymbol{C}^n\right) = \sum_{j=1}^{t} \operatorname{codim}(X_{i_j}, \boldsymbol{C}^n)$$

for all subsequences $i_1 < i_2 < \cdots < i_t$ of $\{1, \cdots, n\}$ then we shall call it a very proper subsequence.

In the case of germs of determinantal varieties we can obtain

COROLLARY 1. *Let $X_1, \cdots, X_r$ be a proper sequence of germs of determinantal subvarieties of $\boldsymbol{C}^n$ of type (m_i, n_i, l_i) respectively. For each i_j such that X_{i_j} is not a complete intersection, set $k_j = m_{i_j} + n_{i_j} - 2l_{i_j} + 3$. Let $X = \bigcap_{j=1}^{r} X_{i_j}$. Then if* dim $X < \min_j k_j$, *X is smoothable.*

COROLLARY 2. *Let $X_1, \cdots, X_r$ be a proper sequence of germs of analytic subvarieties of $\boldsymbol{C}^n$. Suppose each X_i is either a complete intersection or a perfect subvariety of codimension* 2. *Let $X = \bigcap X_i$. Then* dim $X \leqq 3$ *implies X is smoothable.*

PROOF. By [**9**], [**11**] if X_j is perfect of codimension 2 it is determinantal of type $(n, n-1, n-1)$. Thus $k_j = 4$. Now apply the previous corollary.

We now turn to the versal deformation spaces of intersections of the above type.

3. Versal deformation spaces. For a germ of an analytic subvariety X of $\boldsymbol{C}^n$ let Θ_X denote the sheaf of tangent vectors of X. Then the $\mathcal{O}_X$ module of isomorphism classes of first order infinitesimal deformations of X, T_X^1, is defined by the exact sequence

$$0 \longrightarrow \Theta_X \longrightarrow \Theta_{\boldsymbol{C}^n}|_X \xrightarrow{\ \rho\ } N_X \longrightarrow T_X^1 \longrightarrow 0$$

where $N_X = \operatorname{Hom}_{\mathcal{O}_{\boldsymbol{C}^n}}(\mathscr{I}(X), \mathcal{O}_X)$ and ρ is the mapping taking $\sum_j \theta_j(x) \cdot \partial/\partial x$ to the homomorphism $f_i \mapsto \sum_j \theta_j \cdot \partial f_i/\partial x_j$. See [**12**] for further details.

Now let X_1, X_2 be germs of analytic subvarieties at the origin and set $X = X_1 \cap X_2$. Consider $\mathcal{O}_X$ to be a module over $\mathcal{O}_{X_i}$ and let $\alpha_i: N_{X_i} \otimes \mathcal{O}_{X_i}$

$\to \mathrm{Hom}_{\mathcal{O}_{C^n}}(\mathscr{I}(X_i), \mathcal{O}_X)$ be the map $\alpha_i(T \otimes f)(g) = fT(g)$ for $f \in \mathcal{O}_X$, $g \in \mathscr{I}(X_i)$, $T \in N_{X_i}$. We define $N_{X,X_1} = \mathrm{Im}\ \alpha_1$; $N_{X,X_2} = \mathrm{Im}\ \alpha_2$.

The following can then be proven:

LEMMA 3.1. *Let X_1, X_2 be perfect germs of analytic subvarieties at the origin in $\boldsymbol{C}^n$ which we assume to be defined by ideals J_1, J_2 respectively. Let $X = X_1 \cap X_2$ and suppose* codim $(X, \boldsymbol{C}^n) =$ codim$(X_1, \boldsymbol{C}^n) +$ codim $(X_2, \boldsymbol{C}^n)$. *Then if X_i, $i = 1, 2$, is either a complete intersection or of codimension 2, then (1) α_i is onto, $i = 1, 2$, and (2) $N_X = N_{X,X_1} \oplus N_{X,X_2}$.*

Using induction we then obtain

COROLLARY 3.2. *Let $X_1, \cdots, X_r$ be a very proper sequence of perfect germs of analytic subvarieties at the origin in $\boldsymbol{C}^n$ and suppose each X_i is either a complete intersection or of codimension 2. Let $Y_i = X_1 \cap \cdots \cap X_{i-1} \cap X_{i+1} \cap \cdots \cap X_r$. Then if $X = \bigcap_{i=1}^r X_i$ we have $N_X = \bigoplus_{i=1}^r N_{X,Y_i}$.*

We now state:

THEOREM 2. *Let $X_1, \cdots, X_r$ be a very proper sequence of perfect germs of analytic subvarieties at the origin in $\boldsymbol{C}^n$ with $X = \bigcap_{i=1}^r X_i$. Suppose each X_i is either a complete intersection or of codimension 2. Then every element of T_X^1 lifts to a flat analytic deformation of X.*

PROOF. Let $g \in N_X$ represent $[g] \in T_X^1$. Then by Corollary 3.2 we have $g = \bigoplus g_i$, $g_i \in N_{X,Y_i}$, and by definition $g_i = \alpha_i(h_i \otimes 1)$ for some infinitesimal deformation h_i of X_i. By **[6]**, **[9]**, h_i lifts to a flat analytic deformation H_i of X_i and, by Lemma 2.3, $\bigcap H_i$ is then a flat analytic deformation of X inducing g.

We now clearly have

COROLLARY 3.3. *Let $X, X_1, \cdots, X_r$ be as in Theorem 2. Suppose* $\dim_C T_X^1 = N < \infty$ *so that, by* **[2]**, *X has an analytic versal deformation space $V \to S$. Then $S \approx \boldsymbol{C}^N$.*

REMARK. N_{X,Y_i} in the above theorem and corollary consists of the space of all infinitesimal first-order deformations of X obtained by holding Y_i fixed and moving only X_i. Thus by Corollary 3.2 and Theorem 2 every deformation of X can be written as a combination (intersection) of movements of X in Y_i obtained by holding Y_i fixed and moving only X_i. Note that even movements of X_i which are trivial deformations may induce nontrivial deformations of X. For example let X_1 be the perfect analytic subvariety of codimension two in $\boldsymbol{C}^4$ given by the vanishing of the maximal minors of the relation matrix

$$R = \begin{bmatrix} z_1 & z_2^2 \\ z_2 & z_3 \\ z_3^2 & z_4 \end{bmatrix}.$$

Let X_2 be the nonsingular hypersurface with defining equation $h = z_2^2 - z_3^3 + z_1 + z_4$. The deformation space $\tilde{X}$ of X is then given by intersecting the variety $\tilde{X}_1$ in $\boldsymbol{C}^4(z_1, \cdots, z_4) \times \boldsymbol{C}^{10}(t_1, \cdots, t_{10})$ defined by the relation matrix

$$\begin{bmatrix} z_1 & z_2^2 + t_3 z_2 + t_4 \\ z_2 & z_3 \\ z_3^2 + t_1 z_3 + t_2 & z_4 \end{bmatrix}$$

with the variety $\tilde{X}_2$ defined by

$$H(z,t) = z^2 - z^3 + z_1 + z_4 + t_5 z_3^2 + t_6 z_3 + t_7 z_2^2 + t_8 z_2 + t_9 z_2 z_3 + t_{10}.$$

Note that the first four parameters $t_1, \cdots, t_4$ correspond to moving X in X_2 while holding X_2 fixed while the last six parameters correspond to moving X in X_1 holding X_1 fixed. Note also that T_X^1 is not the direct sum of $T_{X_1}^1$ and $T_{X_2}^1$ since X_2 being rigid has $T_{X_2}^1 = \{0\}$ and dim $T_{X_1}^1 = 4$. Also as deformations of X_2, all the $\tilde{X}_{2,t}$ are isomorphic to X_2 and $\tilde{X}_2 \approx X_2 \times \boldsymbol{C}^{10}$. However these trivial deformations of X_2 induce nontrivial deformations of $X_1 \cap X_2 = X$.

REFERENCES

1. A. Grothendieck and J. Dieudonné, *Éléments de géométrie algébriques*. IV, Inst. Hautes Études Sci. Publ. Math. No. 24 (1964). MR **33** #7330.

2. H. Grauert, *Über die Deformationen isolierter Singularitäten analytischer Mengen*, Invent. Math. **15** (1972), 171–198. MR **45** #2206.

3. R. C. Gunning, *Lectures on complex analytic varieties*. I. *The local parametrization theorem*, Math. Notes, Princeton Univ. Press, Princeton, N. J.; Univ. of Tokyo Press, Tokyo, 1970. MR **42** #7941.

4. ———, *Lectures on complex analytic varieties*. II. *Finite analytic mappings*, Math. Notes, no. 14, Princeton Univ. Press, Princeton, N. J., 1974. MR **50** #7570.

5. R. Hartshorne, *Topological conditions for smoothing algebraic singularities*, Topology **13** (1974), 241–253. MR **50** #2170.

6. A. Kas and M. Schlessinger, *On the versal deformation of a complex space with isolated singularity*, Math. Ann. **196** (1972), 23–29. MR **45** #3769.

7. G. Kempf and D. Laksov, *The determinantal formula of Schubert calculus*, Acta Math. **132** (1974), 153–162. MR **49** #2773.

8. S. Kleiman, *The transversality of a general translate*, Compositio Math. **28** (1974), 287–297. MR **50** #13063.

9. M. Loday, *Deformation des germes d'espaces analytiques*, Séminaire F. Norguet, Lecture Notes in Math., vol. 409, Springer-Verlag, Berlin and New York, 1973, pp. 140–164.

10. D. Mumford, *A remark on the paper of M. Schlessinger*, Rice Univ. Studies **59** (1973), no. 1, 113–117. MR **50** #319.

11. M. Schaps, *Deformations of Cohen-Macaulay schemes of codimension two*, Tel Aviv University.

12. M. Schlessinger, *On rigid singularities*, Rice Univ. Studies **59** (1973), no. 1, 147–162. MR **49** #9258.

WEIZMANN INSTITUTE

TEL AVIV UNIVERSITY

Proceedings of Symposia in Pure Mathematics
Volume 30, 1977

THE STRUCTURE OF WEIGHTED HOMOGENEOUS POLYNOMIALS

P. ORLIK AND R. RANDELL*

Let $f\colon C^{n+1} \to C$ be a complex polynomial and $(w_0, \cdots, w_n)$ positive rational numbers. Call f *weighted homogeneous* with weights $(w_0, \cdots, w_n)$ if each monomial of f, $\alpha z_0^{r_0} z_1^{r_1} \cdots z_n^{r_n}$ satisfies $r_0/w_0 + r_1/w_1 + \cdots + r_n/w_n = 1$. A *critical point* of f, $z \in C^{n+1}$, is a point where all the partials $\partial f/\partial z_i$, $i = 0, \cdots, n$, vanish. The aim of this paper is to describe the authors' recent work on the structure of weighted homogeneous polynomials with an isolated critical point.

1. Introduction. Let $V_t = \{z \in C^{n+1} | = t\}.f(z)$ For a weighted homogeneous polynomial (whp hereafter) with isolated critical point at $\mathbf{0}$, the variety V_0 is singular at $\mathbf{0}$. It has been the object of extensive study [**4**], [**26**]–[**31**], [**38**]. According to Milnor [**26**] if S_ε^{2n+1} is a small sphere centered at $\mathbf{0}$ and $K^{2n-1} = S_\varepsilon \cap V_0$, then $f/|f|: S_\varepsilon - K \to S^1$ is a smooth fiber bundle with fiber diffeomorphic to V_1, the open, nonsingular variety, a real $2n$-dimensional parallelizable manifold of the homotopy type of a wedge of μ copies of S^n. The multiplicity μ is computed according to Milnor and Orlik [**27**],

$$\mu = \prod_{i=0}^{n} (w_i - 1).$$

The diffeomorphism $h\colon V_1 \to V_1$ by which the two ends of $V_1 \times [0, 1]$ are identified to obtain $S_\varepsilon - K$ is called the characteristic map, given by [**26**] and [**31**] as follows. Express the rational numbers $w_i = u_i/v_i$ in irreducible form and let $d = [u_0, \cdots, u_n]$ be the least common multiple of the numerators. Define $q_i = d/w_i$ and $\xi = \exp(2\pi i/d)$. Then

AMS (*MOS*) *subject classifications* (1970). Primary 57D45; Secondary 14B05.
Key words and phrases. Singularities of polynomial maps, Milnor fibration, weighted homogeneous polynomials, monodromy.
*The authors were partially supported by NSF.

© 1977, American Mathematical Society

$$h(z_0, \cdots, z_n) = (\xi^{q_0} z_0, \xi^{q_1} z_1, \cdots, \xi^{q_n} z_n).$$

The corresponding map in the crucial homology dimension, $h_*^A: H_n(V_1; A) \to H_n(V_1; A)$, is called the monodromy map. Its explicit computation for $A = \boldsymbol{Z}$ is still an open problem. Partial results have been obtained by Pham [**33**], Brieskorn [**6**], Hirzebruch and Mayer [**20**], and others [**2**], [**3**], [**11**], [**12**], [**14**], [**15**], [**19**], [**30**]. The eigenvalues of the corresponding map with $A = \boldsymbol{C}$ are much easier to obtain, as in Milnor and Orlik [**27**] and Brieskorn [**7**].

Call the whp f *decomposable* if $f(z_0, \cdots, z_n) = f_1(z_0, \cdots, z_k) + f_2(z_{k+1}, \cdots, z_n)$ where $f_1: \boldsymbol{C}^{k+1} \to \boldsymbol{C}$ is a whp of type $(w_0, \cdots, w_k)$ with an isolated critical point at $\boldsymbol{0} \in \boldsymbol{C}^{k+1}$ and $f_2: \boldsymbol{C}^{n-k} \to \boldsymbol{C}$ is a whp of type $(w_{k+1}, \cdots, w_n)$ with an isolated critical point at $\boldsymbol{0} \in \boldsymbol{C}^{n-k}$. Then a result of Oka [**28**] asserts that the fiber $V_1(f)$ is homotopy equivalent to the reduced join of the fibers $V_1(f_1)$ and $V_1(f_2)$. The matrix of the monodromy map for f is therefore the tensor product of the matrices of the monodromies for f_1 and f_2. See also Sebastiani and Thom [**39**].

In this note we consider weighted homogeneous polynomials which are indecomposable in the above sense. By associating a weighted graph $G(\boldsymbol{w})$ to the weights $\boldsymbol{w} = (w_0, \cdots, w_n)$ and looking for certain subgraphs $\Gamma(\boldsymbol{w})$, we obtain new necessary conditions for the set $(w_0, \cdots, w_n)$ to occur as the weights of a whp with isolated singularity. This reduces the integral monodromy problem to a few specific questions.

There has been much recent interest in monodromy problems, and for the convenience of the reader we have included an extensive bibliography.

2. Graphs. Let $\boldsymbol{w} = (w_0, \cdots, w_n)$ be an $(n + 1)$-tuple of positive rationals.

DEFINITION 2.1. The *weighted graph* $G(\boldsymbol{w})$ is associated to $\boldsymbol{w}$ as follows.

(i) $G(\boldsymbol{w})$ has vertices A_i, $i = 0, \cdots, n$.

(ii) For each integer w_i, add a 1-cell $E_{i,i}$ with endpoints at A_i, oriented arbitrarily and labeled w_i.

(iii) Consider the rational numbers $a_{j,k} = w_j(w_k - 1)/w_k$, $j \neq k$. For each pair (j, k) with integer $a_{j,k}$, add a 1-cell $E_{j,k}$ with endpoints A_j and A_k, oriented so that $\partial E_{j,k} = A_k - A_j$ and labeled $a_{j,k}$.

REMARK. The 1-cell $E_{i,i}$ corresponds to a monomial $z_i^{w_i}$, while the 1-cell $E_{j,k}$ corresponds to $z_k z_j^{a_{j,k}}$.

EXAMPLE 2.2. The weights $\boldsymbol{w} = (7, 35/2, 35/3)$ give rise to $G(\boldsymbol{w})$ below:

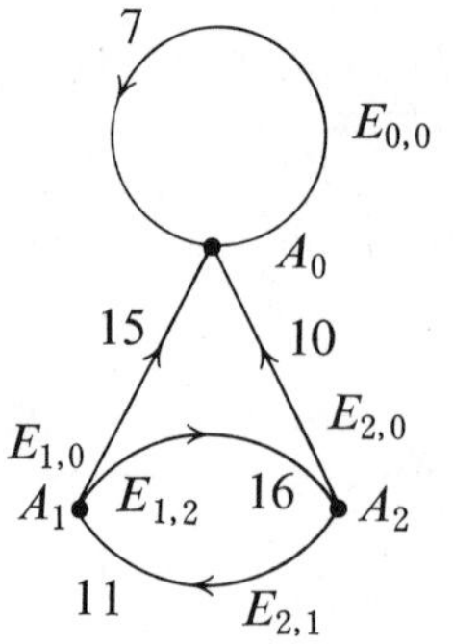

Note. In order to simplify notation we shall suppress the labels A_i and $E_{j,k}$ when this leads to no misunderstanding.

DEFINITION 2.3. An *admissible subgraph* $\Gamma(\boldsymbol{w})$ of $G(\boldsymbol{w})$ has the following properties.

(i) $A_i \in \Gamma(\boldsymbol{w})$, $i = 0, \cdots, n$.

(ii) Each component of $\Gamma(\boldsymbol{w})$ has the homotopy type of a circle oriented by the 1-cells it contains. Furthermore, a retraction to this circle may be given so that points in each 1-cell move only in the positive direction of the orientation of that cell.

DEFINITION 2.4. Given any subgraph of $G(\boldsymbol{w})$ of the form below

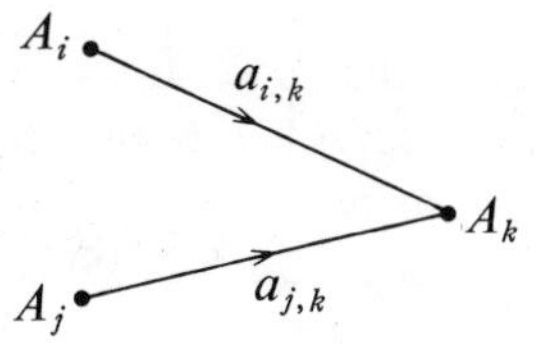

we define the *vertex condition* at A_k of A_i and A_j;

> vc(k; i, j): there exist positive integers b_i, b_j, and $\varepsilon_l = 0$ or 1 so that $\varepsilon_l/w_l + b_i/w_i + b_j/w_j = 1$, for some $l \neq j, k$. If $\varepsilon_l = 1$, $l \neq i$ also.

EXAMPLE 2.5. The admissible subgraphs of (2.2) are

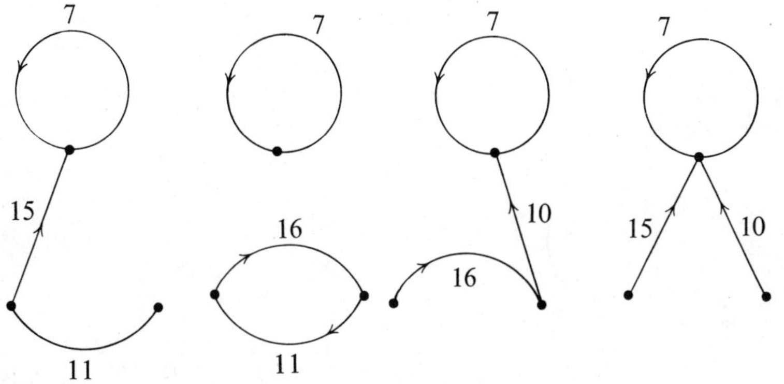

where in the last graph the vertex condition vc(0; 1, 2) is satisfied by $\varepsilon = 0$, $b_1 = 10$, $b_2 = 5$. Note that the choice of b_1 and b_2 is not unique in general.

EXAMPLE 2.6. The graph of $\boldsymbol{w} = (10, 10/3, 10/3)$

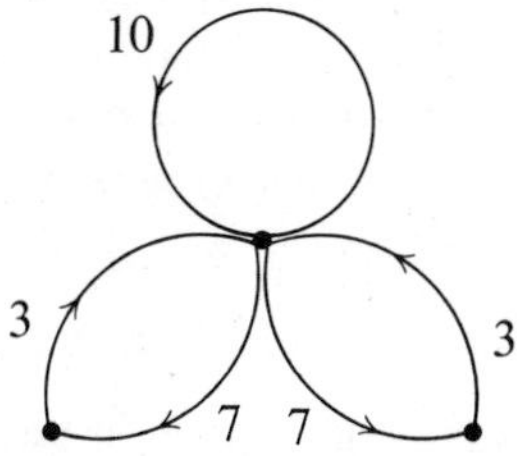

has three admissible subgraphs;

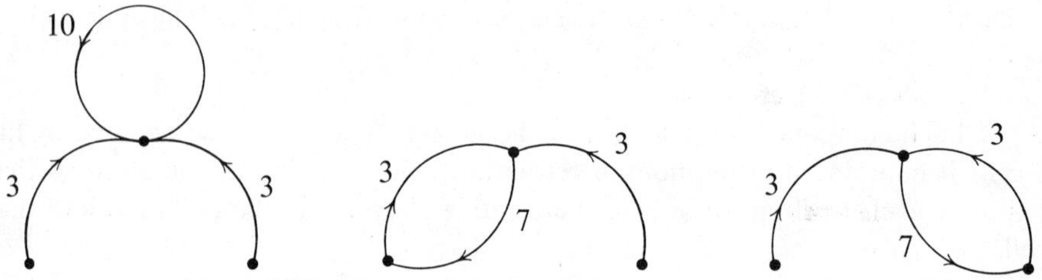

but vc(0; 1, 2) is not satisfied. Note that $\mu = (10 - 1)(10/3 - 1)(10/3 - 1) = 49$ is an integer, but it follows from (2.9) below that there is no whp of type (10, 10/3, 10/3) with an isolated singularity.

EXAMPLE 2.7. The graph of $\boldsymbol{w} = (3, 9/2, 15/2, 45)$ is

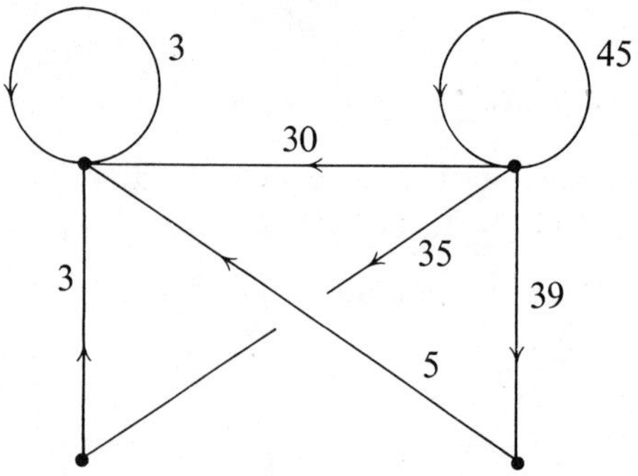

with admissible subgraph

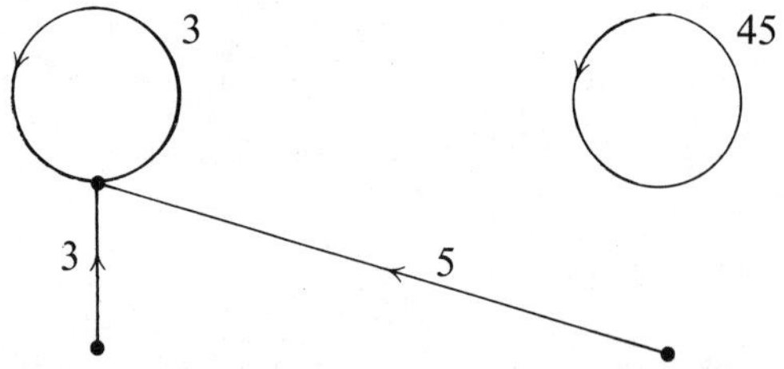

In fact, any admissible subgraph contains the left-hand component above. Here vc(0; 1, 2) is satisfied *only* by $\varepsilon_3 = 1$, $b_1 = 2$, $b_2 = 4$, and $f(z_0, z_1, z_2, z_3) = z_0^3 + z_0 z_1^3 + z_0 z_2^5 + z_3^{45} + z_3 z_1^2 z_2^4$ has an isolated critical point.

Now suppose that f is a whp with isolated singularity and weights $\boldsymbol{w} = (w_0, \cdots, w_n)$. Form the graph of f, $G(f)$ with edges $E_{i,i}$ for monomials $\alpha z_i^{a_i}$ and $E_{j,k}$ for $\alpha z_k z_j^{a_{j,k}}$, in the obvious way.

PROPOSITION 2.8. *$G(f)$ satisfies all possible vertex conditions.*

PROOF. An investigation of partial derivatives shows directly that if some vertex condition is not satisfied, the singularity of f is not isolated.

THEOREM 2.9. *Suppose* $\boldsymbol{w}$ *gives the weights of some whp* f *with isolated singularity. Then* $G(\boldsymbol{w})$ *contains an admissible subgraph in which all possible vertex conditions are satisfied.*

PROOF. Define f to be minimal if whenever any monomial is deleted from f, the resulting polynomial no longer has an isolated critical point. Then it is clear that there is a minimal whp $\hat{f}$ with the same weights as f. Then $G(\hat{f})$ is admissible and, by (2.8), $G(f)$ satisfies all vertex conditions.

The converse of 2.9 does *not* hold, even if we add the hypothesis that $\mu = \prod_{i=0}^{n} (w_i - 1)$ is an integer.

EXAMPLE 2.10. Let $f_a(z_0, \cdots, z_4) = z_0^3 + z_0z_1^3 + z_0z_2^5 + z_3^{45} + z_3z_1^2z_2^4 + z_3z_4^a$. f_a has weights $\boldsymbol{w}_a = (3, 9/2, 15/2, 45, 45a/44)$. $G(f_a)$ is clearly admissible and the only vertex condition, vc(0; 1, 2), is satisfied (see (2.7)). If $a = 3$, $\prod (w_i - 1)$ is not an integer, so there can be no whp with weights $\boldsymbol{w}_3$ and an isolated critical point. If $a = 2$, $\prod (w_i - 1)$ *is* an integer, but a more detailed study shows that still there is no whp with weights $\boldsymbol{w}_2$ and isolated critical point.

Example (2.10) suggests that the appearance of the monomials $z_3z_1^2z_2^4$ and $z_3z_4^2$ in f_2 means that some monomial involving z_1, z_2 and z_4 should appear in f in order to insure that the singularity is isolated. Thus, whenever a "primary" vertex condition vc(i; j, k) is satisfied by monomial $z_l z_j^{b_j} z_k^{b_k}$, and $z_l z_m^{a_m}$ also appears in f, it seems that a "secondary" vertex condition must be satisfied in order to guarantee the existence of the desired whp with isolated singularity.[1]

3. Integral monodromy. The components of an admissible subgraph consist of various trees attached to either a single vertex with a 1-cell,

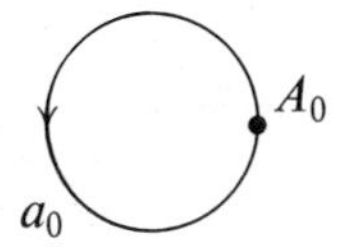

corresponding to the monomial $z_0^{a_0}$; or to a loop

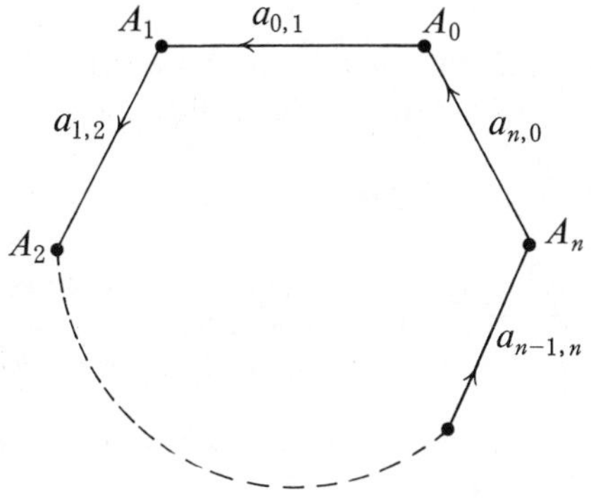

---

[1]ADDED IN PROOF. We now have a necessary and sufficient condition in terms of the weights, see [30]. A. Kouchnirenko has informed us that he has also obtained such a condition.

corresponding to the polynomial $z_0^{a_{0,1}}z_1 + z_1^{a_{1,2}}z_2 + \cdots + z_n^{a_{n,0}}z_0$. This suggests that one should investigate the monodromy corresponding to

(i)

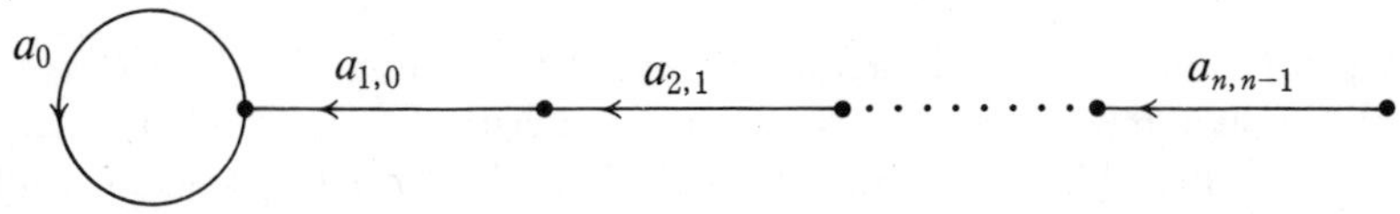

$$z_0^{a_0} + z_0 z_1^{a_{1,0}} + \cdots + z_{n-1} z_n^{a_{n,n-1}}$$

(ii)

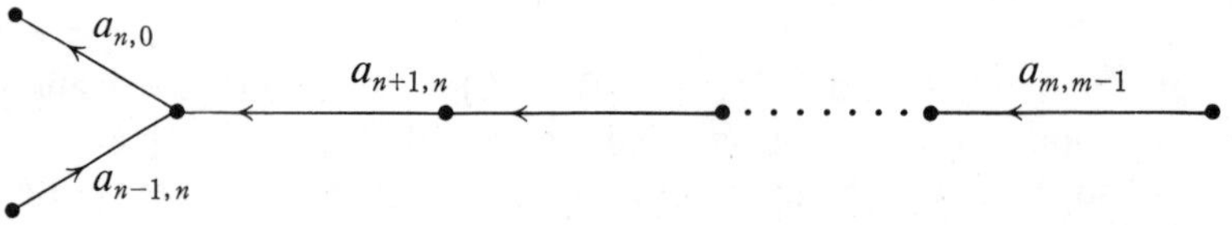

$$z_0^{a_{0,1}}z_1 + z_1^{a_{1,2}}z_2 + \cdots + z_n z_{n-1}^{a_{n-1,n}} + z_0 z_n^{a_{n,0}} + z_n z_{n+1}^{a_{n+1,n}} + \cdots + z_{m-1} z_m^{a_{m,m-1}} + g(z)$$

where vc(n; $n-1$, $n+1$) is satisfied, so that $g(z)$ contains a monomial $z_{n-1}^{b_1} z_{n+1}^{b_2}$ or $z_l z_{n-1}^{b_1} z_{n+1}^{b_2}$.

(iii) The "amalgamation of two graphs along a subgraph", e.g., how to obtain the monodromy of

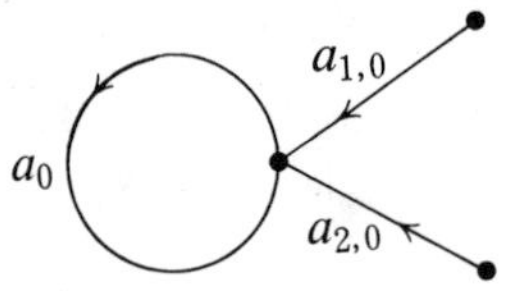

satisfying vc(0; 1, 2) from the monodromies of

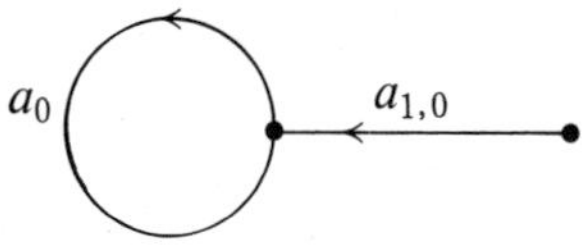

and

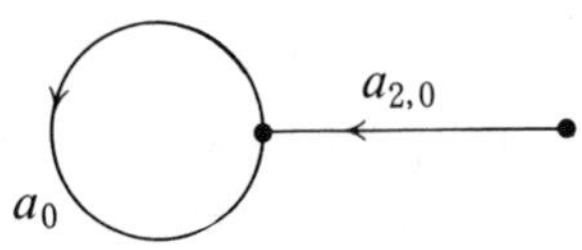

The authors have studied (i) in a somewhat more general setting. The results of that investigation will be published elsewhere [**30**].

BIBLIOGRAPHY

1. N. A'Campo, *Le nombre de Lefschetz d'une monodromie*, Nederl. Akad. Wetensch. Proc. Ser. A **76** = Indag. Math. **35** (1973), 113–118. MR **47** #8903.

2. ———, *Le groupe de monodromie du déploiement des singularités isolées de courbes planes*, Math. Ann. **213** (1975), 1–32.

3. ———, *Sur la monodromie des singularités isolées d'hypersurfaces complexes*, Invent. Math. **20** (1973), 147–169. MR **49** #3201.

4. V. I. Arnol'd, *Normal forms of functions in neighborhoods of degenerate critical points*, Russian Math. Surveys **29** (1974), 10–50.

5. ———, *Classification of unimodal critical points of functions*, Funkcional. Anal. i Priložen. **7** (1973), no. 3, 75–76 = Functional Anal. Appl. **7** (1973), 230–231. MR **48** #2431.

6. E. Brieskorn, *Beispiele zur Differentialtopologie von Singularitäten*, Invent. Math. **2** (1966), 1–14. MR **34** #6788.

7. ———, *Die Monodromie der isolierten Singularitäten von Hyperflächen*, Manuscripta Math. **2** (1970), 103–161. MR **42** #2095.

8. ———, *Vue d'ensemble sur les problèmes de monodromie*, Singularités à Cargèse, Astérisque, nos. 7/8, Soc. Math. France, 1973, pp. 393–413.

9. E. Brieskorn and G.-M. Greuel, *Singularities of complete intersections*, Proc. Internat. Conf. on Manifolds, Tokyo, 1973, pp. 123–129.

10. A. Durfee, *Fibered knots and algebraic singularities*, Topology **13** (1974), 47–59. MR **49** #1523.

11. A. M. Gabrielov, *Bifurcations, Dynkin diagrams and modality of isolated singularities*, Functional Anal. Appl. **8** (1974), 94–98.

12. ———, *Intersection matrices for certain singularities*, Funkcional. Anal. i Priložen. **7** (1973), no. 3, 18–32 = Functional Anal. Appl. **7** (1973), 182–193. MR **48** #2418.

13. G.-M. Greuel, *Der Gauss-Manin-Zusammenhang isolierter Singularitäten von vollständigen Durchschnitten*, Math. Ann. **214** (1975), 235–266.

14. S. M. Gusein-Zade, *Intersection matrices of some singularities of functions of two variables*, Funkcional. Anal. i Priložen. **8** (1974), no. 1, 11–15 = Functional Anal. Appl. **8** (1974), 10–13. MR **49** #3202.

15. ———, *Dynkin diagrams for singularities of functions of two variables*, Functional Anal. Appl. **8** (1974), 295–300.

16. H. Hamm, *Lokale topologische Eigenschaften Komplexer Räume*, Math. Ann. **191** (1971), 235–252. MR **44** #3357.

17. ———, *Exotische Sphären als Umgebungsränder in speziellen komplexen Räumen*, Math. Ann. **197** (1972), 44–56. MR **47** #2625.

18. ———, *Ein Beispiel zur Berechnung der Picard-Lefschetz Monodromie für nichtisolierte Hyperflächensingularitäten*, Math. Ann. **214** (1975), 221–234.

19. A. Hefez and F. Lazzeri, *The intersection matrix of Brieskorn singularities*, Invent. Math. **25** (1974), 143–158.

20. F. Hirzebruch and K. H. Mayer, *0(n)-Mannigfaltigkeiten, exotische Sphären und Singularitäten*, Lecture Notes in Math., no. 57, Springer-Verlag, Berlin and New York, 1968. MR **37** #4825.

21. L. Kauffman, *Branched coverings, open books, and knot periodicity*, Topology **13** (1974), 143–160.

22. K. Lamotke, *Die Homologie isolierter Singularitäten*, Math. Z. **143** (1975), 27–44.

23. Lê Dũng Tráng, *Les théorèmes de Zariski de type de Lefschetz*, Centre de Math. Ecole Poly. Paris, 1971.

24. ———, *Calcul du nombre de cycles évanouissants d'une hypersurface complexe*, Ann. Inst. Fourier (Grenoble) **23** (1973) fasc. 4, 261–270. MR **48** #8838.

25. ———, *Computation of the Milnor number of isolated singularity of complete intersection*, Funkcional. Anal. i Priložen. **8** (1974), no. 2, 45–49 = Functional Anal. Appl. **8** (1974), 127–131. MR **50** #2557.

26. J. Milnor, *Singular points of complex hypersurfaces*, Ann. of Math. Studies, no. 61, Princeton Univ. Press, Princeton, N. J.; Univ. of Tokyo Press, Tokyo, 1968. MR **39** #969.

27. J. Milnor and P. Orlik, *Isolated singularities defined by weighted homogeneous polynomials*, Topology **9** (1970), 385–393. MR **45** #2757.

28. M. Oka, *On the homotopy types of hypersurfaces defined by weighted homogeneous polynomials*, Topology **12** (1973), 19–32.

29. P. Orlik, *On the homology of weighted homogeneous manifolds*, Proc. Second Conf. Transformation Groups, I, Lecture Notes in Math., vol. 298, Springer-Verlag, Berlin and New York, 1972, pp. 206–269.

30. P. Orlik and R. Randell, *The classification and monodromy of weighted homogeneous singularities* (preprint).

31. P. Orlik and P. Wagreich, *Isolated singularities of algebraic surfaces with C*-action*, Ann. of Math. (2) **93** (1971), 205–228. MR **44** #1662.

32. V. P. Palamodov, *Multiplicity of a holomorphic transformation*, Funkcional. Anal. i

Priložen. **1** (1967), no. 3, 54–65 = Functional Anal. Appl. **1** (1967), 218–226. MR **38** #4720.

33. F. Pham, *Formules de Picard-Lefschetz généralisées et ramification des intégrales*, Bull. Soc. Math. France **93** (1965), 333–367. MR **33** #4064.

34. ———, *Introduction à l'étude topologique des singularités de Landau*, Mém. Sci. Math., Fasc. 164, Gauthier-Villars, Paris, 1967. MR **37** #4837.

35. R. Randell, *Generalized Brieskorn manifolds*, Bull. Amer. Math. Soc. **80** (1974), 111–115. MR **49** #4007.

36. R. Randell, *The homology of generalized Brieskorn manifolds*, Topology **14** (1975), 347–355.

37. K. Saito, *Einfach-elliptische Singularitäten*, Invent. Math. **23** (1974), 289–325. MR **50** #7147.

38. ———, *Quasihomogene isolierte Singularitäten von Hyperflächen*, Invent. Math. **14** (1971), 123–142. MR **45** #3767.

39. M. Sebastiani and R. Thom, *Un résultat sur lamonodromie*, Invent. Math. **13** (1971), 90–96. MR **45** #2201.

40. D. Siersma, *Classification and deformation of singularities*, Thesis, Amsterdam, 1974.

UNIVERSITY OF WISCONSIN-MADISON

UNIVERSITY OF MICHIGAN-ANN ARBOR

Proceedings of Symposia in Pure Mathematics
Volume 30, 1977

DEFORMATIONS OF QUOTIENT SURFACE SINGULARITIES

H. C. PINKHAM*

Let the finite group G act on C^2, leaving the origin 0 fixed and acting freely elsewhere. Call the quotient of this action $\pi: C^2 \to X_G$. The singularities that occur have been classified by Prill [5]. Brieskorn determined the dual graph of the minimal resolution, proved they are taut, i.e., uniquely determined by their graph, and computed their embedding dimension [1].

We determine, at least in principle, their infinitesimal deformations. We are mainly interested in the case of large embedding dimension, the hypersurface case being well known. We conclude with a more explicit result in the case G is cyclic, already studied by Riemenschneider [6] by different methods.

Note. The quotients of C^n, $n \geqq 3$, by finite groups are rigid (Schlessinger [7]). For $n = 1$ the quotient is smooth, hence trivially rigid.

Notation. $V = C^2 \setminus (0)$, $U = X_G \setminus \pi(0)$; $X_G \hookrightarrow^{i} C^n$ some embedding. θ = tangents, N = normal bundle, T^1 = classifying space of infinitesimal deformations of X_G, i.e., deformations of X_G over $C[\varepsilon]$, $\varepsilon^2 = 0$. Superscript G means invariants under G.

Schlessinger [7] has shown (in a more general context) that

$$0 \to T^1 \to H^1(U, \theta_U) \xrightarrow{\varphi} H^1(U, i^*\theta_{C^n})$$

is exact, where the right-hand map is induced by the usual map φ:

$$0 \to \theta_U \xrightarrow{\varphi} i^*\theta_{C^n} \to N_{U/C^n} \to 0.$$

Schlessinger also showed $\theta_U = (\pi_*\theta_V)^G$ ($g \in G$ acts on $D \in \theta_V$ by $D \to gDg^{-1}$). Since $(\pi_*\theta_V)^G$ is a direct summand of $\pi_*\theta_V$, $H^1(U, \theta_U) = H^1(V, \theta_V)^G$. Also $H^1(U, \mathcal{O}_U) = H^1(V, \mathcal{O}_V)^G$.

AMS (MOS) subject classifications (1970). Primary 14D15, 14B05; Secondary 14J15.

*Partially supported by NSF grant GP 32843.

© 1977, American Mathematical Society

Let $z_1, \cdots, z_n$ be generators of the ring of X_G giving the embedding in $\boldsymbol{C}^n$. We may assume G acts linearly on $\boldsymbol{C}^2$, so if x, y are coordinates on $\boldsymbol{C}^2$, $z_j = f_j(x, y)$, f_j homogeneous of degree d_j. $i^*\theta_{C^n}$ is free on $\partial/\partial z_j$, θ_V is free on $\partial/\partial x, \partial/\partial y$.

THEOREM. *φ is obtained by taking invariants of the map*

$$(*) \qquad \Phi\colon H^1(V, \mathcal{O}_V\, \partial/\partial x \oplus \mathcal{O}_V\, \partial/\partial y) \to H^1\Big(V, \bigoplus_j \mathcal{O}_V\, \partial/\partial_{z_j}\Big)$$

given on the jth component by $\bigcup \partial f_j/\partial x \oplus \bigcup \partial f_j/\partial y$, $\bigcup$ = cup product, $\partial f_j/\partial x$ viewed as an element of $H^0(V, \mathcal{O}_V)$.

Note. A similar result is proved in [**4**].

Let $\boldsymbol{P} = \boldsymbol{P}^1$. $V = \operatorname{Spec}(\bigoplus \mathcal{O}_{\boldsymbol{P}}(\nu))$, $-\infty < \nu < \infty$, $\boldsymbol{P}$ with homogeneous coordinates x and y. Hence both sides of $(*)$ are graded. Since Φ is also graded $(\partial f_j/\partial x) \in H^0(\mathcal{O}_{\boldsymbol{P}}(d_j - 1))$ we can write Φ as the sum, $-\infty < \nu < \infty$, of

$$H^1\Big(\boldsymbol{P}, \mathcal{O}_{\boldsymbol{P}}(\nu + 1)\frac{\partial}{\partial x} \oplus \mathcal{O}_{\boldsymbol{P}}(\nu + 1)\frac{\partial}{\partial y}\Big) \xrightarrow{\Phi_\nu} \bigoplus_{j=1}^{n} H^1\Big(\boldsymbol{P}, \mathcal{O}_{\boldsymbol{P}}(d_j + \nu)\frac{\partial}{\partial z_j}\Big).$$

Taking invariants, we get a grading on $T^1 = \bigoplus T^1(\nu) = \bigoplus(\ker \Phi_\nu)^G$. We could have concluded this directly by noting that the ring of X_G has $\boldsymbol{C}^*$ action, z_j having degree d_j. The $T^1(\nu)$ correspond to the eigenspaces of the induced action of $\boldsymbol{C}^*$ on T^1; see [**3**].

Note that $T^1(\nu) = 0$ for $\nu \geqq 0$. This is what we call "negative grading" [**3**]. According to Wahl's work on equisingularity this should correspond to the tautness of these singularities.

The theorem shows that to compute T^1 we need only find a basis for the invariants of the group action. It is always convenient, however, to have a minimal one.

We now carry out the computation explicitly in the case G cyclic. This is the only case worked out so far.

DEFINITION. Let q and n be relatively prime integers, $0 < q < n$. $G_{q,n}$ denotes the cyclic group of order n acting on $\boldsymbol{C}^2$ by $(x, y) \mapsto (\zeta x, \zeta^q y)$, ζ an nth root of unity. The corresponding singularity is denoted $X_{q,n}$.

Clearly the invariants are generated by the monomials $x^i y^j$, $i + qj \equiv 0 \bmod n$. Hence $X_{q,n}$ admits a $\boldsymbol{C}^* \times \boldsymbol{C}^*$ action.

PROPOSITION (RIEMENSCHNEIDER [**6**]). *A minimal set of generators for the invariants of $\boldsymbol{C}[x, y]$ under $G_{q,n}$ is $x^{i_\varepsilon} y^{j_\varepsilon}$, $1 \leqq \varepsilon \leqq e$, where i_ε, j_ε, e and auxiliary integers a_ε are defined as follows*: $i_1 = n$, $i_2 = n - q$, $i_{\varepsilon-1} = a_\varepsilon i_\varepsilon - i_{\varepsilon+1}$, $0 \leqq i_{\varepsilon+1} < i_\varepsilon$, $i_{e-1} = 1$, $i_e = 0$, $j_1 = 0$, $j_2 = 1$, $j_{\varepsilon+1} = a_\varepsilon j_\varepsilon - j_{\varepsilon-1}, \cdots, j_e = n$.

The i_ε increase strictly and the j_ε decrease strictly. This turns out to be the crucial fact. Note that $n/(n - q)$ has the continued fraction expansion

$$\frac{n}{n-q} = a_2 - \cfrac{1}{a_3 - \cfrac{1}{a_4 - \ddots - \cfrac{1}{a_{e-1}}}}, \qquad a_\varepsilon \geqq 2.$$

THEOREM. *$T^1(X_{q,n})$ has as basis inside $H^1(V, \mathcal{O}_V\partial/\partial x \oplus \mathcal{O}_V\partial/\partial y)^{G_{q,n}}$ the elements*

$$\frac{x}{x^{i_\varepsilon}y^{j_\varepsilon}}\frac{\partial}{\partial x} \in T^1(-i_\varepsilon - j_\varepsilon), \qquad 2 \leqq \varepsilon \leqq e-2,$$

$$\frac{y}{x^{i_\varepsilon}y^{j_\varepsilon}}\frac{\partial}{\partial y} \in T^1(-i_\varepsilon - j_\varepsilon), \qquad 3 \leqq \varepsilon \leqq e-1,$$

$$\left(\frac{1}{x^{i_\varepsilon}y^{j_\varepsilon}}\right)^a\left(j_\varepsilon x\frac{\partial}{\partial x} - i_\varepsilon y\frac{\partial}{\partial y}\right) \in T^1(-ai_\varepsilon - aj_\varepsilon),$$

$$\begin{aligned} 2 \leqq \varepsilon &\leqq e-1, \\ 2 \leqq a &\leqq a_\varepsilon - 1, \quad \text{if } e \neq 3, \\ &\leqq a_2, \qquad\quad \text{if } e = 3. \end{aligned}$$

Hence T^1 has

$$\begin{aligned} \dim &= 2(e-3) + \sum_{\varepsilon=2}^{e-1}(a_\varepsilon - 2) = \left(\sum_{\varepsilon=2}^{e-1} a_\varepsilon\right) - 2, \quad \text{if } e > 3, \\ &= a_2 - 1, \quad \text{if } e = 3. \end{aligned}$$

Note. The dimension formula was obtained by Riemenschneider in [**6**, p. 237]. He gives no indication of proof for $e \geqq 5$ but claims it can be established by an explicit computation using the equations between the z_j and their syzygies. This seems difficult without an a priori bound for dim T^1. For $X_{1,n}$, the cone over the rational curve of degree n in $\boldsymbol{P}^n$, the formula reduces to Mumford's [**2**]. See also [**3**, Chapter 2, §§6–8].

It is not difficult to identify the cokernel of Φ_ν:

$$0 \to \mathcal{O}_{\boldsymbol{P}}(1)\frac{\partial}{\partial x} \oplus \mathcal{O}_{\boldsymbol{P}}(1)\frac{\partial}{\partial y} \xrightarrow{\Phi_0} \bigoplus_{\varepsilon=1}^{e} \mathcal{O}_{\boldsymbol{P}}(i_\varepsilon + j_\varepsilon) \xrightarrow{\Psi_0} \bigoplus_{\varepsilon=1}^{e-2} \mathcal{O}_{\boldsymbol{P}}(i_\varepsilon + j_{\varepsilon+2}) \to 0$$

where Ψ_0 has εth component $(1 \leqq \varepsilon \leqq e-2)$:

$$(0, \cdots, 0, \underset{\substack{\uparrow \\ \varepsilon\text{th place}}}{y^{j_{\varepsilon+2}-j_\varepsilon}}, -a_{\varepsilon+1}x^{i_\varepsilon - i_{\varepsilon+1}}y^{j_{\varepsilon+2}-j_{\varepsilon+1}}, x^{i_\varepsilon - i_{\varepsilon+2}}, 0, \cdots, 0).$$

This allows us to compute the normal bundle of U in $\boldsymbol{C}^n$, which is of help in understanding the deformations of $X_{q,n}$. For $X_{1,n}$, $i_\varepsilon + j_{\varepsilon+2} = n+2$ for all ε, checking Mumford's computation [**2**].

References

1. E Brieskorn, *Rationale Singularitäten komplexer Flächen*, Invent. Math. **4** (1967/68), 336–358. MR **36** #5136.
2. D. Mumford, *A remark on the paper of M. Schlessinger*, Rice Univ. Studies **59** (1973), no. 1, 113–117. MR **50** #319.
3. H. Pinkham, *Deformations of algebraic varieties with* $\boldsymbol{G}_m$ action, Astérisque, no. 20, Soc. Math. France, 1974.
4. H. Pinkham, *On a result of Riemenschneider*, Manuscripta Math. **16** (1975), 137–144.
5. D. Prill, *Local classification of quotients of complex manifolds by discontinuous groups*, Duke Math. J. **34** (1967), 375–386. MR **35** #1829.
6. O. Riemenschneider, *Deformationen von Quotientensingularitäten (nach zyklischen Gruppen)*, Math. Ann. **209** (1974), 211–248.
7. M. Schlessinger, *Rigidity of quotient singularities*, Invent. Math. **14** (1971), 17–26. MR **45** #1912.

COLUMBIA UNIVERSITY

Proceedings of Symposia in Pure Mathematics
Volume 30, 1977

SIMPLE ELLIPTIC SINGULARITIES, DEL PEZZO SURFACES AND CREMONA TRANSFORMATIONS

H. C. PINKHAM*

At the 1975 Summer Institute on Several Complex Variables, Brieskorn asked the following question: What singularities can appear in the fibres of the versal deformation of a simple elliptic surface singularity of degree 4? For degrees 1, 2 and 3, the only hypersurface cases, this question has been answered by Saito [**11**]. Degree 4 is the only remaining complete intersection case. In this note we announce the result for arbitrary degree: We actually determine all configurations of singularities in the fibres. Our method also yields an abundance of information concerning fine invariants of the singularity, usually available only in the complete intersection case: monodromy, discriminant locus, fundamental group of the complement of the discriminant, etc. For simplicity we restrict to the case degree $\geqq 3$. A simple elliptic singularity of degree k is then just the cone C_E over a smooth curve E of genus 1 embedded in $\boldsymbol{P}^{k-1}$ by a complete linear system of degree k. For details concerning these singularities and their deformations see [**11**] and [**10**, Chapter 9]. The former treats only the hypersurface case; the latter shows that C_E can be smoothed by deformation iff $k \leqq 9$.

THEOREM. *The only configurations of singularities appearing in fibres of the versal deformation of a simple elliptic singularity of degree* $k \geqq 3$ *are*:

(i) *For all* k, *simple elliptic singularities of the same degree*,

(ii) *if* $3 \leqq k \leqq 7$, *configurations of rational double points for which the dual graph of the minimal resolution is, in the standard notation* (*see* [1]):

$k = 7$	A_1
$k = 6$	*any subgraph of* $A_1 \times A_2$
$k = 5$ (*resp.* 4, 3)	*any proper subgraph of the extended Dynkin diagram* $\tilde{A}_4$ (*resp.* $\tilde{D}_5$, $\tilde{E}_6$) *in the sense of* [**1**].

AMS (*MOS*) *subject classifications* (1970). Primary 14D15, 14J25, 14E05; Secondary 14B05.
*Partially supported by NSF grant GP 32843.

© 1977, American Mathematical Society

Using 6.6 and 5.1 of [**10**] we see that deformations of C_E lift to projective deformations of the projective cone $\bar{C}_E$ in $\boldsymbol{P}^k$; see [**10**, Chapter 9]. Using Del Pezzo's classification of surfaces of degree k in $\boldsymbol{P}^k$ (see [**9**] for a modern treatment), we see that a singular deformation of $\bar{C}_E$ must be either a cone over an elliptic curve or a degenerate Del Pezzo surface. By definition a Del Pezzo surface is the image of $\boldsymbol{P}^2$ by a system of (generically smooth) cubics with $9-k$ not necessarily distinct base points. If the base points are in sufficiently general position the surface is nonsingular; in certain special positions, e.g., when two base points come together, when three lie on a line, when six lie on a conic, the surface is singular. We then call the Del Pezzo surface degenerate. An elementary analysis of the singularities that occur then yields (ii). These singularities, especially those appearing on the cubic surface, have been studied classically; see [**12**] and [7, Part III].

This naturally leads us to consider sets of $9-k$ points on a smooth plane cubic and the Del Pezzo surfaces obtained by blowing them up. This situation was first studied by Coble [**5**], then Du Val [**6**] and [7, Parts II and III], and finally in the smooth case by Manin [**8**]. One is led to a finite group of Cremona transformations on $\boldsymbol{P}^2$ which for $k = 7$ (resp. 6, 5,$\cdots$, 1) is isomorphic to the finite reflection group A_1 (resp. $A_1 \times A_2$, A_4, D_5, E_6, E_7, E_8). A certain "complexified affine Weyl group" then plays for C_E a similar role to that played by the finite reflection group in the theory of simultaneous resolution of rational double points: Brieskorn [**4**] and Tjurina [**13**]. (Note that Tjurina's construction of the versal deformation space of E_6, E_7 and E_8 is anticipated by Bramble [**2**] who also shows how the finite reflection group arises by Cremona transformations.) In turn this yields the fine invariants mentioned above, and suggests a proof of the theorem that does not appeal to Del Pezzo's classification, hence also valid for $k = 1$ and 2.

Finally one can ask if there is a deeper connection between the elliptic singularities and the affine Weyl group, as in the rational double point case; see [**4**].

ADDED IN PROOF. E. Looijenga has informed me he has obtained similar results in the hypersurface case,but by a different method.

REFERENCES

1. N. Bourbaki, *Éléments de mathématique*, Fasc. XXXIV. *Groupes et algèbres de Lie.* Chaps. 4, 5, 6, Actualitiés Sci. Indust., no. 1337, Hermann, Paris, 1968. MR **39** #1590.

2. C. C. Bramble, *A collineation group isomorphic with the group of double tangents of the plane quartic*, Amer. J. Math. **40** (1918), 351–365.

3. E. Brieskorn, *Die Auflösung der rationalen Singularitäten holomorpher Abbildungen*, Math. Ann. **178** (1968), 255–270. MR **38** #2140.

4. ———, *Singular elements of semi-simple algebraic groups*, Proc. Internat. Congress Math. (Nice, 1970), vol. 2, Gauthier-Villars, Paris, 1971, pp. 279–284.

5. A. B. Coble, *Point sets and allied Cremona groups.* I, II, III, Trans. Amer. Math. Soc. **16** (1915), 155–198; ibid. **17** (1916), 345–385; **18** (1917), 331–372.

6. P. Du Val, *On the directrices of a set of points in a plane*, London Math. Soc. (2) **35** (1933), 23–74.

7. ———, *On isolated singularities which do not affect the conditions of adjunction.* I, II, III, Proc. Cambridge Philos. Soc. **30** (1934), 453–465, 483–491.

8. Ju. I. Manin, *Cubic forms: Algebra, geometry, arithemetic,* "Nauka", Moscow, 1972; English transl., North-Holland Math. Library, vol. 4, North-Holland, Amsterdam, 1973.

9. M. Nagata, *On rational surfaces.* I. *Irreducible curves of arithmetic genus* 0 *or* 1, Mem. Coll. Sci. Univ. Kyoto Ser. A Math. **32** (1960), 351–370. MR **23** #A3739.

10. H. C. Pinkham, *Deformations of algebraic varieties with G_m action*, Astérisque, no. 20, Soc. Math. France, 1974.

11. K. Saito, *Einfach-elliptsiche Singularitäten*, Invent. Math. **23** (1974), 289–325. MR **50** #7147.

12. G. Timms, *The nodal cubic surfaces and the surfaces from which they are derived by projection*, Proc. Royal Soc. A **119** (1928), 213–248.

13. G. N. Tjurina, *Resolution of singularities for flat deformations or rational double points*, Funkcional. Anal. i Priložen. **4** (1970), no. 1, 77–83 = Functional Anal. Appl. **4** (1970), 68–73. MR **42** #2031.

COLUMBIA UNIVERSITY

Proceedings of Symposia in Pure Mathematics
Volume 30, 1977

SIMULTANEOUS RESOLUTIONS OF SINGULARITIES

OSWALD RIEMENSCHNEIDER

Let X be an isolated normal complex-analytic singularity of dimension $\geqq 2$, and let $\sigma: \tilde{X} \to X$ be a (fixed) resolution of X. We are interested in connections between (complex-analytic, flat) deformations of X and deformations of $\tilde{X}$. Details will appear in [5].

THEOREM [4], [5]. *Let $\tilde{Z} \to^{\tilde{\pi}} T$ be a deformation of $\tilde{X} = \tilde{Z}_{t_0} = \tilde{\pi}^{-1}(t_0)$, $t_0 \in T$, and assume (without loss of generality) that $\tilde{\pi}$ is* 1*-convex. Let*

(i) $H^1(\tilde{X}, \mathcal{O}_{\tilde{X}}) = 0$, *or*

(ii) $\dim_C H^1(\tilde{Z}_t, \mathcal{O}_{\tilde{Z}_t}) = \text{const}$, *$T$ reduced at t_0. Then, in the reduction diagram*

$$\begin{array}{ccc} \tilde{Z} & \xrightarrow{\tilde{\sigma}} & Z \\ & {}_{\tilde{\pi}}\searrow \quad \swarrow_{\pi} & \\ & T & \end{array}$$

π is a deformation of $Z_{t_0} \cong X$.

Conversely, if the reduction diagram gives a deformation π of $Z_{t_0} \cong X$, and if $\dim H^2(\tilde{Z}_t, \mathcal{O}_{\tilde{Z}_t}) = \text{const}$, *then* $\dim H^1(\tilde{Z}_t, \mathcal{O}_{\tilde{Z}_t}) = \text{const}$.

COROLLARY [5]. *For each deformation $\tilde{Z} \to^{\tilde{\pi}} T$ of $\tilde{X}$ there exists a maximal reduced subspace $T_a \subset T$ such that the induced deformation $\tilde{Z}_a = \tilde{Z} | T_a \to^{\tilde{\pi}_a} T_a$ can be blown down to a deformation of X (provided that* $\dim H^2(\tilde{Z}_t, \mathcal{O}_{\tilde{Z}_t}) = \text{const}$).

Now, suppose that $\dim H^2(\tilde{X}, \mathcal{O}_{\tilde{X}}) = 0$ and that there exists a deformation $\tilde{\pi}: \tilde{Z} \to T$ of $\tilde{\pi}^{-1}(t_0) = \tilde{X}$ which is versal with respect to deformations of the germ of $\tilde{X}$ along its exceptional set. [So far, the existence of such a family is not proved in general. However, it seems to be possible to prove a completeness theorem that

AMS (MOS) subject classifications (1970). Primary 32C40, 32C45, 32G05, 32G13; Secondary 14B05, 14D15, 14D20.

Key words and phrases. Deformations of complex analytic singularities, deformations of resolutions, blowing down fibre by fibre.

© 1977, American Mathematical Society

gives the existence in special cases.][1] Then construct $\tilde{Z}_a \to T_a$ as in the corollary above. One can prove

THEOREM [5]. *The reduction diagram*

$$\tilde{Z}_a \xrightarrow{\tilde{\sigma}_a} Z'_a$$
$$\tilde{\pi}_a \searrow \swarrow \pi'_a$$
$$T_a$$

is versal for deformations of X (with reduced base space) together with a simultaneous resolution (without base change and special fibre $\tilde{X} \to X$).

Denote the versal deformation of X by $\pi : Z \to S$. Then there exists a pull-back diagram

$$
\begin{array}{ccc}
Z'_a & \longrightarrow & Z \\
\pi'_a \downarrow & & \downarrow \pi \\
T_a & \xrightarrow{\tau} & S
\end{array}
$$

Question 1. Is τ finite?

Question 2. If the answer to Question 1 is yes, denote by S_a the image of T_a under τ (with reduced structure) and let $\pi_a = \pi | Z_a : Z_a = Z | S_a \to S_a$ be the restriction to S_a. Is then the diagram

$$
\begin{array}{ccc}
\tilde{Z}_a & \xrightarrow{\tilde{\sigma}_a} & Z_a \\
\tilde{\pi}_a \downarrow & & \downarrow \pi_a \\
T_a & \xrightarrow{\tau_a} & S_a
\end{array}
$$

versal for deformations of X together with a simultaneous resolution after finite surjective base change and special fibre $\tilde{X} \to X$ (all base spaces assumed to be reduced)?

REMARKS. The answer to the first question is "yes" for dim $X = 2$ and $\tilde{X}$ = minimal resolution of X [1]. If, in that case, X is a rational singularity, S_a is an irreducible component of S, the "Artin component" of S [1]. If, moreover, X is a rational double point, the answer to the second question is "yes" [2].

EXAMPLES. (1) Let $X_{n,q}$ be the 2-dimensional singularity with dual graph

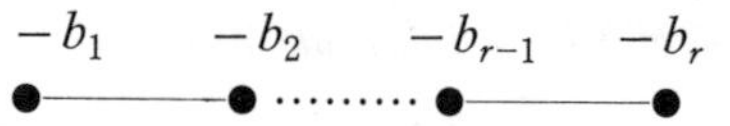

$$-b_1 \quad\quad -b_2 \quad\quad -b_{r-1} \quad -b_r$$

●————●·········●————●   ● $\cong P_1(C)$,

where [3]

$$n/q = b_1 - 1 \vert b_2 - \cdots - 1 \vert b_r.$$

Since $X_{n,q}$ is rational, $T_a = T$; T_a is smooth, dim $T_a = \Sigma(b_\rho - 1)$. The deformation $Z_a \to S_a$ is the "special family" given in [3]. A direct computation shows that $\tau_a : T_a \to S_a$ is a Galois covering with group the direct product of the Weyl groups associated to the maximal configurations of -2 curves (cf. Wahl, these PROCEEDINGS, pp. 91–94, Conjecture 2).

---

[1]NOTE ADDED IN PROOF. D. Leistner, University of Münster, has given a proof of the versality theorem.

(2) Let X be a simple elliptic singularity with dual graph

$$\overset{-b}{\circ}$$

$\circ$ = elliptic curve ($b \geqq 1$). Then T is smooth of dimension $b+1$. For $b = 1, 2, 3$, S is smooth of dimension $11 - b$ [**6**], S_a is smooth of dimension 1 and $\tau_a: T_a \to S_a$ is an isomorphism. Hence $T_a \neq T$, and S_a is not an irreducible component of S.

REFERENCES

1. M. Artin, *Algebraic construction of Brieskorn's resolutions*, J. Algebra **29** (1974), 330–348. MR **50** #7143.

2. F. Huikeshoven, *On the versal resolutions of deformations of rational double points*, Invent. Math. **20** (1973), 15–34.

3. O. Riemenschneider, *Deformationen von Quotientensingularitäten (nach zyklischen Gruppen)*, Math. Ann. **209** (1974), 211–248.

4. ———, *Bemerkungen zur Deformationstheorie nichtrationaler Singularitäten*, Manuscripta Math. **14** (1974), 91–99.

5. ———, *Familien komplexer Räume mit streng pseuaokonvexer spezieller Faser*, Comment. Math. Helv. (submitted).

6. K. Saito, *Einfach-elliptische Singularitäten*, Invent. Math. **23** (1974), 289–325. MR **50** #7147.

UNIVERSITÄT HAMBURG

Proceedings of Symposia in Pure Mathematics
Volume 30, 1977

EQUISINGULARITY AND LOCAL ANALYTIC GEOMETRY

JOHN STUTZ

In an early paper [**1**], Zariski presented in summary form his results concerning the equivalence of various definitions of equisingularity in the case of a hypersurface V whose singular locus Sg V lies in codimension 1. As he pointed out in the introduction of [**1**], Zariski viewed these results as a first step toward the definition of an equisingular stratification of any analytic set; that is, roughly speaking, a division of any analytic set W into manifolds M_i such that at each point of M_i "V has a singularity not worse than the singularity at a generic point of M_i." In fact, what one wants, at least in the simple case we will consider, is that the singularity is in some sense "locally constant." Viewed in this context, Zariski's results provide both a standard and a point of departure for the further development of the notion of equisingularity. Thus, it seems desirable to have as simple proofs of these results as is possible. Now, simplicity, like beauty, is to some extent in the eye of the beholder. Zariski's conditions are of many different types: topological, algebraic, analytic, etc. Thus, there are many different avenues of approach. All of the results in [**1**] are stated without proof. Zariski's proofs are found in [**2**], and are quite algebraic. Here we will adopt the viewpoint of local analytic geometry. Thus for us equisingularity means that V can be represented locally as a branched covering, branching over a manifold. This is consistent with Zariski's approach in [**1**]. Using only a few simple results about branched coverings, we can give complete, and hopefully simple, proofs of the equivalence of Zariski's various conditions. In §1, we will recall, with proofs, a few results on branched coverings which are necessary for our approach and are not found in a text such as [**3**]. In the next two sections, we will develop Zariski's results following the order of exposition in [**1**]. Because our aim is to simplify things as much as possible, we will deal only with the case when V is irreducible. However, all of our arguments go over to the general

AMS (MOS) subject classifications (1970). Primary 32C40, 32C45.

© 1977, American Mathematical Society

case with only technical modifications. In the last section, we will discuss the connection between the various conditions and the structure of the singularity of V.

1. Let L be an analytic subset of U, a domain in $\boldsymbol{C}^{n+1}$ with $\dim V \equiv n$. Suppose $0 \in \operatorname{Sg} V$ and V is irreducible at 0. Let $(x, y, z) = (x, \cdots, x_{n-1}, y, z)$ be coordinates in U with origin 0, define a map $\pi: U \to \boldsymbol{C}^n$ via $\pi(x, y, z) = (x, y)$, and set $\pi(0) = 0'$. If $\pi^{-1}(0') \cap V$ is discrete, we obtain a branched covering (V, π, D). Here and below we will assume that U is replaced, when necessary, by a sufficiently small polydisk centered at 0. This is acceptable since all of Zariski's conditions actually refer to the germ of V at some fixed point of Sg V. Also, we will write π for $\pi|V$ and will assume that $\pi^{-1}(0') = 0$. Let $\Delta \subset D$ be the critical locus of (V, π, D) and let ν be the degree of the topological covering $\pi: V - \pi^{-1}(\Delta) \to D - \Delta$. Associated to a branched covering there is $P \in \mathcal{O}(D)[Z]$, of degree ν, whose germ at $0'$ is a Weierstrass polynomial. See [**3**, Theorem 12, p. 104] for its construction. Let $F \in \mathcal{O}(U)$ such that F_x, the germ of F at $x \in U$, generates ${}_V I_x$, the ideal of V at x. F_0 divides P_0 so $\nu \geqq$ order of F at 0. In order to know when we obtain equality we introduce the Zariski tangent cone.

DEFINITION 1. $C_3(V, 0) = \{r \in \boldsymbol{C}^{n+1}$: there is $(p_i) \subset V$, $(a_i) \in \boldsymbol{C}$ with $(p_i) \to 0$ and $(a_i p_i) \to r\}$.

Expand F about 0 and set G equal the sum of the lowest order terms.

PROPOSITION 1. (a) $C_3(V, 0) = \{r \in \boldsymbol{C}^{n+1}: G(r) = 0\}$ [**4**, *Theorem* 10.4].
(b) $\nu =$ *order of* F *at* $0 \Leftrightarrow (0, 0, 1) \notin C_3(V, 0)$.

PROOF. (a) For $s \in U$, $\lambda \in D(0, 1) = \{\lambda \in \boldsymbol{C}: |\lambda| \leqq 1\}$, we write $K(\lambda, s) = F(\lambda \cdot s) = G(\lambda \cdot s) + \lambda H(\lambda, s)$ for $K, H \in \mathcal{O}(D(0, 1) \times U)$. Now suppose $r \in C_3(V, 0)$ and $(a_i p_i) \to r$. Set $\lambda = 1/a_i$ and $s = a_i p_i$; then F is identically zero and H converges as $i \to \infty$, so we get $G(r) = 0$. Conversely, suppose $r \notin C_3(V, 0)$; then for W a sufficiently small polydisk about r we have $K(\lambda, s) \neq 0$ for $(\lambda, s) \in D(0, 1) \times W$, $\lambda \neq 0$. Since $K(0, s) \equiv 0$, we can write $K(\lambda, s) = \lambda^m L(\lambda, s)$ with $L(0, r) \neq 0$. Since F is of order ν, K is of order ν in λ, so $m = \nu$ and $G(0, r) \neq 0$.

(b) Let P be the polynomial associated to (V, π, D). Near 0, F divides P so we have $\deg P = \nu \geqq$ order F. Since P is Weierstrass, G can contain a term of the form KZ^m, $K \neq 0$, if and only if $\nu =$ order F so in particular $m = \nu$. The existence of such a term is equivalent to the condition that $G(0, 0, 1) \neq 0$, so we are finished.

The numbering of C_3 and C_4, to be introduced below, comes from [**5**].

The order of F at 0 is the same for all functions F such that F_0 generates ${}_V I_0$. It is generally called the multiplicity of V at 0 because it can be shown to be the multiplicity of ${}_V\mathcal{O}_0$, the local ring of V at 0. For any covering one has $\pi(\operatorname{Sg} V) \subset \Delta$. In order to see when in fact $\Delta = \pi(\operatorname{Sg} V)$ we need to introduce the cone $C_4(V, 0)$.

DEFINITION 2. Let $T(V)$ be the tangent space to V and set

$$C_4(V, 0) = \{r \in \boldsymbol{C}^{n+1}: (0, r) \in \overline{T(V - \operatorname{Sg} V)}\}.$$

PROPOSITION 2. *For any branched covering constructed as above, if* $(0, 0, 1) \notin C_4(V, 0)$, *then* $\pi(\operatorname{Sg} V) = \Delta$.

PROOF. If $\pi(\operatorname{Sg} V) \notin \Delta$ then there is $(p_i) \subset V - \operatorname{Sg} V$, $(p_i) \to 0$, such that $\operatorname{rank} \pi_*(p_i) < n$. Because $\pi(x, y, z) = (x, y)$ this is only possible if $(0, 0, 1) \in T(V, p_i)$ so we get $(0, 0, 1) \in C_4(V, p)$.

From what we have done so far it is easy to see that $C_3(V, 0)$ and $C_4(V, 0)$ are analytic subsets of $\boldsymbol{C}^{n+1}$ and that $\dim C_3(V, 0) = n$. One can also show that these spaces are independent of the coordinates used in their definition.

2. From here on we will assume that $\dim \operatorname{Sg} V = n - 1$.

DEFINITION 3. V is equisingular along $\operatorname{Sg} V$ at 0 if there is a branched covering (V, π, D) such that Δ is nonsingular at $0' = \pi(0)$.

Equisingularity is a much stronger condition than it appears at first. In order to see this we need to change the form of our covering slightly. First of all, shrinking D if necessary, and changing (x, y), we may assume that $\Delta = \{(x, y) \in D : y = 0\}$. Since V is irreducible at p, we may also assume that $\pi : V - \pi^{-1}(\Delta) \to D - \Delta$ is a connected covering of degree v. Define $\alpha : D \to D$ via $\alpha(x, y) = (x, y^v)$. $\alpha|D - \Delta$ is also a connected covering of degree v. Because the fundamental group of $D - \Delta$ is the integers, any two such coverings are homeomorphic so we obtain $\beta : D - \Delta \to V - \pi^{-1}(\Delta)$ such that $\alpha = \pi \circ \beta$. α and π are locally biholomorphic so β is holomophic. Further, for any compact set $K \subset D$, $\beta(K - \Delta) \subset \pi^{-1}(\alpha(K))$, so π proper implies that β is locally bounded and so extends to all of D. Thus, we can write $\beta = (x, y^v, f)$ for f holomorphic on D. Next observe that since π has discrete fibers we must have $\dim \pi^{-1}(\Delta) < n$ and so, since V is of pure dimension n, we have $V = \overline{V - \pi^{-1}(\Delta)}$. Since $\beta(D - \Delta) = V = \pi^{-1}(\Delta)$, we get $\operatorname{Sg} V \subset \pi^{-1}(\Delta) = \beta(\Delta)$. $\beta(\Delta)$ is an $(n - 1)$-dimensional manifold and $\dim \operatorname{Sg} V = n - 1$, so we obtain our first result on equisingularity.

PROPOSITION 3. *If V is equisingular at 0, then $\operatorname{Sg} V$ is nonsingular. V has nonsingular normalization $\beta : D \to V$ and the normalization map is a weakly holomorphic homeomorphism.*

PROOF. $(\beta|D - \Delta)^{-1}$ is holomorphic. Its boundedness near Δ follows from the properness of α and π. From the fact that β is onto it follows immediately that V is irreducible at every point of $\operatorname{Sg} V$, so $(\beta|D - \Delta)^{-1}$ extends continuously to $\operatorname{Sg} V$. This completes the proof.

In order to go further we must simplify the function f. We expand f about 0 writing $f = f_0(x) + \sum_{i=k} f_i(x) y^i$ with f_k not identically zero. Changing coordinates via $(x, y, z) \mapsto (x, y, z - f_0(x))$ we may assume $f_0 = 0$. Next, if $k = v$, we change via $(x, y, z) \mapsto (x, y, z - f_k(x)y)$ and so we can assume either $k < v$ or $k > v$. The case $k < v$ needs a little further work.

DEFINITION 4. Given ε a vth root of unity set $S_\varepsilon = f(x, y) - f(x, \varepsilon y)$.

PROPOSITION 4. *Given ε there is an integer l_ε and $g_{\varepsilon,0} \in {}_D\mathcal{O}_0$ such that $S_{\varepsilon,0} = y_{\varepsilon,0} \cdot y_0^{l_\varepsilon}$ and $g_\varepsilon(0, 0) \neq 0$.*

PROOF. $\beta|D - \Delta$ is a homeomorphism and $\beta(x, \varepsilon y) = (x, y^v, f(x, \varepsilon y))$ so S_ε must vanish precisely on Δ. The result now follows from the Nullstellensatz.

When $k < v$ we can take a primitive vth root of unity and write

$$S_\varepsilon = (1 + \varepsilon^k) f_k(x) \cdot y^k + y^{k+1} k(x, y).$$

Immediately, one sees that $f_k(0) \neq 0$. This allows us to obtain C_3 and C_4 explicitly at an equisingular point. In our present coordinates the result is as follows:

$$C_3(V, 0) = C_4(V, 0) = \{(x, y, z): y = 0\} \quad \text{if } k < v,$$
$$= \{(x, y, z): z = 0\} \quad \text{if } k > v.$$

The expressions for C_3 become obvious if one writes the sequence (p_i) used to obtain an element of C_3 in the form $p_i = \beta(x_i, y_i)$. For C_4 let (e_j) be the vector space basis associated with the coordinates (x, y, z). At $p = \beta(q, r) \in V - \text{Sg } V$, $T(V, p)$ is spanned by:

$$W_i = \beta_*(q, r)(e_i) = e_i + \frac{\partial f}{\partial x_i} e_{n+1}, \quad i < n,$$
$$W_n = \beta_*(q, r)(e_n) = v y^{v-1} e_n + \frac{\partial f}{\partial y} e_{n+1}.$$

From this the result for C_4 is immediate. Now, we are in a position to obtain Zariski's result in transversal coordinates.

DEFINITION 5. (a) (x, y, z) is transversal if it gives (V, π, D) of minimal degree at 0.

(b) (x, y, z) is equisingular if it gives (V, π, D) with $\pi(\Delta)$ nonsingular at $0'$.

PROPOSITION 5. *If V is equisingular at* 0, *then all transversal coordinates are equisingular.*

PROOF. Since Sg V is nonsingular at 0, we have $T(\text{Sg } V, 0) \subset C_3(\text{Sg } V, 0) \subset C_3(V, 0)$. (V, π, D) of minimal degree implies π is transverse to $C_3(V, 0)$ so also to $T(\text{Sg } V, 0)$ and $C_4(V, 0)$. The first fact implies $\pi(\text{Sg } V)$ is nonsingular at $0'$ and the second that $\Delta = \pi(\text{Sg } V)$ so we are done.

Despite this result not all transversal coordinates are equally useful in studying equisingularity. The following definition picks out the ones which are best for our purposes.

DEFINITION 6. (x, y, z) are good coordinates at an equisingular point $0 \in V$ if $\text{Sg } V = \{(x, y, z): y = z = 0\}$ and $C_3(V, 0) = \{(x, y, z): z = 0\}$.

It is clear from what we have done that good coordinates always exist at an equisingular point. Defining π as usual we see that, since $(0, 0, 1) \notin C_3(V, 0)$, we must have $\pi^{-1}(0') \cap V$ discrete so π defines (V, π, D). Good coordinates are transversal and so equisingular. Further, if one looks over the construction and modification of β given above one sees that the β associated with good coordinates has $v <$ order of f. Before going on to consider equivalent criteria for equisingularity, we give one further application of these simple calculations.

PROPOSITION 6. *If V is equisingular at $0 \in \text{Sg } V$, then the multiplicity of V at p is constant for $p \in \text{Sg } V$ near p.*

PROOF. Simply note that our calculation of C_3 could be repeated at any $p \in \text{Sg } V$ where $v <$ order of f at $\beta^{-1}(p)$, so at all p near 0 in Sg V. Thus $(0, 0, 1) \notin C_3(V, p)$ for $p \in \text{Sg } V$ near 0, so by Proposition 1 the multiplicity is constantly equal v.

3. We will now prove the equivalence of the various criteria for equisingularity beginning with the Jacobian condition. As above, let $F \in \mathcal{O}(U)$ such that F_x generates ${}_V I_x$ for all $x \in U$ and let ${}_V\bar{\mathcal{O}}_0$ be the germs of weakly holomorphic functions on U at 0. We consider the ideal $J \subset {}_V\bar{\mathcal{O}}_0$ generated by $\{\partial f/\partial x_i, \partial f/\partial y, \partial f/\partial z\}$ for some coordinates.

PROPOSITION 7. *The following are equivalent*:
(a) V *is equisingular along* Sg V *at* 0.
(b) Sg V *is nonsingular at* 0 *and* dim $C_4(V, 0) = n$.
(c) Sg V *is nonsingular at* 0 *and* J *is principal*.

PROOF. (a) $\Rightarrow$ (b) was done above. Conversely, if dim $C_4(V, 0) = n$, then one can take coordinates defining a branched cover (V, π, D) with $(0, 0, 1) \notin T(\text{Sg } V, 0) \cup C_4(V, 0)$. By Proposition 2 such coordinates will be equisingular.

(a) $\Rightarrow$ (c). Take good coordinates at 0. $F \circ \beta \equiv 0$ all its derivatives must be zero. Computing one finds:

$$\frac{\partial}{\partial x_i}(F \circ \beta) = \frac{\partial F}{\partial x_i} \circ \beta + \frac{\partial f}{\partial x_i}\frac{\partial F}{\partial z} \circ \beta \equiv 0,$$
$$\frac{\partial}{\partial y}(F \circ \beta) = v y^{v-1}\frac{\partial F}{\partial y} + \frac{\partial f}{\partial y}\frac{\partial F}{\partial z} \circ \beta \equiv 0.$$

y^{v-1} divides $\partial f/\partial y$ so we can rewrite the last equation as

$$\frac{\partial F}{\partial y} \circ \beta + g\frac{\partial F}{\partial y} \circ \beta \equiv 0.$$

Now $(\beta | D - \Delta)^{-1}$ is weakly holomorphic so composing we see that $\partial F/\partial z$ generates J. Conversely, if J is principal, we can write $dF(p) = (h_1(p)g(p), \cdots, h_{n+1}(p)g(p))$ for $p \in V - \text{Sg } V$ near 0 and $g, h, \cdots, h_{n+1}$ representatives of elements of ${}_V\bar{\mathcal{O}}_0$. Since V is irreducible at p, we know $s = \lim_{p \to 0} dF(p)$ exists. However, since $T(V, p) = \{(p, r)\colon \langle dF(p), r\rangle = 0\}$, this implies that $C_4(V, 0)$ is the plane orthogonal to s. So we are done.

Next we have the Whitney a, b conditions.

DEFINITION 7. Suppose Sg V is nonsingular at 0. We say V is a regular over Sg V at 0 if given $(p_i) \subset V - \text{Sg } V$, $(p_i) \to 0$, such that $T(V, p_i) \to T$, an n-plane, we have $T(\text{Sg } V, 0) \subset T$. We say that V is b regular if in addition to (p_i) as above we have $(q_i) \subset \text{Sg } V$, $(q_i) \to 0$, and $a_i\,(p_i - q_i) \to r$; then $r \in T$.

PROPOSITION 8. *V is equisingular at* 0 *if and only if* Sg V *is nonsingular near* 0, *the multiplicity of V is constant near* 0 *in* Sg V, *and V is a, b regular over* Sg V *at* 0.

PROOF. Suppose V is equisingular at 0. We know that Sg V is nonsingular and that V is equimultiple along Sg V. Take good coordinates, and let (p_i), (q_i), (a_i), etc., be as in Definition 7 above. In general, one has $T \subset C_4(V, 0)$ so in this case $T = C_4(V, 0)$ and a regularity is immediate. Write $p_i = \beta(b_i, c_i)$ and $q_i = \beta(d_i, 0)$; then $a_i(p_i - q_i) = (a_i(b_i - d_i), a_i c_i^v, a_i f(c_i))$. Since $(c_i) \to 0$, $(a_i c_i^v)$ converges, and $v <$ order f, it follows that $(a_i f(c_i)) \to 0$ and $r \in C_4(V, 0)$. This proves b regularity. From Proposition 7 one sees that to prove the converse it suffices to show that our conditions imply that $C_4(V, 0)$ is an n-plane. Take transversal coordinates such that Sg $V = \{(x, y, z) \subset U\colon y = z = 0\}$ and let P be the associated polynomial. Write $P = z^v + \sum P_i(x, y)\, z^{v-i}$ where $P_i = \sum P_{i,j}(x)y^j$. At $p \in \text{Sg } V$ we must have order P at $p \geqq$ order F at $p = v$, so we must have $P_{i,j} = 0$ for all $i < j$. Now, essentially by dividing P by y^v, one obtains a polynomial $Q \in \mathcal{O}(D)(Z)$ such that $Q(z/y)$ vanishes identically on $\{(x, y, z) \in V\colon y \neq 0\}$. This implies that z/y is weakly holomorphic in V, so using the irreducibility of V at 0 we see that z/y

has a limit r as $(x, y, z) \in V$ approaches 0. We also see that for $p = (x, y, z) \in V$ we cannot have $y = 0$ unless $z = 0$ also. Thus $\{(x, y, z) \in V: y = 0\} = \text{Sg } V$. Now, take any sequence $p_i = (x_i, y_i, z_i) \in V - \text{Sg } V$ with $(p_i) \to 0$ and set $q_i = (x_i, 0, 0)$ and $a_i = 1/y_i$. If $T(V, P_i)$ converges, then by a, b regularity its limit is $\{(x, y, z): ry = z\}$. This means $C_4(V, 0)$ is an n-plane so the proof is complete.

Next, we come to the criteria involving resolution via local quadratic transformations. Rather than formulating the condition initially, we will simply resolve V in a neighborhood of an equisingular point and then point out the relevant aspects of the process. Let V be equisingular at 0 and select good coordinates. Define t via $t(x, y, z) = (x, y, yz)$ and set $v^1 = t^{-1}(v)$. Such a map is called a local quadratic transformation. Since $v <$ order of f, we get the following commutative diagram

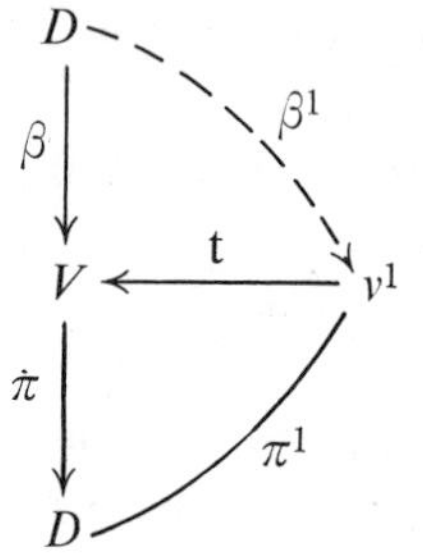

where $\beta(x, y, z) = (x, y^v, f^1)$ with f^1 satisfying $y^v f^1 = f$, and π projection as usual. It is easy to see that (V^1, π^1, D) branches over $\pi(\text{Sg } V)$ so V^1 is equisingular at $0' = t^{-1}(0)$. Further, β^1 is exactly the map that would be constructed if one applied the methods of §2 to (V^1, π^1, D). Clearly, we may continue "dividing by y" in this way until $v \geqq$ order of f^1 so let us consider that case. As in §2 we can change coordinates to kill off the 0th order terms in f^1, so $\text{Sg } V^1 = \{(x, y, z): y = z = 0\}$ and then interchange y and z. The resulting coordinates will be good coordinates for V^1, so we are back essentially in our original situation, except that the multiplicity of V^1 at $0' = t^{-1}(0)$ is strictly lower than that of V at 0. What is involved here is simply the observation made following Proposition 4 that, if $0 <$ order $f = k < v$, then $f_k(0, 0) \neq 0$. By continuing this process, one obtains a sequence

$$V \overset{t^1}{\leftarrow} V^1 \overset{t^2}{\leftarrow} V^2 \leftarrow \cdots \overset{t(m)}{\leftarrow} V^m$$

with the following properties:

(1) Each $t^{(i)}$ is a coordinate change on a local quadratic transformation.

(2) V^m is nonsingular.

(3) $t^i | \text{Sg } V^i: \text{Sg } V^i \to \text{Sg } V^{i-1}$ is biholomorphic for $i < m$ as is t^m restricted to m, the inverse image of $\text{Sg } V^{m-1}$.

(4) Define p via $p(x, y, z) = (x)$; then $p \circ t^i = p$.

(5) The composite $t: V^m \to V$ is a resolution.

(6) Each V^i is equisingular along $\text{Sg } V^i$ at 0^i for $i < m$. Notice that $\rho: V \to \Delta$ exhibits V as a family of plane curve singularities. For each $q \in \Delta$ our resolution restricts to a resolution of ${}_qV = \rho^{-1}(q)$ of the type usually considered for such singularities. Further, these resolutions are essentially the same; in particular the multiplicity of ${}_qV^i$ is the same for a given i for all q. It is at this point that our assumption of irreducibility allows a real rather than a technical simplification. Let

X_0, Y_0 be the germs of curves at the origin in C^2 with $X_0 = \bigcup X_{0,i}$ and $Y_0 = \bigcup Y_{0,j}$ their decomposition into irreducible components. We say that X_0 and Y_0 are equivalent if there is a bijective map $\varphi: \{X_{0,i}\} \to \{Y_{0,j}\}$ such that:

(1) $X_{0,i}$ and $\varphi(X_{0,i})$ have the same multiplicity at 0.

(2) $C_3(X_{0,i_1}, 0) = C_3(X_{0,i_2}, 0) \Leftrightarrow C_3(\varphi(X_{0,i_1}), 0) = C_3(\varphi(X_{0,i_2}), 0)$.

In the general case $V = \bigcup V_i$ with each V_i being the type of variety we have considered and $V_{i_1} \cap V_{i_2} = \operatorname{Sg} V$ for all i_1, i_2. ${}_qV$ decomposes via ${}_qV_i = {}_pV \cap V_i$ and for $q_1, q_2 \in \Delta$ the natural map ${}_{q_1}V_i \mapsto {}_{q_2}V_i$ satisfies (1) and (2). This extends to each stage of the resolution. See [**6**] for details. In treating the equivalence of singularities, Zariski actually introduces three apparently distinct equivalence relations. All, however, reduce to a correspondence of the type we have described.

PROPOSITION 9. *Let* 0 *be a regular point of* Sg V. *If there are coordinates so that locally* Sg $V = \{(x, y, z) \in U : y = z = 0\}$ *and so that a sequence as above satisfying* (1), (2), (3), (4), *and* (5) *exists, then* V *is equisingular at* 0.

PROOF. We may assume that V^{m-1} is singular so t^m must be a quadratic transformation. One sees inductively from (3) and (4) that rank $\rho(0^i) = n - 1$ for all i. If $e_n \in T(V^m, 0^m)$ one would have rank $t^m_*(0^m) = n$ contradicting our assumption that V^{m-1} is singular. Thus, $\tau: V^m \to C^n$, defined via $\tau(x, y, z) = (x, z)$, is locally biholomorphic. One can define $\bar{\tau} = f^{(2)} \circ \cdots \circ f^{(m)} \circ \tau^{-1}$ on a neighborhood N of $\tau(0^m)$. It can be written in the form (x, h, k). Clearly, we can assume that t^1 is a quadratic transformation so $t^1 \circ \bar{\tau} = (x, h, hk)$. Now $t^1 \circ \bar{\tau} = t \circ \tau^{-1}$ so since t is a resolution we may compute $T(V, p)$ for $p \notin \operatorname{Sg} V$ using our expression for the coordinate functions of $t^1 \circ \bar{\tau}$. To make this simpler, observe that it follows inductively from (3) and (4) that $\rho | \operatorname{Sg} V^i$ is locally biholomorphic and so also is $\rho | n$. We can change coordinates so that $(t \circ \tau^{-1})^{-1}(\operatorname{Sg} V) = \{(x, z) \in N : z = 0\}$ locally. This implies that h expands as $h = \sum_{\alpha \geqq 1} h_\alpha(x) z^\alpha$ so one computes easily as in §2 that $C_4(V, 0)$ is in fact an n-plane. Our result now follows from Proposition 7.

4. Fix $0 \in \operatorname{Sg} V$ at which V is equisingular, take good coordinates, and define $\rho: V \to \Delta$ as above. We have $\rho(0) = (0, 0)$. One way to interpret the "local constancy" of V is to compare V with $\tilde{V} = {}_0V \times \Delta$, that is to compare ρ with the trivial family $\tilde{\rho}$. In this section, we will discuss the type of equivalence between V and $\tilde{V}$ given by each of the equisingularity criteria. Projection in the usual way gives covers (V, π, D) and $(\tilde{V}, \tilde{\pi}, D)$ both branched over Δ and of the same degree. Arguing as we did at the beginning of §3, we can produce a homeomorphism $\delta: V \to \tilde{V}$ with $\tilde{\pi} \circ \delta = \pi$. Using an interpolation technique due to Whitney [**5**], we can extend δ to a homeomorphism $\delta: U \to U$ still satisfying $\tilde{\pi} \circ \delta = \pi$. Thus, working with just the definition of equisingularity, one can show that V, $\tilde{V}$ are homeomorphic as embedded via a map which preserves the structure of the family ρ. One can obtain the same result in a second manner as follows: In [**7**] we showed that $\{p \in \operatorname{Sg} V : \dim C_4(V, p) > n\}$ is an analytic subset of Sg V of dimension $< n - 1$. Thus, from the Jacobian condition, we can see that if V is equisingular at $0 \in V$, then we can take U such that V is equisingular at every point of $\operatorname{Sg} V \cap U$. Thus, from our results in a, b regularity we see that in U, $(V - \operatorname{Sg} V, \operatorname{Sg} V)$ is an a, b regular stratification of V. Further, from our calculation of $T(V, p)$ for $p \in \operatorname{Sg} V$, it is easy to see that $\rho | V - \operatorname{Sg} V$ is a submersion. All of this remains true without the assumption of irreducibility. Further, if we take $U - V$ as an additional stratum, we obtain a stratification of U with the same properties since a, b regularity and rank $\rho = n - 1$

are obvious on $U - V$. Now, our result follows from Thom's first isotopy lemma [**8**]. This approach does not depend in an essential way on the fact that Sg V is in codimension 1. In [**1**] Zariski presented a generalization of the definition of equisingularity based on branched coverings which is applicable in the case where dim Sg $V < n - 1$. Recently, Speder [**9**] has shown that a version of equisingularity close to Zariski's implies Whitney conditions, so one has results of this ambient homeomorphism type in higher codimension. In another direction Pham and Tessier [**10**] showed as a consequence of their work on saturation that in the hypersurface case δ is actually Lipschitz. The meaning of the condition on the resolution is clear in this context; what we showed is that the elements of the family ρ have resolutions, of the type usually associated with singularities of plane curves, which are essentially identical.

We should point out that if one keeps the condition that Sg V is codimension 1 but allows V to have arbitrary embedding dimension, Zariski's system of equivalencies breaks down. In [**10**] we show, for example, that this condition on resolutions and the ambient Lipschitz homeomorphism conditions are completely different. The reader wishing to pursue the study of equisingularity further should consult [**12**] and [**13**] for more information on the general types of equisingularity being studied now and the relationships between them. For a treatment of the topics covered in this paper following Zariski's approach, but less difficult than his *Studies in Equisingularity* series, one should consult [**14**].

References

1. O. Zariski, *Equisingular points on algebraic varieties*, Seminari 1962/63, Anal. Alg. Geom. e Topol., vol. 1, Ist. Naz. Alta Mat., Edizioni Cremonese, 1965, pp. 164–177. MR **33** #7336.

2. ———, *Studies in equisingularity*. II. *Equisingularity in codimension* 1 (*and characteristic zero*), Amer. J. Math. **87** (1965), 972–1006. MR **33** #125.

3. R. C. Gunning and H. Rossi, *Analytic functions of several complex variables*, Prentice-Hall, Englewood Cliffs, N. J., 1965. MR **31** #4927.

4. H. Whitney, *Tangents to analytic variety*, Ann. of Math. (2) **81** (1965), 496–549. MR **33** #745.

5. ———, *Local properties of analytic varieties*, Differential and Combinatorial Topology (A Sympos. in Honor of Marston Morse), Princeton Univ. Press, Princeton, N. J., 1965, pp. 205–244. MR **32** #5924.

6. J. Becker and J. Stutz, *Resolving singularities via local quadratic transformations*, Rice Univ. Studies, **59** (1973), no. 1, 1–9. MR **49** #627.

7. J. Stutz, *Analytic as branched coverings*. Trans. Amer. Math. Soc. **166** (1972), 241–259. MR **48** #2420.

8. J. Mather, *Notes on topological stability*, Harvard, 1970.

9. J. P. Speder, *Equisingularity et conditions de Whitney*, Amer. J. Math. (to appear).

10. F. Pham and B. Tessier, *Fractions Lipschitziennes d'une algèbra analytic complex et saturation de Zariski* (preprint).

11. J. Stutz, *Equisingularity and equisaturation in codimension* 1, Amer. J. Math. **94** (1972), 1245–1268. MR **48** #11565.

12. J. Brainçon and J. P. Speder, *Families equisingular de surfaces à singularité isolée*, C. R. Acad. Sci. Paris Sér. A **280** (1975).

13. O. Zariski, *Some open questions in the theory of singularities*, Bull. Amer. Math. Soc. **77** (1971), 481–491. MR **43** #3266.

14. ———, *Contributions to the problem of equisingularity*, Questions on Algebraic Varieties (C.I.M.E., III Ciclo, Varenna, 1969), Edizioni Cremonese, Rome, 1970, pp. 261–343. MR **43** #1987.

STATE UNIVERSITY OF NEW YORK AT ALBANY

Proceedings of Symposia in Pure Mathematics
Volume 30, 1977

DEFORMATIONS OF COHERENT ANALYTIC SHEAVES WITH ISOLATED SINGULARITIES

GÜNTHER TRAUTMANN

In this note some results on isolated singularities of coherent analytic sheaves are announced. Details will be given in two forthcoming papers at some other place. For isolated singularities of coherent sheaves in $\boldsymbol{C}^n$ spaces of moduli are given, and at the same time the deformation problem is solved for all types of isolated singularities. These types are described by the sheaves $\mathscr{L}^l(A)$ (cf. (4.2)) and for any sheaf $\mathscr{F}$ with only finitely many isolated singularities there is an isomorphism $\mathscr{F} \oplus \mathcal{O}^p \cong \mathscr{L}^l(A) \oplus \mathcal{O}^q$. Further, for any type-index l there is an explicitly defined algebraic subvariety S^l of moduli in some $\boldsymbol{C}^{N_l}$ and an explicitly defined S^l-flat sheaf $\mathscr{L}^l$ on $\boldsymbol{C}^n \times S^l$ such that the fibers $\mathscr{F}^l(A)$ for $A \in S^l$ on $\boldsymbol{C}^n$ form a *complete* family of deformations everywhere, which is semi-universal at $0 \in S^l$. The varieties S^l of moduli and the sheaves $\mathscr{F}^l$ are given in terms of canonical matrices, which generalize the syzygy matrices of Hilbert [**4**]. The varieties S^l naturally also show that all isolated singularities are of algebraic nature.

1. Deformations of sheaves. Let X and S be complex spaces and $\mathscr{G}$ be a coherent $\mathcal{O}_{X\times S}$-sheaf on $X \times S$. $\mathscr{G}$ is called S-flat, if for every $(x, s) \in X \times S$ the stalk $\mathscr{G}_{(x,s)}$ is a flat $\mathcal{O}_{S,s}$-module by the canonical homomorphism from $\mathcal{O}_{S,s}$ to $\mathcal{O}_{X\times S,(x,s)}$. For any $s \in S$ let $j_s : X \to X \times S$ be the canonical injection $x \to (x, s)$. The sheaf $j_s^*\mathscr{G}$ is denoted by $\mathscr{G}(s)$ and called the fiber of $\mathscr{G}$ at s. It is identified with $\mathscr{G}/\mathfrak{m}_S(s)\mathscr{G}$, where $\mathfrak{m}_S(s)$ is the maximal ideal sheaf of $s \in S$. Let now $\mathscr{F}$ be a given coherent sheaf on X. A deformation of $\mathscr{F}$ is a quadruple $(\mathscr{G}, S, s_0, \alpha)$, consisting of a germ of an analytic space (S, s_0), a germ of an S-flat sheaf $\mathscr{G}$ on $X \times S$, and an isomorphism $\alpha: \mathscr{G}(s_0) \to \mathscr{F}$. The collection of all deformations of $\mathscr{F}$ is made into a category $\boldsymbol{D}(X, \mathscr{F})$, if we define morphisms

$$(\mathscr{G}, S, s_0, \alpha) \xrightarrow[(f,\varphi)]{} (\mathscr{G}', S', s_0', \alpha')$$

AMS (MOS) subject classifications (1970). Primary 32G13, 32L10.

© 1977, American Mathematical Society

between two objects to be pairs (f, φ), where $S \xleftarrow{f} S'$ is a morphism of germs of spaces and $\varphi\colon (\mathrm{id} \times f)^* \mathscr{G} \to \mathscr{G}'$ is an isomorphism which is compatible with the α's. For simplification we write $\mathscr{G}_S$ for $(\mathscr{G}, S, s_0, \alpha)$ and identify $\mathscr{G}_S$ with some representative of the sheaf germ $\mathscr{G}$. If $K \subset X$ is a compact subset, the category of deformations of $\mathscr{F}|K$ is defined similarly and denoted by $\boldsymbol{D}(K, \mathscr{F})$. A deformation $\mathscr{G}_S$ in $\boldsymbol{D}(X, \mathscr{F})$ or $\boldsymbol{D}(K, \mathscr{F})$ is called complete, if for any other deformation $\mathscr{G}'_{S'}$, there exists a morphism $\mathscr{G}_S \to \mathscr{G}'_{S'}$. If $\mathscr{G}_S$ is complete, any deformation of $\mathscr{F}$ is obtained by pulling back $\mathscr{G}_S$ via a morphism $S' \to^f S$, such that the collection of fibers $\mathscr{G}'(s')$, $s' \in S'$, is contained in the collection of fibers $\mathscr{G}(s)$, $s \in S$. The deformation $\mathscr{G}_S$ is called *semi-universal* or *versal*, if it is complete and if in addition the tangential map $T_{s_0}(S) \leftarrow T'_{s_0}(S')$ given by $\mathscr{G}_S \to \mathscr{G}'_{S'}$ is unique. $\mathscr{G}_S$ is called a deformation of order e if $\mathfrak{m}_S(s_0)^{e+1} = 0$. The deformations of order e form a category $\boldsymbol{D}_e(X, \mathscr{F})$ or $\boldsymbol{D}_e(K, \mathscr{F})$, and the notions of "$e$-complete" and "$e$-versal" are defined as above.

2. Existence of semi-universal deformations. The singular locus $S(\mathscr{F})$ of a coherent sheaf $\mathscr{F}$ on a complex space X is defined by $S(\mathscr{F}) = \{x \in X, \mathscr{F}_x \text{ not free and } \mathscr{F}_x \neq 0\}$. It is well known that $S(\mathscr{F})$ is an analytic subvariety of X. Let now X be a closed analytic subspace in some domain of $\boldsymbol{C}^N$, $0 \in X$, and let $\mathscr{F}$ be coherent on X. *If K is a compact polydisc of $\boldsymbol{C}^N$, such that $S(\mathscr{F}) \subset \mathring{K}$ and such that K is $\mathscr{F}$-privileged in some sense* (*cf.* **[1]**) *then there exists a semi-universal deformation $\mathscr{G}_S$ of $\mathscr{F}$ in* $\boldsymbol{D}(K, \mathscr{F})$, *such that the tangential space $T_{s_0}(S)$ is isomorphic to* $\mathrm{Ext}^1(K, \mathscr{F}, \mathscr{F})$. Moreover:

(2.1) If $\mathcal{O}_X^r \to \mathcal{O}_X^q \to \mathcal{O}_X^p \to \mathscr{F} \to 0$ is a resolution of $\mathscr{F}$ by holomorphic matrices B_0 and A_0 in a neighborhood of K and if $m = \dim_C \mathrm{Ext}^1(K, \mathscr{F}, \mathscr{F})$, then $\mathscr{G}_S$ is given by an exact sequence

$$\mathcal{O}_{X\times S}^r \xrightarrow{B} \mathcal{O}_{X\times S}^q \xrightarrow{A} \mathcal{O}_{X\times S}^p \longrightarrow \mathscr{G}_S \longrightarrow 0,$$

where S is the germ corresponding to an ideal $\mathfrak{a} \subset \boldsymbol{C}\langle t_1,\cdots, t_m\rangle$ of the ring of convergent power series in $t_1,\cdots, t_m$, and where $B = B_0 + \sum_{0<|\mu|} B_\mu t^\mu$ and $A = A_0 + \sum_{0<|\mu|} A_\mu t^\mu$ are convergent series of matrices with holomorphic coefficients B_μ, A_μ in a neighborhood of K. The ideal $\mathfrak{a}$ and the matrices B_μ and A_μ are constructed by using the methods of reduction developed by H. Grauert **[3]**, such that $BA = 0 \bmod \mathfrak{a}$ and such that for every order e the matrices B mod $\mathfrak{a} + \mathfrak{m}^{e+1}$ and A mod $\mathfrak{a} + \mathfrak{m}^{e+1}$ define an e-versal deformation of $\mathscr{F}$, where $\mathfrak{m} \subset \boldsymbol{C}\langle t_1,\cdots, t_m\rangle$ is the maximal ideal.

(2.2) For $|\mu| = 1$ the classes $[A_\mu] \in \mathrm{Ext}^1(K, \mathscr{F}, \mathscr{F})$, defined by the homomorphisms $\mathcal{O}_X^q \to_{A_\mu} \mathcal{O}_X^p \to \mathscr{F}$, form a basis of $\mathrm{Ext}^1(K, \mathscr{F}, \mathscr{F})$.

(2.3) The *obstructions* for *extending deformations* of $\mathscr{F}$ are given by the group $\mathrm{Ext}^2(K, \mathscr{F}, \mathscr{F})$ in the following way:

Let S and S' be germs of analytic spaces such that $\mathcal{O}_{S'}$ is an extension of $\mathcal{O}_S$ by an exact sequence $0 \to \mathscr{T} \to \mathcal{O}_{S'} \to \mathcal{O}_S \to 0$, where $\mathfrak{m}_{S'}\mathscr{T} = 0$ and $\mathscr{T} = \sum_{i=1}^k f_i \mathcal{O}_{S'}$. Then every $g \in \Gamma(K, \mathscr{T}\mathcal{O}_{X\times S'})$ has a unique representation $g = \sum_i g_i f_i$ with $g_i \in \Gamma(K, \mathcal{O}_X)$. If $\mathscr{G}_S$ is a deformation of $\mathscr{F}$ with a resolution $\mathcal{O}_{X\times S}^r \to_B \mathcal{O}_{X\times S}^q \to_A \mathcal{O}_{X\times S}^p \to \mathscr{G}_S \to 0$ such that $B|K \times \{s_0\} = B_0$, $A|K \times \{s_0\} = A_0$, and if B', A' are holomorphic matrices on $K \times S'$ such that $B'|K \times S = B$ and $A'|K \times S = A$, the pair (B', A') defines a deformation $\mathscr{G}_{S'}$ of $\mathscr{F}$ with $\mathscr{G}_{S'}|K \times S = \mathscr{G}_S$ if and only if $B'A' = 0$. In general we only have $B'A' = \sum_i f_i R_i$ with holomorphic matrices

R_i on K, since $B'A'|K \times S = 0$. There exists a deformation $\mathscr{G}_{S'}$ on $K \times S'$ with $\mathscr{G}_{S'}|K \times S = \mathscr{G}_S$ if and only if the classes $[R_i] \in \mathrm{Ext}^2(K, \mathscr{F}, \mathscr{F})$, defined by the homomorphisms $\mathcal{O}_X^r \to_{R_i} \mathcal{O}_X^p \to \mathscr{F}$, are zero for $i = 1, \cdots, k$.

3. The Ext **groups of the Hilbert syzygies.** To simplify the statements below we also consider matrices of type $p \times q$ (i.e., with p rows and q columns), where $p = 0$ or $q = 0$ (empty matrices). The usual algebra of matrix addition and multiplication is maintained, if we agree that the product of a matrix of type $p \times 0$ with a matrix of type $0 \times q$ is the zero matrix of type $p \times q$.

(3.1) Let t and $n \geqq 1$ be integers. We write $A_{t,0}(n)$ for a column $(a_{\nu_0 \cdots \nu_t})$, $1 \leqq \nu_0 < \cdots < \nu_t \leqq n$, in $C\binom{n}{t+1}$ and $A_{t,0}(n-1)$, resp. $A^n_{t-1,0}(n-1)$, for its subcolumns $(a_{\nu_0 \cdots \nu_t})$, $\nu_t < n$, resp. $(a_{\nu_0 \cdots \nu_{t-1} n})$. For every integer s we assign to $A_{t,0}(n)$ a matrix $A_{t,s}(n)$ of type $\binom{n}{t+s+1} \times \binom{n}{s}$ with coefficients in C by using induction on n, t, s in the following way: $A_{t,s}(n)$ is "empty" for $t \leqq -2$, for in the case where $0 \leqq t + s + 1$, $s \leqq n$ is not satisfied. Let $A_{-1,0}(0) = A_{-1,0}(1)$. By

$$A_{t,s}(n) := \left[\begin{array}{c|c} A_{t,s}(n-1) & 0 \\ \hline A^n_{t-1,s}(n-1) & (-1)^t A_{t,s-1}(n-1) \end{array}\right]$$

the induction is complete and $A_{t,s}(n)$ is well defined for any $n \geqq 1$ and any $t, s \in \boldsymbol{Z}$. Moreover:

(a) The entries of $A_{t,s}(n)$ are 0 or $\pm a$ where a is an element of $A_{t,0}(n)$.

(b) If $t = -1$ the matrix $A_{-1,0}(n)$ is an element $a \in C$ and $A_{-1,s}(n)$ is the diagonal matrix $(-1)^s aE$ with $\binom{n}{s}$ rows.

(c) If $t = 0$ we write $A_s(n) = A_{0,s}(n)$.

(3.2) If $z_1, \cdots, z_n$ are the coordinate functions of C^n we denote by $Z_s = Z_s(n)$ the matrix which is built up by the elements $z_1, \cdots, z_n$ in the same way as $A_s(n)$. It is the sth syzygy matrix of Hilbert (cf. [4]), which defines a homomorphism

$$\mathcal{O}\binom{n}{s+1} \xrightarrow[Z_s]{} \mathcal{O}\binom{n}{s},$$

where $\mathcal{O}$ is the sheaf of holomorphic functions in C^n. We denote its cokernel by Cok Z_s.

(3.3) We write $A_{t,s} = A_{t,s}(n)$ and define $\mathfrak{A}_{t,s}$ to be the C vector space of all such $A_{t,s}$ such that $\dim_C \mathfrak{A}_{t,s} = \binom{n}{t+1}$ for every s. The matrices $A_{t,s}$ are normal coordinates for the Ext classes of the Z_i as stated in the following propositions.

(3.4) For any Stein domain $X \subset C^n$ with $0 \in X$ and for any $0 \leqq i, j \leqq n-1$, $\mathrm{Ext}^1_{\mathcal{O}}(X, \mathrm{Cok}\, Z_i, \mathrm{Cok}\, Z_j) = \mathfrak{A}_{i-j,j}$, where addition of Ext classes is identical with usual addition of matrices and where the isomorphism is interpreted by the relation $Z_{i+1} A_{i-j,j} + A_{i-j,j+1} Z_j = 0$.

(3.5) If t_1, t_2, s are integers and if $B_{t_2,t_1+s+1} \in \mathfrak{A}_{t_2,t_1+s+1}$ and $A_{t_1,s} \in \mathfrak{A}_{t_1,s}$, then the product $B_{t_2,t_1+s+1} A_{t_1,s}$ is in $\mathfrak{A}_{t_1+t_2+1,s}$, which can be interpreted as a Yoneda pairing.

4. Matrices $A_{t,s}$ and $Z_s(A)$. Let $l = (l_0, \cdots, l_{n-1})$ be a row of integers $l_i \geqq 0$. We write $|l|$ for $\sum_i l_i$ and denote by L_i the interval of all $\lambda \in \boldsymbol{N}$ such that $l_{n-1} + \cdots + l_i$

$\geqq \lambda > l_{n-1} + \cdots + l_{i+1}$, which is empty, if $l_i = 0$. For $t, s \in \boldsymbol{Z}$ we consider matrices $A_{t,s}$ with coefficients in $\boldsymbol{C}$ and matrices Z_s^l, which are built up as follows:

$$A_{t,s}^l = \begin{array}{|ccc|} A_{(t,s)}^{|l|,|l|} & \cdots & A_{(t,s)}^{|l|,1} \\ \vdots & & \vdots \\ A_{(t,s)}^{1,|l|} & \cdots & A_{(t,s)}^{1,1} \end{array}, \quad Z_s^l = \begin{array}{|cccc|} Z_{(s)}^{|l|} & 0 & \cdots & 0 \\ 0 & \ddots & & \vdots \\ \vdots & & \ddots & 0 \\ 0 & \cdots & 0 & Z_{(s)}^1 \end{array},$$

where $A_{(t,s)}^{\mu\nu} \in \mathfrak{A}_{i-j+t,j+s}$ if $\mu \in L_i$ and $\nu \in L_j$, and where $Z_{(s)}^\lambda = Z_{i+s}$, if $\lambda \in L_i$, is a Hilbert matrix. Let $\mathfrak{A}_{t,s}^l$ be the $\boldsymbol{C}$ vector space of all $A_{t,s}^l$. As in §3, we use the index s to indicate that any $A_{t,s}^l$ is determined by $A_{t,0}^l$. (Each $A_{(t,s)}^{\mu\nu} = A_{i-j+t,j+s}^{\mu\nu}$, $\mu \in L_i$, $\nu \in L_j$ is determined by $A_{i-j+t,0}^{\mu\nu}$ and vice versa, if $0 \leqq j + s \leqq n - 1$.) For $t = 0$ we write $A_s^l = A_{0,s}^l$ and $\mathfrak{A}_s^l = \mathfrak{A}_{0,s}^l$.

(4.1) As in (3.4) there is a pairing $\mathfrak{A}_{t_2,t_1+s+1}^l \times \mathfrak{A}_{t_1,s}^l \to \mathfrak{A}_{t_1+t_2+1,s}^l$ such that

$$C_{t_1+t_2+1,s}^l = (-1)^s B_{t_2,t_1+s+1}^l A_{t_1,s}^l \quad \text{if } C_{t_1+t_2+1,0}^l = B_{t_2,t_1+1}^l A_{t_1,0}^l.$$

(4.2) Let $S^l = \{A_0^l \in \mathfrak{A}_0^l,\ A_1^l A_0^l = 0\}$, which is an algebraic subvariety of $\boldsymbol{C}^N$, where $N = \dim_{\boldsymbol{C}} \mathfrak{A}_0^l = \sum_{i,j=1}^n l_i l_j \binom{n}{i-j+1}$. If $A_0^l \in S^l$, then $A_{s+1}^l A_s^l = 0$ for any s by (4.1). For $A_0^l \in S^l$ we write $Z_s^l(A) = Z_s^l + A_s^l$. Then $Z_s^l(A)$ defines a sheaf homomorphism $\mathcal{O}^{q_{s+1}} \to \mathcal{O}^{q_s}$, where $q_s = \sum_i l_i \binom{n}{i+s}$, and we denote by $\mathscr{Z}_s^l(A)$ its image sheaf. Let $\mathscr{C}^l(A)$ be the cokernel of $Z_0^l(A)$. The following statements hold:

(4.3) For every $A_0^l \in S^l$ the following sequence is exact:

$$\cdots \longrightarrow \mathcal{O}^{q_{s+1}} \xrightarrow[Z_s(A)]{} \mathcal{O}^{q_s} \longrightarrow \cdots \xrightarrow[Z_0(A)]{} \mathcal{O}^{q_0} \longrightarrow \mathscr{C}^l(A) \longrightarrow 0.$$

(4.4) Each $\mathscr{Z}_s^l(A)$ has only a finite number of isolated singularities. If $A_{(0,s)}^{\mu\nu} = 0$ for $\mu \geqq \nu$ then only $0 \in \boldsymbol{C}^n$ is a singular point.

(4.5) If H_c^* denotes cohomology with compact supports, then for $i \leqq n - 1$, $\dim_{\boldsymbol{C}} H_c^i(\boldsymbol{C}^n, \mathscr{C}^l(A)) \leqq l_i$, and $\dim_{\boldsymbol{C}} H_c^i(\boldsymbol{C}^n, \mathscr{C}^l(A)) = l_i$ for all $i \leqq n - 1$ if and only if A_0^l contains no diagonal matrices of type $A_{-1,j}^{\mu\nu}$ for $\nu \in L_j$, which are not zero.

5. Completeness of S^l at every point. Letting $T_0^l \in S^l$ vary in S^l, we obtain an exact sequence

$$\cdots \longrightarrow \mathcal{O}_{\boldsymbol{C}^n \times S}^{q_{s+1}} \xrightarrow[Z_s(T)]{} \mathcal{O}_{\boldsymbol{C}^n \times S}^{q_s} \longrightarrow \cdots \xrightarrow[Z_0(T)]{} \mathcal{O}_{\boldsymbol{C}^n \times S}^{q_0} \longrightarrow \mathscr{C}^l \longrightarrow 0$$

where $\mathscr{C}^l$ denotes the cokernel of $Z_0^l(T)$. If $\mathscr{Z}_s^l$ denotes the image sheaf of $Z_s^l(T)$, then $\mathscr{C}^l$ and any $\mathscr{Z}_s^l$ are S^l-flat, and for any $A_0^l \in S^l$ the sheaf $\mathscr{C}^l(A)$, resp. $\mathscr{Z}_s^l(A)$, is the fiber of $\mathscr{C}^l$, resp. $\mathscr{Z}_s^l$ at A_0^l.

(5.1) *Let $A_0^l \in S^l$ and $X \subset \boldsymbol{C}^n$ be a Stein domain such that $S(\mathscr{C}^l(A)) \subset X$. Then $\mathscr{C}^l$ is a complete deformation of $\mathscr{C}^l(A)$ in the category $\boldsymbol{D}(X, \mathscr{C}^l(A))$. In $0 \in S^l$ the deformation $\mathscr{C}^l$ is semi-universal, i.e.,* $\mathfrak{A}_0^l = \operatorname{Ext}^1(X, \mathscr{C}^l(0), \mathscr{C}^l(0))$. *The same holds for every $\mathscr{Z}_s^l$, $s \geqq 0$.*

This theorem can be proved by calculating the groups $\operatorname{Ext}^i(X, \mathscr{C}^l(A), \mathscr{C}^l(A))$ in terms of $A_{t,s}$, which can be done by a theorem of type *GAGA* for Ext groups.

6. Classifying isolated singularities of coherent sheaves in $\boldsymbol{C}^n$. Let X be a convex domain in $\boldsymbol{C}^n$ and let $\mathscr{F}$ be a coherent analytic sheaf on X with only finitely many singularities in X. One can prove by using methods similar to those in **[5]** that there are coherent sheaves $\mathscr{F}_1, \cdots, \mathscr{F}_k$ in $\mathbf{C}^n$ having only one singular point in X such that for some $p, q \geqq 0$ there is an isomorphism $\mathscr{F} \oplus \mathcal{O}^p = \mathscr{F}_1 \oplus \cdots \oplus \mathscr{F}_k \oplus \mathcal{O}^q$ on X. Therefore the singularities of $\mathscr{F}$ behave like independent single singularities. Moreover, as in **[5]**, each $\mathscr{F}_i$ is equivalent (i.e., isomorphic up to free direct summands $\mathcal{O}^p$) to a sheaf $\mathscr{C}^{l^{(i)}}(A^{(i)})$ where $l^{(i)}$ is determined by $\dim_{\mathbf{C}} H_c^j(\mathbf{C}^n, \mathscr{F}_i) = l_j^{(i)}$ for $j \leqq n-1$. By "adding up" the sheaves $\mathscr{C}^{l^{(i)}}(A^{(i)})$ we obtain a matrix $A_0^l \in S^l$, where $l = \sum_i l^{(i)}$, such that $\mathscr{F}$ is equivalent to $\mathscr{C}^l(A)$ and $\dim_{\mathbf{C}} H_c^j(X, \mathscr{F}) = l_j$. Therefore l is uniquely determined by $\mathscr{F}$ by this property. However, there may be larger l's such that an equivalence with some $\mathscr{C}^l(A)$ still holds (cf. (4.5)). But we can characterize each S^l by the following property:

(6.1) Let l be given. Then S^l is the space of moduli of all equivalence classes of coherent sheaves $\mathscr{F}$ on $\mathbf{C}^n$ with finite $S(\mathscr{F})$ such that, for $0 \leqq i \leqq n-1$, $\dim_{\mathbf{C}} H_c^i(\mathbf{C}^n, \mathscr{F}) = l_i - \lambda_i - \lambda_{i+1}$ where $\lambda_i \geqq 0$ are integers for $i = 0, \cdots, n$ with $\lambda_0 = \lambda_n = 0$ depending on $\mathscr{F}$.

7. Theorem of isomorphisms. If $C_{-1,0}^l \in \mathfrak{A}_{-1,0}^l$ is an invertible matrix, then any $C_{-1,s}^l$ is also invertible and its inverse $\check{C}_{-1,s}^l$ is also in $\mathfrak{A}_{-1,s}^l$. Therefore the invertible elements in $\mathfrak{A}_{-1,s}^l$ form a group $\mathfrak{G}_s^l$ which is determined by $\mathfrak{G}_0^l$. If $C_{-1,0}^l \in \mathfrak{G}_0^l$, by $A_0^l \to -\check{C}_{-1,1}^l A_0^l C_{-1,0}^l$ we obtain an operation $S^l \times \mathfrak{G}_0^l \to S^l$, such that $\mathfrak{G}_0^l$ is a subgroup of all automorphisms of S^l. The equation $C_{-1,1}^l B_0^l + A_0^l C_{-1,0}^l = 0$ is equivalent to $C_{-1,s+1}^l Z_s^l(A) + Z_s^l(A) C_{-1,s}^l = 0$ for $s \geqq 0$ and thereby we obtain induced isomorphisms $\mathscr{C}^l(A) \gtrapprox \mathscr{C}^l(B)$ and $\mathscr{Z}_s^l(A) \gtrapprox \mathscr{Z}_s^l(B)$, $s \geqq 0$. Conversely the following holds:

(7.1) Let $A_0^l, B_0^l \in S^l$ and let $\mathscr{C}^l(A) \oplus \mathcal{O}^p \cong \mathscr{C}^l(B) \oplus \mathcal{O}^q$. Then $p = q$ and there exists an automorphism $C_{-1,0}^l \in \mathfrak{G}_0^l$ of S^l, which induces an isomorphism $\mathscr{C}^l(A) \gtrapprox \mathscr{C}^l(B)$ in the above way.

References

1. A. Douady, *Le problème des modules pour les sous-espaces analytiques compacts d'un espace analytique donné*, Ann. Inst. Fourier (Grenoble) **16** (1966), fasc. 1, 1–95. MR **34** #2940.
2. O. Forster and K. Knorr, *Über die Deformationen von Vektorraumbündeln auf kompakten komplexen Räumen*, Math. Ann. **209** (1974), 291–346.
3. H. Grauert, *Über die Deformationen isolierter Singularitäten analytischer Mengen*, Invent. Math. **15** (1972), 171–198. MR **45** #2206.
4. D. Hilbert, *Über die Theorie der dlgebraischen Formen*, Math. Ann. **36** (1890), 473–534.
5. G. Trautmann, *Darstellung von Vektorraumbündeln über $C^n - \{0\}$*, Arch. Math. (Basel) **24** (1973), 303–313. MR **50** #5010.
6. ———, *Deformation von isolierten Singularitäten kohärenter analytischer Garben*. I, mimeographed notes, University Kaiserslautern, 1974, pp. 1–100; Math. Ann. (to appear).

UNIVERSITÄT KAISERSLAUTERN

Proceedings of Symposia in Pure Mathematics
Volume 30, 1977

LOCAL COHOMOLOGY GROUPS FOR RESOLUTIONS OF SINGULARITIES

JONATHAN M. WAHL

1. Let Spec R be a normal n-dimensional affine variety over $\boldsymbol{C}$ ($n \geqq 2$), with isolated singularity at $P \in \operatorname{Spec} R$. A resolution $f: X \to \operatorname{Spec} R$ is a proper map, where X is smooth, which is an isomorphism off $E = f^{-1}(P)$. We are interested in computing the local cohomology group $H^1_E(X, F)$, for certain locally free sheaves F on X; basically, we want to compare cohomology of X with that of Spec $R - \{P\}$. We will state some theorems, especially in the 2-dimensional case, and give applications to equisingularity and to simultaneous resolution of rational singularities.

2. An example of this type is the Grauert-Riemenschneider theorem [**6**], which states that $H^i_E(\mathcal{O}_X) = 0$, $i < n$. This is usually stated in the form $R^i f_* K_X = 0$, $i > 0$, where K_X = canonical line bundle of X; one uses the duality of $H^i_E(F)$ and $H^{n-i}(X, K_X \otimes F^\vee)$ (cf. [**8**]). Ogus has used this to observe that if Spec R has a resolution satisfying $H^i(\mathcal{O}_X) = 0$, $i > 0$, then R is Cohen-Macaulay; for the Cohen-Macaulay property of R may be interpreted in terms of local cohomology $H^i_P(\operatorname{Spec} R)$, and one can relate $H^i_E(\mathcal{O}_X)$ to $H^i_P(\operatorname{Spec} R)$ by a spectral sequence [**7**].

In another direction, Burns and Wahl [**4**] have shown that $H^1_E(\theta_X)$ (θ_X = tangent bundle) is a term which, in case $n = 2$, has consequences for the deformation theory of a *projective* smooth surface containing a formal neighborhood of $E \subset X$. In particular, every smooth surface containing a complete nonsingular rational curve E of self-intersection -2 has a first-order deformation in which E "disappears". We state two equivalent conjectures, whose truth would imply that this phenomenon does not obviously generalize to higher dimensions.

CONJECTURE 1. *Suppose $f: X \to \operatorname{Spec} R$ is as above, and $E = f^{-1}(P)$ is a smooth projective variety. Suppose further that the conormal bundle L of E in X is ample (i.e., positive). Then $H^1_E(\theta_X) = 0$, unless (E, L) is $(\boldsymbol{P}^n, \mathcal{O}(1))$ or $(\boldsymbol{P}^1, \mathcal{O}(2))$.*

AMS (*MOS*) *subject classifications* (1970). Primary 14B05, 32C40; Secondary 14J15.

© 1977, American Mathematical Society

CONJECTURE 1′. *Let E be a nonsingular projective variety, L an ample line bundle. Then $H^0(E, \theta_E \otimes L^\vee) = 0$, unless (E, L) is $(\boldsymbol{P}^n, \mathcal{O}(1))$ or $(\boldsymbol{P}^1, \mathcal{O}(2))$.*

THEOREM 1 [**11**]. *The conjectures are true if* $\dim E \leqq 2$ (*i.e., if* $\dim R \leqq 3$).

The case of $\dim E = 1$ follows trivially from the Riemann-Roch theorem; for $\dim E = 2$, one uses classification of surfaces. The conjectures describe a (possible) simple cohomological description of $\boldsymbol{P}^n$; perhaps this question is related to ampleness of the tangent bundle.

In the situation $f: X \to \operatorname{Spec} R$ as above, Grauert has an example [**5**] with E nonsingular, but the conormal bundle L is not ample; in this case, $H^1_E(\theta_X) \neq 0$, and one presumably gets local contributions to global deformations, as in [**4**].

3. If $\dim R = 2$, we may speak of the (unique) *minimal good resolution*, or MGR, of Spec R; it is the minimal resolution for which E is a union of nonsingular curves intersecting transversally, no three through a point. We also have a subbundle $S \subset \theta_X$ of vector fields preserving each exceptional component of E; the dual of S is the sheaf of "1-forms with logarithmic poles along E" [**9**]. S may be defined as the kernel of the natural map of θ_X into the direct sum of the normal bundles of all the irreducible components of E. Our results (see [**12**]) are:

THEOREM 2. *If $X \to$ Spec R is the MGR of a rational singularity* (*i.e.*, $H^1(\mathcal{O}_X) = 0$), *then* $H^1_E(\mathcal{O}_X(E)) = 0$.

THEOREM 3. *If $X \to$ Spec R is the MGR, then* $H^1_E(S) = 0$.

THEOREM 4. *If $X \to$ Spec R is the MGR of a rational singularity, then* $H^1_E(S(E)) = 0$.

Here, $F(E) = F \otimes \mathcal{O}_X(E)$, where E is the reduced divisor $f^{-1}(P)$. Theorem 2 is rather straightforward, but Theorems 3 and 4 require long and delicate arguments. From Theorem 4 follows easily:

THEOREM 5. *If $X \to$ Spec R is the MGR of a rational singularity, then* $\dim H^1_E(\theta_X) = \#$ *of rational* -2 *curves in E.*

4. The applications arise from the cohomology sequence

$$H^0(X, F) \to H^0(X - E, F) \to H^1_E(F) \to H^1(X, F) \to H^1(X - E, F).$$

First, Theorem 3 implies that a certain subspace of the moduli space of X ("the equisingular deformations of X") maps *injectively* into the moduli space of the singularity Spec R (see [**11**]). For, $H^1(X, S) \subset H^1(X, \theta_X)$ consists of first-order deformations of X for which none of the irreducible components of E disappear; Theorem 3 says that such a deformation is trivial if it is trivial on $X - E$, i.e., gives a trivial deformation of Spec R. Actually, the existence of a blowing-down map (to relate deformations of X to those of Spec R) is rather subtle [**11**]; one may consider only those deformations of X for which $H^1(\mathcal{O}_X)$ does not jump.

Second, recall the Artin-Brieskorn map Φ: Res $\to$ Def arising from the MGR $X \to$ Spec R of a rational singularity [**1**]:

(i) Def is the local moduli space for Spec R.

(ii) Res is the moduli space representing simultaneous resolutions of deformations of Spec R.

(iii) Res is smooth, and Φ maps Res finitely onto an irreducible component (the Artin component) of Def.

The kernel of the tangent space map of Φ is easily identified with $H^1_E(\theta_X)$. So, one can use Theorems 2 and 5 to prove [**12**]

THEOREM 6. *If the MGR of a rational singularity contains no* -2 *curves, then* Φ: Res $\to$ Def *is a closed immersion, the Artin component of* Def *is smooth, and the locus of singular deformations there has codimension* $\geqq 2$, *whence simultaneous resolution takes place without base change.*

Brieskorn has shown that for the rational double points, Φ is a Galois cover, whose group is the Weyl group of the semisimple Lie algebra associated to the "Dynkin diagram" of the singularity [**2**]. This and Theorems 5 and 6 give good evidence for

CONJECTURE 2 (SEE ALSO [**3**, 7.3]). Φ *is a Galois cover of its image, with group the direct product of Weyl groups associated to the maximal configurations of* -2 *curves in E. In particular, the Artin component is smooth.*

There is an action of the group on Res defined functorially as in [**4**]; unfortunately, we know of no explicit action. In addition to the above cases, the conjecture is known to be true for cyclic quotient singularities (follows from [**10**]), and for configurations containing no -3 curves for which the -2 curves are disjoint.

5. The proofs of Theorems 2, 3, and 4 start with

$$H^1_E(F) = \xrightarrow{\lim} H^0(F \otimes \mathcal{O}_Z(Z)),$$

the limit taken over effective divisors supported on E. One shows that every such Z contains an exceptional component E_i so that α: $H^0(F \otimes \mathcal{O}_Z(Z)) \to H^0(F \otimes \mathcal{O}_{E_i}(Z))$ is the zero map, as this implies that

$$H^0(F \otimes \mathcal{O}_{Z-E_i}(Z - E_i)) \xrightarrow{\sim} H^0(F \otimes \mathcal{O}_Z(Z)),$$

and one can induct. The easiest way to do this is to find an E_i with $H^0(F \otimes \mathcal{O}_{E_i}(Z)) = 0$; this is always possible in the situation of Theorem 2, and makes it an "easy" result. However, this is not possible for Theorems 3 and 4 (even for the rational double points!). One must prove that any cycle Z causing difficulty contains a very special subcycle; and that, using this subcycle, the map α is 0 for an appropriate E_i. Each of these two steps is lengthy.

BIBLIOGRAPHY

1. M. Artin, *Algebraic construction of Brieskorn's resolutions*, J. Algebra **29** (1974), 330–348. MR **50** #7143.

2. E. Brieskorn, *Singular elements of semi-simple algebraic groups*, Proc. Internat. Congress Math. (Nice, 1970), vol. 2, Gauthier-Villars, Paris, 1971, pp. 279–284.

3. D. Burns and M. Rapaport, *On the Torelli problem for Kählerian K–3 surfaces*, Ann. Sci. École Norm. Sup. **8** (1975), 235–274.

4. D. Burns and J. Wahl, *Local contributions to global deformations of surfaces*, Invent. Math. **26** (1974), 67–88. MR **50** #2168.

5. H. Grauert, *Über Modifikationen und exzeptionelle analytische Mengen*, Math. Ann. **146** (1962), 331–368. MR **25** #583.

6. H. Grauert and O. Riemenschneider, *Verschwindungssätze für analytische Kohomologiegruppen auf komplexen Räumen*, Invent. Math. **11** (1970), 263–292. MR **46** #2081.

7. A. Grothendieck, *Local cohomology*, Lecture Notes in Math., vol. 41, Springer-Verlag, Berlin and New York, 1967.

8. R. Hartshorne and A. Ogus, *On the factoriality of local rings of small embedding codimension*, Comm. Algebra **1** (1974), 415–437. MR **50** #322.

9. N. M. Katz, *Algebraic solutions of differential equations* (*p-curvature and Hodge filtration*), Invent. Math. **18** (1972), 1–118. MR **49** #2728.

10. O. Riemenschneider, *Deformationen von Quotientensingularitäten*, Math. Ann. **209** (1974), 211–248.

11. J. Wahl, *Equisingular deformations of normal surface singularities*. I, Ann. of Math. (to appear).

12. ———, *Vanishing theorems for resolutions of surface singularities*, Invent. Math. **31** (1975), 17–41.

UNIVERSITY OF NORTH CAROLINA, CHAPEL HILL

PRINCIPAL LECTURE

MASATAKE KURANISHI

APPLICATION OF $\bar{\partial}_b$ TO DEFORMATION OF ISOLATED SINGULARITIES

Proceedings of Symposia in Pure Mathematics
Volume 30, 1977

APPLICATION OF $\bar{\partial}_b$ TO DEFORMATION OF ISOLATED SINGULARITIES

MASATAKE KURANISHI*

1. Introduction. The purpose of the present paper is to give the outline of deformation theory of isolated singularities based on tangential Cauchy-Riemann equations. The deformation theory of singularities has already been developed by several mathematicians. (For a historical note see the article of O. Forster in these PROCEEDINGS [**1**].) However the methods so far are algebraic. That is to say, deformations are regarded as deformations of defining equations of singularities. As for the analytic approach, besides the one exposed here, which follows the idea suggested by Hugo Rossi several years ago, Richard Hamilton constructed a theory which relies on the $\bar{\partial}$ operator on a tubular neighborhood of the boundary. This approach leads to a nonlinear boundary value (noncoercive) problem of Cauchy-Riemann equations [**2**].

2. Isolated singularities and CR structures. Let V be a Stein analytic space (of complex dimension n) smooth except at a point p such that there is a relatively compact open neighborhood U of p in V with the property that the boundary of U is smooth and strongly pseudoconvex. Denote by M the boundary of U. M being a real submanifold of codimension one in the complex manifold $V - \{p\}$, the complex tangent vector bundle CTM has a distinguished subbundle $^\circ T''_V M$. Namely, it consists of all elements in CTM which are of type (0, 1) when we regard CTM canonically as a subbundle of $CTV|M$. Setting $E'' = {}^\circ T''_V M$, the equation

$$(1) \qquad Xf = 0 \quad \text{for all } X \in E''$$

is called the tangential Cauchy-Riemann equation on M induced by the ambient complex space V. By the construction it is obvious that the subbundle $E'' = {}^\circ T''_V M$ satisfies the following conditions:

AMS (MOS) subject classifications (1970). Primary 32H15, 32H10.

*The research was partially supported by NSF Grant MPS 71–03442.

© 1977, American Mathematical Society

(2) If L and L' are sections of E'', so is $[L, L']$.

We refer to any subbundle E'' of CTM of fiber dimension $n-1$ with the property $E' \cap E'' = \{0\}$, $E' = \bar{E}''$, as an almost CR structure on M. If it further satisfies the condition (2), we call E'' a CR structure. The notion of pseudoconvexity of a hypersurface in a complex manifold can be formulated solely in terms of the almost CR structure induced in the hypersurface. Namely, pick sections $Y_1, \cdots, Y_{n-1}$ of E'' on an open set G of M such that they generate $E''|G$ and also a real vector field T on G complementary to $E' + E''$. Write

$$[Y_j, \bar{Y}_k] \equiv c_{jk} iT \pmod{Y_1, \cdots, Y_{n-1}, \bar{Y}_1, \cdots, \bar{Y}_{n-1}}.$$

We say that E'' is strongly pseudoconvex when the hermitian matrix (c_{jk}) obtained in this way is always nonsingular and its eigenvalues are of the same sign. If M is strongly pseudoconvex in V, ${}^{\circ}T''_V M$ is strongly pseudoconvex. Conversely, for any strongly pseudoconvex CR structure E'' on M and for any point p in M, Boutet de Monvel [6] showed recently that there are $f_1, \cdots, f_n \in C^\infty(M)$ such that $Xf_j = 0$ for any section X of E'' and the map $x \to (f_1(x), \cdots, f_n(x)) \in \boldsymbol{C}^n$ is an embedding on a neighborhood of p. This means in particular that any strongly pseudoconvex CR structure E'', when restricted to small open sets, is induced by an ambient complex manifold. Pick an open covering $\{G_\alpha\}$ of M together with an ambient complex manifold W_α of G_α such that $W_\alpha - G_\alpha$ consists of two components and $E''|G_\alpha$ is the induced CR structure on G_α by W_α. Then by the theorem of H. Lewy [3] there is a unique component W'_α of $W_\alpha - G_\alpha$ such that any solution of the equation (1) on W_α extends uniquely to a holomorphic function on W'_α, provided we choose W_α sufficiently thin. This will allow us to piece together W'_α (shrinking G_α a little if necessary) and construct a complex manifold N with boundary M (regarding the pseudoconcave part of the boundary of N as open), even though the complex structure may not extend beyond M. We may say that E'' is induced by the ambient complex manifold N. Since $f_1, \cdots, f_n$ in the theorem of Boutet de Monvel are defined everywhere in M, we may conclude that we can construct N as above such that the holomorphic functions on N separate points. Then it is a theorem of H. Rossi [8] that we can fill in the hole on N. He showed that the set of the maximal ideals of the algebra of the holomorphic functions on N, say S, has the natural structure of normal Stein analytic space. The obvious injection $N \to S$ is holomorphic and the image is open. In this way we can replace the deformations of normal isolated singularities by the deformations of CR structures.

Deformations of isolated singularities may be viewed in two steps. Namely, the first is the deformations of the smooth part of the analytic set and the second is the way singular points are added to complete it. Now the second step is not unique. Because of the blowing up and its inverse, this step is very complicated. Our contention is that the CR structure induced on the boundary completely controls the first step and also gives one a definite way of doing the second step.

3. Integrability conditions. We develop the deformation theory of CR structures following the pattern established in the deformation theory of complex structures. Let us recall the first step of the latter. We fix a reference complex structure on a manifold, say N, which is considered as a subbundle T'' of the complex tangent vector bundle CTN. T'' consists of complex tangent vectors of type (0, 1). We note the direct sum decomposition

(3) $$CTN = T'' + T', \qquad T' = \bar{T}''.$$

Then any almost complex structure sufficiently close to T'' is considered as a subbundle which is the graph of a bundle map

(4) $$\omega : T'' \to T'.$$

We denote this almost complex structure by T''_ω. Thus almost complex structures sufficiently close to T'' are parametrized by T'-valued differential forms of type (0, 1). T'' has parameter 0. If T''_ω is a complex structure, it follows that

(5) $$\bar{\partial}\omega - \tfrac{1}{2}[\omega, \omega] = 0$$

where $[\omega, \omega]$ is the type (0, 2) form constructed by means of the bracket of vector fields and exterior product. It is the famous theorem of Newlander and Nirenberg that the converse is true. These considerations are the ground on which we can apply the theory of elliptic differential operators to construct the versal family of deformations of compact complex structures.

As for the reference CR structure ${}^\circ T''$, there is no canonical decomposition like (3). We are forced to choose one. We pick a subbundle F of CTM of fiber dimension 1 such that

(6) $$CTM = {}^\circ T'' + {}^\circ T' + F, \qquad \overline{{}^\circ T''} = {}^\circ T', \qquad \bar{F} = F.$$

Then any almost CR structure sufficiently close to ${}^\circ T''$ is the graph of a bundle map

(7) $$\varphi' : {}^\circ T'' \to {}^\circ T' + F.$$

It is a little awkward to use a bundle like ${}^\circ T'' + F$ to parametrize almost CR structures. We avoid this by observing that the restriction to CTM of the canonical projection map $CTV|M \to T'V|M$ has the kernel ${}^\circ T''$ and hence this map induces an isomorphism of ${}^\circ T' + F$ to $T'V|M$. Denote by

(8) $$\tau : T'V|M \to {}^\circ T' + F$$

the inverse of the isomorphism. Then we can write

(9) $$\varphi' = \tau \circ \varphi$$

where

(10) $$\varphi : {}^\circ T'' \to T'V|M$$

is a bundle map. Thus almost CR structures on M sufficiently close to ${}^\circ T''$ are parametrized by $T'V|M$-valued differential forms of type $(0, 1)_b$. Namely, ${}^\circ T''_\varphi$ is the graph of the bundle map $\tau \circ \varphi : T'V|M \to {}^\circ T' + F$. Therefore

(11) $${}^\circ T''_\varphi = \{X - \tau \circ \varphi(X); X \in {}^\circ T''\}.$$

In other words we have the isomorphism:

(11′) $${}^\circ T'' \ni X \mapsto X - \tau \circ \varphi(X) \in {}^\circ T''_\varphi.$$

The next problem is to decide which of the almost CR structures ${}^\circ T''_\varphi$ are CR structures. This will lead to an equation like (5) for φ. Now we can rewrite the integrability condition (2) in terms of differential forms as follows:

$$\text{(2}'\text{)}\qquad \begin{array}{l}\text{If } \theta \text{ is a differential form of degree 1 such that } \theta(X) = 0 \\ \text{for all } X \in E'', \text{ then } d\theta(X, X') = 0 \text{ for all } X, X' \in E''.\end{array}$$

Thus we can write down the condition for $°T''_\varphi$ being a CR structure when we can find a generator for differential forms of degree 1 which annihilate $°T''_\varphi$. To do this we use a local chart and introduce several notations. Before proceeding, we pause here to note that the condition (2′) can be reformulated in the following more suggestive way: Consider the diagram

$$\text{(12)}\qquad \begin{array}{ccc} \Lambda^1(M, \boldsymbol{C}) & \rightarrow & \Lambda^2 M, \boldsymbol{C}') \\ \downarrow & & \downarrow \\ C^\infty(M, (E'')^*) & \dashrightarrow & C^\infty(M, \Lambda^2(E'')^*) \end{array}$$

where the vertical arrows are induced by the injection $E'' \rightarrow \boldsymbol{C}TM$. Then the condition (2′) is equivalent to the following:

(2″) There is a unique dotted arrow which makes the diagram (2) commutative.

When E'' is a CR structure, we denote by $\bar\partial_{E''}$ the dotted arrow obtained in (2″). Once this can be done it follows easily that we can construct similarly the differential operator

$$\text{(13)}\qquad \bar\partial_{E''}: C^\infty(M, \Lambda^p(E'')^*) \rightarrow C^\infty(M, \Lambda^{p+1}(E'')^*).$$

Since we always have the differential operator $\bar\partial_{E''}: C^\infty(M, \boldsymbol{C}) \rightarrow C^\infty(M, (E'')^*)$ as the composition of the exterior derivative and the restriction map, we conclude that E'' is a CR structure if and only if we have the $\bar\partial_{E''}$-complex

$$\text{(14)}\qquad C^\infty(M, \boldsymbol{C}) \rightarrow C^\infty(M, \Lambda^1(E'')^*) \rightarrow C^\infty(M, \Lambda^2(E'')^*) \rightarrow \cdots.$$

When $E'' = °T''$, this is the $\bar\partial_b$-complex.

Now we come back to the problem of writing down the generator mentioned above. Let our reference CR structure $°T''$ be induced locally by a real submanifold G in an open ball in $\boldsymbol{C}^n$. Denote by $z = (z^1, \cdots, z^n)$ the general elements in the ball, and let

$$\text{(15)}\qquad h = 0$$

be the equation of G, where h is a real-valued C^∞ function in z. The choice of h is not unique. But we pick one and preserve it throughout. By the injection i of G into the ball in $\boldsymbol{C}^n$ we identify $\boldsymbol{C}TG$ as a subbundle of $\boldsymbol{C}T\boldsymbol{C}^n|G$. Since F in (6) is preserved under conjugation, we can write

$$\text{(16)}\qquad F = \boldsymbol{C}(P' - P''), \qquad P' \in T'\boldsymbol{C}^n|G, \qquad P'' = \bar{P}'.$$

We normalize the choice of P' by the requirement

$$\text{(17)}\qquad \langle dh,\ P'\rangle = 1.$$

Set

$$\text{(18)}\qquad d'h = \sum_k h_k dz^k, \qquad h_{\bar k} = \bar h_k,$$

$$\text{(19)}\qquad P' = \sum_k p^k \partial/\partial z^k, \qquad p^{\bar k} = \overline{p^k}.$$

By (17),

$$\sum_k p^k h_k = 1. \tag{17'}$$

Set

$$Z^{\bar{k}} = i^*(d\bar{z}^k - p^{\bar{k}}\, d''h), \tag{20}$$

$$Z_{\bar{k}} = \partial/\partial\bar{z}^k - h_{\bar{k}} P''. \tag{21}$$

By (17) we see easily that $Z_{\bar{k}} \in {}^\circ T''$. $Z^{\bar{k}}$ generates all differential forms of degree one which annihilate ${}^\circ T' + F$. $Z_{\bar{k}}$ generates ${}^\circ T''$. They have the relations:

$$\sum_k h_{\bar{k}} Z^{\bar{k}} = 0, \qquad \sum_k p^{\bar{k}} Z_{\bar{k}} = 0. \tag{22}$$

Now a $TV|M$-valued differential form of type $(0, 1)_b$, say φ, can be expressed on G as

$$\begin{aligned} \varphi &= \sum_k \varphi^k \otimes \frac{\partial}{\partial z^k}, \\ \varphi^k &= \sum_l \varphi^k_{\bar{l}} Z^{\bar{l}} \quad \text{with } \sum p^{\bar{l}} \varphi^k_{\bar{l}} = 0. \end{aligned} \tag{23}$$

Because of (11), the differential forms on G of degree 1 which annihilate ${}^\circ T''_\varphi$ are generated by

$$\theta^k = i^* dz^k + \varphi^k \qquad (k = 1, \cdots, n). \tag{24}$$

Since ${}^\circ T''_\varphi$ is a CR structure (assuming that φ is sufficiently small) if and only if $E'' = {}^\circ T''_\varphi$ satisfies the condition (2′), it follows that ${}^\circ T''_\varphi$ is a CR structure if and only if (since $d\theta^k = d\varphi^k$)

$$d\varphi^k \equiv 0 \qquad (\text{mod } \theta^1, \cdots, \theta^n). \tag{25}$$

Set

$$\partial^\tau/\partial z^k = \tau(\partial/\partial z^k) = \partial/\partial z^k - h^k P''. \tag{26}$$

We calculate the condition (25) more explicitly using the expression (23). Then we arrive at the following conclusion: For a sufficiently small $T'V|M$-valued differential form φ of type $(0, 1)_b$, ${}^\circ T''_\varphi|G$ is a CR structure if and only if

$$\begin{aligned} P(\varphi) = {} & \bar{\partial}_b \varphi - \sum_{j,k,l} (\partial^\tau \varphi^k_{\bar{l}}/\partial z^i)\varphi^i \wedge Z^{\bar{l}} \otimes \partial/\partial z^k \\ & + \Big(\sum_i h_i \varphi^i\Big) \wedge \sum_{l,k} \Big(\bar{\partial}_b p^l - \sum_i \varphi^i \partial^\tau p^{\bar{l}}/\partial z^i\Big) \varphi^k_{\bar{l}} \otimes \partial/\partial z^k \end{aligned} \tag{27}$$

vanishes identically.

$P(\varphi)$ is constructed depending on the chart z of the ambient complex manifold inducing ${}^\circ T''$ and of the function h in (15). However one can show that $P(\varphi)$ is independent of such choice. This can be done by explicitly calculating the right-hand side of (27) when we make changes in the choice. Recently D.C. Spencer and H. Goldschmidt found an intrinsic defining formula of $P(\varphi)$.

4. Heuristic argument for the construction of versal families. Let us recall the basic idea in the construction of the versal families of deformations of compact complex manifolds.

A diffeomorphism of N transforms an almost complex structure to an almost complex structure. Thus the diffeomorphism group of N acts on the set of almost complex structures on N. This action sends complex structures to complex structures. Two structures on the same orbit are isomorphic structures. Since we are interested in deformations we consider only almost complex structures sufficiently close to the reference complex structure T'' and actions of diffeomorphisms sufficiently close to the identity map. Hence we may describe our situation roughly as follows: A sufficiently small open neighborhood of 0 in $A^{(0,1)}(N, T')$ is fibered into orbits by the action of small open neighborhoods of the identity in the diffeomorphism group of N. We consider the subset of this fiber space consisting of all ω such that ${}^\circ T''_\omega$ is a complex structure. This is a fiber subspace, say B. If we can find a cross-section passing through 0, say C, of fibers of B, $\{T''_\omega; \omega \in C\}$ will be considered as a universal family of deformations of N. However, it can happen (for some N) that it is impossible to find a decent such C. This is due to the fact that the dimension of the complex automorphism group of T''_ω which acts as the isotopy group at ω may change with ω. To avoid this difficulty we fiber B instead into orbits by action of diffeomorphisms which are complementary to the automorphism group of T''. To be more precise, we parametrize first a small neighborhood of the identity in the diffeomorphism group of N by a small neighborhood of 0 in $C^\infty(N, T')$ by an exponential map. For a small $\xi \in C^\infty(N, T')$ denote by g_ξ the diffeomorphism parametrized by ξ. Written in a complex chart z of N

$$g_{t\xi}(z)^k \equiv z^k + t\xi^k \ (\mathrm{mod}\ t^2), \qquad \xi = \sum_k \xi^k \,\partial/\partial z^k. \tag{28}$$

Denote by ${}^\perp C^\infty(N, T')$ the subspace of $C^\infty(N, T')$ orthogonal to the subspace of holomorphic sections of T' (with respect to a hermitian metric in N). Let $G^\perp N$ be the set of diffeomorphisms of N parametrized by elements in a small neighborhood of 0 in ${}^\perp C^\infty(N, T')$. Now, instead of fibering B into orbits by small neighborhoods of the identity in the diffeomorphism group of N, we fiber B into orbits by small neighborhoods of the identity in $G^\perp N$. Then it is possible to find a cross-section. A family of deformations of N constructed in this way is the versal family of deformations of N. Before we proceed further, we insert here a notation. For a small $\omega \in A^{(0,1)}(N, T')$ and a diffeomorphism g sufficiently close to the identity map of N, the transform of T''_ω by f is equal to T''_θ. We set $\theta = \omega \circ g$. Then we find that

$$\omega \circ g_\xi = \omega + \bar\partial \xi + \cdots \tag{29}$$

where $\cdots$ includes all terms which are not linear in (ω, ξ). This formula plays an important role in the construction of the versal family.

Now we start to carry over the above consideration to deformations of isolated singularities viewed as deformations of CR structures. Then we notice a new phenomenon due to the fact that we can wiggle the boundary. Let ${}^\circ T''_\varphi$ be a CR structure on M induced by an ambient complex manifold N_1. Since N_1 is diffeomorphic to the ambient complex manifold N of ${}^\circ T''$, we may write $N_1 = N_\omega$. Let $f\colon M \to N$ be a C^∞ injective map sufficiently close to the injection $i\colon M \to N$. The complex manifold N_ω induces a CR structure on $f(M)$, which we transplant to a CR structure on M via f. We call it the transform of ${}^\circ T''_\varphi$ by f. Since the above process is nothing but a wiggling of the boundary it is obvious that ${}^\circ T''_\varphi$ and its transform give rise to isomorphic singularities. Thus we find that in the deformation theory of isolated singularities the set of injections $M \to N$ sufficiently close to i plays the role of diffeomorphism group in the deformation theory of complex structures. This is the only modification we have to make.

Let us go over the fibering we consider more explicitly. We first parametrize injections of M into N sufficiently close to i by a small open neighborhood of $C^\infty(M, T'N|M)$, say f_ξ for $\xi \in C^\infty(M, T'NM)$, by an exponential map. In a complex analytic chart z of N it means that

$$f_\xi^k(z) = z^k + \xi^k(z) + \cdots, \qquad \xi = \sum \xi^k \partial/\partial z^k, \tag{30}$$

for $z \in M$. Denote by I the set of all f_ξ where ξ are sufficiently small and orthogonal to the vector space of $\bar\partial_b$ closed sections of $T'N|M$. Consider the set B of φ such that φ is sufficiently small and ${}^\circ T''_\varphi$ is induced by a complex structure N_ω. The elements in I act on the elements in B. Consider the fibering of B into orbits of the action by small neighborhoods of i in I. We shall try to find a cross-section of the fiber space B and show that a cross-section represents a versal family of deformations of the isolated singularity out of which we obtained ${}^\circ T''$.

We might feel that our picture is a little blurred because we considered only φ such that ${}^\circ T''_\varphi$ is induced by a complex manifold N_ω which lies on both sides of M, whereas in §2 we constructed for any integrable φ a complex manifold which induces ${}^\circ T''_\varphi$ but lies only in one side of M. However this does not stop us from constructing the versal family by the following reasoning: To construct a family which we wish to be the versal family we do not need ω's, and in order to show that the family we constructed is versal we are allowed to consider only φ's such that the ambient complex manifolds lie on both sides of M. The last is due to the fact that we start from an analytic set V with an isolated singularity so that ${}^\circ T''$ is induced by an ambient complex manifold which lies on the both sides of M and that any small deformation of V induces a CR structure with the same property.

As before we define $\varphi \circ f$ so that the transform of ${}^\circ T''_\varphi$ by f is ${}^\circ T''_{\varphi\circ f}$. Then we find after a little calculation that

$$\varphi \circ f_\xi = \varphi + \bar\partial_b \xi + \cdots \tag{31}$$

where $\cdots$ includes all terms not linear in (ω, ξ).

5. The construction of versal families. We recall first how the versal family of deformations of a compact complex manifold, say N, was constructed. As was explained in the preceding section, we are to find a decent set C of $\omega \in \Lambda^{(0,1)}(N, T'N)$ satisfying the condition:

$$\bar\partial\omega - \tfrac{1}{2}[\omega, \omega] = 0 \tag{5}$$

such that it contains 0 and it cuts transversally the set

$$^\perp\{\omega \circ g_\xi\,;\ \xi \in {}^\perp C^\infty(N, T'), \xi \text{ small}\}. \tag{32}$$

When we linearize the problem, we see by (29) that we are asked to find a complete set of representatives of the cohomology classes in the $T'N$-valued differential forms of type (0, 1). The standard way is to solve the equation

$$\bar\partial\omega = 0, \qquad \bar\partial^*\omega = 0. \tag{33}$$

This observation suggests that a good candidate for our C is the set of sufficiently small solutions of the equation

$$(34)\qquad \bar\partial\omega - \tfrac{1}{2}[\omega, \omega] = 0, \qquad \bar\partial^*\omega = 0.$$

Actually it can be shown that such C forms the versal family. The equation (34) is solved as follows: Since solutions of (34) satisfy the condition

$$(35)\qquad G\bar\partial^*(\bar\partial\omega - \tfrac{1}{2}[\omega, \omega]) + G\bar\partial\bar\partial^*\omega + H(\omega) \text{ is harmonic,}$$

where G is the Green's operator and H is the harmonic projection, we first solve the equation (35) and decide which of the solutions of (35) are solutions of (34). Since the sufficiently small solutions of (35) form a finite-dimensional complex manifold, it can be shown that solutions of (34) in this manifold form an analytic set. Now we can solve the equation (35) when we can invert the map

$$(36)\qquad \omega \mapsto H(\omega) + G\bar\partial^*(\bar\partial\omega - \tfrac{1}{2}[\omega, \omega]) + G\bar\partial\bar\partial^*\omega = \omega - \tfrac{1}{2}G\bar\partial^*[\omega, \omega].$$

Because of the ellipticity of the Laplacian $\Delta = \bar\partial^*\bar\partial + \bar\partial\bar\partial^*$, the map (36) induces an analytic map of the Banach manifold obtained by completing with respect to Sobolev norm. Therefore to find the inverse of (36), we check that the differential at 0 is the identity map and apply the Banach inverse mapping theorem.

To find the CR analog of the above construction we merely have to replace the equation (5) by the equation (cf. (27))

$$(37)\qquad P(\varphi) = 0$$

and the formula (29) by (31). Thus our problem is to invert the map

$$(38)\qquad \varphi \mapsto H(\varphi) + N\bar\partial_b^* P(\varphi) + N\bar\partial_b\bar\partial_b^*\varphi$$

where N is the Neumann operator. (For subellipticity and Neumann operators, see Kohn's article [**4**] in these PROCEEDINGS.) However, the analogy fails here because the Laplacian $\Delta_b = \bar\partial_b^*\bar\partial_b + \bar\partial_b\bar\partial_b^*$ is not elliptic and hence the map (38) does not induce the map of a Sobolev Banach manifold into itself. The way to get around the difficulty is to note that Δ_b is subelliptic and use the Nash-Moser inverse mapping theorem. The theorem says that when we have a map like (38), if the differentials at points near 0 are subelliptic with a uniform estimate and invertible then the map is invertible. However, the differentials of the map (38) at nonzero points do not appear to be subelliptic. Thus we are forced to modify our construction: In order to obtain the subellipticity of differentials we have to bring the boundary Cauchy-Riemann operators of each ${}^\circ T''_\varphi$ into our picture. Now it is necessary to introduce a number of operators. Before proceeding further we note that the inverse in the Nash-Moser theorem is constructed by ingeniously combining Newton's algorithm and the smoothing operators (cf. [**7**]).

We recall that the $\bar\partial_{E''}$-complex was constructed in (14) by means of the diagram (12) in the case $E'' = {}^\circ T''_\varphi$. However, we need such an operator for all sufficiently small φ (not merely for φ for which ${}^\circ T''_\varphi$ is a CR structure). We obtain this by picking cross-sections of the vertical arrows in (12). Such cross-sections which are natural from our standpoint are induced by the decomposition (cf. (6))

$$CTM = E'' + E' + F, \qquad E'' = {}^\circ T''_\varphi.$$

We denote by $\bar\partial_b^\varphi$ the sequence of differential operators thus obtained:

$$(39)\qquad \bar\partial_b^\varphi : \Lambda_b^{(0,p)}(M, C) \to \Lambda_b^{(0,p+1)}(M, C).$$

They form a complex if and only if ${}^\circ T''_\varphi$ is a CR structure. In terms of a complex analytic chart $z = (z^1, \cdots, z^n)$ of N (and using the notations introduced near the end of §3),

$$\bar\partial^\varphi_b f = \sum_l (Z^\varphi_{\bar l} f) Z^{\bar l}, \quad f \in C^\infty(G, C), \tag{40}$$

$$\bar\partial^\varphi_b Z^{\bar l} = \sum_k \varphi^k \wedge \bar\partial^\varphi_b (h_k p^{\bar l}). \tag{41}$$

The formulae (40) and (41) determine uniquely the operator $\bar\partial^\varphi_b$ because of the linearity and the rule $\bar\partial^\varphi_b(\theta \wedge \psi) = (\bar\partial^\varphi_b \theta) \wedge \psi + (-1)^l \theta \wedge \bar\partial^\varphi_b \psi$ where l is the degree of θ. It is interesting to note here the formula

$$\bar\partial^\varphi_b \circ \bar\partial^\varphi_b f = -\sum_k (\partial^\tau f/\partial z^k) P(\varphi)^k.$$

This could be used to show that $P(\varphi)$ is well defined independent of choices in the defining formula (27). Since in our construction of versal families we work on $T'N|M$-valued differential forms of type $(0, p)_b$ we have to define $\bar\partial^\varphi_b$ for such forms. We do this by explicitly writing down the definition in terms of the complex analytic chart z in N and showing that it is well defined globally. For

$$\mu \in \Lambda^{(0,p)}_b(M, T'N|M)$$

write

$$\mu = \sum_k \mu^k \otimes \partial/\partial z^k, \qquad \mu^k \in \Lambda^{(0,p)}_b(G, \boldsymbol{C}).$$

Then we define $\bar\partial^\varphi_b \mu$ on G by

$$\bar\partial^\varphi_b \mu = \sum_k \Big(\bar\partial^\varphi_b \mu^k + \sum_l \Gamma^k_{\bar l}(\varphi) \wedge \mu^l \Big) \otimes \partial/\partial z^k$$

where

$$\Gamma^k_{\bar l}(\varphi) = \sum_j ((\partial^\tau \varphi^k_{\bar j}/\partial z^l) Z^{\bar j} - \varphi^k_{\bar j} \alpha^{\bar j}_l(\varphi)),$$
$$\alpha^{\bar k}_l(\varphi) = h \bar\partial_l{}^\varphi_b p^{\bar k} + \sum_j h_j \varphi^j \partial^\tau p^{\bar k}/\partial z^l.$$

By means of a hermitian metric we introduce $(\bar\partial^\varphi_b)^*$. The Laplacian $\Delta^\varphi_b = (\bar\partial^\varphi_b)^* \bar\partial^\varphi_b + \bar\partial^\varphi_b (\bar\partial^\varphi_b)^*$ is still subelliptic. The dimension of the kernel of Δ^φ_b may depend on φ. However, we can show that, for φ sufficiently small, the dimension of the sum of the eigenspaces of eigenvalues sufficiently small, say H'_φ, is independent of φ. Denote by ρ^φ the orthogonal projection (in L_2 norm) to H'_φ. Denote by N^φ the composition of $I - \rho^\varphi$ with the Neumann operator of Δ^φ_b where I is the identity map. We have the formula:

$$\rho^\varphi + N^\varphi \Delta^\varphi_b = \text{the identity map}.$$

We use ρ^φ and N^φ instead of the harmonic projection and the Neumann operator of Δ^φ_b because the latter do not depend smoothly on φ. When we write down $(\bar\partial^\varphi_b)^*$ in terms of a local chart, we find that partial derivatives of $\varphi^k_{\bar l}$ appear in the coefficients of the zeroth order terms of the expression. For a technical reason these coefficients cause some trouble in the construction of the universal family. Therefore we subtract the terms which contain partial derivatives of $\varphi^k_{\bar l}$ and piece the remaining parts together by means of a partition of unity. In this way we construct $(\bar\partial^\varphi_b)^\star$. It is a differential operator having the same principal part as $(\bar\partial^\varphi_b)^*$.

We are ready to state what we will do instead of trying to find the inverse of the map (38). For each sufficiently small harmonic (with respect to Δ_b) $T'N|M$-valued differential form of type $(0,1)_b$, say t, solve the equation

$$\rho^\varphi \varphi + N^\varphi((\bar\partial_b^\varphi)^* P(\varphi) + \bar\partial_b^\varphi(\bar\partial_b^\varphi)^* \varphi) = \rho^\varphi t.$$

It can be shown by means of the Nash-Moser theorem that the equation has a unique solution, say $\varphi(t)$, which is sufficiently small. Then we write down the equation for t so that ${}^\circ T''_{\varphi(t)}$ is a CR structure. In this way we construct a family of CR structures. By analyzing $\varphi \circ f_\xi$ more closely than was done in (31) we can show that the family induces the versal family of deformations of the isolated singularity we started with.

References

1. O. Forster, Proc. Sympos. Pure Math., vol. 30, part 2, Amer. Math. Soc., Providence, R. I., 1977, pp. 199–217.
2. Richard Hamilton, *Deformation of complex structures on manifolds with boundary*. I–III, Cornell University (preprint).
3. L. Hörmander, *An introduction to complex analysis in several variables*, Van Nostrand, Princeton N. J., 1966. MR **34** #2933.
4. J. J. Kohn, Proc. Sympos. Pure Math., vol. 30, part 1, Amer. Math. Soc., Providence, R. I., 1977, pp. 215–237.
5. M. Kuranishi, *Deformations of compact complex manifolds*, Séminaire Mathématiques Supérieures, no. 39 (Été 1969), Press Univ. Montréal, Montréal, Que., 1971. MR **50** #7588.
6. Boutet de Monvel, *Integration des equations de Cauchy-Riemann induites formelles*, Seminaire Goulaouic-Lions-Schwartz, 1974/75.
7. J. Moser, *A rapidly convergent iteration method and non-linear partial differential equations*, I, II, Ann. Scuola Norm. Sup. Pisa (3) **20** (1966), 265–315, 499–535. MR **33** #7667; **34** #6280.
8. H. Rossi, *Attaching analytic spaces to an analytic space along a pseudoconcave boundary*, Proc. Conf. on Complex Analysis (Minneapolis, 1964), Springer, Berlin, 1965, pp. 242–256. MR **31** #381.

Columbia University

FUNCTION THEORY AND REAL ANALYSIS

Proceedings of Symposia in Pure Mathematics
Volume 30, 1977

THE DIRICHLET PROBLEM FOR A COMPLEX MONGE-AMPÈRE EQUATION

ERIC BEDFORD* AND B. A. TAYLOR**

1. On C^n, write $d = \partial + \bar{\partial}$, $d^c = i(\bar{\partial} - \partial)$ so that $dd^c u = 2i\partial\bar{\partial}u$, and let $\beta_n = (i/2)^n \prod_{j=1}^n dz_j \wedge d\bar{z}_j$ be the usual volume form. We study here the nonlinear Dirichlet problem,

$$(1) \qquad \begin{aligned} (dd^c u)^n = dd^c u \wedge \cdots \wedge dd^c u &= f\beta_n \quad \text{on } \Omega, \\ u \text{ plurisubharmonic on } &\Omega, \\ u = \phi \text{ on } &\partial\Omega, \end{aligned}$$

where Ω is a bounded open set in C^n, $f \geqq 0$, and ϕ is a continuous function on $\partial\Omega$. For arbitrary plurisubharmonic functions u, it is known that $dd^c u$ is a positive current of type (1, 1) [**5**, p. 70]; but, it is not clear that the higher exterior powers of $dd^c u$ are well defined. In fact, examples indicate that it is probably not possible to define $(dd^c u)^n$ as a distribution for all plurisubharmonic functions u [**8**]. However, for bounded C^2 plurisubharmonic functions, Chern, Levine, and Nirenberg have given in [**3**] an estimate which makes it clear how to define $(dd^c u)^n$ when u is a continuous plurisubharmonic function. If $\| u \|_\Omega = \sup\{|u(z)| : z \in \Omega\}$, then they prove that for each compact subset K of Ω there is a constant $C = C(K)$ such that

$$\int_K (dd^c u)^n \leqq C\{\| u \|_\Omega\}^n$$

for all C^2 plurisubharmonic functions u on Ω. With this result (and its proof), it is easy to show that the operator $(dd^c u)^n$, thought of as a mapping from the C^2 plurisubharmonic functions on Ω to the space of nonnegative Borel measures on Ω,

---

AMS (MOS) subject classifications (1970). Primary 32F05, 35D05; Secondary 32E99.

*Research supported in part by a Sloan Foundation Grant to the Courant Institute of Mathematical Sciences, New York University, and by the Army Research Office grant DAHC04–75–G–0149.

**Research supported in part by the National Science Foundation grant GP 37628.

© 1977, American Mathematical Society

has a continuous extension to the space of all continuous plurisubharmonic functions on Ω. With this definition of $(dd^c u)^n$ as a nonnegative Borel measure on Ω, we prove the following existence and uniqueness theorem for the Dirichlet problem (1).

THEOREM 1. *Let Ω be a bounded, strongly pseudoconvex set in $\mathbf{C}^n$ with C^2 boundary. Let $\phi \in C(\partial\Omega)$ and $f \geqq 0, f \in C(\bar{\Omega})$. Then there exists a unique solution to the Dirichlet problem* (1) *in the class of all functions continuous on $\bar{\Omega}$ and plurisubharmonic in Ω.*

The differential equation of (1) bears a strong resemblance to the real Monge-Ampère equation (see, e.g., **[1]**, **[6]**, **[7]**). For example, when $n = 2$ and (z, w) are the variables for $\mathbf{C}^2$, the equation of (1) is

$$u_{z\bar{z}}u_{w\bar{w}} - u_{z\bar{w}}u_{w\bar{z}} = f(z, w)$$

while the real Monge-Ampère equation for $u = u(x, y)$ is

$$u_{xx}u_{yy} - (u_{xy})^2 = f(x, y).$$

The real Monge-Ampère equations have been studied extensively in connection with problems in real differential geometry, and seem to be very difficult to solve in a completely satisfactory way. For example, interior regularity for the analogue of the Dirichlet problem (1) was only recently shown by A. V. Pogorelov **[7]** (when $n \geqq 3$), and regularity up to the boundary is apparently not yet known. However, the existence of weak solutions has been known for some time. A. D. Aleksandrov introduced an elegant theory of convex surfaces, and his techniques showed the existence and uniqueness of convex solutions of certain real Monge-Ampère equations. The book of Pogorelov **[6]** contains an extensive treatment of these questions when $n = 2$, and the recent survey article by Gluck **[4]** gives an account of some related work.

From our point of view, the approach outlined here for the study of the Dirichlet problem (1) is analogous to the early work on the real Monge-Ampère equations. We are able to prove the existence and uniqueness of continuous solutions to the problem (although in some special cases we can do better, i.e., Theorem 3). However, in contrast to the situation for the real Monge-Ampère equations, the problem (1) does not in general have unique solutions if the plurisubharmonic function u fails to be continuous. There are examples of one-parameter families $\{u_\lambda\}$ of plurisubharmonic functions on $\mathbf{C}^n$ which are smooth except at the origin, have logarithmic singularities at the origin, and satisfy

$$(dd^c u_\lambda)^n = \delta_0, \qquad u_\lambda(z) = 0 \text{ if } |z| = 1$$

where δ_0 is the point mass measure at the origin.

Our method of proof for Theorem 1 is the familiar Perron method. The solution is exhibited as the upper envelope of a family of subsolutions. The minimum principle on which this construction is based, and which gives the uniqueness part of Theorem 1, is the following.

THEOREM 2. *Let Ω be a bounded open set in $\mathbf{C}^n$. If u, v are continuous on $\bar{\Omega}$, plurisubharmonic in Ω, and if in Ω*

$$(dd^c u) \leqq (dd^c v)^n \quad (\textit{as nonnegative measures})$$

then

$$\min\{u(z) - v(z) : z \in \bar{\Omega}\} = \min\{u(z) - v(z) : z \in \partial\Omega\}.$$

We remark that this minimum principle may fail if u, v are not continuous. In a special case, we also prove a regularity result.

THEOREM 3. *If Ω is the unit ball in C^n, if $\phi \in C^2(\partial\Omega)$, and if $f^{1/n} \in C^2(\bar{\Omega})$, $f \geqq 0$, then the unique, continuous plurisubharmonic solution of* (1) *has locally bounded second partial derivatives.*

In general, the solution u of problem (1) will not be of class C^2 when $f \equiv 0$, even if ϕ is real analytic.

Under stronger conditions, the Laplacian of u also satisfies a maximum principle.

THEOREM 4. *If Ω is a bounded open set in C^n, if $u \in C^2(\bar{\Omega})$ and is plurisubharmonic on Ω and solves* (1) *with $0 \leqq f \in C(\bar{\Omega})$ and $f^{1/n}$ plurisubharmonic on Ω, then for any $\zeta = (\zeta_1, \cdots, \zeta_n) \in C^n$,*

$$\max\left\{\sum_{i,j=1}^{n} \frac{\partial^2 u(z)}{\partial z_i \partial \bar{z}_j} \zeta_i \bar{\zeta}_j : z \in \bar{\Omega}\right\} = \max\left\{\sum_{i,j=1}^{n} \frac{\partial^2 u(z)}{\partial z_i \partial \bar{z}_j} \zeta_i \bar{\zeta}_j : z \in \partial\Omega\right\}.$$

2. The operator $\Phi(u)$. As an important ingredient in our work, we introduce an operator $\Phi(u)$ which is defined for all plurisubharmonic functions and is nicely behaved under smoothing. This operator is essentially $[(dd^c u)^n]^{1/n}$. Precisely, since $dd^c u$ is a positive (1, 1) current when u is plurisubharmonic, the partial derivatives $\partial^2 u/\partial z_i \partial \bar{z}_j$ are Borel measures. Choose a positive Borel measure λ so that all the $\partial^2 u/\partial z_i \partial \bar{z}_j$ are absolutely continuous with respect to λ. Then write $\partial^2 u/\partial z_i \partial \bar{z}_j = h_{i,j} d\lambda$, and define $\Phi(u)$ to be the Borel measure

$$\Phi(u) = [4^n n! \det [h_{i,j}]]^{1/n}\, d\lambda.$$

It is easy to verify that $\Phi(u)$ is well defined since $[\det A]^{1/n}$ is homogeneous of degree 1 in the nonnegative $n \times n$ Hermitian matrix A. If the derivatives $\partial^2 u/\partial z_i \partial \bar{z}_j$ are all absolutely continuous with respect to Lebesgue measure dV on C^n, then $\Phi(u) = g dV$ and $\Phi(u)^n = g^n dV = (dd^c u)^n$. In general, $\Phi(u)$ is absolutely continuous with respect to dV and $\Phi(u)^n \leqq (dd^c u)^n$ for continuous plurisubharmonic functions u.

THEOREM 5. (i) *Φ is a concave function of u, i.e., if u_1, u_2 are plurisubharmonic, $0 < \lambda < 1$, then*

$$\Phi(\lambda u_1 + (1 - \lambda) u_2) \geqq \lambda \Phi(u_1) + (1 - \lambda)\Phi(u_2).$$

(ii) *Φ is upper semicontinuous; that is, if u, u_j, $j \geqq 1$, are plurisubharmonic, and if $u_j \to u$ locally in L^1, and if the measures $\Phi(u_j)$ converge weakly, then $\Phi(u) \geqq \lim_{j \to \infty} \Phi(u_j)$.*

Define, for $g \in C(\bar{\Omega})$, $g \geqq 0$, $\phi \in C(\partial\Omega)$, the class $\mathscr{B}(g, \phi)$ to be the class of all plurisubharmonic functions v on Ω such that $\Phi(v) \geqq g dV$ (dV = Lebesgue measure), $\limsup_{\zeta \to z} v(\zeta) \leqq \phi(z)$, all $z \in \partial\Omega$. Utilizing the properties of Φ and the automorphism group of the unit ball, we prove the following regularity result.

THEOREM 6. *Let Ω be the unit ball in $\mathbf{C}^n$. If $u(z) = \sup\{v(z): v \in \mathscr{B}(g, \phi)\}$ is the upper envelope of the family $\mathscr{B}(g, \phi)$, and if $\phi \in C^2(\partial\Omega)$, $g \in C^2(\bar{\Omega})$, then $u \in C(\bar{\Omega})$ and has locally bounded second partial derivatives.*

With this much regularity for u, we can then prove that $\Phi(u) = gdV$, and then that u also solves the Dirichlet problem (1), with Ω the unit ball in $\mathbf{C}^n$ and $f = g^n$. The general existence result of Theorem 1 follows easily from this special case.

3. The Bremermann function and regularity of envelopes of holomorphy. One interest in the Dirichlet problem (1) results from the work of Bremermann [**2**], who introduced the following function. Given $\phi \in C(\partial\Omega)$, let $\mathscr{B}(\phi, \Omega)$ denote the Perron-Bremermann family of all plurisubharmonic functions v on Ω such that $\limsup_{z\to\zeta} v(z) \leqq \phi(\zeta)$ for all $\zeta \in \partial\Omega$, and then define

$$(S\phi)(z) = \sup\{v(z) \mid v \in \mathscr{B}(\phi, \Omega)\}.$$

Bremermann showed that $S\phi$ is plurisubharmonic on Ω and $S\phi(z) \to \phi(\zeta)$ as $z \to \zeta \in \partial\Omega$, provided that Ω is strongly pseudoconvex. It was later proved by J. B. Walsh [**9**] that $S\phi$ is continuous on $\bar{\Omega}$. Bremermann also noted that if $S\phi \in C^2(\Omega)$, then $(dd^c u)^n = 0$ in Ω. Now, in general, $S\phi$ is not a C^2 function, but in any case, Theorem 1 shows that $S\phi$ is characterized as the unique continuous solution of the special case of (1) with $f \equiv 0$.

Bremermann also proved that the envelope of holomorphy of the Hartogs' domain in $\mathbf{C}^{n+1}$, $\{(z, w): z \in \partial\Omega, |w| \leqq \exp(-\phi(z))$, or $z \in \Omega, |w| \leqq e^{-M}\}$, $M = \max\{\phi(z): z \in \partial\Omega\}$, is described as $\{(z, w) \in \mathbf{C}^{n+1}: |w| \leqq \exp(-S\phi(z)), z \in \bar{\Omega}\}$, when Ω is strictly pseudoconvex. Thus, the boundary of this envelope of holomorphy is characterized as the solution to the partial differential equation $(dd^c u)^n = 0$. In particular, if Ω is the unit ball in $\mathbf{C}^n$ and ϕ is smooth, then Theorem 3 yields a smoothness property of the boundary of this envelope of holomorphy.

BIBLIOGRAPHY

1. A. D. Aleksandrov, *Dirichlet's problem for the equation* Det $\|z_{ij}\| = \phi(z_1, \cdots, z_n, z, x_1, \cdots, x_n)$. I, Vestnik Leningrad. Univ. Ser. Mat. Meh. Astr. **13** (1958), no. 1, 5–24. (Russian) MR **30** #3385.

2. H. J. Bremermann, *On a generalized Dirichlet problem for plurisubharmonic functions and pseudo-convex domains, characterization of Šilov boundaries*, Trans. Amer. Math. Soc. **91** (1959), 246–276. MR **25** #227.

3. S. S. Chern, Harold I. Levine and L. Nirenberg, *Intrinsic norms on a complex manifold*, Global Analysis (Papers in Honor of K. Kodaira), Univ. of Tokyo Press, Tokyo, 1969, pp. 119–139. MR **40** #8084.

4. H. Gluck, *Manifolas with preassigned curvature—a survey*, Bull. Amer. Math. Soc. **81** (1975), 313–329.

5. P. Lelong, *Plurisubharmonic functions and positive differential forms*, Gordon and Breach, New York, 1968. MR **39** #4436.

6. A. V. Pogorelov, *Monge-Ampère equations of elliptic type*, Izdat. Har'kovsk. Univ., Kharkov, 1960; English transl., Noordhoff, Groningen, 1964. MR **23** #A1137; **31** #4993.

7. ———, *The Dirichlet problem for the n-dimensional analogue of the Monge-Ampère equation*, Dokl. Akad. Nauk SSSR **201** (1971), 790–793 = Soviet Math. Dokl. **12** (1971), 1727–1731. MR **45** #2305.

8. Y.-T. Siu, *Extension of meromorphic maps*, Ann. of Math. (to appear).

9. J. B. Walsh, *Continuity of envelopes of plurisubharmonic functions*, J. Math. Mech. **18** (1968/69), 143–148. MR **37** #3049.

COURANT INSTITUTE OF MATHEMATICAL SCIENCES

UNIVERSITY OF MICHIGAN

Proceedings of Symposia in Pure Mathematics
Volume 30, 1977

TYPE CONDITIONS FOR REAL SUBMANIFOLDS OF C^n

THOMAS BLOOM

Introduction. This paper is a report on joint work with Ian Graham some of which is to appear in [**1**].

The notion of point of type m on a real hypersurface in C^2 was introduced by J. J. Kohn [**2**]. His definition (see 1.2) involves properties of commutators of holomorphic tangential vector fields. We give a geometric characterization of this condition (Theorem 1.4) and consider various generalizations.

The main motivation in studying these questions comes from their relation to subelliptic estimates for $\bar\partial$ and $\bar\partial_b$. These may be found in the references [**2**], [**3**], [**4**] and some recent work of Rothschild-Stein and J. J. Kohn reported on at the Williamstown conference (these PROCEEDINGS).

1. Hypersurfaces.

1.1. Let M be a real C^∞ submanifold of codimension one in an open subset $U \subset C^n$. Let r be a C^∞ defining function for M, that is

$$M = \{z \in U \mid r(z) = 0\} \quad \text{and } dr \neq 0 \text{ on } M.$$

A holomorphic tangential vector field is a C^∞ vector field F on U such that

(1) $F = \sum_{i=1}^n a_i\, \partial/\partial z_i$ where $a_i \in C^\infty(U)$, and

(2) the value of F at any point $Q \in M$ is an element of the holomorphic tangent space to M at Q.

We let $\mathscr{L}_0$ denote the module (over $C^\infty(U)$) spanned by tangential holomorphic vector fields and their conjugates. For each integer $\mu \geqq 1$ we let $\mathscr{L}_\mu$ denote the module of vector fields spanned by commutators of order $\leqq \mu$ of elements in $\mathscr{L}_0$.

1.2. DEFINITION. Let $P \in M$. Then P is of type m (m is an integer $\geqq 1$ or $+\infty$) if for every $G \in \mathscr{L}_{m-1}$, $\langle \partial r(P), G(P)\rangle = 0$ while for some $G \in \mathscr{L}_m$ we have $\langle \partial r(P), G(P)\rangle \neq 0$.

Here $\langle\ ,\ \rangle$ denotes contraction between a cotangent and a tangent vector.

AMS (MOS) subject classifications (1970). Primary 32F99.

© 1977, American Mathematical Society

We use the notation $t(P) = m$.

Let X be an $(n-1)$-dimensional complex submanifold of a neighbourhood of P tangent to M at P.

1.3. DEFINITION. We define X to be tangent to M at P to order S if r restricted to X vanishes to order $S + 1$ at P.

For S an integer $\geqq 1$ or $+\infty$ we use the notation $a(P) = S$ if there exists a complex $(n - 1)$-dimensional submanifold tangent to M at P to order S but none tangent to order $S + 1$. We write $a(P) = +\infty$ if either:

(1) there is a complex $(n - 1)$-dimensional submanifold tangent to M at P to order $+\infty$, or

(2) for every integer N however large, there is a complex $(n - 1)$-dimensional submanifold tangent to M at P to order N.

1.4. THEOREM. $a(P) = t(P)$.

This result may also be phrased as follows. A point $P \in M$ is of type m if and only if there exist local coordinates centered at P such that the defining function for M has the form

$$r = \operatorname{Re}(z_n) + \phi$$

where ϕ vanishes at P to order $\geqq 2$ and the lowest order of a term in the Taylor series for ϕ at P not involving z_n is $m + 1$. Furthermore it is not possible to find local coordinates such that the corresponding term in ϕ vanishes to order $> m + 1$.

Thus, a point on a hypersurface where the Levi form is nondegenerate is of type 1.

2. Generic submanifolds.

2.1. Let M be a generic real C^∞ submanifold of codimension $k < n$ in an open set $U \subset C^n$. Let $r_1, \cdots, r_k$ be C^∞ defining functions for M. That is

$$M = \{z \in U \mid r_1 = \cdots = r_k = 0\};$$

$dr_1 \wedge \cdots \wedge dr_k \neq 0$ on M and $\partial r_1 \wedge \cdots \wedge \partial r_k \neq 0$ on M.

With $\mathscr{L}_\mu$ as in 1.1, we let $\mathscr{L}_\mu(P)$ $(\mu = 0, 1, 2, \cdots)$ denote the vector space of values at P of vector fields in $\mathscr{L}_\mu$.

2.2. DEFINITION. We define P to be a point of type $(m_1, \cdots, m_k)$ where $m_1, \cdots, m_k$ are integers $\geqq 1$ or $+\infty$ if the following conditions hold:

(1) $\dim_C \mathscr{L}_\mu(P) = 2n - 2k$ for $0 \leqq \mu < m_1$,

(2) $\dim_C \mathscr{L}_\mu(P) = 2n - 2k + i$ for $m_i \leqq \mu < m_{i+1}$.

We note that $m_1 \leqq m_2 \leqq \cdots \leqq m_k$. Of course if $m_i = m_{i+1}$ then (2) is vacuous. We will use the notation $t(P) = (m_1, \cdots, m_k)$.

2.3. DEFINITION. A weighted coordinate system consists of the following:

(1) A local coordinate system $z_1, \cdots, z_{n-k}; w_1, \cdots, w_k$.

(2) The assignment of weights to the coordinate functions and their conjugates. Any z_i or $\bar{z}_i$ is assigned the weight 1 whereas w_i and $\bar{w}_i$ are assigned the weight α_i where α_i is an integer $\geqq 2$ or $+\infty$.

2.4. DEFINITION. Any monomial in $z, \bar{z}, w, \bar{w}$ is assigned the weight the sum of the weights of its factors. A C^∞ function ϕ is assigned the weight the minimum of the weights of the (nonzero) monomials which occur in its Taylor series expansion at P. We use the notation $\mathrm{wt}_P(\phi)$.

Consider a weighted coordinate system at P such that, for $i = 1, \cdots, k$,
(1) $r_i = \operatorname{Re}(w_i) + \phi_i$ where ϕ_i vanishes at P to order $\geqq 2$,
(2) $\operatorname{wt}_P(w_i) = \operatorname{wt}_P(r_i) = l_i + 1$.

Let $\mathfrak{A}$ denote the set of all $(l_1, \cdots, l_k)$ which arises in this manner with $l_1 \leqq \cdots \leqq l_k$ (i.e., we consider all possible sets of defining functions and weighted coordinates). Let $\mathfrak{N}$ denote the ordered set of all k-tuples $(n_1, \cdots, n_k)$ where n_i is an integer $\geqq 1$ or $+\infty$ and $n_1 \leqq \cdots \leqq n_k$ and the order relation is given by the rule $(n_1, \cdots, n_k) > (n_1', \cdots, n_k')$ if $n_i \geqq n_i'$ for all i, and for some value of i we have strict inequality.

2.5. CONJECTURE. *$\mathfrak{A}$ has a unique least upper bound in $\mathfrak{N}$. Furthermore denoting that least upper bound by $a(P)$ we have $a(P) = t(P)$.*

We have proved the following.

2.6. THEOREM. *Let $P \in M$ be of type $(m_1, \cdots, m_k)$. Then there exists (if $m_1 < +\infty$) $(l_1, \cdots, l_k) \in \mathfrak{A}$ with $l_1 = m_1$. For any $(l_1, \cdots, l_k) \in \mathfrak{A}$ we have $l_1 \leqq m_1$.*

3. Single vector fields.

3.1. Let M, r be as in 1.1. With $P \in M$ let L be a holomorphic tangential vector field such that $L(P) \neq 0$. Let $\mathscr{L}_0$ denote the $C^\infty(U)$-module spanned by L, $\bar{L}$, and $\mathscr{L}_\mu$ the module spanned by commutators of order $\leqq \mu$ of elements of $\mathscr{L}_0$.

3.2. DEFINITION. L is of type m at P if there exists $G \in \mathscr{L}_m$ such that $\langle \partial r(P), G(P) \rangle \neq 0$ while for every $G \in \mathscr{L}_{m-1}$ we have $\langle \partial r(P), G(P) \rangle = 0$.

We use the notation $t(L, P) = m$.

3.3. PROPOSITION. *Suppose these exists a one-dimensional complex submanifold X of a neighbourhood of P tangent to M at P to order S; then there exists a tangential holomorphic vector field L such that $L(P)$ is tangent to X at P and $t(L, P) \geqq S$.*

In particular if M contains a complex manifold passing through P then there exists an L with $t(L, P) = +\infty$.

The converse is not true, even if M is real analytic as the following example shows. Let $M \subset C^3$ be given by the defining function

$$r = 2\operatorname{Re}(z_3) + \tfrac{1}{2}(z_1^3\bar{z}_1 + \bar{z}_1^3 z_1) + \tfrac{3}{4}z_1^2\bar{z}_1^2 - z_2\bar{z}_2.$$

Let

$$L = 2\frac{\partial}{\partial z_1} + 2(z_1 + \bar{z}_1)\frac{\partial}{\partial z_2} - (3z_1^2\bar{z}_1 + \bar{z}_1^3 + 3\bar{z}_1^2 z_1 - 2z_1\bar{z}_2 - 2\bar{z}_1\bar{z}_2)\frac{\partial}{\partial z_3};$$

then $t(L, P) = +\infty$ where $P = (0, 0, 0)$.

The maximal order of tangency of a one-dimensional complex submanifold with M at P is 3.

REFERENCES

1. T. Bloom and I. Graham, *A geometric characterization of points of type m on real submanifolds of C^n*, J. Differential Geometry (to appear).

2. J. J. Kohn, *Boundary behaviour of $\bar{\partial}$ on weakly pseudo-convex manifolds of dimension two*, J. Differential Geometry **6** (1972), 523–542. MR **48** #727.

3. P. Greiner, *Subelliptic estimates for the* $\bar{\partial}$*-Neumann problem in* $\boldsymbol{C}^2$, J. Differential Geometry **9** (1974), 239–250. MR **49** #9441.

4. I. Naruki, *An analytic study of a pseudo-complex structure*, Proc. Internat. Conf. Functional Analysis and Related Topics (Tokyo, 1969), Univ. of Tokyo Press, Tokyo, 1970, pp. 72–82. MR **43** #4069.

UNIVERSITÉ DE PARIS VI

UNIVERSITY OF TORONTO

Proceedings of Symposia in Pure Mathematics
Volume 30, 1977

FACTORIZATION THEOREMS FOR HARDY SPACES OF COMPLEX SPHERES

R. R. COIFMAN, R. ROCHBERG AND GUIDO WEISS*

The purpose of this paper is to announce the extension to Hardy spaces in several variables certain of well-known factorization theorems on the unit disk. Proofs of these results and related results for Hardy spaces defined by Riesz systems on $\boldsymbol{R}^n$ are presented in **[1]**.

Let $B = B_n$ be the unit ball in complex n-space, $B_n = \{z = (z_1, \cdots, z_n) \in \boldsymbol{C}^n;$ $|z| < 1\}$. Denote the sphere by $\partial B = \partial B_n = \{z \in \boldsymbol{C}^n; |z| = 1\}$. Denote Lebesgue measure on the sphere by $d\sigma(z)$ and solid Lebesgue measure by $d\nu(z)$. The Hardy space $H^p(\partial B_n)$, $1 \leqq p < \infty$, is defined to be the space of functions F which are holomorphic on B_n and for which

$$\|F\|_p = \sup_{r<1} \left(\int_{\partial B_n} |F(rz)|^p \, d\sigma(z) \right)^{1/p} < \infty.$$

General references for the theory of these spaces are **[6]** and **[7]**.

THEOREM I. *Given F in $H^1(\partial B_n)$ there are G_i, H_i in $H^2(\partial B_n)$ such that*

$$F = \sum_{i=1}^{\infty} G_i H_i \quad \textit{and} \quad \sum_{i=1}^{\infty} \|G_i\|_2 \, \|H_i\|_2 \leqq c\|F\|_1.$$

By restricting this result to functions that do not depend on all of the variables, we obtain results on Hardy type spaces defined with respect to weighted solid Lebesgue measure. For $1 \leqq p < \infty$, $k = 0, 1, 2, \cdots$, let $H^p(B_n, (1 - |z|^2)^k d\nu)$ be the subspace of $L^p(B_n, (1 - |z|^2)^k d\nu)$ consisting of holomorphic functions.

THEOREM II. *Given F in $H^1(B_n, (1 - |z|^2)^k d\nu)$ there are G_i, H_i in $H^2(B_n, (1 - |z|^2)^k d\nu)$ such that*

AMS (MOS) subject classifications (1970). Primary 32A10, 44A25; Secondary 32A30, 47B47, 46E15.

*Research supported in part by the National Science Foundation grants MPS75–02411 and MPS75–06367.

© 1977, American Mathematical Society

$$F = \sum_{i=1}^{\infty} G_i H_i \quad \text{and} \quad \sum_{i=1}^{\infty} \|G_i\|_2 \|H_i\|_2 \leqq c\|F\|_1.$$

REMARK. The spaces $H^p(B_1, d\nu)$ are sometimes called Bergman spaces and denoted A^p. Horowitz has recently shown [**5**] that the strong version of the classical factorization theorem, i.e., that if F is in A^1 then $F = GH$ for G, H in A^2 and G nonvanishing, is *not* true.

For z, ζ in ∂B, let $d(z, \zeta) = |1 - \bar{\zeta}\cdot z|^{1/2}$ where $\bar{\zeta}\cdot z = \sum_{i=1}^{n} \bar{\zeta}_i z_i$. It is easy to verify that d is a metric on ∂B_n. Let $S_{\zeta,r}$ denote the sphere on ∂B of radius r in the d metric, $S_{\zeta,r} = \{z \in \partial B; d(z, \zeta) < r\}$ and let $|S|$ denote the $d\sigma$ measure of S. It is easy to see that $|S_{\zeta,r}| \simeq r^{2n}$ (for small r). The crucial property of these spheres is the inequality $|S_{\zeta,r}| > c|S_{\zeta,2r}|$. This property makes it possible to prove the standard covering lemmas and to develop a real variable version of H^1 theory. The fundamental objects in the real variable theory which is developed in [**2**] are atoms, functions $a(\zeta)$ defined on ∂B which are either identically one or are supported on a d sphere S and satisfy

$$|a(\zeta)| \leqq \frac{1}{|S|}, \qquad \int_{\partial B} a(\zeta)\, d\sigma(\zeta) = 0.$$

The holomorphic analogs of atoms are holomorphic atoms, that is functions $A(z)$ which are the holomorphic projection of atoms $a(\zeta)$, i.e.,

$$A(z) = c\int_{\partial B} \frac{a(\zeta)}{(1 - \bar{\zeta}\cdot z)^n}\, d\sigma(\zeta), \qquad |z| < 1.$$

A fundamental result is

THEOREM III. *Every F in $H^1(\partial B)$ can be written as $F = \sum_{i=1}^{\infty} \lambda_i A_i$ where the A_i are holomorphic atoms and the λ_i are complex scalars with $\sum|\lambda_i| \leqq c\|F\|_1$.*

A locally integrable function $b(z)$ on ∂B is said to be of bounded mean oscillation (with respect to the metric d) of

$$\sup_S \frac{1}{|S|}\int_S \left| b - \frac{1}{|S|}\int_S b \right| d\sigma = \|b\|_{\mathrm{BMO}} < \infty$$

where the supremum is over all spheres in the d metric. The space of all such functions is denoted BMO. Theorem III is equivalent to the result that the dual space of $H^1(\partial B)$ is the space of analytic functions in BMO. For $n = 1$ this is a result of C. Fefferman [**3**]. In fact, the proof of Theorem III is essentially the adaptation to the d metric of the proof given in Fefferman and Stein [**4**] of the analogous result in $\boldsymbol{R}^n$. Using Theorem III, the proof of Theorem II is reduced to proving the following proposition.

PROPOSITION IV. *Every holomorphic atom A can be written as $A = \sum_{i=1}^{N} B_i C_i$ with B_i, C_i in $H^2(\partial B)$ and $\sum\|B_i\|_2 \|C_i\|_2 \leqq C$ and both C and N depend only on the dimension n.*

In proving Proposition IV we may assume without loss of generality that an atom a is supported in a sphere S centered at $\mathbf{1} = (0, 0, \cdots, i)$ (the "North pole"). We let $\varphi_s = \chi_S/|S|$ and write

$$1 - \bar{w}\cdot z = -\bar{w}_0\cdot z_0 + (1 + iz_n)(i\bar{w}_n) + (1 - i\bar{w}_n)$$

for $z = (z_1, z_2, \cdots, z_n)$, $w = (w_1, w_2, \cdots, w_n) \in \partial B_n$, where $z_0 = (z_1, z_2, \cdots, z_{n-1}, 0)$ and $w_0 = (w_1, w_2, \cdots, w_{n-1}, 0)$. We have

$$(1 - \bar{w} \cdot z)^n = \sum_{|J|+k+l=n} d_{J,k,l}\, \bar{w}_0^J(1 - i\bar{w}_n)^k\, \bar{w}_n^l z_0^J(1 + iz_n)^l,$$

where $J = (j_1, \cdots, j_{n-1})$, $|J| = j_1 + j_2 + \cdots + j_{n-1}$ and $z_0^J = z_1^{j_1} z_2^{j_2} \cdots z_{n-1}^{j_{n-1}}$. We write

$$1 = \int_{\partial B_n} \varphi_s(w)\, dw = \sum_{|J|+k+l=n} z_0^J(1 + iz_n)^l\, D_{J,k,l}(z),$$

where

$$D_{J,k,l}(z) = d_{J,k,l} \int_{\partial B_n} \frac{\bar{w}_0^J(1 - i\bar{w}_n)^k\, \bar{w}_n^l\, \varphi_S(w)}{(1 - z \cdot \bar{w})^n} dw.$$

Thus,

$$A(z) = \sum_{|J|+k+l=n} z_0^J(1 + iz_n)^l\, D_{J,k,l}(z) A(z).$$

The proof is completed by showing that each summand can be split into a product $B(z)C(z)$ with $\|B\|_2\, \|C\|_2 \leqq K$ with K a constant not depending on the atom a. This is done by making explicit estimates on the size of $A(z)$ and $D_{j,k,l}(z)$ using standard singular integral estimates on the Szegö kernel; see **[1]**, **[2]**.

Theorem I is closely related to a general result concerning commutators of singular integral operators and multiplication operators. One instance of this general result is the following. Let P be the projection of $L^p(\partial B_n)$ to boundary values of functions in $H^p(\partial B_n)$ given by

$$(Pf)(z) = \tfrac{1}{2} f(z) + c_n \text{P.V.} \int_{\partial B_n} \frac{f(\zeta)}{(1 - \bar{\zeta} \cdot z)^n}\, d\sigma(\zeta).$$

THEOREM V. *Let $b(z)$ be a locally integrable function on ∂B_n. Let B be the operator multiplication by b. The commutator operator $BP - PB$ is a bounded map of $L^2(\partial B_n)$ to $L^2(\partial B_n)$ if and only if b is in BMO. In this case the operator norm of $BP - PB$ is equivalent to the BMO norm of b.*

REFERENCES

1. R. R. Coifman, R. Rochberg and Guido Weiss, *Factorization theorems for Hardy spaces in several variables*, Ann. of Math. **103**(1976).

2. R. R. Coifman and G. Weiss, *H^p spaces and harmonic analysis*, Bull. Amer. Math. Soc. (to appear).

3. C. Fefferman, *Characterizations of bounded mean oscillation*, Bull. Amer. Math. Soc. **77** (1971) 587–588. MR **43** #6713.

4. C. Fefferman and E. M. Stein, *H^p spaces of several variables*, Acta Math. **129** (1972), 137–193.

5. C. Horowitz, *Zeros of functions in the Bergman spaces*, Bull. Amer. Math. Soc. **80** (1974), 713–714. MR **49** #3164.

6. A. Korányi and S. Vági, *Singular integrals on homogeneous spaces and some problems of classical analysis*, Ann. Scuola Norm. Sup. Pisa Cl. Sci. **25** (1971), 575–648.

7. E. M. Stein, *Boundary behavior of holomorphic functions of several complex variables*, Princeton Univ. Press, Princeton, N. J., 1972.

WASHINGTON UNIVERSITY, ST. LOUIS

Proceedings of Symposia in Pure Mathematics
Volume 30, 1977

GEVREY REGULARITY UP TO THE BOUNDARY FOR THE $\bar{\partial}$ NEUMANN PROBLEM

M. DERRIDJ

I. Introduction. Let Ω be a strictly pseudoconvex domain in C^n. In [4], J. J. Kohn considered the problem

$$(1) \qquad (\bar{\partial}\bar{\partial}^* + \bar{\partial}^*\bar{\partial})u = \square\, u = f, \qquad f \in L^2(\Omega),$$

where $\bar{\partial}^*$ is the adjoint of $\bar{\partial}$.

He showed that if $\partial\Omega$ is in C^∞ class and f is a (p, q) form, $q \geqq 1$, C^∞ up to the boundary in Ω, then a solution u of (1) which is in Dom ($\square$) is also C^∞ up to the boundary.

At the interior, this problem was also solved in [3]. We wish to know further regularity up to boundary for u, when we know further regularity in $\bar{\Omega}$ for f.

In the interior of Ω, we know that if f is real analytic then u is also real analytic (theory of elliptic equations). But at the boundary we have only subellipticity.

Here, we prove that if f is a (p, q) form, $q \geqq 1$, which is in the class G^2 in $\bar{\Omega}$ (i.e., up to the boundary), where $\partial\Omega$ is defined by a function r which is in G^2 class, then u is also in the class G^2 in $\bar{\Omega}$.

This result was proved by D. Tartakoff when $q \geqq 2$ [5]. In general fashion we use a method from Morrey and Nirenberg [6] and some other techniques [1]. Note that for $q = 1$ then $\bar{\partial}^* u$ is a function and we know that Kohn's inequality fails for functions.

II. Definitions and notations. We recall first some notations about $\bar{\partial}$ which are in the book of G. B. Folland and J. J. Kohn [2]. The space of (p, q) forms in $C^\infty(\bar{\Omega})$ is denoted by $\Lambda^{p,q}(\bar{\Omega})$.

If $u, v \in \Lambda^{p,q}(\bar{\Omega})$, i.e.,

$$u = \sum_{|I|=p;|J|=q} u_{IJ} dz_I \wedge d\bar{z}_J, \qquad v = \sum,$$

AMS (MOS) subject classifications (1970). Primary 35H05, 35N15, 35B99.

© 1977, American Mathematical Society

then $(u, v) = \sum(u_{IJ}, v_{IJ})_{L^2(\Omega)}$, $\|u\|^2 = (u, u)$. So the adjoint $\bar\partial^*$ of $\bar\partial$ is defined by:
$v \in \mathrm{Dom}(\bar\partial^*) \Leftrightarrow \exists\, C > 0$ such that $|(v, \bar\partial u)| \leq C\|u\|$,
$u \in \Lambda^{p,q}(\bar\Omega)$, and
$(\bar\partial^* v, u) = (v, \bar\partial u)$, $u \in \Lambda^{p,q}(\bar\Omega)$, $v \in \mathrm{Dom}(\bar\partial^*)$.
Note $D^{p,q} = \Lambda^{p,q}(\bar\Omega) \cap \mathrm{Dom}(\bar\partial^*)$.

Similarly, we may define (see [2]) the same things when we restrict to a neighborhood $V \cap \bar\Omega$ of a point $P \in \Omega$, in $\bar\Omega$.

Now, we recall the basic estimate of J. J. Kohn

$$(2.1)\quad |||Du|||^2_{-1/2} \leqq C(\|\bar\partial u\|^2 + \|\bar\partial u\|^2 + \|u\|^2), \qquad u \in D^{p,q} \cap \Lambda_0^{p,q}(V \cap \bar\Omega)$$

where $\|\ \ |||_s$ denotes the s-Sobolev norm in the tangential direction. Locally, we denote by x the tangential variables and by y the normal variable.

Let us denote

$$Q(u, v) = (\bar\partial u, \bar\partial v) + (\bar\partial^* u, \bar\partial^* v) = (u, v).$$

Gevrey classes. If Ω is an open set in $\boldsymbol{R}^N$, we define $G^s(\Omega)$ for $s \geqq 1$ by $u \in G^s(\Omega) \Leftrightarrow \forall K \subset\subset \Omega$, $\exists C_k$ such that $\sup_k |D^\alpha u| \leqq C^{|\alpha|+1}|\alpha|!^s$ where

$$D^\alpha = \partial^{|\alpha|}u/\partial x_1^{\alpha_1}\cdots\partial x_N^{\alpha_N}, \qquad \alpha = (\alpha_1, \cdots, \alpha_N).$$

If K is a compact set, then we say that u is in $G^S(K)$ if us extends to $\tilde u \in G^S(\tilde\Omega)$, $\tilde\Omega$ being a neighborhood of K. Now if $\Omega \subset \boldsymbol{C}^n$, we say that has the G^S regularity if Ω is defined by a G^S function r.

THEOREM 3.1. *Let Ω be a strictly pseudoconvex domain in $\boldsymbol{C}^n$ such that $\partial\Omega$ has the G^2 regularity. Let $f \in L^2_{p,q}(\Omega)$ and u be a solution of $\Box u = f$ in $D(\Box)$.*
If f is in the class $G^2(\bar\Omega)$, then u is also in $G^2(\bar\Omega)$.

REMARKS. (1) The theorem is true for $s \geqq 2$.

(2) The theorem is in fact local (i.e., if f has the G^S regularity, $s \geqq 2$ near a point $P \in \partial\Omega$, then u has the G^s regularity near this point). We have written the theorem in the global form, for simplicity.

SKETCH OF THE PROOF. We have the following proposition:

PROPOSITION 3.2. *Let V_0 be a small neighborhood of $P \in \partial\Omega$. Then there exists a constant C_0 such that:*

$$(3.1)\quad \|\bar\partial u\|^2 + \|\bar\partial^* u\|^2 + |||Du|||^2_{-1/2} \leqq C_0\, Q(u, u), \qquad u \in D^{p,q} \cap \Lambda_0^{p,q}(V_0 \cap \bar\Omega),$$

$$(3.2)\quad |||\bar\partial u|||^2_{1/2} + |||\bar\partial^* u|||^2_{1/2} + \|u\|_1^2 \leqq C_0\{Q(v, u) + Q(\psi T_{1/2} u, \psi T_{1/2} u)\}$$

where ψ is a function in $D(V \cap \bar\Omega)$, $\psi = 1$ on $V_0 \cap \bar\Omega$, V a neighborhood of $\bar V_0$, and $T_{1/2}$ is the pseudodifferential operator in the x variables whose symbol is $(1 + |\xi|^2)^{1/4}$.

The proposition is a consequence of Kohn's inequality. In fact (3.1) is Kohn's inequality. And (3.2) is obtained by applying (3.1) to $\psi T_{1/2} u$, using the fact that a tangential operator keeps invariant $D^{p,q}$, and a property on brackets of pseudodifferential operators.

Before the next proposition, we need some notations.

Let V_0 be a cube and $\omega = V_0 \cap \bar\Omega$. For $\varepsilon > 0$ small we define $V_\varepsilon = \{z \in V_0, d(z, (V_0)) > \varepsilon\}$ and $\omega_\varepsilon = V_\varepsilon \cap \bar\Omega$. We suppose, for simplicity, that $\omega_\varepsilon = \varphi$ if $\varepsilon > 1$.

Now, if ε and ε_1 are two small numbers, $\varepsilon > 0$, $\varepsilon_1 > 0$, $\varepsilon + \varepsilon_1 \leqq 1$, there exists a function $\varphi_{\varepsilon,\varepsilon_1}$ such that

(a) $\varphi_{\varepsilon,\varepsilon_1} \in D(\omega_{\varepsilon_1})$, $0 \leqq \varphi_{\varepsilon,\varepsilon_1} \leqq 1$,

(b) $\varphi_{\varepsilon,\varepsilon_1} = 1$ on $\omega_{\varepsilon+\varepsilon_1}$,

(c) $\sup_{\omega\varepsilon_1}|D^\alpha \varphi_{\varepsilon,\varepsilon_1}| \leqq C_\alpha\, \varepsilon^{-|\alpha|}$, C_α independent of ε and ε_1.

We denote by $\|u\|_{s,\omega_\varepsilon}$ the s-Sobolev norm in ω_ε and by $|||u|||_{s,\omega_\varepsilon}$ the s-tangential Sobolev norm in ω_ε.

Then we have the main proposition.

PROPOSITION 3.3. *There exists a constant $B > 0$ such that, for any integer j, any number ε, with $j\varepsilon \leqq 1$, we have*

$$(3.4)\qquad \varepsilon^{2(2|\alpha|+2)}\{\|\bar\partial D_x^\alpha u\|^2_{\omega_{j\varepsilon}} + \|\bar\partial^* D_x^\alpha u\|^2_{\omega_{j\varepsilon}} + |||DD_x^\alpha u|||^\varepsilon_{-1/2,\omega_j}\} \leqq B^{2(|\alpha|+1)},$$

$$(3.5)\qquad \varepsilon^{2(2|\alpha|+3)}\{|||\bar\partial D_x^\alpha u|||^2_{1/2,\omega_{j\varepsilon}} + |||\bar\partial^* D_x^\alpha u|||^2_{1/2,\omega_{j\varepsilon}} + \|D_x^\alpha u\|^2_{1,\omega_{j\varepsilon}}\} \leqq B^{2(|\alpha|+1)}$$

for any α such that $|\alpha| \leqq j - 1$.

OUTLINE OF THE PROOF. We suppose (3.4) and (3.5) true for j and we show that if B is suitable (not depending on j) then we have (3.4) and (3.5) for $j + 1$. For that we consider $\varphi_{\varepsilon,j\varepsilon}$ and from the properties (3.3) it is sufficient to estimate, if we note $\varphi = \varphi_{\varepsilon,j\varepsilon}$,

$$\|\bar\partial\varphi D_x^\alpha u\|^2 + \|\bar\partial^*\varphi D_x^\alpha u\|^2 + |||D\varphi D_x^\alpha u|||^2_{-1/2},$$

if we want to prove (3.4) at the $(j + 1)$ step.

For this, by applying inequality (3.1) to the form $\varphi D_x^\alpha u$ (which we can do because φD_x^α is a tangential operator), it remains to estimate the expression $Q(\varphi D_x^\alpha u, \varphi D_x^\alpha u)$.

In the proof, one has to be very careful to make suitable commutations and integrations by parts. We omit the details which are long. It is important to keep the function φ in all the proof; in fact we prove precisely the following:

There exists a constant $C_0 > 0$ such that, for any $\lambda > 0$ we have

$$(3.6)\qquad \begin{aligned}&\varepsilon^{2(2|\alpha|+2)}\{\|\bar\partial\varphi D_x^\alpha u\|^2 + \|\bar\partial^*\varphi D_x^\alpha u\|^2 + |||D\varphi D_x^\alpha u|||^2_{-1/2}\}\\ &\leqq C_0\,\lambda\, \varepsilon^{2(2|\alpha|+2)}(\|\bar\partial\varphi D_x^\alpha u\|^2 + \|\bar\partial^*\varphi D_x^\alpha u\|^2 + |||D\varphi D_x^\alpha u|||^2_{-1/2}) + (C_0/\lambda)B^{2|\alpha|}\end{aligned}$$

and

$$(3.7)\qquad \begin{aligned}&\varepsilon^{2(2|\alpha|+3)}\{|||\bar\partial\varphi D_x^\alpha u|||^2_{1/2} + |||\bar\partial^*\varphi D_x^\alpha u|||^2_{1/2} + \|\varphi D_x^\alpha u\|^2_1\}\\ &\leqq C_0\lambda\, \varepsilon^{2(2|\alpha|+3)}(|||\bar\partial\varphi D_x^\alpha u|||^2_{1/2} + |||\bar\partial^*\varphi D_x^\alpha u|||^2_{1/2} + \|\varphi D_x^\alpha u\|^2_1) + (C_0/\lambda)B^{2|\alpha|}.\end{aligned}$$

Now, by choosing λ suitably ($\lambda = 1/2C_0$) we obtain the inequalities (3.4) and (3.5).

COROLLARY 3.4. *There exists a constant $B_1 > 0$ such that (if $\omega_c \subset \omega$)*

$$(3.8)\qquad \|D_x^\alpha u\|_{1,\omega_c} \leqq B_1^{|\alpha|+2}(|\alpha| + 1)!^2.$$

PROOF. (3.8) follows from (3.5) if we choose ε such that $j\varepsilon = c$ (here $\omega_c \subset \omega$, $c \leqq 1$).

Corollary 3.4 says that we have a good estimation of the tangential derivatives.

PROPOSITION 3.5. *There exists a constant $B_2 > 0$ such that*

$$\left\| D_y^\beta D_x^\alpha u \right\|_{\omega_\varepsilon} \leqq B_2^{|\alpha|+|\beta|+1} \beta!^2 |\alpha|!^2. \tag{3.9}$$

PROOF. We make an induction on β. (3.5) says that it is true for $\beta \leqq 1$ and α any multi-index. To make the induction we return to the equation $\square u = f$ in the local coordinate and use that $y = 0$ is not characteristic for $\square$. The proof is finished if we recall that (3.9) implies that

$$\sup_{\omega_\varepsilon} \left| D_y^\beta D_x^\alpha u \right| \leqq B_3^{|\alpha|+|\beta|+1} \beta!^2 |\alpha|!^2,$$

by the Sobolev theorem.

REFERENCES

1. M. Derridj and C. Zuily, *Regularité Gevrey pour les opérateurs de L. Hörmander*, J. Math. Pures Appl. **52** (1973).
2. G. B. Folland and J. J. Kohn, *The Neumann problem for the Cauchy-Riemann complex*, Ann. of Math. Studies, Princeton Univ. Press, Princeton, N. J., 1972.
3. L. Hörmander, *L^2-estimates and existence theorems of the $\bar{\partial}$ operator*, Acta Math. **113** (1965), 89–152. MR **31** #3691.
4. J. J. Kohn, *Harmonic integrals on strongly pseudoconvex manifolds.* I, Ann. of Math (2) **78** (1963), 112–148. MR **27** #2999.
5. D. Tartakoff, *Gevrey hypoellipticity for subelliptic boundary value problems*, Comm. Pure Appl. Math. **26** (1973), 251–312. MR **49** #7586.
6. G. B. Morrey and L. Nirenberg, *On the analyticity of the solutions of linear elliptic systems of partial differential equations*, Comm. Pure Appl. Math. **10** (1957), 271–290. MR **19**, 654

UNIVERSITÉ DE PARIS-SUD

Proceedings of Symposia in Pure Mathematics
Volume 30, 1977

SOME RECENT DEVELOPMENTS IN THE THEORY OF THE BERGMAN KERNEL FUNCTION: A SURVEY

KLAS DIEDERICH

I. Basic definitions and facts.

1. The Bergman kernel function of complex analysis can be looked at as a special example of a very general theory of function spaces with reproducing kernels (cf. Aronszajn [1]). One possible definition for the general situation is:

DEFINITION. Let X be a set and H a separable $\mathbf{C}$-Hilbert space of functions on X with scalar product $(\cdot, \cdot)$ and norm $\|\cdot\|$. H is called a function space with reproducing kernel on X if

(R) for all $x \in X$ the evaluation functional $K_x: H \to \mathbf{C}$, $x \mapsto f(x)$, is bounded on H.

This becomes somewhat more concrete if we make the stronger assumption

(R′) X is a Hausdorff space, $H \subset C(X)$, and for every compact subset $M \subset X$ there is a constant $c_M > 0$ such that $\|K_x\| \leqq c_M$ for all $x \in M$.

If H is a function space with reproducing kernel, the Riesz representation theorem gives:

for all $x \in X$ there exists $K_x(y) \in H$ such that $f(x) = K_x f = (f(y), K_x(y))$.

DEFINITION 1.1. The (uniquely determined) function

$$K(y, x) = K_x(y) : X \times X \to \mathbf{C}$$

is called the reproducing kernel of X with respect to H.

From these definitions, it follows quite easily that the reproducing kernel has the following properties:

PROPOSITION 1.1. (a) *K is hermitian, i.e., $K(y, x) = \overline{K(x, y)}$ for all $x, y \in X$.*

AMS (MOS) subject classifications (1970). Primary 32H10; Secondary 32H15, 32F15.

© 1977, American Mathematical Society

(b) $K(x) = K(x, x) = \|K(\cdot, x)\|^2 = \|K_x\|^2$ *for all* $x \in X$.

(c) If $\{\phi_n\}_1^\infty$ *is an orthonormal basis of* H, *then one has, for all* $x \in X$, $K(\cdot, x) = \sum_{n=1}^\infty \overline{\phi_n(x)}\, \phi_n$, *the series being convergent with respect to* $\|\cdot\|$. *As a series of functions in both variables* $(x, y) \in X \times X$ *it is uniformly convergent on every compact subset* $M \subset X \times X$.

2. The following two additional points of view shall be mentioned here because of their importance for the Bergman kernel:

(a) If for every $x \in X$ there is a function $f \in H$ with $f(x) \neq 0$, one has

$$1/K(x, x) = \min \{\|f\|^2 \mid f \in H, f(x) = 1\}$$

and if we denote by $N_x(y)$ the solution of this "variational" problem, we have $K(y, x) = N_x(y)/\|N_x\|^2$.

(b) $K(y, x)$ can be looked at as defining a mapping $K: X \to H$, $x \mapsto K_x$. If H separates the points of X, K is injective. Let $\pi_1 : H - \{0\} \to S(H) = \{f \in H \mid \|f\| = 1\}$ be defined by $f \to f/\|f\|$. Then by setting

$$\delta(x, y) := \inf_{\alpha, \beta \in \mathbf{R}} \{\|e^{i\alpha}\pi_1(K_x) - e^{i\beta}\pi_1(K_y)\|\} \tag{1.1}$$

we get a pseudometric δ on X.

Another pseudometric on X can be defined as follows. Call π_2 the projection of H onto its projective space PH. On PH one has the differential metric induced by the differential form on $S(H)$

$$d\sigma^2 = \sum d\zeta^i d\bar{\zeta}^i - |\sum \bar{\zeta}^j d\zeta^j|^2.$$

Now one defines

$$\Delta(x, y) = \sigma(\pi_2(K_x), \pi_2(K_y)) \tag{1.2}$$

(cf. Kobayashi [**13**]).

3. One of the most important cases, to which the general theory of function spaces with reproducing kernels applies, is the case of square-integrable holomorphic functions on bounded domains in $\mathbf{C}^n$.

DEFINITION 1.2. Let $\Omega \subset \mathbf{C}^n$ be a bounded domain; we write

$$\mathscr{H}^2(\Omega) = \left\{ f \text{ holomorphic on } \Omega \,\middle|\, \int_\Omega |f|^2 \, d\lambda < \infty \right\}$$

($d\lambda$ = Lebesgue measure on Ω).

It is easy to prove

LEMMA. *For* $z \in \Omega$ *and* $f \in \mathscr{H}^2(\Omega)$ *one has*

$$|f(z)| \leqq (c_n/d^n(z)) \|f\|_\Omega$$

with $\|f\|_\Omega = (\int_\Omega |f|^2 \, d\lambda)^{1/2}$ *and* $d(z) = \operatorname{dist}(z, b\Omega)$ *and some constant* c_n *depending only on* n.

An immediate consequence is:

PROPOSITION 1.2. $\mathscr{H}^2(\Omega)$ *is a separable Hilbert space with respect to* $(f, g)_\Omega = \int_\Omega f\bar{g}\, d\lambda$ *and is a function space with reproducing kernel* $K_\Omega(z, w)$ (*Bergman kernel of* Ω).

Some properties of the Bergman kernel are:

PROPOSITION 1.3. (a) *$K_\Omega(z, w)$ is holomorphic in z and antiholomorphic in w, $K_\Omega(z) = K_\Omega(z, z) > 0$ for all $z \in \Omega$.*

(b) *The orthogonal projection $\kappa : L^2(\Omega) \to \mathscr{H}^2(\Omega)$ is given by*

$$\kappa(f)(w) = \int_\Omega f(z)\, \overline{K_\Omega(z, w)}\, d\lambda_z.$$

(c) *If $z \in \Omega_1 \subset \Omega_2 \subset\subset \boldsymbol{C}^n$, one has $K_{\Omega_1}(z) \geqq K_{\Omega_2}(z)$.*

(d) *If $\Phi\colon \Omega_1 \to \Omega_2$ is a biholomorphic map, one has*

$$K_{\Omega_1}(z, w) = K_{\Omega_2}(\Phi(z), \Phi(w)) \det \Phi'(z)\, \overline{\det \Phi'(w)}.$$

REMARK. The transformation rule in Proposition 1.3(d) shows that one can think of the Bergman kernel function as the coefficient of an $(n, 0) \times (0, n)$-form on $\Omega \times \Omega$ and this is the way in which the theory can be generalized to certain complex manifolds.

Next, we want to ask what we can say about the two pseudometrics defined in (1.1) and (1.2) in the case of the space $\mathscr{H}^2(\Omega)$.

PROPOSITION 1.4. (a) *For $\mathscr{H}^2(\Omega)$ both δ_Ω and Δ_Ω become metrics on Ω, which are invariant under biholomorphic mappings.*

(b) *The function $\log K_\Omega(z)$ is strictly plurisubharmonic on Ω, i.e., the form*

$$ds^2 = \sum \frac{\partial^2 \log K_\Omega(z)}{\partial z_i\, \partial \bar{z}_j}\, dz_i\, d\bar{z}_j$$

is a positive definite hermitian differential metric on Ω (Bergman metric). The induced metric is Δ_Ω.

REMARK. It is not known whether in general every isometry of the Bergman metric is biholomorphic.

Propositions 1.3(b) and (d) and 1.4 give the basic properties that are the main reasons why the Bergman kernel function has been studied in complex analysis. In particular, the hope has always been that a careful study of the geometry of the Bergman metric would give some insight into the equivalence problem for bounded domains with respect to biholomorphic mappings. And, in fact, a well-known proof shows that the hyperball and the polycylinder in $\boldsymbol{C}^n$, $n \geqq 2$, are not biholomorphically equivalent by using the explicitly known Bergman metrics of the two domains. In 1974 M. Skwarczynski **[15]** proved, by using the metric δ:

PROPOSITION 1.5. *Let $\Omega_1, \Omega_2 \subset\subset \boldsymbol{C}^n$ be domains with complete Bergman metrics. Assume there is a holomorphic mapping $\phi\colon V_1 \to V_2$ of an open subset $V_1 \subset \Omega_1$ into an open subset $V_2 \subset \Omega_2$, such that the transformation rule for K_{Ω_1} and $K_{\Omega_2}(\phi(\cdot), \phi(\cdot))$ is satisfied on $V_1 \times V_1$. Then ϕ can be extended to a biholomorphic map Φ from Ω_1 onto Ω_2.*

On the other hand, in only a "few" cases is it possible to determine the Bergman kernel function explicitly; and, in general, almost nothing is known about how properties of the Bergman kernel K_Ω in the interior of Ω depend on Ω. This is one of the main reasons that one has to study the boundary behavior of the kernel function and related objects with the aim of expressing the limits, which one eventually gets, in terms of differential-geometric objects on $b\Omega$ (if $b\Omega$ is smooth).

Another reason for this can be that one wants to study the properties of the projection operator $\kappa : f \to \int f(z)\, \overline{K(z, w)}\, d\lambda_z$ of Proposition 1.3(b). In this case, however, it is not quite clear what the right order of argumentation finally will be: to study κ by using results on the boundary behavior of K or, vice versa, to study κ first and then use it to describe K.

II. The boundary behavior of the Bergman kernel.

1. *Basic results.* Since the first papers of St. Bergman on this subject, the boundary behavior of the kernel has been studied for several classes of domains. The best results are known for strictly pseudoconvex domains with smooth boundaries (and, of course, in those cases, where the kernel can be explicitly computed). Let us restrict our attention here to strictly pseudoconvex domains with smooth boundary.

The first results in this direction are contained in Bergman [**2**]. But they are valid only under strong additional geometric hypotheses on Ω. The first general results were

THEOREM 2.1. *Let $\Omega \subset\subset \mathbf{C}^n$ be a strictly pseudoconvex domain with smooth boundary and ψ a strictly plurisubharmonic defining function for $b\Omega$. Then one has*

(a) *Hörmander* (1965), [**11**]:

$$\lim_{z \to z_0 \in b\Omega} \psi^{n+1} K_\Omega(z) = \frac{n!}{\pi^n} \det \begin{pmatrix} \psi_1 & & & \\ \vdots & & & \\ \psi_n & \varphi_{i\bar j} & & \\ \psi & \psi_{\bar 1} & \cdots & \varphi_{\bar n} \end{pmatrix} (z_0).$$

(*In the following, this determinant is always denoted by Δ_ψ.*)

(b) *Diederich* (1970, 1973), [**6**]: *Analogous formulas for the first and mixed second derivatives of K_Ω and the coefficients of the Bergman metric hold.*

Both proofs proceed essentially as follows:

(1) A localization theorem is proved saying that the boundary behavior of certain objects (e.g., K_Ω) depends only on the shape of $b\Omega$ near the point $z_0 \in b\Omega$.

(2) The boundary $b\Omega$ is approximated at z_0 by a biholomorphic image of the hyperball.

(3) The objects aie explicitly computed for the hyperball.

Hörmander proves (1) by using his $\bar\partial$ methods. Diederich uses an approximation theorem for $\mathscr{H}^2$ that follows from classical Stein theory. The investigation of the derivatives of K_Ω is reduced to the study of certain closed subspaces of $\mathscr{H}^2$ and their reproducing kernels.

The next large progress, in a slightly different direction, was made by N. Kerzman in [**12**]. His main result is

THEOREM 2.2. *Let Ω be as above. Then $K_\Omega(z, w) \in C^\infty(\bar\Omega \times \bar\Omega - D)$ with $D = \{(z,w) \in \partial\Omega \times \partial\Omega \mid z = w\}$.*

The methods in the proof of N. Kerzman are totally different from those used to prove Theorem 2.1. The main tool is the following formula of J. J. Kohn for the projection operator κ:

$$\kappa f = f - \bar\partial^* N \bar\partial f. \tag{2.1}$$

Here $N: L^2_{(0,1)}(\Omega) \to L^2_{(0,1)}(\Omega)$ is the $\bar\partial$ Neumann operator of J. J. Kohn and $\bar\partial^*$ is the adjoint of $\bar\partial$, i.e., $\bar\partial^*(\sum_1^n f_j d\bar z_j) = -\sum_1^n(\partial f_i/\partial z_j)$. (For details on these operators see also [10].) The proof of N. Kerzman gives the following statement.

PROPOSITION. *If $\Omega \subset\subset \mathbf{C}^n$ is an arbitrary (weakly) pseudoconvex domain with smooth boundary, such that* (2.1) *holds, and if N is pseudolocal on Ω, then $K_\Omega(z, w) \in C^\infty(\bar\Omega \times \bar\Omega - D)$.*

Since strictly pseudoconvex domains $\Omega \subset\subset \mathbf{C}^n$ with smooth boundary satisfy the hypothesis of this proposition, this gives also Theorem 2.2.

2. *The asymptotic formula of Ch. Fefferman.* After these results were obtained, a critical point in the study of the kernel function was reached. Theorem 2.1(a) says that, for any defining function ψ of $b\Omega$, the function

$$L_\psi(z) := (-\psi)^{n+1}(z)K_\Omega(z)$$

is continuous up to the boundary and the results of (b) are a hint for $L_\psi(z)$ to be even C^2 up to the boundary. This led to the conjecture

$$L_\psi \in C^\infty(\bar\Omega). \tag{2.2}$$

Furthermore, I. Naruki showed in [12] how this would have the important consequence that biholomorphic mappings between strictly pseudoconvex domains with smooth boundary extend smoothly to the boundary. Unfortunately, the situation is much more complicated, as can be seen from

THEOREM 2.3 (CH. FEFFERMAN, 1974). *There are functions $\Phi, \tilde\Phi \in C^\infty(\bar\Omega)$ such that*

$$L_\psi(z) = (-\psi)^{n+1}(z)K_\Omega(z) = \Phi(z) + \tilde\Phi(z)(-\psi)^{n+1}(z)\log(-\psi(z)),$$
$$\Phi\big|\partial\Omega = c_n\Delta(\psi)\big|\partial\Omega,$$

and, in general, one has $\tilde\Phi\big|\partial\Omega \neq 0$.

REMARK. (a) One always has $L_\psi \in C^n(\bar\Omega)$, but it can happen that $L_\psi \notin C^{n+1}(\bar\Omega)$.

(b) Fefferman's full result is stronger, because he also describes what happens to $K_\Omega(z, w)$ if z and w approach independently from one another the same point $z_0 \in b\Omega$.

(c) Recently, L. Boutet de Monvel and J. Sjöstrand have slightly improved the result of Ch. Fefferman and given an independent new proof by using their techniques of Fourier integral operators with complex-valued phase functions.

The proof involves a very complicated machinery of integral operators, especially developed for this purpose. But the main procedure is as follows:

Fix a point $w \in \Omega$ near $b\Omega$ and call p the projection of w on $b\Omega$. Approximate $b\Omega$ at p up to third order by a biholomorphic image $\tilde\Omega$ of the hyperball. We assume for simplicity that $\tilde\Omega \subset \Omega$. Now we can write the Dirac δ-function with center w as

$$\delta_w = K_{\tilde\Omega}(\cdot, w) + (\delta_w - K_{\tilde\Omega}(\cdot, w)) \quad \text{on } \tilde\Omega$$
$$\Rightarrow \delta_w = K_{\tilde\Omega}(\cdot, w)\cdot\chi_{\tilde\Omega} + (\delta_w - K_{\tilde\Omega}(\cdot, w))\,\chi_{\tilde\Omega} \quad \text{on } \Omega$$

and because $\tilde\Omega \subset \Omega$, one has

$$K^+ = (\delta_w - K_{\tilde\Omega}(\cdot, w))\,\chi_{\tilde\Omega} \in \mathscr{H}^{2\perp}(\Omega).$$

Now, the important observation is that the known explicit formula for the Bergman kernel of the hyperball shows that its pull-back to $\tilde\Omega$, $K_{\tilde\Omega}(\cdot, w)$, continues analytically across $b\tilde{\Omega}$, and since $\tilde{\Omega}$ approximates Ω near p it extends to Ω near p. If we make the additional simplification that $K_{\tilde{\Omega}}(\cdot, w)$ extends to all of Ω, we get

$$\delta_w + K_{\tilde{\Omega}}(\cdot, w)\, \chi_{\Omega-\tilde{\Omega}} = K_{\tilde{\Omega}}(\cdot, w) + K^+(\cdot, w).$$

Now let us interpret the terms of this equality as operators on $L^2(\Omega)$:

$$I + R = K_{\tilde{\Omega}} + K^+.$$

We have $K_{\tilde{\Omega}} f \in \mathscr{H}^2(\Omega)$, $K^+ f \in \mathscr{H}^{2\perp}(\Omega)$. R, $K_{\tilde{\Omega}}$, K^+ are "explicitly" known. Furthermore, R is small in the operator norm because $\Omega - \tilde{\Omega}$ is very thin near p. Therefore we get, for $f \in L^2(\Omega)$,

$$\begin{aligned} f &= (I + R)(I + R)^{-1} f \\ &= (K_{\tilde{\Omega}} - K_{\tilde{\Omega}} R + K_{\tilde{\Omega}} R^2 - \cdots) f \\ &\quad + (K^+ - K^+ R + K^+ R^2 - \cdots) f = f_1 + f_2 \end{aligned}$$

with $f_1 \in \mathscr{H}^2(\Omega)$ and $f_2 \in \mathscr{H}^{2\perp}(\Omega)$. This implies that

$$K_\Omega = K_{\tilde{\Omega}} - K_{\tilde{\Omega}} R + K_{\tilde{\Omega}} R^2 - + \cdots.$$

This shows that the operator R has to be analyzed if one wants to find the properties of K_Ω. (For details, see Ch. Fefferman **[8]**.)

III. Applications.

1. *The extension theorem for biholomorphic mappings.* Let $\Omega_1, \Omega_2 \subset\subset \mathbf{C}^n$ be strictly pseudoconvex with smooth boundaries and $F: \Omega_1 \to \Omega_2$ a biholomorphic mapping. What are the regularity properties of F at the boundary?

In 1973, N. Vormoor proved, by using the boundary behavior of the Carathéodory metric, that F extends to a homeomorphism of $\bar{\Omega}_1 \to \bar{\Omega}_2$.

Ch. Fefferman showed, by using Theorem 2.3 and the Bergman metric:

THEOREM 3.1. *F extends to a diffeomorphism of $\bar{\Omega}_1 \to \bar{\Omega}_2$* **[8]**.

REMARK. This result gives the possibility of using differential geometric objects on the boundaries to study biholomorphic mappings.

Again, the technical details of the proof are very complicated, whereas the basic ideas can be explained quite easily: At first we look at an arbitrary strictly pseudoconvex domain $\Omega \subset\subset \mathbf{C}^n$ with smooth boundary. Let $t \to X(t)$ by any geodesic of the Bergman metric which does not remain within a compact subset of Ω. One wants to show that, for all t_0 large enough, the following happens:

(a) There exists a neighborhood U of $w_0 = \dot{X}(t_0)$ in the unit sphere S^{2n-1} in the tangent space at $z_0 = X(t_0)$, such that

$$\pi_{z_0} : U \to b\Omega, \quad w \mapsto \lim_{x \to \infty} X(t, w, z_0),$$

is well defined and even a diffeomorphism onto a small neighborhood of $\pi_{z_0}(w_0) \in b\Omega$.

Here $X(t, w, z_0)$ is the geodesic starting at z_0 in the direction of w.

Furthermore, one wants to show that

(b) $b\Omega$ is totally covered by such images.

Let us suppose for the moment that these two facts can be proved and let us denote by $\hat{F}$ the continuous extension of F according to Vormoor. Take a geodesic

$X_1(t)$ in Ω_1 which goes to $b\Omega_1$. Then $F \circ X_1(t) = X_2(t)$ is a geodesic in Ω_2. Pick a $t_0 \gg 0$ such that (a) holds both for $X_1(t_0) = z_0$ and $X_2(t_0) = \zeta_0$. Because F is an isometry with respect to the Bergman metric, one obtains the following commutative diagram:

$$\begin{array}{ccc} U_1 & \xrightarrow{dF} & U_2 \\ \downarrow{\scriptstyle \pi_{z_0}} & & \downarrow{\scriptstyle \pi_{\zeta_0}} \\ b\Omega_1 & \xrightarrow{\hat{F}} & b\Omega_2 \end{array}$$

where U_i is a neighborhood of $\dot{X}_i(t_0)$ in S^{2n-1}. This, together with (b), gives that $\hat{F}|b\Omega_1$ is smooth, and because the coordinate functions of $\hat{F}$ are harmonic in Ω_1 it follows from potential theory that $\hat{F}$ is smooth on $\bar{\Omega}_1$.

This reduces the problem to the proof of statements (a) and (b), and that turns out to be very hard because the differential equations describing the geodesics of the Bergman metric become highly singular at the boundary $b\Omega$ if the logarithmic term in the asymptotic formula of Theorem 2.3 is nonzero. Therefore the main step in deriving Theorem 3.1 from Theorem 2.3 is finding the right changes of variables that destroy these singularities. (In fact, statements (a) and (b) are only proved for C^k-differentiability, with an arbitrary positive integer k fixed.)

2. *Regularity properties of the canonical solution of* $\bar{\partial}u = \alpha$. Let α be a $\bar{\partial}$-closed (0, 1)-form on a strictly pseudoconvex domain $\Omega \subset\subset \boldsymbol{C}^n$ with smooth boundary, and assume that α is at least square-integrable on Ω with respect to the Lebesgue measure. Then J. J. Kohn's theory of the $\bar{\partial}$ Neumann problem gives an $f \in L^2(\Omega)$ with $\bar{\partial}f = \alpha$ and $f \in \mathscr{H}^{2\perp}(\Omega)$ and, obviously, f is uniquely determined by these conditions. We call it the canonical solution of the equation $\bar{\partial}u = \alpha$. On the other hand, several different solution operators for this equation have been constructed in the last years (for a bibliography see **[10]**), for which much better regularity properties are known than for the canonical solution. It may also be possible to derive these properties for the canonical solution by using the following consequence of Proposition 1.3(b):

If $f \in L^2(\Omega)$ is any solution of $\bar{\partial}u = \alpha$, then

$$f_0 = f - \int_\Omega f(z)\,\overline{K(z, w)}\,d\lambda_z$$

is the canonical solution of this equation. Therefore, a good knowledge of the boundary behavior of $K(z, w)$ can give regularity properties for the canonical solution of the $\bar{\partial}$ problem.

Based on this method together with Ch. Fefferman's result on the Bergman kernel, N. Kerzman announced the following result (Proc. Internat. Congr. Math. (Vancouver, 1974), Canad. Math. Congr., 1975).

THEOREM 3.2. *The canonical solution operator is a bounded operator from* $L^\infty_{0,1}(\Omega) \to L^\infty(\Omega)$.

3. *The Fefferman metric.* In their recent paper **[4]** S.S. Chern and J. Moser solved the equivalence problem with respect to locally biholomorphic mappings for nondegenerate hypersurfaces in $\boldsymbol{C}^n$ by constructing a complete system of differential

geometric invariants. In this section, we want to describe how to get a conformal Lorentz metric on the trivial circle bundle over the boundary of a smooth strictly pseudoconvex domain Ω by using the kernel function.

To avoid difficulties with signs, we denote by ψ the negative of a strictly plurisubharmonic C^∞-defining function for Ω, i.e., $\psi > 0$ in Ω. Theorem 2.3 says that the function

$$L_\psi = \psi^{n+1} K_\Omega(z) \in C^n(\bar{\Omega})$$

and has strictly positive boundary values. Therefore, the function

$$u_\Omega = \frac{\psi}{L_\psi^{1/(n+1)}} \in C^{n+1}(\bar{\Omega}) \tag{3.1}$$

and vanishes of first order at $b\Omega$. Furthermore, $u_\Omega | \Omega = K^{-1/(n+1)}$ shows that it does not depend on the choice of ψ.

Following I. Naruki [**14**], we introduce an additional variable $z_0 \in \boldsymbol{C}^*$ and look at the function

$$U_\Omega(z_0, z) = |z_0|^{2/(n+1)} u_\Omega(z) \quad \text{on } \boldsymbol{C}^* \times \Omega. \tag{3.2}$$

If $F: \Omega \to \tilde{\Omega}$ is a biholomorphic mapping onto the strictly pseudoconvex smooth domain $\tilde{\Omega}$, we know already that we can extend it to a diffeomorphism from Ω to $\tilde{\Omega}$ and we define

$$\begin{aligned} \mathscr{F}: \boldsymbol{C}^* \times \bar{\Omega} &\to \boldsymbol{C}^* \times \bar{\tilde{\Omega}}, \\ (z_0, z) &\to (z_0/\det F'(z), F(z)). \end{aligned} \tag{3.3}$$

$\mathscr{F}$ is holomorphic on $\boldsymbol{C}^* \times \Omega$ and $\det \mathscr{F}' \equiv 1$. An immediate consequence of Proposition 1.3(d) is that the functions U_Ω are invariant under such mappings $\mathscr{F}$, i.e.,

$$U_\Omega = U_{\Omega} \circ \mathscr{F}. \tag{3.4}$$

Together with $\det \mathscr{F}' \equiv 1$, this gives

LEMMA 3.1. *The indefinite hermitian metric*

$$ds^2 = \sum_0^n \frac{\partial^2 U_\Omega}{\partial z_j \partial \bar{z}_k} dz_j \, d\bar{z}_k \quad \text{on } \boldsymbol{C}^* \times \bar{\Omega}$$

is invariant under the mappings $\mathscr{F}$. It is nondegenerate.

Up to this point, the main ideas of this construction are contained in I. Naruki [**14**]. Now, following Ch. Fefferman [**9**], we want to study what happens if we restrict ds^2 to $\boldsymbol{C}^* \times b\Omega$, because ds^2 cannot be computed at the points lying over Ω whereas the restriction to $\boldsymbol{C}^* \times b\Omega$ can be computed explicitly and even quite easily, as we will see later.

If one introduces polar coordinates $z_0 = re^{i\theta}$ on $\boldsymbol{C}^*$ and restricts ds^2 to the tangent space of $\boldsymbol{C}^* \times b\Omega = \boldsymbol{R}^+ \times S^1 \times b\Omega$, one gets

$$ds^2 = r^{2/(n+1)} \left\{ \frac{-i}{n+1} (\partial u_\Omega - \bar{\partial} u_\Omega) d\theta + \sum_{j,k} \frac{\partial^2 u_\Omega}{\partial z_j \partial \bar{z}_k} dz_j \, d\bar{z}_k \right\}.$$

Unfortunately, this is now degenerate because $\partial/\partial r$ does not appear in this expression. But we can throw out the variable r by simply putting $r = 1$. Altogether, one gets

THEOREM 3.3 (FEFFERMAN). *The metric*

$$ds^2 = \frac{-i}{n+1}(\partial u_\Omega - \bar\partial u_\Omega)d\theta + \sum_{j,k} \frac{\partial^2 u_\Omega}{\partial z_j \partial \bar z_k} dz_j\, d\bar z_k \tag{3.5}$$

is a nondegenerate Lorentz metric on $S^1 \times b\Omega$. Under the mappings $\mathscr{F}$ of (3.3), *restricted to $S^1 \times b\Omega$, it splits off a factor $|\det F'|^{2/(n+1)}$, i.e.,*

$$F^*(ds^2_{\tilde\Omega}) = |\det F'|^{2/(n+1)}\, ds^2_\Omega. \tag{3.6}$$

Therefore, the conformal class of ds^2 is a biholomorphic invariant.

We recall that a geodesic of a Lorentz metric is called a light ray if its tangent vectors always have length zero. The following is a well-known fact:

LEMMA. *Two conformally equivalent Lorentz metrics have the same light rays* (*as sets*).

DEFINITION. The projections of the light rays of ds^2 from $S^1 \times b\Omega$ to $b\Omega$ are called the Fefferman chains on $b\Omega$.

The Fefferman chains are biholomorphically invariant.

4. *Explicit computation of the Fefferman metric.* If one tries to compute ds^2 explicitly, one sees at once that, in the notation of Theorem 2.3, it is not enough to know $\Phi|b\Omega$. One also has to know the first order term of Φ on $b\Omega$. More exactly: let us denote by π the orthogonal projection of Ω on $b\Omega$ (near $b\Omega$). Then one can write

$$\begin{aligned}\Phi(z) = \Phi^{(0)}(\pi(z)) + \Phi^{(1)}(\pi(z))\psi(z) + \cdots + \Phi^{(n)}(\pi(z))\psi^n(z)\\ + O(\psi^{n+1}).\end{aligned} \tag{3.7}$$

On the other hand, one sees at once that

$$(\partial^2 u_\Omega/\partial z_j \partial \bar z_k)|b\Omega = \psi_j \Phi_{\bar k} + \psi_{\bar k}\Phi_j + \psi_{j\bar k}\Phi.$$

This shows that one can compute ds^2 if and only if one knows $\Phi^{(1)}$ (on $b\Omega$).

How to determine $\Phi^{(1)}$. Let us first make the following remarks:

(a) The functions $\Phi^{(0)}, \cdots, \Phi^{(n)}$ depend on the choice of ψ.

(b) $$\Phi^{(0)}(z) = c_n \Delta_\psi(z) = c_n \det\begin{pmatrix} \psi_1 & & \\ \vdots & \psi_{i\bar j} & \\ \psi_n & & \\ \psi & \psi_{\bar 1} \cdots & \psi_{\bar n}\end{pmatrix}(z)$$

(Monge-Ampère determinant) for all $z \in b\Omega$.

(c) For $\Omega = \{\psi(z) = 1 - \sum_{j=1}^n |z_j|^2 > 0\}$ we have $\Delta_\psi \equiv 1$ on Ω and $K_\Omega(z) = c_n/\psi^{n+1}(z)$, $z \in \Omega$.

(d) We have the following transformation rule for Δ_ψ:

LEMMA 3.2. *Let $F\colon \Omega \to \tilde\Omega$ be biholomorphic and $\tilde\psi$ a smooth function on $\tilde\Omega$. Define $\psi = |\det F'|^{-2/(n+1)}\, \tilde\psi \circ F$. Then one has $\Delta_\psi = \Delta_{\tilde\psi} \circ F$.*

PROOF. If one defines $\tilde\Psi(\tilde z_0, \tilde z) = |\tilde z_0|^{2/(n+1)}\, \tilde\psi$ on $\boldsymbol{C}^* \times \tilde\Omega$, $\Psi(z_0, z) = |z_0|^{2/(n+1)}\, \psi$ on $\boldsymbol{C}^* \times \Omega$ and $\mathscr{F}$ as in (3.3), one gets at once:

$$\tilde\Psi \circ \mathscr{F} = \Psi$$

and

$$\Delta_{\tilde\Psi} = (-1)^n \det(\tilde\Psi_{i\bar j})^n_0, \qquad \Delta_\Psi = (-1)^n \det(\Psi_{i\bar j})^n_0.$$

Because of $\det \mathscr{F}' \equiv 1$ and the known rule

$$\det (\tilde{\Psi}_{ij})_0^n = \left|\det \mathscr{F}'\right|^2 \det (\Psi_{ij})_0^n \circ \mathscr{F},$$

this gives what we want.

All these observations lead to the following questions:

(1) Let Ω be a smooth, strictly pseudoconvex domain. Is there always a function $\psi \in C^k(\bar{\Omega})$, k large enough, such that $\psi > 0$ on Ω, $\psi | b\Omega = 0$ and $\Delta_\psi \equiv 1$?

(2) Which regularity properties does ψ have?

(3) What is the relation between ψ and u_Ω (cf. (3.1))?

The answers to these questions are unknown. But the important fact here is that the equation $\Delta_\psi \equiv 1$ can quite easily be formally solved up to the order $n + 1$ at the boundary. One has

THEOREM 3.4. *Let ψ be a strictly superharmonic function on $\bar{\Omega}$ with $\Omega = \{\psi > 0\}$ and $d\psi \neq 0$ on $b\Omega$. Define*

$$\psi_{(1)} = \psi \cdot \Delta_\psi^{-1/(n+1)}$$

and, for $2 \leqq s \leqq n + 1$,

$$\psi_{(s)} = \psi_{(s-1)} \left(1 + \frac{1 - \Delta(\psi_{(s-1)})}{(n + 2 - s)s}\right).$$

Then one has $\Delta_{\psi_{(s)}} = 1 + O(\psi^s)$. Each such $\psi_{(s)}$ is uniquely determined up to terms of the form $O(\psi^{s+1})$.

Fortunately, these formal solutions of the Monge-Ampère equation $\Delta_\psi \equiv 1$ already solve the problem of computing the term $\Phi^{(1)}$ in (3.7). H. Christoffers [5] and K. Diederich [7] proved independently

THEOREM 3.5. *For the functions $\psi_{(2)}$ of Theorem* 3.4, *one has*

$$(\psi_{(2)})^{n+1} \cdot K_\Omega(z) - c_n = O((\psi_{(2)})^2),$$

i.e., $\Phi^{(1)} \equiv 0$ *for this defining function.*

REMARK. It seems most likely that the proofs of Theorem 3.5 can be pushed further to show that $(\psi_{(s)})^{n+1} K_\Omega(z)$ for $s > 2$ approximates c_n at $b\Omega$ even better than for $s = 2$.

5. *A new definition for Fefferman's metric.* We have seen above that the Fefferman metric can be explicitly computed from the function $\psi_{(2)}$. On the other hand, $\psi_{(2)}$ is uniquely determined up to terms of order $O(\psi^3)$, which together with Lemma 3.2 and Theorem 3.1 shows the following: If $F: \Omega \to \tilde{\Omega}$ is biholomorphic, $\tilde{\Omega} = \{\tilde{\psi} > 0\}$, $d\tilde{\psi} \neq 0$, $\tilde{\psi}$ strictly plurisuperharmonic on $\bar{\tilde{\Omega}}$ and if one defines

$$\hat{\psi} = \left|\det F'\right|^{-2/(n+1)} \tilde{\psi}_{(2)} \circ F,$$

then $\hat{\psi} = \psi_{(2)} + O(\psi^3)$ on Ω and can therefore be used to define ds^2 on $S^1 \times b\Omega$. Therefore, we can give the following new definition for ds^2, which no longer refers to the kernel function.

DEFINITION. Take $u = \psi_{(2)}$ as in Theorem 3.4 and define on $S^1 \times b\Omega$

$$ds^2 = -\frac{i}{n+1}(\partial u - \bar{\partial} u)\, d\theta + \sum_{j,k} \frac{\partial^2 u}{\partial z_j \partial \bar{z}_k} dz_j\, d\bar{z}_k.$$

REMARKS. (a) One can of course express ds^2 directly in terms of an arbitrary strictly plurisuperharmonic defining function of Ω. The formula is:

$$
\begin{aligned}
ds^2 = & -i(\partial\psi - \partial\bar{\psi})\, d\theta + (n+1) \sum_{j,k} \psi_{j\bar{k}}\, dz_j d\bar{z}_k - \Delta_{\psi}^{-1} \cdot (\partial\Delta_{\psi}\bar{\partial}\psi + \partial\psi\bar{\partial}\Delta_{\psi}) \\
& + \frac{1}{n} \operatorname{trace}\left[\begin{pmatrix} 0 & 0 \\ 0 & \Delta_{\psi}^{-1}(\Delta_{\psi})_{j\bar{k}} - \Delta_{\psi}^{-2}(\Delta_{\psi})_j(\Delta_{\psi})_{\bar{k}} \end{pmatrix} \cdot \begin{pmatrix} 0 & \psi_{\bar{k}} \\ \psi_j & \psi_{j\bar{k}} \end{pmatrix}^{-1}\right] (\partial\psi)\,(\bar{\partial}\psi).
\end{aligned} \tag{3.8}
$$

(b) By using the characterization of D. Burns and St. Shnider [3] of the Chern-Moser chains, one can show that the Fefferman chains agree as sets with the Chern-Moser chains. It is very difficult to compute these chains with the methods of Chern and Moser even for simple examples, whereas the explicit formula (3.8) for ds^2 and a formula for the Hamiltonian H of ds^2, which can be derived from it (see [**9**]), makes it possible actually to compute the Fefferman chains in many special cases or, at least, to describe certain global properties of them (for details see [**9**]).

References

1. N. Aronszajn, *Theory of reproducing kernels*, Trans. Amer. Math. Soc. **68** (1950), 337–404. MR **14**, 479.

2. St. Bergman, *Über die Kernfunktion und ihr Verhalten am Rande*. I, II, J. Reine Angew. Math. **169** (1933), 1–42; ibid. **172** (1934), 89–128.

3. D. Burns, Jr., K. Diederich and S. Shnider, *Distinguished curves on strictly pseudoconvex boundaries in* C^{n+1} (to appear).

4. S. Chern and J. Moser, *Real hypersurfaces in complex manifolds*, Acta Math. **133** (1974), 219–271.

5. H. Christoffers, (to appear).

6. K. Diederich, a) *Das Randverhalten der Bergmanschen Kernfunktion und Metrik in streng pseudokonvexen Gebieten*, Math Ann. **187** (1970), 9–36. MR **41** #7149.

b) *Über die* 1. *und* 2. *Ableitungen der Bergmanschen Kernfunktion und ihr Randverhalten*, Math. Ann. **203** (1973), 129–170. MR **48** #6472.

7. ———, (to appear).

8. Ch. Fefferman, *The Bergman kernel and biholomorphic equivalence of pseudoconvex domains*, Invent. Math. **26** (1974), 1–65. MR **50** #2562.

9. ———, *Monge-Ampere equations, the Bergman kernel and geometry of pseudoconvex domains* (to appear).

10. G. B. Folland and J. J. Kohn, *The Neumann problem for the Cauchy-Riemann complex*, Ann. of Math. Studies, no. 75, Princeton Univ. Press, Princeton, N.J., 1972.

11. L. Hörmander, L^2 *estimates and existence theorems for the* $\bar{\partial}$*-operator*, Acta Math. **113** (1965), 89–152. MR **31** #3691.

12. N. Kerzman, *The Bergman kernel function. Differentiability at the boundary*, Math. Ann. **195** (1972), 149–158. MR **45** #3762.

13. S. Kobayashi, *Geometry of bounded domains*, Trans. Amer. Math. Soc. **92** (1959), 267–290. MR **22** #3017.

14. I. Naruki, *On the equivalence problem for bounded domains*, Res. Inst. Math. Sci. Kyoto **118** (1972).

15. M. Skwarczyński, *The Bergman function and semiconformal mappings*, Bull. Acad. Polon. Sci. Sér. Sci. Math. Astronom. Phys. **22** (1974), 667–673. MR **50** #7592.

Westfalische Wilhelms-Universität Münster

Proceedings of Symposia in Pure Mathematics
Volume 30, 1977

THE MAXIMUM MODULUS PRINCIPLE FOR CR FUNCTIONS ON SMOOTH REAL EMBEDDED SUBMANIFOLDS OF C^n

DAVID ELLIS AND C. DENSON HILL*

Let M be a smooth real embedded submanifold of C^n. We assume that M is a CR submanifold. Let $\bar{\partial}_M$ denote the tangential Cauchy-Riemann operator on M. Let $CR(M)$ denote the algebra of functions u differentiable on M for which $\bar{\partial}_M u = 0$. Such a function $u \in CR(M)$ is called a *CR function* on M.

We would like to state necessary and sufficient conditions for $\bar{\partial}_M$ to obey a local maximum modulus principle on M.

DEFINITION. $\bar{\partial}_M$ obeys the *local maximum modulus principle* on M if given any open connected set $\mathscr{D}$ in M and any $u \in CR(\mathscr{D})$, then $|u|$ cannot have a (weak) local maximum at any point of $\mathscr{D}$ unless u is constant on $\mathscr{D}$.

These necessary and sufficient conditions can be stated in terms of intrinsic objects called extreme points. But in more useful analytic terms, we can also obtain conditions in terms of the Levi form of M.

DEFINITION. We call a point $p \in M$ an *extreme point* of M if there exists a local holomorphic coordinate system $z = (z_1, z_2, \cdots, z_n)$ in a neighborhood U of p such that $z(p) = 0$ and $M \cap U \subset \{z : y_1 \geqq 0\}$. We assume that, locally near p, M is not contained in any C^k for $k < n$, i.e., that C^n is locally minimal for M near p.

Let $\mathscr{H}T_p(M)$ denote the holomorphic tangent space to M at p. The Levi form of M can be expressed as a map assigning a normal vector to each holomorphic tangent vector:

DEFINITION. (i) For any $p \in M$ and $X \in \mathscr{H}T_p(M)$ set $\boldsymbol{Z} = X - iJX$. Then the *Levi orm* of M at p assigns to $\boldsymbol{Z}$ the normal vector $L_p(\boldsymbol{Z})$ defined by

$$L_p(\boldsymbol{Z}) = B_p(X, X) + B_p(JX, JX)$$

AMS (MOS) subject classifications (1970). Primary 32A30.

*Research supported by an Alfred P. Sloan Research Fellowship and by the National Science Foundation grants NSF PO43 957X00 and MPS 72–05055 AO2.

© 1977, American Mathematical Society

where B_p is the second fundamental form of M at p.

(ii) For any $\xi \in N_p(M)$, the normal space of M at p, the map $L_p^\xi\colon \mathscr{H}T_p(M) \to \boldsymbol{R}$ defined by $L_p^\xi(\boldsymbol{Z}) = \langle L_p(\boldsymbol{Z}), \xi \rangle$ is called the *Levi form in the ξ-direction.* Here $\langle\, ,\rangle$ denotes the real inner product in $\boldsymbol{R}^{2n}$. It can be shown that L_p^ξ is a Hermitian form on $\mathscr{H}T_p(M)$.

Statement of results. (1) If p is an extreme point of M then there exists a normal direction $\xi \in N_p(M)$ such that L_p^ξ is positive semidefinite.

REMARK. The converse of Statement (1) is, in general, not true. Kohn and Nirenberg [**3**] have given an example of a bounded pseudoconvex domain $\Omega \subset \boldsymbol{C}^2$ with smooth boundary $b\Omega$ (so that the Levi form of $b\Omega$ is positive semidefinite with respect to the inner normal at each point of $b\Omega$), but such that at a certain point $p \in b\Omega$ any one-dimensional complex-analytic variety which passes through p must intersect Ω and $C\Omega$. Thus p cannot be an extreme point of M.

(2) If for a point $p \in M$ there exists a $\xi \in N_p(M)$, $\xi \neq 0$, such that L_p^ξ is positive definite on $\mathscr{H}T_p(M)$, then p is an extreme point of M.

(3) If $\bar\partial_M$ obeys the local maximum modulus principle on M, then M can contain no extreme points.

Let $m = \dim_{\boldsymbol{C}} \mathscr{H}T_p(M)$ and let $\lambda_1(\xi) \geqq \lambda_2(\xi) \geqq \cdots \geqq \lambda_m(\xi)$ be the eigenvalues of the Hermitian form L_p^ξ.

(4) If $\bar\partial_M$ obeys the local maximum modulus principle on M, then for every $p \in M$ and every $\xi \in N_p(M)$,

$$(*) \qquad \lambda_1(\xi)\, \lambda_m(\xi) \leqq 0.$$

It is easily seen that the necessary condition $(*)$ is, in general, not sufficient. However we have:

(5) If for every $p \in M$ and every $\xi \in N_p(M) \cap JT_p(M)$, $\xi \neq 0$,

$$\lambda_1(\xi)\, \lambda_m(\xi) < 0,$$

then $\bar\partial_M$ obeys the local maximum modulus principle on M.

The necessary conditions described above appear in [**1**]. The sufficient condition appears in [**2**].

REFERENCES

1. D. Ellis, C. D. Hill and C. Seabury, *The maximum modulus principle.* I. *Necessary conditions,* Indiana J. Math. (to appear).

2. D. Ellis and C. D. Hill, *The maximum modulus principle.* II. *Sufficient conditions,* Indiana J. Math. (to appear).

3. J. Kohn and L. Nirenberg, *A pseudo-convex domain not admitting a holomorphic support function,* Math. Ann. **201** (1973), 265–268. MR **48** #8850.

HERBERT H. LEHMAN COLLEGE (CUNY)

STATE UNIVERSITY OF NEW YORK AT STONY BROOK

Proceedings of Symposia in Pure Mathematics
Volume 30, 1977

REAL SUBMANIFOLDS WITH DEGENERATE LEVI FORM*

MICHAEL FREEMAN

1. Introduction. A real submanifold M of $\mathbf{C}^n$ is *straightened* near a point z on M if there exist open sets U and V, a biholomorphic map $f = (f_1, \cdots, f_n)\colon U \to V$, and a real submanifold M_0 of $\mathbf{C}^q$ such that $z \in U$ and

$$f(U \cap M) = (M_0 \times \mathbf{C}^{n-q}) \cap V. \tag{1.1}$$

Note that f induces a local foliation of M by the complex submanifolds

$$f^{-1}(\{c\} \times \mathbf{C}^{n-q}) = f_1^{-1}(c_1) \cap \cdots \cap f_q^{-1}(c_q), \qquad c \in M_0. \tag{1.2}$$

In fact, finding a straightening map amounts to finding independent holomorphic functions $f_1, \cdots, f_q$ such that M is exhausted by the common level sets indicated explicitly in (1.2).

A nontrivial straightening is usually not possible, even for a real-analytic manifold admitting local foliations. For it turns out that foliations of the form (1.2) are very special. At least, their leaves must be tangent to null directions of the Levi form of M, and hence no straightening is possible if the Levi form is non-degenerate. There are still further constraints, of higher order. This paper will describe an invariant sequence of obstructions to straightening carried by M, which under mild regularity conditions provides a precise description of the manner in which M may be straightened. These obstructions can be regarded from two basically dual viewpoints: as modules of differential one-forms on M, or as modules of vector fields tangent to M. The former view is taken here, and the main results are summed up as Theorem 2.1. This paper is only an announcement and brief exposition. A discussion taking both points of view, with more examples and complete proofs, will appear elsewhere [**3**].

AMS (MOS) subject classifications (1970). Primary 53A55, 32C05; Secondary 53B25.

Key words and phrases. Straightening, local biholomorphic invariant, Pfaffian system, module of one-forms.

*Research supported by NSF GP43247 at the University of Kentucky.

© 1977, American Mathematical Society

A simple example which admits complex foliation but no local straightening is furnished by the real-analytic hypersurface $M^5 = \{(z_1, z_2, z_3): x_1^3 + x_2^3 + x_3^3 = 0\} \subset \boldsymbol{C}^3 - \{0\}$, where $z_j = x_j + iy_j$ with x_j and y_j real. It is a *tube manifold*, independent of the y_j. This hypersurface has a complex foliation whose leaves are in fact complex affine lines described in $\boldsymbol{C}^3$ by holding a, b, and c fixed in the parametrization of M^5 given by

$$(1.3) \qquad (a, b, c, t) \to p(a, b, c) + t \operatorname{Re} p(a, b, c), \qquad t \in \boldsymbol{C},\ a, b, c \in \boldsymbol{R}.$$

Here $p(a, b, c)$ is the point on M^5 given by $p(a, b, c) = (a + ib, -(1 + a^3)^{1/3} + ic, 1)$. Later it will be shown that M^5 *cannot* be locally straightened. In particular this foliation cannot arise in the manner given by (1.2) from a straightening map (1.1).

The first q coordinates $f_1, \cdots, f_q$ of a straightening map (1.1) can be regarded as solutions of a certain type of Pfaffian differential system. This viewpoint is developed in the first part of this paper, which will describe how such systems can be expressed as modules of one-forms of a special class that is invariant under local biholomorphic maps. It is stated in Theorem 2.1 that M always carries an increasing sequence of these special Pfaffian systems, each of which is a local biholomorphic invariant of M. Each of these systems is contained in any Pfaffian system defined by a straightening map (1.1). The same is of course true for their union $\mathscr{J}$, which thus acts as an obstruction to straightening. Moreover, there are many instances when there exists a straightening map defining $\mathscr{J}$ itself, which is then the complete obstruction to straightening.

This study is purely local, and henceforth all objects are to be interpreted as germs at z. Thus $\mathscr{E}^0$ is the ring of (germs at z of) smooth complex-valued functions and $T\boldsymbol{C}^n$ is the complexified $\mathscr{E}^0$-module of smooth vector fields. $M = \{\rho = 0\}$ where $\rho = (\rho_1, \cdots, \rho_m)$ is a smooth $\boldsymbol{R}^m$-valued map of rank m. All geometric objects on M are regarded as extended to $\boldsymbol{C}^n$. Thus the model for the space of smooth sections of the tangent bundle of M is $T = \{X \in T\boldsymbol{C}^n: X\rho_j = d\rho_j(X) \in O(M), \text{all } j\}$, the $\mathscr{E}^0$-module of vector fields tangent to M. The $\mathscr{E}^0$-module of complex tangent vector fields is

$$(1.4) \qquad H = \{X \in T\boldsymbol{C}^n : \partial\rho_j(X) \text{ and } \bar{\partial}\rho_j(X) \in O(M), \text{all } j\}.$$

Here $O(M)$ stands for the subspace of objects vanishing on M. It is assumed that M is a CR manifold, so that H is (the extension to $\boldsymbol{C}^n$ of) the space of sections of a bundle on M whose fibres are defined by the pointwise version of (1.4). An object is *invariant* if it is independent of a choice of defining functions ρ_j and preserved by local biholomorphic maps.

The symbol $\mathscr{E}^1$ denotes the $\mathscr{E}^0$-module of one-forms on $\boldsymbol{C}^n$. It has the usual decomposition $\mathscr{E}^1 = \mathscr{E}^{1,0} + \mathscr{E}^{0,1}$ defined by the complex structure of $\boldsymbol{C}^n$. For convenience it is desired to express a Pfaffian differential system on M as a submodule $\mathscr{M}$ of $\mathscr{E}^1$; that is, defined on $\boldsymbol{C}^n$ and not just on M. This is permissible provided that $\mathscr{M}$ can be recovered completely from its pull-back $j^*\mathscr{M}$ defined by the inclusion $j: M \to \boldsymbol{C}^n$. In other words, submodules $\mathscr{M}$ are required such that $\mathscr{M} = j^{*-1}(j^*\mathscr{M})$, or equivalently $\mathscr{M} \supset \ker j^*$.

Note that $\ker j^* = \{\beta \in \mathscr{E}^1: \beta = \sum_{j=1}^m a_j d\rho_j + \rho_j b_j,\ a_j \in \mathscr{E}^0,\ b_j \in \mathscr{E}^1\}$ which is denoted by $\mathrm{sp}\{d\rho, \rho\}$. In general $\mathrm{sp}\{\tau, \rho\}$denotes the submodule of $\mathscr{E}^1$ spanned by the subset τ and $O(M)$.

Accordingly the Pfaffian system induced on M by $\mathscr{E}^{1,0}$ is $j^{*-1}(j^* \mathscr{E}^{1,0}) = \mathscr{E}^{1,0} + \mathrm{sp}\{d\rho, \rho\}$. It is routine to see that this coincides with

$$\mathscr{E}^{1,0} + \mathscr{I}_1, \tag{1.5}$$

where

$$\mathscr{I}_1 = \mathrm{sp}\{\partial\rho, \bar\partial\rho, \rho\} \tag{1.6}$$

is the module which defines H. Elements of (1.5) are called *relative* (1, 0)*-forms*. They are precisely the forms in $\mathscr{E}^1$ whose restrictions to H are linear with respect to the complex structure induced by $\mathbf{C}^n$. Since $\mathscr{I}_1$ is invariant, this definition is also invariant. Note that a function g is CR (satisfies the tangential Cauchy-Riemann equations) if and only if $dg \in \mathscr{E}^{1,0} + \mathscr{I}_1$.

It is clear that the Pfaffian system

$$\mathscr{S} = j^{*-1}(j^* \,\mathrm{sp}\{df_1, \cdots, df_q\})$$

induced on M by a straightening map (1.1) is a submodule of (1.5). But it is more special than that. Since ρ_i is constant on each leaf (1.2) it follows from linear algebra that

$$\partial\rho_i = \sum_{j=1}^{q} a_{ij}\, df_j \mod O(M), \tag{1.7}$$

which implies $\partial\rho \subset \mathscr{S}$. Since $d\rho = \partial\rho + \bar\partial\rho$ it follows that

$$\mathscr{S} \supset \mathscr{I}_1 \tag{1.8}$$

and $\mathscr{S} = \mathrm{sp}\{df, \bar\partial\rho, \rho\}$.

$\mathscr{S}$ is the *straightening module* defined by f. Because of (1.8), $\mathscr{I}_1$ is to be regarded as an obstruction to straightening, the first of an invariant chain of obstructions attached to M. Relation (1.8) has a simple geometric interpretation. It is just a nonstandard way of expressing the well-known fact that the module of vector fields tangent to the leaves (1.2) is contained in H:

$$\{X \in T\mathbf{C}^n : df_j(X) \text{ and } d\bar f_j(X) \in O(M), j = 1, \cdots, q\} \subset H. \tag{1.9}$$

Furthermore, $\mathscr{S}$ generates a d-closed ideal in the $\mathscr{E}^0$-algebra of differential forms. For differentiation of (1.7) yields

$$-d\bar\partial\rho_i = d\partial\rho_i = \sum_{j=1}^{q} da_{ij} \wedge df_j \mod \mathscr{I}_1, \tag{1.10}$$

which in conjunction with (1.8) shows that

$$d\mathscr{S} \subset (\mathscr{S}). \tag{1.11}$$

The symbol (E) is used to denote the ideal generated by a set E.

A larger obstruction to straightening is caused by the null space N of the Levi form of M. It is defined as

$$N = \{Y \in T\mathbf{C}^n : L\rho_i(X, Y) \in O(M) \text{ for all } X \in H \text{ and all } i\}, \tag{1.12}$$

where $L\rho_i$ is the complex Hessian of ρ_i.

PROPOSITION 1.1. *If f satisfies* (1.1) *then* $\{X \in TC^n : df_j(X)$ *and* $d\bar{f}_j(X) \in O(M), j = 1, \cdots, q\} \subset N$.

This strengthens (1.9). It can be shown easily by an inspection of (1.1), for the statement is invariant, and it can be verified for $M_0 \times C^{n-q}$ by computing the complex Hessian of a defining function which is dependent only on the variables of C^q. Later this result will be expressed in a dual form, in the same spirit as (1.8).

Proposition 1.1 suggests that the existence of a straightening map (1.1) will depend on the structure of N. This is so, and the results exposed in this paper can be considered as an investigation of this structure. To understand them it will be necessary to quote a preliminary theorem on the structure of N contained in [2]. Before stating this theorem it may be helpful to consider in some detail the example M^5 above, where $\rho = \rho_0 \equiv x_1^3 + x_2^3 + x_3^3$. It is desired to compute the Levi form of M^5 (which is the restriction to $H \times H$ of the complex Hessian of ρ_0), and to find N and relate it to the foliation parametrized by (1.3).

By solving $(2/3)\, \partial\rho_0 = x_1^2 dz_1 + x_2^2 dz_2 + x_3^2 dz_3$ and its conjugate for $x_1^2 dz_1$ and $x_1^2 d\bar{z}_1$ near a point on M^5 where $x_1 \neq 0$ and substituting for $x_1^4 dz_1 \wedge d\bar{z}_1$ in

$$(2/3)\, x_1^3\, \partial\bar{\partial}\rho_0 = x_1^4 dz_1 \wedge d\bar{z}_1 + x_1^3 x_2 dz_2 \wedge d\bar{z}_2 + x_1^3 x_3 dz_3 \wedge d\bar{z}_3$$

one finds

$$\begin{aligned}(2/3)\, x_1^3\, \partial\bar{\partial}\rho_0 = {} & (x_1^3 x_2 + x_2^4) dz_2 \wedge d\bar{z}_2 + x_2^2 x_3^2\, dz_2 \wedge d\bar{z}_3 \\ & + x_2^2 x_3^2 dz_3 \wedge d\bar{z}_2 + (x_1^3 x_3 + x_3^4) dz_3 \wedge d\bar{z}_3 \bmod (\mathscr{I}_1).\end{aligned}$$

The Levi form of M^5 is defined by the 2×2 "matrix" on the right side. This hermitian form can be diagonalized mod $O(M)$ by elementary methods to obtain

$$\partial\bar{\partial}\rho_0 = -\frac{3}{2}\,\frac{x_2 x_3}{x_1^3}\,\tau_0 \wedge \bar{\tau}_0 \bmod (\mathscr{I}_1), \tag{1.13}$$

where $\tau_0 = x_3 dz_2 - x_2 dz_3$. This shows explicitly that the Levi form of M^5 is degenerate but not identically zero. In fact

$$N = \{X \in TC^n : \partial\rho_0(X), \tau_0(X), \bar{\partial}\rho_0(X), \text{ and } \bar{\tau}_0(X) \in O(M)\}. \tag{1.14}$$

Moreover, N coincides with the set of vector fields tangent to the complex foliation defined by (1.3), because it is clear that these vector fields are defined on M^5 by $X_0 = x_1 \partial/\partial z_1 + x_2 \partial/\partial z_2 + x_3 \partial/\partial z_3$ and $\bar{X}_0$ and one sees immediately that $\partial\rho_0(X_0) = (3/2)\rho_0$ and $\tau_0(X_0) \equiv 0$.

Returning to the general situation it is assumed henceforth that N is the space of (extensions to C^n of) smooth sections of the vector bundle whose fibres are defined by the pointwise analog of (1.12) (it is sufficient for this [2] to assume that these fibres have constant dimension). The content of the following theorem is that under this assumption the calculations just made for M^5 can be generalized, and expressions for $\partial\bar{\partial}\rho_i$ and N similar to those above can be obtained for any manifold. Moreover, there will exist a local complex foliation of M with leaves tangent to N. Let $k = \operatorname{codim} H$ and $p = \operatorname{codim} N$.

THEOREM 1.2 [2]. (A) *There exists a set* $\tau^2 = \{\tau^2_{k+1}, \cdots, \tau^2_p\} \subset \mathscr{E}^{1,0}$ *such that* $N = \{X \in H : \tau^2_j(X)$ *and* $\bar{\tau}^2_j(X) \in O(M), j = k+1, \cdots, p\}$. *The set* τ^2 *is a second-order construction.*

(B) *The module of relative* (1, 0) *-forms*

(1.15) $$\mathscr{I}_2 = \mathrm{sp}\{\tau^2, \partial\rho, \bar{\partial}\rho, \rho\}$$

is a second-order invariant; in particular it is independent of a choice of τ^2 *satisfying* (A).

(C) *For each i there exist functions* α^i_{jl} *such that*

(1.16) $$\partial\bar{\partial}\rho_i = \sum_{j,l=k+1}^{p} \alpha^i_{jl}\tau^2_j \wedge \bar{\tau}^2_l \quad \mathrm{mod}\ (\mathscr{I}_1).$$

(D) *The module* $\mathscr{S}_2 = \mathscr{I}_2 + \bar{\mathscr{I}}_2$ *satisfies* $d\mathscr{S}_2 \subset (\mathscr{S}_2)$.

In view of (A) and (B), $\mathscr{I}_2$ can be considered the module of relative (1,0)-forms defining N. For M^5 it was shown in (1.14) that τ^2 can be taken as the single (1, 0)-form $\tau_0 = x_3dz_2 - x_2dz_3$. Of course $k = 1$ and $p = 2$ for this example.

It is now possible to restate Proposition 1.1 in a dual formulation:

(1.17) *If* $\mathscr{S}$ *is the straightening module of a map* (1.1) *then* $\mathscr{S} \supset \mathscr{I}_2$.

This strengthens (1.8) since $\mathscr{I}_2 \supset \mathscr{I}_1$.

Statement (C), which has been verified in (1.13) for M^5, can be interpreted via the identities $d(\bar{\partial}\rho) = \partial\bar{\partial}\rho = -d(\partial\rho)$ as asserting that $\mathscr{I}_1 \cup d\mathscr{I}_1 \subset (\mathscr{I}_2)$. In fact $\mathscr{I}_2$ can be characterized as the smallest module of relative (1,0)-forms whose ideal contains $\mathscr{I}_1$ and $d\mathscr{I}_1$. This is how $\mathscr{I}_2$ appears in Theorem 2.1 below.

The module $\mathscr{S}_2 = \mathrm{sp}\{\tau^2, \partial\rho, \bar{\tau}^2, \bar{\partial}\rho, \rho\}$ of (D) is the full module of one-forms defining N, and (D) is the integrability condition (in dual form) that implies the existence of a local complex foliation of M for which N is the module of tangent vector fields. For M^5 this foliation was found to be the one given by (1.3).

Thus this foliation exists in considerable generality. It will be seen that it may not (in fact usually does not) admit a definition of the form (1.2) by a straightening map. This behavior is displayed by the example M^5. It is clear that a nontrivial straightening of M^5 is possible only if $q = 2$, and then (1.14) shows that the dimensions of $\mathscr{S}$ and $\mathscr{I}_2$ are the same. By (1.17), $\mathscr{S}$ and $\mathscr{I}_2$ must coincide. But $d\mathscr{I}_2 \not\subset (\mathscr{I}_2)$, for by direct calculation one finds $d\tau_0 \wedge \tau_0 \wedge \partial\rho_0 \wedge \bar{\partial}\rho_0 \notin O(M)$. This contradiction of (1.11) shows that *no* biholomorphic map satisfies (1.1). In other words, M^5 is a real-analytic hypersurface which cannot be straightened, even though it admits a local complex foliation tangent to N. By the way, it is about the simplest example of this kind; it is obvious that any linear manifold can be straightened and since the obstructions to straightening are of third and higher order it turns out that any "quadric" manifold can also be straightened. Moreover, in $\mathbf{C}^2$ a manifold with degenerate Levi form is Levi-flat, and if real-analytic is therefore locally biholomorphic to a real linear subspace [**2**].

2. Invariant obstructions to straightening. The example M^5 suggests that there must be other obstructions to straightening besides $\mathscr{I}_1$ and $\mathscr{I}_2$, perhaps something to measure the failure of $\mathscr{I}_2$ to generate a d-closed ideal.

THEOREM 2.1. *There exists a chain* $\mathscr{I}_1 \subset \mathscr{I}_2 \subset \mathscr{I}_3 \subset \cdots$ *of modules of relative* (1,0)*-forms beginning with* $\mathscr{I}_1$ *defined by* (1.6) *and satisfying the following properties*:

(A) *For each* $k \geqq 2$, $\mathscr{I}_k$ *is the smallest module* $\mathscr{K}$ *of relative* (1, 0)*-forms such that* $\mathscr{I}_{k-1} \cup d\mathscr{I}_{k-1} \subset (\mathscr{K})$.

(B) *For each* $k \geqq 2$ *the module* $\mathscr{J}'_k = \mathscr{J}_k + \bar{\mathscr{J}}_2$ *satisfies* $d\mathscr{J}'_k \subset (\mathscr{J}'_k)$.

(C) *The union* $\mathscr{J} = \bigcup_{k=1}^{\infty} \mathscr{J}_k$ *is the smallest module* $\mathscr{K}$ *of relative* (1, 0)-*forms such that* $\mathscr{K} \supset \mathscr{J}_1$ *and* $d\mathscr{K} \subset (\mathscr{K})$.

The theorem is equally true in the real-analytic category.

REMARKS 2.2. Each $\mathscr{J}_k$ generates the "d-closure" of $(\mathscr{J}_{k-1})$; thus $(\mathscr{J}_{k-1})$ is d-closed if and only if $\mathscr{J}_{k-1} = \mathscr{J}_k$. The smallness properties in (A) and (C) determine the $\mathscr{J}_k$ and $\mathscr{J}$ uniquely. (Incidentally, their existence does not appear to follow immediately from these properties; in general d-closure does not behave well with respect to intersection.) As stated earlier, $\mathscr{J}_2$ turns out to be the module defined by (1.15). It follows easily from (A) and (C) that $\mathscr{J}$ and the $\mathscr{J}_k$ are invariants; in fact $\mathscr{J}_k$ is an invariant of order k. Each can be calculated in a finite number of steps. The chain $\{\mathscr{J}_k\}$ terminates at k, meaning $\mathscr{J}_k = \mathscr{J}_{k+1} = \cdots = \mathscr{J}$, if and only if $d\mathscr{J}_k \subset (\mathscr{J}_k)$. In the real-analytic category the chain always terminates, and so all of the $\mathscr{J}_k$, hence $\mathscr{J}$, can be calculated in a finite number of steps.

2.3. The invariants $\mathscr{J}_k$ are independent. In fact for each $n \geqq 2$ there exists a real-analytic hypersurface M^{2n-1} in $\boldsymbol{C}^n$ for which $\mathscr{J}_2 \neq \mathscr{J}_3 \neq \cdots \neq \mathscr{J}_n = \mathscr{E}^{1,0} + \mathscr{J}_1$. These hypersurfaces are easily defined as generalized "tangent developables", but verification of their properties requires a somewhat lengthy calculation. (They are not generalizations of M^5. Regardless of degree, homogeneous functions like $x_1^3 + x_2^3 + x_3^3$ seem to define manifolds with only one-dimensional degeneracy of the Levi form, which is not enough "room" to distinguish the $\mathscr{J}_k$ for $k \geqq 3$.)

2.4. There are two trivial special cases of the theorem:

(i) The Levi form is nondegenerate; that is $N = 0$. Then $\tau^2 \cup \partial\rho$ generates $\mathscr{E}^{1,0}$ so $\mathscr{J}_2 = \mathscr{E}^{1,0} + \mathscr{J}_1$, the entire module of relative (1,0)-forms.

(ii) The *Levi flat* case $N = H$. Then $\tau^2 = \phi$ is the only choice and $\mathscr{J}_2 = \mathscr{J}_1$, i. e., $d\mathscr{J}_1 \subset (\mathscr{J}_1)$, which has the more common expression $\partial\bar{\partial}\rho_j \in (\partial\rho, \bar{\partial}\rho, \rho)$.

In either case the chain terminates at $\mathscr{J}_2 (= \mathscr{J})$.

2.5. Statement (C) follows easily from (A). Moreover, if $\mathscr{S}$ is any straightening module it follows from (1.8) and (1.11) that $\mathscr{S} \supset \mathscr{J}$. Therefore $\mathscr{J}$ is always an obstruction to straightening. In the real-analytic category it can be shown that if $\mathscr{J}$ has constant rank on M it defines an integrable Pfaffian system ($j^*\mathscr{J}$ in fact) whose solutions g_j are real-analytic CR functions (and also define $\mathscr{J}$). By a result of Tomassini [**4**] there exist holomorphic functions f_j such that $f_j|M = g_j|M$, and $\mathscr{J}$ is defined by the df_j. These holomorphic functions can be shown to be the first q coordinates of a straightening map (1.1), so in this case $\mathscr{J}$ is itself a straightening module. Hence in the real-analytic case where it has constant rank, $\mathscr{J}$ provides the complete obstruction to straightening.

As a corollary of this fact, it follows that a real-analytic M such that $\mathscr{J}_2 = \mathscr{J}_3$ admits a straightening whose M_0 has nondegenerate Levi form. Therefore real-analytic hypersurfaces with degenerate Levi form (of constant rank) and this invariant property $\mathscr{J}_2 = \mathscr{J}_3$ are subject to the Chern-Moser classification [**1**].

2.6. Even though $\mathscr{J}_k$ may not generate a d-closed ideal, statement (B) says that $\mathscr{J}'_k$ always does. This fact is useful in part because it permits an inductive construction of the $\mathscr{J}_k$, as will be presently indicated. It also implies that for each $k \geqq 2$ the conjugate-symmetric submodule $\mathscr{S}_k = \mathscr{J}_k + \bar{\mathscr{J}}_k$ satisfies

(2.1) $$d\mathscr{I}_k \subset (\mathscr{I}_k).$$

This is immediate from $d\mathscr{I}_k \subset (\mathscr{I}'_k) \subset (\mathscr{I}_k)$, which follows from (B), and its conjugate. For $k = 2$, (B) is just a restatement of Theorem 1.2(D) since $\mathscr{I}'_2 = \mathscr{I}_2$. The integrability condition (2.1) shows that each $\mathscr{I}_k$ which is of constant rank defines a local complex foliation of M which (since $\mathscr{I}_k \supset \mathscr{I}_2$) is a subfoliation of the one defined by $\mathscr{I}_2$ (that is, the one tangent to N). This chain of subfoliations can be distinguished from the great profusion of arbitrary ones by their possession of a certain relative flatness property which will not be elaborated here. Its definition requires a calculation of the modules of vector fields annihilated on M by the $\mathscr{I}_k$.

SKETCH OF PROOF FOR (A) AND (B). The idea is to adjoin a set σ of (1, 0)-forms to $\mathscr{I}_k$ such that $\mathscr{I}_{k+1} \equiv \mathrm{sp}\{\sigma\} + \mathscr{I}_k$ satisfies (A). This procedure is outlined in the case of a hypersurface M for which N has codimension 1 in H; there a single (1,0)-form will suffice to satisfy Theorem 1.2(A) and generate $\mathscr{I}_2 = \mathrm{sp}\{\tau^2\} + \mathscr{I}_1$. In this case $\mathscr{I}_{k+1}$ can be obtained from $\mathscr{I}_k$ by adjunction of a single (1, 0)-form τ^{k+1}:

It follows from Theorem 1.2(C) that $d\partial\rho_i$ and $d\bar{\partial}\rho_i \in (\mathscr{I}_2)$ and from Theorem 1.2 (D) that

$$d\tau^2 \in \mathscr{I}_2 = (\bar{\tau}^2, \mathscr{I}_2) = (\tau^2, \partial\rho, \bar{\tau}^2, \bar{\partial}\rho, \rho).$$

Therefore there exists a one-form τ^3 such that

(2.2) $$d\tau^2 = \tau^3 \wedge \bar{\tau}^2 \bmod (\mathscr{I}_2).$$

It is easy to show that one can take τ^3 to be a (1, 0)-form in (2.2). Letting $\mathscr{I}_3 = \mathrm{sp}\{\tau^3\} + \mathscr{I}_2$ it follows that $d\mathscr{I}_2 \cup \mathscr{I}_2 \subset (\mathscr{I}_3)$. The proof of minimality requires some algebra which is omitted. Relation (2.2) implies that

$$0 = d^2\tau^2 = d\tau^3 \wedge \bar{\tau}^2 \bmod (\mathscr{I}_3)$$

which, after some more algebra, shows that $d\tau^3 \in (\mathscr{I}'_3) = (\bar{\tau}^2, \mathscr{I}_3)$. Therefore (A) and (B) are verified for $k = 3$. The process can be continued, for (B) shows that

$$d\tau^3 = \tau^4 \wedge \bar{\tau}^2 \bmod (\mathscr{I}_3)$$

for some τ^4, which as before can be taken as a (1,0)-form. Now let $\mathscr{I}_4 = \mathrm{sp}\{\tau^4\} + \mathscr{I}_3$, etc.

REFERENCES

1. S. S. Chern and J. Moser, *Real hypersurfaces in complex manifolds*, Acta. Math. **133** (1974), 219–271.

2. M. Freeman, *Local complex foliation of real submanifolds*, Math. Ann. **209** (1974), 1–30. MR **49** #10911.

3. ———, *Local biholomorphic straightening of real submanifolds* (to appear).

4. G. Tomassini, *Tracce delle funzioni olomorfe sulle sottovarietà analitichà reali d'una varietà complessa*, Ann. Scuola Norm. Sup. Pisa. (3) **20** (1966), 31–43. MR **34** #6808.

UNIVERSITY OF KENTUCKY

Proceedings of Symposia in Pure Mathematics
Volume 30, 1977

BOUNDARY BEHAVIOUR OF THE CARATHÉODORY AND KOBAYASHI METRICS

IAN GRAHAM

We wish to report on two addenda to the results which appear in [8]. One is a generalization of these results to domains in manifolds. The other is an application to an estimate for the Carathéodory metric used by G. Henkin [9].

Let G be a domain with compact closure in $\mathbf{C}^n$ or in a Stein manifold. Let $z \in G$ and let ξ be a tangent vector at z. The Carathéodory metric is defined by

$$F_C(z, \xi) = \sup_{f: G \to \Delta;\, f(z) = 0} |f_*(\xi)|.$$

Here Δ denotes the unit disk and f is of course holomorphic. The Kobayashi metric on G is defined by

$$F_K(z, \xi) = \inf\{\alpha | \alpha > 0, \exists f: \Delta \to G \text{ with } f(0) = z, f'(0) = \xi/\alpha\}.$$

These objects are Finsler metrics, non-Hermitian in general. They are of interest because they are invariant under biholomorphic mappings. The Carathéodory metric has been studied by Reiffen [12], and the Kobayashi metric by Royden [13]. The Kobayashi distance function is of course due to Kobayashi [11]. The relationship between distance functions and metrics is discussed in [8], [12], [13]. In [8] we prove the following theorem, using domains of comparison and related techniques from the study of the boundary behaviour of the Bergman kernel function and associated quantities [2], [5], [6], [10] (these techniques have been carried much further in [7]):

THEOREM. *Let G be a (bounded) strongly pseudoconvex domain in $\mathbf{C}^n$ with smooth boundary. Let $F(z, \xi)$ be either the Carathéodory or Kobayashi metric on G. Let $z_0 \in \partial G$. Let $d(z, \partial G)$ denote the Euclidean distance of a point $z \in G$ to the boundary. Then*

AMS (MOS) subject classifications (1970). Primary 32E30, 32F15, 32H15.

© 1977, American Mathematical Society

(i)
$$\lim_{z\to z_0} F(z, \xi)d(z, \partial G) = \tfrac{1}{2}\|\xi_N(z_0)\|.$$

($\xi_N(z_0)$ denotes the complex normal component of ξ at the boundary point z_0.) If $\xi_N(z_0) = 0$, i.e., ξ is a holomorphic tangent vector to ∂G at z_0, then

(ii)
$$\lim_{z\to z_0, z\in A} (F(z, \xi))^2 d(z, \partial G) = \tfrac{1}{2}\mathscr{L}_{z_0}(\xi).$$

(We must use a function r such that $\|\nabla_z r(z_0)\| = 1$ in defining the Levi form $\mathscr{L}_{z_0}$.) The approach in the second limit must be nontangential unless one does the following: sufficiently near the boundary we may decompose any vector into complex normal and tangential components. $\xi = \xi_N(z) + \xi_T(z)$. Then

(iii)
$$\lim_{z\to z_0}(F(z, \xi_T(z))^2 d(z, \partial G) = \tfrac{1}{2}\mathscr{L}_{z_0}(\xi_T(z_0))$$

with no restrictions.

The addenda are the following:

(1) Let G be a strongly pseudoconvex domain with smooth boundary and compact closure in a Stein manifold M. Let r be a smooth function on M which defines the boundary ∂G, i.e., $r < 0$ in G, $r > 0$ on $M - \bar{G}$, $dr \neq 0$ on ∂G. Let ξ be a holomorphic vector field (with smooth, not holomorphic, coefficients) in a neighborhood of $z_0 \in \partial G$. Using $-r$ in place of the Euclidean distance to the boundary we obtain the following statements:

(i)
$$\lim_{z\to z_0}(-r(z))F(z, \xi(z)) = |\langle \partial r, \xi\rangle_{z_0}|$$

($\langle\ ,\ \rangle$ denotes contraction between cotangent and tangent vectors).

(ii) If ξ is tangential to ∂G in a neighborhood of z_0 then

$$\lim_{z\to z_0}(-r(z))\,(F(z, \xi))^2 = \langle \partial\bar{\partial} r, \xi \wedge \bar{\xi}\rangle_{z_0}.$$

If ξ is tangential at z_0 but not in a neighborhood then z must be restricted to approach z_0 nontangentially here (e.g., along curves which intersect the boundary transversally). Finally, given a Hermitian metric on M, statement (i) and the first part of (ii) are uniform in unit vector fields.

These results are a straightforward extension of the above theorem. As noted in [**8**] the approximation theorem used to localize the boundary behaviour of the Carathéodory metric carries over to this case, with slightly stronger smoothness assumptions on ∂G. This approximation theorem is in the style of Lemma 3.5.2 in [**10**] with an additional idea of N. Kerzman.

(2) In [**9**] Henkin makes use of the following estimate for the Carathéodory metric on a strongly pseudoconvex domain in $\boldsymbol{C}^n$ with smooth boundary:

$$F(z, \xi) \simeq \frac{\|\xi_N(z)\|}{d(z, \partial G)} + \frac{\|\xi_T(z)\|}{(d(z, \partial G))^{1/2}}.$$

The decomposition into complex normal and tangential components is defined for points sufficiently near ∂G by translating the decomposition at the nearest boundary point. This estimate is also used by H. Alexander in [**1**]. To my knowledge a proof of this estimate has not appeared. We shall indicate how it follows from the above theorem.

PROOF OF HENKIN'S ESTIMATE. (a) The Carathéodory metric is subadditive in ξ

[**8**, §2.1]. Thus $F(z, \xi) \leqq F(z, \xi_N(z)) + F(z, \xi_T(z))$. Together with the results of the above theorem this gives

$$F(z, \xi) \leqq C\left[\frac{\|\xi_N(z)\|}{d(z, \partial G)} + \frac{\|\xi_T(z)\|}{(d(z, \partial G))^{1/2}}\right] \quad \text{for some constant } C.$$

(b) We wish to show

$$F(z, \xi) \geqq c\left[\frac{\|\xi_N(z)\|}{d(z, \partial G)} + \frac{\|\xi_T(z)\|}{(d(z, \partial G))^{1/2}}\right] \quad \text{for some constant } c.$$

We shall reduce this to a statement about the unit ball. Let $z_0 \in \partial G$. z_0 is the nearest point on the boundary to points z on the interior normal to ∂G at z_0. If U is a small neighborhood of z_0 we have $F_{G\cap U}(z, \xi) / F_G(z, \xi) \to 1$ as $z \to z_0$ [**8**, Proposition 6]. Hence it suffices to prove the inequality with F replaced by $F_{G\cap U}$ and z near z_0. In fact we shall replace $G \cap U$ by a domain $G' \subset U$ such that

(i) $G' \supset G \cap U$, and in fact $G' \supset (\bar{G} \cap U - \{z_0\})$;

(ii) $z_0 \in \partial G'$, and the decomposition into complex normal and tangential directions at z_0 is the same for G and G';

(iii) G' is biholomorphic via a transformation whose Jacobian matrix at z_0 is the identity followed by a complex linear transformation to the intersection of the unit ball B with a neighborhood V of a boundary point. Calling the combined transformation Φ, we thus have $d(z, \partial G) \simeq d(\Phi(z), \partial B)$ for points z near z_0 on the interior normal. (We may take G' to be defined by the terms up to second order in the function $r(z) - \varepsilon\|z - z_0\|^2$ for sufficiently small ε.)

We have $F_{G\cap U} \geqq F_{G'}$, so it suffices to estimate $F_{G'}$ for points z on the interior normal. In view of property (iii) of G' it suffices to estimate $F_{B\cap V}$, or in fact F_B since $F_{B\cap V} \geqq F_B$. Now the estimate for F_B follows from the explicit expression [**8**, §2.2] for the Carathéodory metric on B.

We may state this result for strongly pseudoconvex domains in Stein manifolds also. Replace $d(z, \partial G)$ by $(-r(z))$, and use the decomposition into complex normal and tangential directions defined by the level sets of r.

BIBLIOGRAPHY

1. H. Alexander, *Proper holomorphic mappings of bounded domains*, Proc. Sympos. Pure Math., vol. 30, part 2, Amer. Math. Soc., Providence, R.I., 1977, pp. 171–174.

2. S. Bergman, *Über die Kernfunktion und ihr Verhalten am Rande*. I, II, J. Reine Angew. Math. **169** (1933), 1–42; **172** (1935), 89–128.

3. C. Carathéodory, *Über das Schwarze Lemma bei analytischen Funktionen von zwei komplexen Veränderlichen*, Math. Ann. **97** (1926), 76–98.

4. ———, *Über die Geometrie der analytischen Abbildungen, die durch analytische Funktionen von zwei Veränderlichen vermittelt werden*, Abh. Math. Sem. Univ. Hamburg **6** (1928), 97–145.

5. K. Diederich, *Das Randverhalten der Bergmanschen Kernfunktion und Metrik in streng pseudokonvexen Gebieten*, Math. Ann. **187** (1970), 9–36. MR **41** #7149.

6. ———, *Über die* 1. *und* 2. *Ableitungen der Bergmanschen Kernfunktion und ihr Randverhalten*, Math. Ann. **203** (1973), 129–170. MR **48** #6472.

7. C. Fefferman, *The Bergman kernel and biholomorphic mappings of pseudoconvex domains*, Invent. Math. **26** (1974), 1–65. MR **50** #2562.

8. I. Graham, *Boundary behavior of the Carathéodory and Kobayashi metrics on strongly pseudoconvex domains in* $\mathbf{C}^n$ *with smooth boundary*, Trans. Amer. Math. Soc. **207** (1975), 219–240.

9. G. M. Henkin, *An analytic polyhedron is not holomorphically equivalent to a strongly pseudo-*

convex domain, Dokl. Akad. Nauk SSSR **210** (1973), 1026–1029 = Soviet Math. Dokl. **14** (1973), 858–862. MR **48** #6467.

10. L. Hörmander, *L^2 estimates and existence theorems for the $\bar{\partial}$ operator*, Acta Math. **113** (1965), 89–152. MR **31** #3691.

11. S. Kobayashi, *Hyperbolic manifolds and holomorphic mappings*, Pure and Appl. Math., 2, Dekker, New York, 1970. MR **43** #3503.

12. H.-J. Reiffen, *Die differentialgeometrischen Eigenschaften der invarianten Distanzfunktion von Carathéodory*, Schr. Math. Inst. Univ. Münster No. 26 (1963). MR **28** #1320.

13. H. L. Royden, *Remarks on the Kobayashi metric*, Several Complex Variables II (Proc. Internat. Conf., Univ. of Maryland, College Park, Md., 1970), Lecture Notes in Math., vol. 185, Springer-Verlag, Berlin and New York, 1971, pp. 125–137. MR **46** #3826.

UNIVERSITY OF TORONTO

Proceedings of Symposia in Pure Mathematics
Volume 30, 1977

FUNCTION THEORY ON TUBE MANIFOLDS

C. DENSON HILL* AND MICHAEL KAZLOW

Let X be a C^∞ embedded CR manifold in $\boldsymbol{C}^n$, and let $CR(X)$ denote the complex-valued CR functions on X, i.e., the functions u which satisfy the tangential Cauchy-Riemann epuations $\bar{\partial}_X u = 0$ on X. In the trivial case of zero codimension, where X is an open set in $\boldsymbol{C}^n$, we have $CR(X) = \mathcal{O}(X)$, the algebra of holomorphic functions on X. By analogy with this trivial case, it would seem desirable to develop also in the case of positive codimension a "function theory" for $CR(X)$. For certain kinds of purely local questions there are already a number of results in the literature which have been referred to in many other reports in these PROCEEDINGS. But it seems that almost nothing is known about global, or semiglobal, questions—especially for codimension greater than one—and these global questions seem to be very difficult.

In order to understand what kinds of global phenomena for $CR(X)$ might be expected, in general, one would like to find a special model class of CR submanifolds X for which a theory becomes feasible, but which is at the same time rich enough to exhibit most of the typical features. By analogy with the classical development of several complex variables, there are two obvious candidates for such a model: Reinhardt manifolds and tube manifolds. But the study of CR functions on a Reinhardt manifold is equivalent to the study of periodic CR functions on a tube manifold. Therefore we consider here the more general case of tubes.

Aside from their interest as a model case for CR manifolds, the tube manifolds are of interest in their own right since they occur, for example, in the theory of group representations.

1. Tube manifolds. Let $z = x + iy$ denote the usual coordinates in $\boldsymbol{C}^n = \boldsymbol{R}^n + i\boldsymbol{R}^n$; the first factor $\boldsymbol{R}^n$ is the "real domain". For a set $A \subset \boldsymbol{R}^n$ in the real domain we define

(i) $\tau A = A + i\boldsymbol{R}^n$ = tube over A,

AMS (MOS) subject classifications (1970). Primary 32D10, 32F99.

*Research supported by an Alfred P. Sloan Research Fellowship and by the National Science Foundation grants NSF PO43 957X00 and MPS 72–05055 AO2.

© 1977, American Mathematical Society

(ii) ch A = convex hull of A in $\boldsymbol{R}^n$,
(iii) ach A = almost convex hull of A = $\{\text{rel int ch } A\} \cup A$,
(iv) τ ach A = $\tau(\text{ach } A)$.

In (iii), rel int ch A means the interior of the convex hull of A in the relative topology of the smallest affine linear subspace of $\boldsymbol{R}^n$ that contains ch A.

Let M be a connected real m-dimensional locally closed C^∞ submanifold in the real domain $\boldsymbol{R}^n$. By a *tube manifold* we mean a manifold of the form τM. A *Reinhardt manifold* is one of the form exp τM where exp denotes the map that sends $z = (z_1, z_2, \cdots, z_n)$ to $\exp z = (e^{z_1}, e^{z_2}, \cdots, e^{z_n})$. Thus the tube manifolds are those invariant under translation in the pure imaginary directions and Reinhardt manifolds are those invariant under multiplication of each coordinate by a different factor of modulus one. CR functions on exp τM pull back to periodic CR functions on τM.

Let $\pi : \tau M \to M$ be projection onto the real domain. The following remarks make it possible to reduce the complex geometry of τM to the real geometry of the base M:

(a) τM is a generic CR manifold of real dimension $n + m$ and of complex CR dimension m.

(b) The holomorphic tangent space to τM at p is isomorphic to the complexified real tangent space to M at $\pi(p)$.

(c) The Levi form of τM at p is equal to the second fundamental form of M at $\pi(p)$.

2. Generalization of the Bochner tube theorem. If M is an open set in $\boldsymbol{R}^n$ (codim zero case) then the main theorem about tubes is the theorem of Bochner which states that the envelope of holomorphy of τM is τ ch M, i.e., that the restriction map $\mathcal{O}(\tau \text{ ch } M) \rightrightarrows \mathcal{O}(\tau M)$ is an isomorphism. For a connected locally closed C^∞ submanifold M of $\boldsymbol{R}^n$ of any dimension, we have the following nontrivial generalization:

THEOREM 1 (M. KAZLOW). (1) *The restriction map $CR(\tau \text{ ach } M) \rightrightarrows CR(\tau M)$ is a Fréchet isomorphism; here CR denotes CR functions that are C^∞ in the sense of Whitney.*

(2) $\sup_{\tau \text{ ach } M} |\tilde{u}| = \sup_{\tau M} |u|$ *for any extension $\tilde{u}$ of u.*

The proof of Theorem 1 can be found in the Ph.D. thesis of Kazlow [3]. The major difficulties in the proof arise when codim $M > 1$. For the much simpler case where codim $M = 1$, the theorem was proved in [1]. Note that one obtains extension of CR functions on τM to holomorphic functions only when ch M has dimension n.

One can obviously use Theorem 1 to extract information pertaining to the global properties of functions in $CR(\tau M)$. For a more complete discussion, see the thesis of Kazlow.

When one imposes some growth condition on the CR functions on τM, e.g., for $L^2(\tau M) \cap CR(\tau M)$, one can obtain extension results analogous to Theorem 1 in a much more elementary way, e. g., by use of the Fourier transform; see [4].

But the advantage of the methods used in the proof of Theorem 1 is that they also yield semiglobal or semilocal results about the extension of CR functions on

τM. These extension theorems have various applications. We give two examples below:

3. The maximum modulus principle. For definitions of the terms used here, see the report by Ellis and Hill [2] in these PROCEEDINGS. Suppose $m \geqq 2$ and let B_p denote the quadratic form associated to the second fundamental form of M at p. For any $\xi \in N_p$ = the normal space to M at p, let $k_1(\xi) \geqq k_2(\xi) \geqq \cdots \geqq k_m(\xi)$ denote the principal curvatures of M in the ξ direction, i.e., the eigenvalues of the scalar quadratic form $\langle \xi, B_p \rangle$. According to the results announced in [2], a necessary condition for the CR function on τM to obey a local maximum modulus principle is that τM should contain no extreme points which, in the case at hand, implies that

$$k_1(\xi) k_m(\xi) \leqq 0, \qquad \forall\, \xi \in N_p,\ \forall\, p \in M.$$

According to [2] the stronger inequality $k_1(\xi)k_m(\xi) < 0$, $\forall\, \xi \in N_p$, $\forall\, p \in M$, was a sufficient condition for the CR functions on τM to obey the local maximum modulus principle. But now for tube manifolds we can give a necessary and sufficient condition: For ω a neighborhood of p in $\boldsymbol{R}^n$ set $M_\omega = M \cap \omega$.

THEOREM 2. *A necessary and sufficient condition for the CR functions on τM to obey the local maximum modulus principle is that for every $p \in M$ there should exist a fundamental sequence of neighborhoods $\{\omega\}$ of p such that*

$$p \in \textit{interior}\ \mathrm{ch}\ M_\omega.$$

COROLLARY. *For a tube manifold τM the local maximum modulus principle for its CR functions is valid if and only if τM contains no extreme points.*

PROOF. If such a fundamental sequence of neighborhoods $\{\omega\}$ exists then for a function $u \in CR(\tau M)$, $|u|$ cannot have a local maximum at any $p' \in \pi^{-1}(p)$ because u extends locally to a holomorphic $\tilde{u}$ whose modulus would also have to have a local maximum of p'. This can happen only if u is constant. If no such $\{\omega\}$ exists then there is some ω such that $p \in$ boundary ch M_ω. It follows that any $p' \in \pi^{-1}(p)$ is an extreme point and that the local maximum modulus principle is not valid at p'. If τM contains no extreme points then for every $p \in M$ the $\{\omega\}$ of Theorem 2 must exist.

4. Hypoellipticity. Let $\bar{\partial}_M$ denote the tangential Cauchy-Riemann operator on τM, i.e., $CR(\tau M) = \ker \bar{\partial}_M$. Instead of smooth CR functions on τM, one could consider distribution CR functions, i.e., $u \in \mathscr{D}'(\tau M)$ which satisfy $\bar{\partial}_M u = 0$ in the sense of distributions. The methods used to prove Theorem 1 seem to indicate that one can concoct a proof of the following proposition, but the authors have not written down an explicit detailed proof:[1]

PROPOSITION. *Let ω be a neighborhood of $p \in M$ such that $p \in$ interior* ch M_ω. *Then the CR distributions in a neighborhood of any point $p' \in \pi^{-1}(p)$ on τM extend to holomorphic functions in an open set in $\boldsymbol{C}^n$ about p'.*

REMARK. The proposition shows that whenever one can expect to have a maxi-

[1] ADDED IN PROOF. This has now been established.

mum modulus principle, distribution CR functions on τM are smooth CR functions. Hence it was not important to be explicit about the amount of differentiability assumed in the statement of Theorem 2.

If f is a form of type (0, 1) on τM we can consider the inhomogeneous equations $\partial_M u = f$ acting, say, on distributions u on τM and ask when does:

(α) $f \in C^\infty$ near a point $p' \in \tau M \Rightarrow u \in C^\infty$ near p' (hypoellipticity for $\bar{\partial}_M$)?

(β) f real-analytic near a point $p' \in \tau M$ (for real-analytic M) $\Rightarrow u$ real-analytic near p' (real-analytic hypoellipticity for $\bar{\partial}_M$)?

Modulo the proof of the proposition, one can obtain the following

THEOREM 3. *The condition of Theorem 2 is a necessary and sufficient condition for* (α) (*in the case of* $C^\infty M$) *or for* (β) (*in the case of real analytic* M).

PROOF. Let $p = \pi(p')$ and assume that there is a fundamental sequence of neighborhoods $\{\omega\}$ of p such that $p \in$ interior ch M_ω.

In the C^∞ case, the proposition shows that the distribution CR functions on τM near p' and the C^∞ CR functions on τM near p' are the same. This means that the sheaf of germs of CR functions on τM near p' has (the beginning of) two fine resolutions: one by forms with C^∞ coefficients and one by forms with coefficients in $\mathscr{D}'$. By the abstract deRham theorem, they have the same cohomology. What we shall use is that $H^1(C^\infty) = H^1(\mathscr{D}')$: If $\bar{\partial}_M u = f$ with $f \in C^\infty$ and $u \in \mathscr{D}'$ then f represents the zero class in $H^1(\mathscr{D}')$, hence also the zero class in $H^1(C^\infty)$. Thus there is a local $w \in C^\infty$ such that $\bar{\partial}_M w = f$ near p'. Then $\bar{\partial}_M(u - w) = 0$ so $u - w$ and hence u must be C^∞.

In the real-analytic case, a local real-analytic solution w of $\bar{\partial}_M w = f$ for real-analytic f exists by the Cauchy-Kowalewski (actually Cartan-Kähler) theorem. Then $\bar{\partial}_M(u - w) = 0$ so $u - w$ and hence u must be real-analytic.

If there is some suitably small neighborhood ω of p such that $p \notin$ interior ch M_ω then at p' there is a supporting hyperplane. Without loss of generality we can assume $p' = 0$ and Re $z_1 = 0$ is the equation of the hyperplane. Then the function $1/z_1$ represents a distribution u on τM near p' which is not equivalent to a C^∞ function nor to a real-analytic function near p'. But it is easy to see that $\bar{\partial}_M(1/z_1) = 0$ near p'. This gives a counterexample to both C^∞ and real-analytic hypoellipticity.

REFERENCES

1. J. A. Carlson and C.D. Hill, *On the maximum modulus principle for the tangential Cauchy-Riemann equations*, Math. Ann. **208** (1974), 91–97. MR **50** #5011.

2. D. Ellis and C. D. Hill, *The maximum modulus principle for CR functions on smooth real embedded submanifolds of* C^n, Proc. Sympos. Pure Math., vol. 30, part 1, Amer. Math. Soc., Providence, R.I., 1977, pp. 139–140.

3. M. Kazlow, *CR functions on tube manifolds*, Ph.D. Thesis, SUNY at Stony Brook.

4. H. Rossi and M. Vergne, *Equations de Cauchy Riemann tangentielles associées à un domaine de Siegel* (to appear).

STATE UNIVERSITY OF NEW YORK AT STONY BROOK

Proceedings of Symposia in Pure Mathematics
Volume 30, 1977

A UNIQUENESS THEORY FOR THE INDUCED CAUCHY-RIEMANN OPERATOR

L. R. HUNT

1. Introduction. Let M be a $\mathscr{C}^\infty$ real k-dimensional CR submanifold of an n-dimensional complex manifold X. Denote by $\bar{\partial}_M$ the induced CR operator on M. We are interested in the uniqueness properties of solutions to the induced Cauchy-Riemann equations $\bar{\partial}_M u = 0$. In particular, we ask under what conditions do solutions to this equation have the property of uniqueness of analytic continuation (both locally and globally) and which CR submanifolds S of M have the property that solutions of $\bar{\partial}_M u = 0$ on M near $p \in S$ and vanishing on S must vanish in an open neighborhood of p in M?

We have obtained almost complete results concerning these problems in the case that M is a hypersurface in X (see **[4]** and **[5]**). In addition, under an assumption on the Levi algebra, we have proved theorems concerning uniqueness of analytic continuation for smooth solutions to $\bar{\partial}_M u = 0$ in the higher codimensional case **[6]**. The precise results together with tools and methods used in obtaining these results are discussed in the remainder of this article.

Historically, most of the published research dealing with the problem of uniqueness and which can be applied to induced Cauchy-Riemann operators occurs only in the case that M is Levi-flat (i.e., the Levi form on M vanishes at each point). In the case of the hypersurface, our recent results in **[4]** and **[5]** make no assumptions concerning the Levi form.

2. Unipueness. Suppose M is a $\mathscr{C}^\infty$ real k-dimensional CR submanifold of an n-dimensional complex manifold X. The CR structure on M is that induced by the complex structure on X. The CR operator on M, which is denoted by $\bar{\partial}_M$, is defined for smooth functions in **[7]** and in a distributional sense in **[8]** (at least in the hypersurface case). A $\mathscr{C}^\infty$ function f which satisfies $\bar{\partial}_M f = 0$ on M is a *CR function* on M,

AMS (*MOS*) *subject classifications* (1970). Primary 35A05, 32D20.
Key words and phrases. Unique continuation, Cauchy-Riemann operator, CR functions.

© 1977, American Mathematical Society

and a distribution u satisfying $\bar{\partial}_M u = 0$ on M is a *CR distribution* on M.

Let $T(M)$ be the real tangent bundle to M, and let $H(M)$ and $\bar{H}(M)$ be the holomorphic tangent bundle and conjugate holomorphic tangent bundle to M, respectively. The Levi algebra of M is the Lie subalgebra of complex vector fields generated by sections of $H(M)$ and $\bar{H}(M)$. If the dimension of this algebra is constant, then it is the algebra of sections of a vector bundle, and the fiber dimension of the bundle is the excess dimension of the Levi algebra (see **[2]**). We can now state a result concerning uniqueness of CR functions.

THEOREM 1 **[6]**. *Let M be a connected $\mathscr{C}^\infty$ real k-dimensional CR submanifold of X, and assume that the excess dimension of the Levi algebra is maximal at every point. If f is a CR function on M such that $f \equiv 0$ on U, where U is a nonempty open subset of M, then $f \equiv 0$ on M.*

The proof of Theorem 1 depends on the general CR extension theory found in **[7]**. If the excess dimension of M is maximal at each point, then locally all CR functions on M extend to holomorphic functions on an open subset of X. Essentially, we then use the uniqueness theory for holomorphic functions to obtain the unique continuation result for the CR functions.

Suppose that M is a $\mathscr{C}^\infty$ real hypersurface in X. A $\mathscr{C}^1$ real hypersurface S of M is *characteristic at* p for $\bar{\partial}_M$ if the real tangent space $T_p(S)$ to S at p is equal to the holomorphic tangent space $H_p(M)$ to M at p. Also S is a *characteristic* (called a complex hypersurface) if it is characteristic at each of its points. We have proved the following theorem concerning local unique continuation.

THEOREM 2 **[5]**. *Let M be a $\mathscr{C}^\infty$ real hypersurface in X, and let Ω be an open subset of M with $p \in \Omega$. Let $r \in \mathscr{C}^1(\Omega)$ satisfy $r(p) = 0$ and $dr(p) \neq 0$. Suppose that the level surface $S = \{x \in \Omega \mid r(x) = 0\}$ is noncharacteristic for $\bar{\partial}_M$ at p. Then there is a neighborhood ω of p such that if u is a CR distribution on Ω and if $u \equiv 0$ in $\Omega^+ = \{x \in \Omega \mid r(x) > 0\}$, then $u \equiv 0$ in ω.*

The theorem is first proved for a $\mathscr{C}^2$ function r and later reduced to the case where r is $\mathscr{C}^1$. The main tools used in the proof are the results from **[8]** which show that locally every CR distribution on Ω is the generalized boundary values of holomorphic functions in the complement of Ω and the classical Kontiniutätssatz. It is also shown that there can be no local unique continuation for CR distributions across characteristic hypersurfaces.

The following result characterizes the boundary of the support of a CR distribution on a real hypersurface of X.

THEOREM 3 **[5]**. *Let M be a $\mathscr{C}^\infty$ real hypersurface in X. Suppose u is a CR distribution on M, and let A denote the boundary of the support of u. If $A \neq \emptyset$ then for every point $p \in A$ there are a neighborhood U of p, a real-valued function $f \in \mathscr{C}^\infty(U)$ with $df(x) \neq 0$ for all $x \in U$, and a closed nowhere dense set $E \subset$ **R** such that*

(1) $A \cap U = \{x \in U \mid f(x) \in E\}$,

(2) *for each $t \in E$, $S_t = \{x \in U \mid f(x) = t\}$ is a complex manifold with $T(S_t) = H(M)|_{S_t}$.*

Let $\Omega = M - \operatorname{supp} u$. Notice that, as a result of Theorem 2, Ω has the following property: If $W \subset \Omega$ is an open subset with $\mathscr{C}^2$ manifold as boundary, then for each

$p \in \partial W \cap \partial\Omega$ we have $T_p(\partial W) = H_p(M)$. Consequently Theorem 3 is a special case of the following result, which is actually a theorem concerning fiber bundles and is of independent interest.

THEOREM 4 [5]. *Let M be a $\mathscr{C}^\infty$ manifold of dimension n. Let H be a subbundle of the tangent bundle of M with fiber dimension $n - 1$. Suppose $\Omega \subset M$ is an open set with the property that if $W \subset \Omega$ is an open subset with $\mathscr{C}^2$ manifold as boundary, then for each $p \in \partial W \cap \partial\Omega$ we have $T_p(\partial W) = H_p$. Then for each point $p \in \partial\Omega$, there are a neighborhood U of p, a real-valued function $f \in \mathscr{C}^\infty(U)$ with $df(x) \neq 0$ for all $x \in U$, and a closed nowhere dense set $E \subset R$ such that*

(1) $\partial\Omega \cap U = \{x \in U \mid f(x) \in E\}$,

(2) *for each $t \in E$, $S_t = \{x \in U \mid f(x) = t\}$ is an integral manifold of H.*

As an application of Theorem 3 we have the following result involving global uniqueness of analytic continuation on a real hypersurface.

THEOREM 5 [5]. *Suppose M is a connected $\mathscr{C}^\infty$ real hypersurface in X which contains no complex hypersurfaces the closure of which is foliated by complex hypersurfaces. Then every CR distribution on M which vanishes on an open subset of M vanishes identically.*

Now we consider the problem of classifying those CR submanifolds S of a $\mathscr{C}^\infty$ real k-dimensional CR manifold $M \subset X$ which have the property that CR distributions (or CR functions) on M near $p \in S$ that vanish on S must vanish in an open neighborhood of p in M. Our first result in this direction is a consequence of Theorem 2.

THEOREM 6 [5]. *Let M be a $\mathscr{C}^\infty$ real hypersurface in X, and let Ω be an open subset of M with $p \in \Omega$. Let $r \in \mathscr{C}^1(\Omega)$ satisfy $r(p) = 0$ and $dr(p) \neq 0$. Suppose that the level surface $S = \{x \in M \mid r(x) = 0\}$ is noncharacteristic for $\bar{\partial}_M$ at p. Then there is a neighborhood ω of p such that if u is a continuous CR distribution on Ω and if $u \equiv 0$ on S, then $u \equiv 0$ in ω.*

To prove Theorem 6, one defines a new function u' which is 0 in Ω^+ and u in $\Omega - \Omega^+$. By a result in [3], u' is a CR distribution on Ω. Two applications of Theorem 2 then yield the desired result.

The following theorem shows that it is possible in some cases to lower the dimension of the submanifold S in Theorem 6. A CR submanifold S of X of real dimension m is *generic* at a point $p \in S$ if the fiber dimension of $H(S)$ is equal to the maximum of 0 and $(m - n)$.

THEOREM 7 [4]. *Let M be a $\mathscr{C}^\infty$ real hypersurface in X, and let $\Omega \subset M$ be open with $p \in \Omega$. Suppose that S is a $\mathscr{C}^\infty$ real m-dimensional submanifold of Ω which is generic at $p \in S$ and suppose $m \geqq n$. Then there is a neighborhood ω of p in M such that if f is a CR function on Ω and $f \equiv 0$ on S, then $f \equiv 0$ in ω.*

The proof of Theorem 7 involves the repeated use of the solution to the additive Riemann-Hilbert problem due to Andreotti and Hill [1] and the CR extension theory found in [7]. Using these tools one finds arbitrarily close to p there is an open subset of Ω on which f is identically zero. Then an argument involving Theorem 4 will show that f actually vanishes in some open neighborhood ω of p in M.

Theorem 7 shows that locally a CR function is uniquely determined on a $\mathscr{C}^\infty$ totally real submanifold (i.e., a generic manifold having no holomorphic tangent vectors) of real dimension n. Examples are given in **[4]** which show that neither the condition that S be generic nor the condition that $m \geqq n$ in Theorem 7 can be removed.

REFERENCES

1. A. Andreotti and C. D. Hill, *E. E. Levi convexity and the Hans Lewy problem.* I. *Reduction to vanishing theorems.* II. *Vanishing theorems*, Ann. Scuola Norm. Sup. Pisa **26** (1972), 325–363; 747–806.

2. S. J. Greenfield, *Cauchy-Riemann equations in several variables*, Ann. Scuola Norm. Sup. Pisa (3) **22** (1968), 275–314. MR **38** #6097.

3. R. Harvey and J. C. Polking, *Removable singularities of solutions of linear partial differential equations*, Acta Math **125** (1970), 39–56. MR **43** #5183.

4. L. R. Hunt, *Uniqueness properties of* CR-*functions*, J. Differential Equations (to appear).

5. L. R. Hunt, J. C. Polking and M. J. Strauss, *Unique continuation for solutions to the induced Cauchy-Riemann equations* (to appear).

6. L. R. Hunt and M. J. Strauss, *Uniqueness of analytic continuation: Necessary and sufficient conditions*, J. Differential Equations (to appear).

7. L. R. Hunt and R. O. Wells, Jr., *Extension of* CR-*functions*, Amer. J. Math. (to appear).

8. J.C. Polking and R.O. Wells, Jr., *Boundary value of Dolbeault cohomology classes*, Abh. Math. Sem. Univ. Hamburg (to appear).

TEXAS TECH UNIVERSITY

Proceedings of Symposia in Pure Mathematics
Volume 30, 1977

A MONGE-AMPÈRE EQUATION IN COMPLEX ANALYSIS

N. KERZMAN*

This is a report on joint work in progress by J. J. Kohn, L. Nirenberg and the author.

0. A. It is an old difficult question in complex analysis to give sufficient conditions for a holomorphic map $T: \Omega_1 \to \Omega_2$, $\Omega_1, \Omega_2 \subset\subset \boldsymbol{C}^n$ to have a continuous or a smooth extension to $\bar{\Omega}_1$. If $n = 1$ there are familiar results of this type about the Riemann mapping function. The case $n \geqq 2$ has seen great progress in recent years **[5]**, **[8]**, **[11]**, at least for strictly pseudoconvex domains and biholomorphic maps.

We will show that there is a nonlinear partial differential equation which has a bearing on this question. See §4. The left-hand side of this equation is

$$\det[\partial^2/\partial z_i \partial \bar{z}_j], \tag{0}$$

i.e., the determinant of the complex Hessian matrix. In §3 we state a Dirichlet problem for this equation which we would like to solve with C^∞ boundary regularity. We describe a method for attacking the problem (the continuity method) which reduces it to obtaining the basic estimate (16). *We have not been able to prove* (16) *for derivatives of order* $\geqq 3$ so far; and hence the problem remains unsolved.

B. *Cases* $n = 1$ *and* $n \geqq 2$. Instead of proving boundary regularity for T (i.e., for its $2n$ real components) our idea is to do it for a *single* function u; $u = v \circ T$ is the pull-back of certain v. If $n = 1$, (0) reduces to $\frac{1}{4}\Delta$, the ordinary Laplace equation. T preserves harmonic functions and we can adopt as u and v certain harmonic functions and resort to potential theory. This we do in §1 as a desirable model to imitate when $n \geqq 2$.

If $n \geqq 2$, T no longer preserves harmonic functions. Pluriharmonic functions are preserved but they are of no use to us. We found that if $n \geqq 2$, boundary regularity for T follows from that of u provided v is *strictly* plurisubharmonic. See §2. (Besides, there is no pluriharmonic function vanishing even on a piece of the boundary of, say, the ball.) Hence, we look for solutions of a more complicated equation in §3.

AMS (MOS) subject classifications (1970). Primary 35Q99, 32H99; Secondary 32F05, 35J60.
*Sloan Fellow.

© 1977, American Mathematical Society

We hope the study of the complex Monge-Ampère equation may be useful in other areas of complex analysis such as value distribution theory.

The expression corresponding to (0) in the $\boldsymbol{R}^n$ case is

$$\det [\partial^2/\partial x_i \partial x_j]. \tag{0'}$$

It is called an equation of Monge-Ampère type. Thus, (0) is referred to as a "complex Monge-Ampère equation". Much more is known about the real case (0′) than about (0). For the former see **[9]**, **[10]** and their references. For the latter see **[2]**, **[3]**.

In §5 we present an integral formula. It is very special but should be useful to understand some peculiarities of the complex Monge-Ampère equation.

1. Harmonic functions and boundary regularity of holomorphic mappings in $\boldsymbol{C}^1$. Here is an illustration of the method we are trying to generalize to $\boldsymbol{C}^n$, $n \geqq 2$.

THEOREM 1. *Let Ω_1 and Ω_2 be bounded smooth C^∞ domains in $\boldsymbol{C}^1$ and let T: $\Omega_1 \to \Omega_2$ be a proper holomorphic map. Then $T \in C^\infty(\bar{\Omega}_1)$. (Note that Ω_j simply connected or T injective are not required.)*

PROOF. Consider the "annuli" $\Omega_{j,a} = \{z \in \Omega_j;\ \delta(z, C\Omega_j) < a\}$, $j = 1, 2$, where $\delta(z, C\Omega_j)$ is Euclidean distance to the complement of Ω_j. For small $a > 0$, $\Omega_{j,a}$ is C^∞ smooth and its boundary $b\Omega_{j,a} = b\Omega_j \cup b_{j,a}$ where $b_{j,a} = \{z \in \Omega_j;\ \delta(z, C\Omega_j) = a\}$.

Part 1. Assume first there is some $a > 0$ and a real function $\psi \in C^\infty(\bar{\Omega}_{2,a})$ such that

$$|\partial\psi/\partial w| \geqq B \quad \text{in } \Omega_{2,a} \tag{1}$$

which pulls back to a function

$$\varphi = \psi \circ T \in C^\infty(\bar{\Omega}_{1,\alpha}). \tag{2}$$

Here $B > 0$ is a constant, $\alpha > 0$ is small, $\alpha = \alpha(a, T)$ and w is the variable in Ω_2 (recall T is proper).

Claim. If (2) holds then, for any q, $|(d^qT/dz^q)(z)| \leqq C_q$, $z \in \Omega_{1,\alpha}$; C_q is a constant.

The claim proves Theorem 1 in view of the smoothness of Ω_1.

Proof of claim. Let $D_z^q = d^q/dz^q$, $w = T(z)$. Chain rule and T holomorphic yield $D_z\varphi = D_zT(z)D_w\psi(w)$, and iteration shows that

$$D_z^q\varphi = D_z^qT\, D_w\psi + P(D_zT, \cdots, D_z^{q-1}T, D_w\psi, \cdots, D_w^q\psi) \tag{3}$$

where $P = P_q$ is a polynomial with integer coefficients.

From (1), (2) and (3) we see that $|D_z^qT(z)| \leqq C_q < \infty$ for $z \in \Omega_{1,\alpha}$ provided $|D_z^jT(z)| \leqq C_j < \infty$ for $z \in \Omega_{1,\alpha}$ and $1 \leqq j \leqq q - 1$. The claim follows by induction.

Part 2. *How to produce the function* ψ. Fix $c > 0$ small. Solve the following Dirichlet problem on $\Omega_{2,c}$: $\Delta\psi = 0$ on $\Omega_{2,c}$, ψ continuous on $\bar{\Omega}_{2,c}$, $\psi = 0$ on $b\Omega_2$ and $\psi = g$ on $b_{2,c}$; $g \in C^\infty(b_{2,c})$ is arbitrary, $g < 0$ (e.g., $g = -1$). The solution ψ satisfies, in addition,

$$\psi \in C^\infty(\bar{\Omega}_{2,c}), \tag{5}$$

$$(\partial\psi/\partial\nu)(p) \gneqq 0, \tag{6}$$

where $p \in b\Omega_2$ is arbitrary and ν is the outer normal at p. See Remark 1 below for (6); (5) is a standard fact from potential theory (see [**4**, Part II, Chapter 4]). Now (1) holds for small a because ψ is real valued and (2) follows from $\varphi = \psi \circ T$ being the solution of a similar problem on $\Omega_{1,\alpha}$. The theorem is proved.

REMARK 1. *A special case of Hopf's lemma.* If $\Omega \subset\subset \boldsymbol{R}^n$ is smooth, if $u \in C^1(\bar{\Omega})$ is subharmonic nonconstant on Ω and if u attains its maximum at $p \in b\Omega$, then $(\partial u/\partial \nu)(p) \gneqq 0$. See [**4**, p. 151], [**8**]. A simple proof follows from Harnack's inequality applied to a ball internally tangent to Ω at p. This proves (6).

2. Strictly plurisubharmonic functions and boundary regularity of holomorphic mappings in C^n. Part 1 of the proof above is generalized as follows (see §1 for notation).

THEOREM 2. *Let Ω_j, $j = 1, 2$, be smooth C^∞ domains, $\Omega_j \subset\subset \boldsymbol{C}^n$ and let $T: \Omega_1 \to \Omega_2$ be a proper holomorphic mapping. Assume there is $a > 0$ and a strictly plurisubharmonic function $\psi \in C^\infty(\bar{\Omega}_{2,a})$ such that its pull-back $\varphi = \psi \circ T \in C^\infty(\bar{\Omega}_{1,\alpha})$ for some $\alpha > 0$, $\alpha = \alpha(a, T)$. Then $T \in C^\infty(\bar{\Omega}_1)$.*

PROOF. It suffices to show that, for any γ, $|\gamma| < \infty$, there is a constant C_γ such that

$$|D^\gamma T^k(z)| \leqq C_\gamma < \infty, \qquad z \in \Omega_{1,\alpha},\ 1 \leqq k \leqq n. \tag{7}$$

Here $n = n_2$, T^k is the k component of T and $D^\gamma = \partial^{|\gamma|}/\partial z_1^{\gamma_1} \cdots \partial z_n^{\gamma_n}$. Let $L > 0$ be a lower bound for the Levi form of ψ, i.e.,

$$\psi_{i\bar{j}}(w)\, \xi^i \bar{\xi}^j \geqq L|\xi|^2, \qquad w \in \Omega_{2,a},\ \xi \in \boldsymbol{C}^n. \tag{8}$$

We use the summation convention. Lower indexes stand for differentiation, e.g., $\psi_{i\bar{j}} = \partial^2\psi/\partial z_i \partial \bar{z}_j$.

Chain rule, T holomorphic and (8) yield

$$\varphi_{i\bar{i}} = \psi_{k\bar{l}} T_i^k \bar{T}_{\bar{i}}^l \geqq L \sum_k |T_i^k|^2 = L|T_i|^2, \tag{9}$$

which proves (7) in case $|\gamma| = 1$.

Assume now that (7) holds for $|\gamma| \leqq q - 1$ and let $|\gamma| = q$. Differentiation of (9) and an induction process show that

$$D_z^\gamma D_{\bar{z}}^\gamma \varphi = M + E_1 + E_2 \tag{10}$$

where the main term is

$$M = \psi_{kl}\, D^\gamma T^k\, D^\gamma \bar{T}^l \geqq L|D^\gamma T|^2. \tag{11}$$

The error term E_1 is the sum of products of derivatives of T of order $\leqq q - 1$; in E_2 terms contain exactly one factor involving $D^\gamma T$, $|\gamma| = q$. Recall that for $a, b > 0$

$$2ab = 2\delta a \frac{1}{\delta} b \leqq \delta^2 a^2 + \frac{1}{\delta^2} b^2. \tag{12}$$

Choosing δ small, using (12) and adding (10) over all γ, $|\gamma| = q$, we obtain (7) for $|\gamma| = q$. (E_2 "has been absorbed by M".) The proof is finished.

From now on Ω stands for a smooth C^∞ strictly pseudoconvex domain $\Omega \subset\subset \boldsymbol{C}^n$.

3. The complex Monge-Ampère equation. The function $u = \lg |z|^2$ vanishes on bB (B = unit ball in C^n) and near bB it satisfies the equation

$$\det [u_{i\bar{j}}] = \det [\partial^2 u/\partial z_i \partial \bar{z}_j] = 0. \tag{13}$$

Other simple calculations show that for any smooth g

$$\det [g_{i\bar{j}}]\, dz_1 \wedge d\bar{z}_1 \wedge \cdots \wedge dz_n \wedge d\bar{z}_n = C_n(\partial\bar{\partial} g)^n, \tag{14}$$

and if T is holomorphic, $T: V \to C^n$, $V \subset C^n$ and $f = g \circ T$, then

$$\det [f_{i\bar{j}}](z) = (\det [g_{i\bar{j}}])(w) \, |T'(z)|^2 \tag{15}$$

where T' is the Jacobian determinant. This suggests posing the following

Problem M (*the complex Monge-Ampère equation*). Given

(M_1) $\Omega \subset\subset C^n$, Ω strictly pseudoconvex smooth C^∞,

(M_2) $\psi \in C^\infty(\bar{\Omega})$, $\psi \geqq 0$ such that

(M_2') $\psi^{1/m} \in C^\infty(\bar{\Omega})$ for any $m = 1, 2, \cdots$ (see remark below),

find

(M_3) $u \in C^\infty(\bar{\Omega})$ such that

(M_4) u is plurisubharmonic in Ω, $u = 0$ on $b\Omega$ and

(M_5) $\det[u_{i\bar{j}}] = \psi$ on Ω.

REMARK. Condition (M_2') enters into the proof of estimate (16) below in cases $|\alpha| = 1$ and $|\alpha| = 2$. The proof of the estimates so far missing, i.e., for $|\alpha| \geqq 3$, may in the future require replacing (M_2') by a stronger condition.

A. *A maximum principle and uniqueness.* See also [3].

THEOREM 3. *If Ω is open $\subset\subset C^n$, if u and $v \in C^0(\bar{\Omega}) \cap C^2(\Omega)$ are plurisubharmonic in Ω and both vanish on $b\Omega$ and if* $\det [u_{i\bar{j}}] \geqq \det [v_{i\bar{j}}]$ *then $u \leqq v$ in Ω.*

PROOF. $\delta(f)$ stands for $\det[f_{i\bar{j}}]$. We have

$$0 \leqq \delta(u) - \delta(v) = \int_0^1 \frac{d}{dt} \delta[tu + (1 - t)\, v]\, dt$$
$$= \int_0^1 \sum_{ij} A_t^{ij}(u - v)_{i\bar{j}}\, dt = \sum_{ij} B^{ij}(u - v)_{i\bar{j}}$$

where $[A_t^{ij}]$ is the cofactor matrix of $[tu_{i\bar{j}} + (1 - t)v_{i\bar{j}}]$ which is positive semidefinite for each t and $[B^{ij}]$ is its integral in t. If $\delta(u) - \delta(v) \gneqq 0$ in Ω then u is *strictly* plurisubharmonic (recall u and v *are* plurisubharmonic) and thus $[B^{ij}]$ is positive definite. Hence $u - v$ has no interior maximum and $u \leqq v$. The theorem follows by replacing u by $u + \varepsilon |z|^2$ and letting $0 < \varepsilon \to 0$. (Recall Ω is bounded.)

COROLLARY. *If Problem* M *has a solution it is unique.*

B. *How to prove existence: The continuity method.* Deform the datum ψ of Problem M into $\psi_t = (1 - t)\varphi + t\psi$ where $\varphi = \delta(v)$ and v is a fixed strictly plurisubharmonic function defining Ω; here $0 \leqq t \leqq 1$. A problem M_t is thus obtained for each t; M_0 has the solution v. Let $W = \{t \in [0, 1]$; Problem M_τ has a solution for all τ, $0 \leqq \tau \leqq t\}$. If $1 \in W$ then Problem M is solved. The method consists in establishing two facts:

(a) W is open in [0, 1].

(b) W is closed in [0, 1].

Actually (a) *is true.* An application of the standard implicit functions theorem in the relevant infinite-dimensional space yields (a). The nondegeneracy condition required holds for any $0 \leqq t < 1$ because the "Jacobian" involved is the linearized operator

$$L_t = \sum_{kl} A_t^{kl}\, \partial^2/\partial z_k \partial \bar{z}_l$$

which is strongly elliptic and can be inverted. (Recall $[A_t^{kl}]$ is the cofactor matrix of $[tu_{i\bar{j}} + (1-t)v_{i\bar{j}}]$ which is positive definite for $0 \leqq t < 1$.

An a priori estimate is required to prove (b). This is the difficult part.

The basic estimate. For all $t \in W$

$$\sup_{z \in \Omega} |D^\alpha u_t(z)| \leqq C_\alpha, \qquad 0 \leqq |\alpha| < \infty, \tag{16}$$

where $C_\alpha = C(\Omega, \psi)$ is a constant independent of t (u_t is the solution of Problem M_t). *We have proved* (16) *only for* $|\alpha| \leqq 2$ so far.

If (16) holds then a diagonal method argument yields convergent subsequences of u_{t_j} whenever $t_j \nearrow t$, $t_j \in W$, and hence $t \in W$. The uniqueness property is repeatedly used.

Proof of (16) *for* $\alpha = 0$. (The cases $|\alpha| = 1$ and $|\alpha| = 2$ follow the same pattern but are much more involved.) For large $C > 0$, $\delta(Cv) = C^n\varphi \geqq \delta(u_t) = (1-t)\varphi + t\psi$ for all $t \in [0, 1]$ (recall $\varphi > 0$ on $\bar{\Omega}$). By the maximum principle $Cv \leqq u_t$; v and u_t are $\leqq 0$. Hence $|u_t| \leqq |Cv|$.

4. Application of the Monge-Ampère equation.

THEOREM 4. *Let Ω_1 and Ω_2 be smooth C^∞ strictly pseudoconvex domains in $\mathbf{C}^n$ and let $T\colon \Omega_1 \to \Omega_2$ be a proper holomorphic map. If Problem* M *can be solved in Ω_1 and in Ω_2, then $T \in C^\infty(\bar{\Omega})$.*

Note. The following proof requires only that Problem M can be solved for *certain ψ*, not for all of them.

PROOF. Let T' be the Jacobian determinant of T and Z = zero set of T'. Then $Z \neq \Omega_1$ and $T(Z)$ is closed $\neq \Omega_2$ (because T is proper). Choose $\psi \in C_0^\infty(\Omega_2)$ as in Problem M, $\psi \not\equiv 0$, and

$$\text{Support of } \psi \cap T(Z) = \varnothing. \tag{17}$$

Let v be the solution to Problem M on Ω_2; $u = v \circ T$ solves a similar problem on Ω_1. Indeed $\delta(u)$ satisfies the required hypothesis by (15) and (17), e.g., (15) yields $\delta(u) \in C_0^\infty(\Omega_1)$. Hence $u \in C^\infty(\bar{\Omega}_1)$ and, by Remark 1, $(\partial u/\partial \nu)(p) \neq 0$, $(\partial v/\partial \nu)(q) \neq 0$ for any $p \in b\Omega_1$, $q \in b\Omega_2$. This implies that, for $C > 0$ large and constant, $\exp(Cv)$ and $\exp(Cu) = (\exp(Cv)) \circ T$ are strictly plurisubharmonic near $b\Omega_2$ and $b\Omega_1$. See the argument, not the statement [7, p. 264]. Apply Theorem 2 and the proof is finished.

One is interested in noninjective T because of the following

THEOREM 5 (FORNAESS [**6**]). *If Ω_1 and Ω_2 are smooth C^∞ strictly pseudoconvex domains in $\mathbf{C}^n$, $n \geqq 2$, and if $T = \Omega_1 \to \Omega_2$ is a proper holomorphic map $T \in C^\infty(\bar{\Omega}_1)$, then $T'(z) \neq 0$ for all $z \in \Omega_1$.*

This theorem has an obvious corollary:

COROLLARY. *If in addition Ω_2 is simply connected then T is biholomorphic* (*because T is a covering map*).[1]

We had been aware of the following independent simple analytic proof of Theorem 5 involving the special case of Hopf's lemma mentioned in Remark 1.

PROOF. Let $v \in C^\infty(\bar{\Omega}_2)$ be a strictly plurisubharmonic defining function for Ω_2. Then $u = v \circ T \in C^\infty(\bar{\Omega}_1)$ is *not* strictly plurisubharmonic where $T' = 0$ (use chain rule formula for $u_{i\bar{j}}$). But, as above, $\exp(Cu) = \exp(Cv) \circ T$ *is* strictly plurisubharmonic near $b\Omega_1$ for large C. (Here we used Remark 1.) Since $n \geqq 2$ a nonempty $Z = \{z \in \Omega_1,\ T'(z) = 0\}$ would approach $b\Omega_1$ (recall Hartogs' theorem). Hence $Z = \emptyset$ and the proof is finished. The proof of Theorem 5 is finished.

5. An integral formula for a very special case. Let Ω be the unit ball in $\boldsymbol{C}^2$.

(a) If u solves Problem M (see §3) in Ω with radial datum ψ then

$$u(z) = -\frac{4}{\sqrt{\omega}} \int_{r=|z|}^{r=1} \left[\int_{B_r} \psi(z)\, dV \right]^{1/2} \frac{dr}{r} \tag{18}$$

where B_r is the ball of radius r, ω is the area of the unit sphere in $\boldsymbol{R}^4$ and dV is Lebesgue measure in $\boldsymbol{R}^4$.

(b) If ψ is an arbitrary *radial* continuous function in $\bar{B}$, $\psi \geqq 0$, then (18) defines a u which is C^2 in $X = \{z \in \bar{\Omega};\ |z| > |w| \text{ for some } w \text{ where } \psi(w) \neq 0\}$. Furthermore, $\det[u_{i\bar{j}}] = \psi$ in X and $u = 0$ on $b\Omega$. (Note $0 \notin X$.)

The proof of (18) follows from explicitly solving a second-order ordinary differential equation in r (the only variable involved). The conditions $u = 0$ on $b\Omega$ and u smooth at 0 determine the two constants of integration.

REFERENCES

1. H. Alexander, *Proper holomorphic mappings in $\boldsymbol{C}^n$* (to appear).
2. E. Bedford and B. A. Taylor, *The Dirichlet problem for a complex Monge-Ampère equation*, Bull. Amer. Math. Soc. **82** (1976), 102–104.
3. ———, *The minimum principle for a complex Monge-Ampère equation* (to appear).
4. L. Bers, F. John and M. Schechter, *Partial differential equations*, Lectures in Appl. Math., vol. 3, Interscience, New York, 1966. MR **29** #346.
5. C. Fefferman, *The Bergman kernel and biholomorphic mappings of pseudo-convex domains*, Invent. Math. **26** (1974), 1–65. MR **50** #2562.
6. J. Fornaess, *Embedding strictly pseudoconvex domains in convex domains*, Amer. J. Math. (to appear).
7. R. C. Gunning and H. Rossi, *Analytic functions of several complex variables*, Prentice-Hall, Englewood Cliffs, N.J., 1965. MR **31** #4927.
8. G. Henkin, *An analytic polyhedron is not holomorphically equivalent to a strongly pseudoconvex domain*, Dokl. Akad. Nauk SSSR **210** (1973), 1026–1029 = Soviet Math. Dokl. **14** (1973), 858–862. MR **48** #6467.
9. L. Nirenberg, *Monge-Ampere equations and some associated problems in geometry*, Proc. Internat. Congr. Math. (Vancouver, 1974), Canad. Math. Congr., Providence, R.I., 1975.

[1]In this connection, H. Alexander [1] has recently answered an old question: *Arbitrary* proper holomorphic maps of the ball in $\boldsymbol{C}^n$, $n \geqq 2$ *are* biholomorphic. His work does not use the Monge-Ampère equation.

10. A. V. Pogorelov, *Monge-Ampere equations of elliptic type*, Izdat. Har'kovsk. Univ., Kharkov, 1960; English transl., Noordhoff, Groningen, 1964. MR **23** #A1137; **31** #4993.

11. N. Voormoor, *Topologische Fortsetzung biholomorpher Funktionen auf dem Rande bei beschränkten streng-pseudokonvexen Gebieten im* $\mathbf{C}^n$ *mit* C^∞*-Rand*, Math. Ann. **204** (1973), 239–261.

12. S. Yau, *The real Monge-Ampere equation*, Personal communication at Williamstown Summer Institute (1975).

MASSACHUSETTS INSTITUTE OF TECHNOLOGY

Proceedings of Symposia in Pure Mathematics
Volume 30, 1977

CR STRUCTURES AND THEIR EXTENSIONS*

GARO K. KIREMIDJIAN

0. Introduction. Let M_0 be a C^∞ manifold of real dimension $2n - 1$ and let CTM_0 be its complexified tangent bundle. A Cauchy-Riemann (or, simply, a CR) structure on M_0 is given by a complex subbundle $E'' \subset CTM_0$ of complex fiber dimension $n - 1$ such that $E'' \cap \bar{E}'' = \{0\}$ and the Lie bracket of any two sections L, L' of E'' over an open set of M_0 is also a section of E''. If M_0 is the boundary of a complex n-dimensional manifold M, then the complex structure of M and its deformations induce CR structures on M_0.

In this report we briefly outline the results of [3] and [4] which show that, conversely, every small deformation of M_0 is induced by a complex structure on M if M_0 is compact and pseudoconvex, M is an open relatively compact subset of a complex manifold M', and certain cohomology groups over M are zero (cf. §3). The solution of this extension problem is in the spirit of the theory developed by M. Kuranishi in [10].

1. An implicit function theorem. Let E, F be Fréchet spaces with a fundamental system of norms $\| \ \|_s$, $s \in \boldsymbol{Z}^+$, where $\boldsymbol{Z}^+$ is the set of positive integers. For $s \geqq r$ we have $\|x\|_s \geqq \|x\|_r$. We also assume that E has smoothing operators $S(t)$, $t \in \boldsymbol{R}^+$, i.e., $S(t): E \to E$ is an endomorphism such that for r $\leqq s$ there exists a constant $c_{r,s} > 0$ for which the following inequalities hold for all $x \in E$:

$$\|S(t)x\|_s \leqq c_{s,r} t^{s-r} \|x\|_r, \tag{1.1}$$

$$\|x - S(t)x\|_r \leqq c_{s,r}\, t^{r-s} \|x\|_s. \tag{1.2}$$

If Y is a compact manifold (with or without boundary) and V is a vector bundle over Y with a metric along the fibers, then the space of C^∞ sections of V over Y is an example of such a Fréchet space where $\| \ \|_s$ is the Sobolev s-norm.

For $s_1 \in \boldsymbol{Z}^+$ we define $U(s_1, a) = \{x \in E: \|x\|_{s_1} < a\}$. Let $G: U(s_1, a) \to F$ be a map for which the derivative $G'(x): E \to F$, $G'(x)(y) = \lim_{u \to 0} u^{-1} [G(x + uy) - G(x)]$,

AMS (MOS) subject classifications (1970). Primary 32G99; Secondary 32G05, 35J60, 32J25.

*This research is supported in part by a National Science Foundation grant MPS75-07513.

© 1977, American Mathematical Society

exists for all $x \in U(s_1, a)$. We now assume that $G(x)$ has some smoothness properties, i.e., there are integers m and q and real numbers d_q and l with $l > 1, d < l$ and for each $s \in \mathbf{Z}^+$ there are continuous functions $\alpha_s(\tau) > 0$, $\beta_s(\tau) > 0$ on the closed interval $[0, a]$ with $\beta_s(\tau) \leqq C_s \tau^d$ for some constant $C_s > 0$ such that the following conditions are satisfied:

(1.3) $\|G(x)\|_s \leqq \alpha_s(\|x\|_k)(\|x\|_{s+m} + 1)$ for all s and k with $2k > s + q$.

(1.4) $\|G(x + y) - G(x) - G'(x)(y)\|_s \leqq \alpha_s(\|x\|_{s+m})\|y\|_{s+m}^l$ for all $s \in \mathbf{Z}^+$ and $x \in U(s_1, a)$, $y \in E$ with $x + y \in U(s_1, a)$.

(1.5) $\|G'(x)(y)\|_s \leqq \alpha_s(\|x\|_{s+m})\|y\|_{s+m}$ for all $s \in \mathbf{Z}^+$, $x \in U(s_1, a)$ and $y \in E$.

(1.6) There is a subspace $B \subset E$ such that for all $x \in U(s_1, a)$ the equation $G'(x)(y) + G(x) = 0$ has a solution $y \in B$ with

$$\|y\|_s \leqq \alpha_s(\|x\|_k)(\|G(x)\|_{s+m} + \beta_s(\|x\|_{s+m})\|G(x)\|_{k-m})$$

for all s and k with $2k > s + q$.

REMARK 1.1. It is easy to see that (1.3)—(1.6) are based on properties satisfied by nonlinear partial differential operators. In this case m serves as a bound on the order of G and the number of derivatives lost by the solutions of the corresponding linearized differential equation; q is the dimension of the manifold over which G is defined, $l = 2$ and k is chosen so large that one can use the Sobolev inequalities. Condition (1.3) in the general case of Fréchet spaces is the definition of a tame map. This concept has been introduced by R. Hamilton.

THEOREM. *If the map G has the properties* (1.3)—(1.6) *then, for each $z \in U(s_1, b)$, $b = \min\{a, c_{k,k}^{-1}a\}$, there exists $w \in U(k, a)$ with $G(w) = 0$ and $w - z \in B$ provided $\|G(z)\|_k$ is sufficiently small and k is a sufficiently large integer depending only on m, q, d, and l.*

The proof is based on an appropriate modification of Moser's method (cf. [**11**]). The details appear in [**5**]. Similar types of implicit function theorems have also been extensively used by R. Hamilton and M. Kuranishi (cf. [**2**] and [**10**]).

We remark that, in terms of partial differential equations, the above theorem represents a useful technique for the construction of solutions of rather general classes of nonlinear boundary value problems.

2. Noncoercive boundary value problems. In this section we discuss the important work [**8**] of J. J. Kohn and L. Nirenberg in a slightly more general situation. We refer to this paper and [**6**] for further details. For the purposes of the present exposition we may consider M, M_0, and M' merely as C^∞ manifolds.

Let $\mathscr{U}$ be a C^∞ vector bundle over M'. We denote by $C^\infty(\bar{M}, \mathscr{U})$ (resp., $C_0^\infty(M, \mathscr{U})$) the space of C^∞ sections extendible to a neighborhood of $\bar{M}$ (resp., with compact support in M). Let $\mathscr{L}: C^\infty(\bar{M}, \mathscr{U}) \to C^\infty(\bar{M}, \mathscr{U})$ be a second-order differential operator, and let $\langle\ ,\ \rangle$ denote the L_2-inner product on $C^\infty(\bar{M}, \mathscr{U})$ with respect to a hermitian metric along the fibers of $\mathscr{U}$. We assume that there exists a hermitian bilinear form $Q(u, v)$ involving derivatives of at most first order in u and v such that $\langle \mathscr{L}u, v\rangle = Q(u, v)$ for all $u, v \in B$ where the subspace $B \subset C^\infty(\bar{M}, \mathscr{U})$ satisfies the following conditions:

(a) $B \supset C_0^\infty(M, \mathscr{U})$.

(b) If U is a boundary coordinate neighborhood, then for every neighborhood V

with $\bar{V} \subset U$ one can find a real-valued function $\zeta \in C_0^\infty(U)$ such that $\zeta \equiv 1$ on $\bar{V} \cap M$ and $\zeta B \subset B$.

(c) For every translation or differentiation T parallel to M_0, $\zeta TB \subset B$.

Let Q_0 and Q' be the hermitian and skew hermitian parts of Q, i.e., $2Q_0(u, v) = Q(u, v) + \overline{Q(v, u)}$ and $2\sqrt{-1}\, Q'(u, v) = Q(u, v) - \overline{Q(v, u)}$. We assume that there is a constant $C > 0$ such that

(2.1) $Q_0(u, u) \geqq C\|u\|^2$ for all $u \in B$ where $\|u\|^2 = \langle u, u\rangle$.

(2.2) $\|Ku\|^2, \|Lu\|^2 \leqq CQ_0(u, u)$ for all $u \in B$ if $Q'(u, v)$ contains terms of the form $\langle Ku, Lv\rangle$ where K and L are first-order differential operators.

(2.3) $|Q'(u,v)|^2 \leqq CQ_0(u, u)Q_0(v, v)$, $u, v \in B$.

(2.4) M_0 is nowhere characteristic with respect to Q_0.

(2.5) $|u|^2 \leqq CQ_0(u, u)$ for all $u \in B$ where $|\ \ |^2$ is the L_2-norm over M_0.

REMARK 2.1. If $\mathscr{L}^*$ is the adjoint of $\mathscr{L}$ with respect to $\langle\ ,\ \rangle$, i.e., $\langle \mathscr{L}u, v\rangle = \langle u, \mathscr{L}^*v\rangle$ for all $u, v \in C_0^\infty(M, \mathscr{U})$, then, by (2.2), $\mathscr{L} - \mathscr{L}^*$ is allowed to contain second-order terms as long as they behave in a prescribed manner. We point out that condition (ii) in [**8**, p. 452] is a little more restrictive, i.e., $\mathscr{L} - \mathscr{L}^*$ can contain terms of order at most one.

THEOREM 2.1. *Assume that the operator $\mathscr{L}$ satisfies conditions* (2.1)—(2.5). *Then for any $f \in C^\infty(\bar{M}, \mathscr{U})$ there exists a unique $u \in B$ such that $\mathscr{L}u = f$. Moreover, if the coefficients of $\mathscr{L}$ depend smoothly on a section ω over $\bar{M}$ of some C^∞ vector bundle and if for some sufficiently large $k \in \boldsymbol{Z}^+$ the Sobolev k-norm* (*over $\bar{M}$*) $\|\omega\|_k$ *is sufficiently small, then for each $s \in \boldsymbol{Z}^+$ there is a smooth positive function α_s on $\boldsymbol{R}$ such that*

$$\|u\|_s \leqq \alpha_s(\|\omega\|_{s+k})\|f\|_s. \tag{2.6}$$

3. Extensions of CR structures in the class $\mathscr{C}$. Throughout the rest of this paper we will use the summation convention.

Let T' be the holomorphic tangent bundle of M'. Set ${}^\circ T' = \boldsymbol{C}TM_0 \cap (T'|M_0)$, ${}^\circ T'' = {}^\circ\bar{T}'$. Then $\boldsymbol{C}TM_0 = {}^\circ T'' \oplus \boldsymbol{C}F$ where $\boldsymbol{C}F$ is the complexification of a real one-dimensional subbundle F of TM_0. If M_0 is defined by a real-valued C^∞ function h on M', i.e., $M_0 = \{h = 0\}$ and $dh \neq 0$ on M_0, then one can choose a purely imaginary generator $P = P' - P''$ of $\boldsymbol{C}F$, $P'' = \bar{P}'$, such that $dh(P') = dh(P'') = 1$. In terms of local coordinates $z = (z^1, \cdots, z^n)$ on a coordinate neighborhood U we have $P' = p^j\partial/\partial z^j$ and $h_jp^j = h_{\bar{j}}p^{\bar{j}} = 1$ on M_0, where $p^{\bar{j}} = \overline{p^j}$, $h_j = \partial h/\partial z^j$, and $h_{\bar{j}} = \overline{h_j}$. A set of local generators of ${}^\circ T'$ is given by $Z_j = \partial/\partial z^j - h_jP'$, $1 \leqq j \leqq n$. Similarly, $Z^j = i^*dz^j - p^ji^*\partial h$, $1 \leqq j \leqq n$, is a set of local generators of the dual bundle ${}^\circ T'^*$. Here $i: M_0 \to M'$ is the natural embedding of M_0 in M' and $\partial h = h_jdz^j$. We have $p^jZ_j = h_jZ^j = 0$.

A differential form ϕ on M_0 of degree r is of type $(0, r)_b$ if $\phi(X_1, \cdots, X_r) = 0$ whenever any one of $X_1, \cdots, X_r$ is a section of ${}^\circ T' \oplus \boldsymbol{C}F$. In terms of $\bar{Z}^j$, $1 \leqq j \leqq n$, we can write

$$\phi = \sum_{j_1<\cdots<j_r} \phi_{\bar{j}_1\cdots\bar{j}_r}\bar{Z}^{j_1} \wedge \cdots \wedge \bar{Z}^{j_r}.$$

For a function g on a neighborhood U in M_0 the tangential Cauchy-Riemann operator is given by $\bar{\partial}_bg = (\partial g/\partial\bar{z}^j)\bar{Z}^j$. We set

$$\bar{\partial}_b\phi = \sum_{j_1<\cdots<j_r} \bar{\partial}_b\phi_{\bar{j}_1\cdots\bar{j}_r} \wedge \bar{Z}^{j_1} \wedge \cdots \wedge \bar{Z}^{j_r}.$$

Let ρ': $CTM' \to T'$ be the projection, and let τ: $T'|M_0 \to {}^\circ T' \oplus CF$ be the isomorphism which is the identity on ${}^\circ T'$ and $\tau(\rho' X) = X$ for $X \in CF$. Locally, $\tau(\partial/\partial z^j) \equiv \partial^\tau/\partial z^j = \partial/\partial z^j - h_j P''$. According to the theory developed in **[10]** (cf. also **[3]**), every CR structure E'' sufficiently close to ${}^\circ T''$ can be represented by a $T'|M_0$-valued form φ of type $(0, 1)_b$ such that $E'' = \{X - \tau \circ \varphi(X) : X \in {}^\circ T''\}$ and $R(\varphi) = 0$ where $R(\varphi)$ is a $T'|M_0$-valued form of type $(0, 2)_b$ which, in terms of local coordinates, can be written as

$$R(\varphi) = \bar{\partial}_b \varphi - [(\partial^\tau \varphi_l^k/\partial z^j)\varphi^j \wedge \bar{Z}^l - h_\mu \varphi^\mu \wedge \varphi_l^k(\bar{\partial}_b p^{\bar{l}} - (\partial^\tau p^{\bar{l}}/\partial z^j)\varphi^j)](\partial/\partial z^k). \tag{3.1}$$

Set $E'' = {}^\circ T''_\varphi$.

Let $\mathscr{C}$ be the set of all φ for which ${}^\circ T''_\varphi$ represents a CR structure on M_0 and $h_\mu \varphi^\mu = 0$. If the C^∞ T'-valued (0, 1) form ω on $\bar{M}$ gives an almost complex structure $T''_\omega = \{X - \omega(X) : X \in T'' = \bar{T}'\}$ on M, then one can easily show that $(T''_\omega | M_0) \cap CTM_0 = {}^\circ T''_\varphi$ for $\varphi \in \mathscr{C}$ if and only if $t(\omega) = \varphi$ where $t(\omega)$ is the complex tangential part of ω on M_0, i.e., $i^*\omega = t(\omega) + \nu(\omega) i^* \bar{\partial} h$. Also, for $\Omega = \bar{\partial}\omega - [\omega, \omega]$ we have $t(\Omega) = 0$. We recall that T''_ω is integrable (i.e., a complex structure) if and only if $\Omega = 0$. Here [,] is the Poisson bracket.

THEOREM 3.1. *Assume that $n \geqq 3$, the Levi form of M_0 has at least two positive eigenvalues and*

$$H^q(M, \underline{\chi \otimes T'^*}) = 0 \quad \text{for } q = n - 2, n - 1, \tag{3.2}$$

where χ is the canonical bundle and $\underline{\chi \otimes T'^*}$ is the sheaf of germs of holomorphic sections of $\chi \otimes T'^*$. Let k be a sufficiently large integer with respect to k (e.g., $k > 2n$). Then for all $\varphi \in \mathscr{C}$ with sufficiently small Sobolev k-norm $|\varphi|_k$ over M_0 there exists a C^∞ T'-valued (0, 1) *form ω on $\bar{M}$ such that $\bar{\partial}\omega - [\omega, \omega] = 0$ and $t(\omega) = \varphi$, i.e., the CR structure ${}^\circ T''_\varphi$ can be extended to a complex structure T''_ω on M. Moreover, $\|\omega\|_k <$ const $|\omega|_k$.*

We briefly sketch the proof which appears in **[3]**. First, consider the Fréchet space $E = C^{0,1}(\bar{M}, T')$ of C^∞ T'-valued forms of type (0, 1) extendible to a neighborhood of $\bar{M}$ and the map $G : E \to E$ given by $G(\omega) = \bar{\partial}^*(\bar{\partial}\omega - [\omega, \omega])$ where $\bar{\partial}^*$ is the adjoint of $\bar{\partial}$ with respect to some hermitian metric on M'. For a fixed ω the differential operator $\mathscr{L}u = \Box u - 2\bar{\partial}^*[\omega, u]$ satisfies all the conditions in §2 with $B = \{u \in C^{0,1}(\bar{M}, T') : t(u) = 0\}$, $Q(u, v) = \langle \bar{\partial}u, \bar{\partial}v \rangle + \langle \bar{\partial}^* u, \bar{\partial}^* v \rangle - 2\langle [\omega, u], \bar{\partial}v \rangle$ and $\|\omega\|_k$ sufficiently small. This is so because $[\omega, u]$ contains only the components of u and their $\partial/\partial z^j$ derivatives and hence by the Sobolev inequality (applied to the components of ω) and the basic Kohn-Morrey estimate (cf. **[7]** and **[9]**) we have

$$\|[\omega, u]\|^2 \leqq \text{const } \|\omega\|_k^2 (\|\bar{\partial}u\|^2 + \|\bar{\partial}^* u\|^2) \quad \text{for all } u \in B$$

(note (3.2)). Hence, for all $f \in E$ there exists a unique $u \in B$ such that $\mathscr{L}u = f$ with the estimate (2.6). In particular, if $f = -G(\omega)$, by integration by parts the term $\bar{\partial}\bar{\partial}^* u$ drops out. Since $G'(\omega)(u) = \bar{\partial}^*(\bar{\partial}u - 2[\omega, u])$, we can solve $G'(\omega)u + G(\omega) = 0$ with $u \in B$ such that (2.6) holds. Now, Theorem 1.1 can be applied and one can find $\omega \in E$ such that $t\omega = \varphi$ and $G(\omega) = 0$. The estimate $\|\omega\|_k \leqq$ const $|\varphi|_k$ follows from the proof of 1.1.

Since $t(\Omega) = 0$ it follows by 3.2 and the basic Kohn-Morrey estimate that $\|\Omega\|^2$

$\leqq$ const $\|\bar{\partial}\Omega\|^2$ ($\bar{\partial}^*\Omega = G(\omega) = 0$). On the other hand, $\bar{\partial}\Omega = 2[\omega, \Omega]$, so $\|\bar{\partial}\Omega\|^2 \leqq$ const $\|\omega\|_k^2\|\bar{\partial}\Omega\|^2$. Thus, if $|\varphi|_k$ is sufficiently small, these estimates imply that $\Omega = 0$.

4. Extensions of arbitrary small deformations of M_0. Let $E'' \subset CTM_0$ be a subbundle which determines a CR structure on M_0. The projection $CTM_0 \to E''$ induces an injective bundle homomorphism $\alpha\colon E''^* \to CT^*M_0$, and hence an injection $\alpha\colon C^\infty(M_0, \Lambda^q E''^*) \to C^\infty(M_0, \Lambda^q CT^*M_0)$ where Λ^q stands for the qth exterior algebra bundle. If $\beta\colon C^\infty(M_0, \Lambda^q CT^*M_0) \to C^\infty(M_0, \Lambda^q E''^*)$ is the restriction map, then the Cauchy-Riemann operator for the structure E'' is given by

$$\bar{\partial}_{E''} = \beta \circ d \circ \alpha\colon C^\infty(M_0, \Lambda^q E''^*) \to C^\infty(M_0, \Lambda^{q+1}E''^*). \tag{4.1}$$

For $E'' = {}^\circ T''_\varphi$ we have the isomorphisms $\lambda^\varphi\colon {}^\circ T'' \to {}^\circ T''_\varphi$ and $(\lambda^\varphi)^*\colon {}^\circ T''^*_\varphi \to {}^\circ T''^*$ given by $\lambda^\varphi(X) = X - \tau \circ \varphi(X)$. The map

$$\bar{\partial}_b^\varphi\colon C^\infty(M_0, \Lambda^q\, {}^\circ T''^*) \to C^\infty(M_0, \Lambda^{q+1}\, {}^\circ T''^*) \tag{4.2}$$

is defined by the formula $\bar{\partial}_b^\varphi = (\lambda^\varphi)^* \circ \bar{\partial}_{E''} \circ ((\lambda^\varphi)^*)^{-1}$. If $Z_{\bar{l}}^\varphi = \lambda^\varphi(Z_{\bar{l}})$, then the following statements can easily be checked by calculation:

$$\bar{\partial}_b^\varphi f = Z_{\bar{l}}^\varphi f \cdot \bar{Z}^l, \qquad f \in C^\infty(M_0, C), \tag{4.3}$$

$$\bar{\partial}_b^\varphi u = (\bar{\partial}_b^\varphi u_{\bar{L}}) \wedge \bar{Z}^L + \sum_{s=1}^{a} (-1)^{s+1} u_{\bar{L}} h_l \varphi^j \wedge \bar{\partial}_b^\varphi p^{\bar{l}_s} \wedge \bar{Z}^{L_s}, \tag{4.4}$$

$u = u_{\bar{L}}\bar{Z}^L \in C^\infty(M_0, \Lambda^q\, {}^\circ T''^*)$, $L = (l_1 < \cdots < l_q)$, $L_s = (l_1 < \cdots < l_{s-1} < l_{s+1} < \cdots < l_q)$, $\bar{Z}^L = \bar{Z}^{l_1} \wedge \cdots \wedge \bar{Z}^{l_q}$.

We remark that we also have $\bar{\partial}_{E''} \circ \bar{\partial}_{E''} = \bar{\partial}_b^\varphi \circ \bar{\partial}_b^\varphi = 0$.

In **[10]**, M. Kuranishi has introduced the connection forms

$$\Gamma_l^i(\varphi) = (\partial^\tau \varphi_{\bar{k}}^i / \partial z^l)\bar{Z}^k - \varphi_{\bar{k}}^i (h_l \bar{\partial}_b^\varphi \bar{p}^k + (\partial^\tau \bar{p}^k/\partial z^l) h_j \varphi^j).$$

With their help one can define the operator $\bar{\partial}_b^\varphi$ on the space $C_b^{0,q}(M_0, T'|M_0)$ of C^∞ $T'|M_0$-valued forms of type $(0, q)_b$ by the formula

$$\bar{\partial}_b^\varphi u = (\bar{\partial}_b^\varphi u^i + \Gamma_l^i(\varphi) \wedge u^l)(\partial/\partial z^i). \tag{4.5}$$

In this case one can again check that $\bar{\partial}_b^\varphi \circ \bar{\partial}_b^\varphi = 0$.

Let $\mathcal{N}$ be a neighborhood of M_0 in M', $\mathcal{N}_\omega$ a complex structure on $\mathcal{N}$ induced by a T'-valued (0, 1) form ω on $\mathcal{N}$, and $f\colon M_0 \to \mathcal{N}$ an embedding sufficiently close to i.

DEFINITION 4.1. The CR structure ${}^\circ T''_\varphi$ on M_0 induced via f by $\mathcal{N}_\omega$ is represented by the subbundle $E''(f, \omega) \subset CTM_0$ whose fiber over $x \in M_0$ is defined by the relation $df(E''(f, \omega)_x) = CT_{f(x)}(f(M_0)) \cap T''_{f(x)}(\mathcal{N}_\omega)$.

We can parametrize f by elements $\xi \in C^\infty(M_0, T'|M_0)$ so that, in terms of local coordinates z on an open set $U \subset M_0$, $f^j(z) = z^j + \xi^j + S^j(z, \xi)$ where $S^j \in C^\infty(U \times C^n)$ and $|S^j(z, \xi)| \leqq$ const $|\xi|^2$. In this case we denote the form φ in the above definition by $\varphi(\xi, \omega)$. The following theorem is proved in **[4]**.

THEOREM 4.2. *Assume that the Levi form of M_0 is nondegenerate. Let k be a sufficiently large positive integer, and let $\mathcal{N}_\omega$ be a complex structure on $\mathcal{N}$ with $\|\omega\|_{2k}$ sufficiently small. Then one can find $\xi \in C^\infty(M_0, T'|M_0)$ such that $\varphi(\xi, \omega) \in \mathscr{C}$.*

SKETCH OF PROOF. Let $| \ |_k$ be the Sobolev k-norm on M_0, and let $\mathscr{U}(k, \varepsilon) = \{\xi \in C^\infty(M_0, T'|M_0): |\xi|_k < \varepsilon\}$. Consider the map $G: \mathscr{U}(k, \varepsilon) \to C^\infty(M_0, {}^\circ T''^*)$, $G(\xi) = h_\mu \varphi^\mu(\xi, \omega)$. One can show that $G'(\xi)(\eta) = h_\mu[\bar\partial_b^{\varphi(\xi,\omega)} Q(\xi, \omega)(\eta)]^\mu$ where $Q(\xi, \omega)$ is invertible for small ξ and ω. Since $h_\mu(\bar\partial_b^{\varphi(\xi,\omega)} \tilde\eta)^\mu = \psi$ is a tangential operator, under the condition of the nondegeneracy of the Levi form one can solve the equation $h_\mu(\bar\partial_b^{\varphi(\xi,\omega)}\tilde\eta)^\mu = \psi$ for $\tilde\eta \in C^\infty(M_0, {}^\circ T')$. Hence $\eta = Q^{-1}(\xi, \omega)(\tilde\eta)$ is a solution of $G'(\xi)(\eta) = \psi$. Furthermore, it can be established that the estimates (1.3)—(1.5) hold for $G(\xi)$ and $G'(\xi)$. By Theorem 1.1 there exists $\xi \in \mathscr{U}(k, a)$ such that $G(\xi) = 0$, i.e., $\varphi(\xi, \omega) \in \mathscr{C}$.

The above theorem shows that, up to equivalence in the sense of Definition 4.1, a CR structure M_0^φ on M_0 represented by ${}^\circ T''_\varphi$ can be extended to M if M_0^φ can be extended to a neighborhood $\mathscr{N}$ of M_0. The last statement easily follows from the local embedding problem, i.e., the question of whether a CR manifold can locally be realized as a hypersurface in complex Euclidean space. L. Boutet de Monvel has recently proved in [1] that the answer is affirmative if one assumes compactness and strong pseudoconvexity. In the case of local deformations of M_0 we can give an independent and simpler solution of the local embedding problem. First, we recall the basic Kohn-Morrey estimate for the operator $\bar\partial_b^\varphi$.

PROPOSITION 4.3. *Let M_0 be strongly pseudoconvex, and let $\bar\partial_b^{*\varphi}$ be the adjoint of $\bar\partial_b^\varphi$ with respect to the L_2-norm on M_0. If $|\varphi|_k$ is sufficiently small with $k > n$, then there exists a constant c, independent of φ, such that for all $u \in C^\infty(M_0, {}^\circ T''^*)$*

$$|u|_{1/2}^2 \leqq c(|u|_0^2 + |\bar\partial_b^\varphi u|_0^2 + |\bar\partial_b^{*\varphi} u|_0^2). \tag{4.6}$$

It is well known that if (4.6) holds, then the kernel $\mathscr{H}_b^\varphi$ of $\Box_b^\varphi = \bar\partial_b^\varphi\bar\partial_b^{*\varphi} + \bar\partial_b^{*\varphi}\bar\partial_b^\varphi$ is finite dimensional, and there exists an operator $N_b^\varphi: C^\infty(M_0, {}^\circ T''^*) \to C^\infty(M_0, {}^\circ T''^*)$ with the following properties:

(4.7) $\Box_b^\varphi N_b^\varphi u = u - H_b^\varphi u$ for all $u \in C^\infty(M_0, {}^\circ T''^*)$, where H_b^φ is the projection on $\mathscr{H}_b^\varphi$.

$$\bar\partial_b^\varphi N_b^\varphi = N_b^\varphi \bar\partial_b^\varphi, \qquad \bar\partial_b^{*\varphi} N_b^\varphi = N_b^\varphi \bar\partial_b^{*\varphi}, \qquad N_b^\varphi H_b^\varphi = H_b^\varphi N_b^\varphi = 0. \tag{4.8}$$

(4.9) $|N_b^\varphi u|_{s+1} \leqq \text{const}\ |u|_s$ for all u and with a constant independent of φ.

PROPOSITION 4.4. *Assume that M_0 is strongly pseudoconvex. Let M_0^φ be a CR structure on M_0 represented by ${}^\circ T''_\varphi$ with a sufficiently small norm $|\varphi|_k$. Then M_0^φ can locally be realized as a hypersurface in C^n, i.e., for each $x_0 \in M_0$ there exists a neighborhood U of x_0 and C^∞ functions $w^1, \cdots, w^n$ on U such that $\bar\partial_b^\varphi w^j = 0$, $i \leqq j \leqq n$, and $dw^1 \wedge \cdots \wedge dw^n \neq 0$ on U.*

PROOF. Since M_0 is compact and strongly pseudoconvex, we can find a Stein neighborhood $\mathscr{N}$ and r holomorphic functions $f^1, \cdots, f^r$ on $\mathscr{N}$ such that if $x_0 \in M_0$, then, for some n-tuple $\{i_1, \cdots, i_n\} \in \{1, \cdots, r\}$, $f^{i_1}, \cdots, f^{i_n}$ are holomorphic coordinates at x_0. For simplicity we assume that $i_j = j$. Set

$$w^j = f^j - \bar\partial_b^{*\varphi} N_b^\varphi \bar\partial_b^\varphi f^j, \qquad 1 \leqq j \leqq n. \tag{4.10}$$

By (4.7) and (4.8) we have that $\bar\partial_b^\varphi w^j = 0$. Since $Z_{\bar l} f^j = 0$, it follows from (4.3) and (4.9) that $dw^1 \wedge \cdots \wedge dw^n \neq 0$ at x_0 if $|\varphi|_k$ is sufficiently small. Q.E.D.

References

1. L. Boutet de Monvel, Séminaire École Polytechnique, 1974 (preprint).

2. R. Hamilton, *Deformations of complex structures on manifolds with boundary*. I, II, III, IV, Cornell Univ. (preprint).

3. G. K. Kiremidjian, *Extendible pseudo-complex structures*, Advances in Math. (to appear).

4. ———, *Extendible pseudo-complex structures*. II (to appear).

5. ———, *A Nash-Moser type implicit function theorem and nonlinear boundary value problems*, Stanford Univ. (preprint).

6. ———, *On perturbations of linear non-coercive boundary value problems*, Stanford Univ. (preprint).

7. J. J. Kohn, *Harmonic integrals on strongly pseudo-convex manifolds*. II, Ann. of Math. (2) **79** (1964), 450–472. MR **34** #8010.

8. J. J. Kohn and L. Nirenberg, *Non-coercive boundary value problems*, Comm. Pure Appl. Math. **18** (1965), 443–492. MR **31** #6041.

9. J. J. Kohn and H. Rossi, *On the extension of holomorphic functions from the boundary of a complex manifold*, Ann. of Math. (2) **81** (1965), 451–472. MR **31** #1399.

10. M. Kuranishi, *Deformations of isolated singularities and* $\bar{\partial}_b$, Columbia Univ. (preprint).

11. J. Moser, *A new technique for the construction of solutions of nonlinear differential equations*, Proc. Nat. Acad. Sci. U.S.A. **47** (1961), 1824–1831. MR **24** #A2695.

STANFORD UNIVERSITY

Proceedings of Symposia in Pure Mathematics
Volume 30, 1977

UNIFORM BOUNDS ON DERIVATIVES FOR THE $\bar{\partial}$-PROBLEM IN THE POLYDISK

MARIO LANDUCCI*

1. Introduction. In a recent paper [2] G.M. Henkin proved the following theorem:

THEOREM *Let* $f = f_1 d\bar{z}_1 + f_2 d\bar{z}_2$ *be a smooth bounded* (0, 1) $\bar{\partial}$ *-closed form in the unitary polydisk D of* C^2. *Then there exists v smooth in D such that*

(1) $$\bar{\partial} v = f,$$

(2) $$\sup_D |v| \leqq c(D) \sup_D |f|.$$

Hence, in particular, such a solution is in $L^2(D)$.

In a more recent paper [5] it has been shown, under the same hypothesis, that

THEOREM. *The canonical solution u of* (1) (*i.e., the solution belonging to the space orthogonal to holomorphic functions in* $L^2(D)$) *has, for a suitable constant, c, the same property* (2).

The same result was proved in the case of the ball [4] and for a general strictly pseudoconvex domain (Proc. Internat. Congr. Math. (Vancouver, 1974), Canad. Math. Congr., 1975) by N. Kerzman.

The aim of this paper is to give estimates for the derivatives of the canonical solution in terms of the derivatives of f, i.e.,

THEOREM 1. *Let f be as above. Then if u is the canonical solution of* (1), *there exist constants C(s, D) such that*

(3) $$|u|_{s,D} \leqq C(s, D) |f|_{s,D}$$

where as usual

$$|u|_{s,D} = \sup_{D, \alpha+\beta \leqq s} |D_w^\alpha D_{\bar{w}}^\beta u| \text{ and } |f|_{s,D} = \max_{i=1,2} |f_i|_{s,D}.$$

AMS (*MOS*) *subject classifications* (1970). Primary 32H10; Secondary 35C15.

Key words and phrases. Holomorphic functions, $\bar{\partial}$-closed differential forms, $\bar{\partial}$-problem, integral representation formula, orthogonal projection.

*The author was supportd by CNR, GNSAGA.

© 1977, American Mathematical Society

An analogous result was shown, for strictly pseudoconvex domains with smooth boundary, by Y.-T. Siu **[6]**. The author was informed at the 1975 Summer Institute on Several Complex Variables (Williamstown) that results of Greiner-Stein and Rothschild-Stein, not yet published, would give, in the case of *strictly pseudoconvex domain* and with metrics different from the euclidean, results of this type.

2. Sketch of the proof of Theorem 1. To show the statement of Theorem 1, we shall consider as in **[4]** the differential form

$$f^\varepsilon(z) = f_1((1-\varepsilon)z)\,d\bar z_1 + f_2((1-\varepsilon)z)\,d\bar z_2$$

with $0 < \varepsilon < 1$.

Then under the hypothesis of Theorem 1, f^ε is $\bar\partial$-closed and smooth in a polydisk $D' \supset D$; so, if we consider the $\bar\partial$-equation

$$\bar\partial\gamma_\varepsilon = f^\varepsilon \tag{4}$$

in D', this admits a solution u_ε (see, for example, **[3]**) smooth up to the boundary of D.

By the integral representation formula proved in **[4]** (see also the next section), we get, for every $w \in D$,

$$h_\varepsilon(w) = u_\varepsilon(w) - c\int_{|z_1|=1;|z_2|=1} \frac{u_\varepsilon(z)}{(z_1 - w_1)(z_2 - w_2)}\,dz_1\,dz_2 \tag{5}$$

where h_ε is Henkin's solution of (3) in D (see **[2]**).

If now we denote by $P: L^2(D) \to H(D)$ the usual projection of $L^2(D)$ into $H(D)$, and if we put $\varepsilon = 1/k$, it is sufficient, for the proof of Theorem 1, to show the following proposition:

PROPOSITION 1. *Let* $h_k = h((1-1/k)z)$ *defined by* (5). *Then there exists a constant* $C = C(D, s)$ *such that.*

$$|h_k - P(h_k)|_{s,D} < C|f|_{s,D}. \tag{6}$$

In fact if we assume the proposition true, then using Banach-Saks' theorem (see **[1]**) the sequence,

$$\sum_1^n k\frac{1}{n} D^\alpha_{w_1} D^\beta_{w_2}[h_k - P(h_k)] \overset{L^2}{\mapsto} D^\alpha_{w_1} D^\beta_{w_2} u \qquad \forall\, \alpha, \beta, \alpha + \beta \leqq s,$$

when $k \to +\infty$, where $u(w)$ is exactly the canonical solution of (1) in D.[1]

The estimate (3) is, by Proposition 1, obvious.

3. Proof of Proposition 1. We recall that in the case of the polydisk $D \subset C^2$, the orthogonal projection with respect to the scalar product in L^2 is given by,

$$Pu(w) = c\int_{z\in D} \frac{u(z)}{(1 - \bar z_1 w_1)^2 (1 - \bar z_2 w_2)^2}\,d\bar z_1 \wedge dz_1 \wedge d\bar z_2 \wedge dz_2 \tag{7}$$

where $u \in L^2(D)$ and $c = 1/(2\pi i)^2$.

We also recall the integral formula stated in **[4]**; let $u \in C^2(\bar D)$; then for every $w \in D$, the following integral representation holds:

[1]Note that we have denoted again by $\{h_k - P(h_k)\}$ the subsequence extracted from $\{h_k - P(h_k)\}$

$$
\begin{aligned}
u(w) = {} & c\int_D \frac{1}{|z-w|^4}\left[\frac{\partial u}{\partial \bar z_1}(\bar z_1 - \bar w_1) + \frac{\partial u}{\partial \bar z_2}(\bar z_2 - \bar w_2)\right] d\bar z_1 \wedge dz_1 \wedge d\bar z_2 \wedge dz_2 \\
& - c\left[\int_{|z_1|=1;|z_2|\leq 1} \frac{\partial u}{\partial \bar z_2}\frac{\bar z_1 - \bar w_1}{(z_2 - w_2)|z-w|^2} d\bar z_2 \wedge dz_2 \wedge dz_1\right. \\
& \left. + \int_{|z_1|<1;|z_2|=1} \frac{\partial u}{\partial \bar z_1}\frac{\bar z_2 - \bar w_2}{(z_1 - w_1)|z-w|^2} d\bar z_1 \wedge dz_1 \wedge dz_2\right] \\
& - 2\pi i c\left[\int_{|z_1|<1} \frac{\partial u}{\partial \bar z_1}(z_1, w_2)\frac{1}{(z_1 - w_1)} d\bar z_1 \wedge dz_1\right. \\
& \left. + \int_{|z_2|<1} \frac{\partial u}{\partial \bar z_2}(w_1, z_2)\frac{1}{(z_2 - w_2)} d\bar z_2 \wedge dz_2\right] \\
& + c\int_{|z_1|=1;|z_2|=1} \frac{u(z)}{(z_1 - w_1)(z_2 - w_2)} dz_1 dz_2.
\end{aligned} \tag{8}
$$

LEMMA 1. *Let* $u \in C^2(\bar D)$; *then we have*

$$\sup_{\bar D} |P(u)| \leqq C(D)|u|_{1, D}. \tag{9}$$

PROOF. By (8) we immediately get

$$\sup_{\bar D}\left|\int_{|z_1|=1;|z_2|=1} \frac{u(z)}{(z_1 - w_1)(z_2 - w_2)} dz_1\, dz_2\right| \leqq C'(D)|u|_{1,D},$$

and so to obtain (9) it is sufficient to apply the remark to Proposition 1 in **[4]**.

LEMMA 2. *Let* $u \in C^\infty(\bar D)$; *then we have the following identity*:

$$D^\alpha_{w_1} D^\beta_{w_2} P(u)(w) = P(D^\alpha_{z_1} D^\beta_{z_2} u(z)) + G(w), \qquad w \in D, \tag{10}$$

where $G(w)$ *is bounded by* $|\bar\partial u|_{\alpha+\beta, D}$.

PROOF. We shall start proving (10) in the case $\alpha = 1$, $\beta = 0$. By differentiation of (7) we have

$$
\begin{aligned}
\frac{\partial P(u)}{\partial w_1} & = c\int_{z\in D} u(z)\frac{\partial}{\partial \bar z_1}\left[\frac{\bar z_1^2}{(1 - \bar z_1 w_1)^2}\right]\frac{1}{(1 - \bar z_2 w_2)^2} d\bar z_1 \wedge dz_1 \wedge d\bar z_2 \wedge dz_2 \\
& = -P\left(\frac{\partial u}{\partial \bar z_1}\bar z_1^2\right)(w) + c\int_{|z_1|=1;|z_2|\leq 1} \frac{u(z)\bar z_1^2}{(1 - \bar z_1 w_1)^2}\frac{1}{(1 - \bar z_2 w_2)^2} dz_1 \wedge d\bar z_2 \wedge dz_2 \\
& = -P\left(\frac{\partial u}{\partial \bar z_1}\bar z_1^2\right)(w) + I.
\end{aligned}
$$

On the other hand

$$
\begin{aligned}
I & = -c\int_{|z_1|=1;|z_2|<1} u(z)\frac{\partial}{\partial z_1}\left[\frac{1}{(z_1 - w_1)}\right]\frac{1}{(1 - \bar z_2 w_2)^2} dz_1 \wedge d\bar z_2 \wedge dz_2 \\
& = c\int_{|z_1|=1;|z_2|<1} -\frac{\partial u}{\partial z_1}\frac{1}{(z_1 - w_1)}\frac{1}{(1 - \bar z_2 w_2)^2} dz_1 \wedge d\bar z_2 \wedge dz_2 \\
& \quad + c\int_{|z_1|=1;|z_2|<1} \frac{\partial u}{\partial \bar z_1}\frac{1}{(z_1 - w_1)}\frac{1}{(1 - \bar z_2 w_2)^2} d\bar z_1 \wedge d\bar z_2 \wedge dz_2 \\
& = P\left(\frac{\partial u}{\partial z_1}\right)(w).
\end{aligned}
$$

So we get (acting in the same manner with $\partial Pu/\partial w_2$),

$$(11)\qquad \begin{aligned} \frac{\partial P(u)}{\partial w_1} &= P\left(\frac{\partial u}{\partial z_1}\right) - P\left(\frac{\partial u}{\partial \bar{z}_1}\bar{z}_1^2\right),\\ \frac{\partial P(u)}{\partial w_2} &= P\left(\frac{\partial u}{\partial z_2}\right) - P\left(\frac{\partial u}{\partial \bar{z}_2}\bar{z}_2^2\right),\end{aligned}$$

and for the proof it is sufficient to observe that, when we differentiate successively $P(u)$, we find, making use again and again of (11), precisely (10).

The bound of $G(w)$ is immediate using Lemma 1.

Now we are able to prove Proposition 1 and so the main theorem.

Let h_k be as in §2; then (it is smooth up to the boundary of D) by (8) applied to $D^\alpha_{w_1} D^\beta_{w_2} h_k$, we immediately get

$$(12)\qquad D^\alpha_{w_1} D^\beta_{w_2} h_k(w) = F_1(w) + c\int_{|z_1|=1;|z_2|=1} D^\alpha_{z_1} D^\beta_{z_2} h_k \frac{1}{(z_1 - w_1)(z_2 - w_2)}\, dz_1\, dz_2,$$

where $F_1\ (w)$ is clearly bounded by $|f|_{\alpha+\beta, D}$.

On the other hand (see [**4**, Proposition 1]),

$$(13)\qquad \begin{aligned} c\int_{|z_1|=1;|z_2|=1} D^\alpha_{z_1} D^\beta_{z_2} h_k \frac{1}{(z_1 - w_1)(z_2 - w_2)}\, dz_1\, dz_2 \\ = P(D^\alpha_{z_1} D^\beta_{z_2} h_k)(w) + F_2(w)\end{aligned}$$

where $F_2(w)$ is bounded by $|f|_{\alpha+\beta, D}$; so using (10), (12) and (13), we get

$$D^\alpha_{w_1} D^\beta_{w_2} (h_k - P(h_k)) = F_1(w) + F_2(w) + G(w),$$

and hence Proposition 1.

References

1. L. Bers, F. John and M. Schechter, *Partial differential equations*, Lectures in Appl. Math., vol. 3, Interscience, New York, 1966. MR **29** #346.

2. G.M. Henkin, *A uniform estimate for the solution of the $\bar{\partial}$-problem in a Weil region*, Uspehi Mat. Nauk **26** (1971), no.3 (159), 211–212. (Russian) MR **45** #3753.

3. L. Hörmander, *An introduction to complex analysis in several variables*, Van Nostrand, Princeton, N.J., 1966. MR **34** #2933.

4. N. Kerzman, *Remarks on estimates for $\bar{\partial}$-equation*, Lecture Notes in Math., vol. 336, Springer-Verlag, Berlin and New York, 1973, pp. 111–124.

5. M. Landucci, *On the projection of $L^2(D)$ into $H(D)$*, Duke Math. J. **42** (1975).

6. Y.-T. Siu, *The $\bar{\partial}$-problem with uniform bounds on derivatives*, Math. Ann. **207** (1974), 163–176.

Instituto Matematico G. Vitale (Modena)

Proceedings of Symposia in Pure Mathematics
Volume 30, 1977

DIFFUSION ESTIMATES IN COMPLEX ANALYSIS

PAUL MALLIAVIN

This report is a short resume of results obtained by several authors in this area which can be summarized as the study of semielliptic systems via the behaviour of the paths of some associated diffusion: global estimates (resp. estimates for the parametrix singularity) come then from global study (resp. infinitesimal study) of the generic path, Poisson formulae and explicit construction of Green kernels from integration in some path space. (Cf. also **[14]** for another survey.)

1. Poisson formulae and boundaries. The exit measure of a diffusion, that is the statistic of the *end* of the path, will provide a Poisson formula. In complex analysis it is possible to choose arbitrarily a *Kählerian* Laplacian or to mix together several Kählerian Laplacians to obtain a *compound process.* The qualitative study of the paths at infinity will be then done by an approach parallel to the study of the solution of ordinary differential equations.

1.1. *Šilov boundary for a pseudoconvex domain* **[3]**. Denote by D a bounded pseudoconvex domain in C^n, $D = \{z; p(z) < 0\}$ where p is a C^3 plurisubharmonic function, $dp \neq 0$ on the boundary of D; denote by $\partial_s(D)$ the points of the boundary where the Levy form restricted to the complex tangent plane is positive definite.

1.1.1. THEOREM. *For every $z_0 \in D$, there exists a positive Radon measure μ_{z_0}, with support contained in $\overline{\partial_s(D)}$, such that $f(z_0) = \int f\, d\mu_{z_0}$.*

1.1.2. COROLLARY. *Denote by $A(D)$ the algebra of holomorphic functions in D, continuous on $\overline{D}$; then Šilov $(A(D)) = \overline{\partial_s(D)}$.*

This corollary answers a problem raised in **[2]** and also discussed in **[11]**.

The method of proof consists of introducing the Kählerian metric

$$ds^2 = |dz|^2 + \partial\bar{\partial} \log(-p^{-1})$$

AMS (MOS) subject classifications (1970). Primary 60J60, 60J45, 60G17, 35J99; Secondary 32F15, 32E25, 47D05, 32D10, 35K05.

© 1977, American Mathematical Society

and to show that the associated diffusion leaves D in an infinite time, and then that the exit is one necessarily on $\overline{\partial_s D}$.

1.2. *Boundary of Hua's system on a Siegel upper half-plane* **[8]**. The upper half-plane considered will consist of complex 2×2 symmetric matrices:

$$W = \begin{pmatrix} 2^{1/2} w_1 & w_3 \\ w_3 & 2^{1/2} w_2 \end{pmatrix}$$

where $V = \mathrm{Im}(W)$ is positive definite. Denote by ∂_w the following 2×2 matrix with differential operator coefficients:

$$\partial_w = \begin{pmatrix} 2^{1/2} \partial_{w_1} & \partial_{w_3} \\ \partial_{w_3} & 2^{1/2} \partial_{w_2} \end{pmatrix}.$$

The formal matrix product $\bar{\partial}_w \cdot V \cdot \partial_w$ defines a 2×2 matrix whose coefficients are second-order differential operators; the Hua system consists of the four second-order differential operators so obtained. We will extract from this system the two following elliptic operators:

$$\Delta_1 = \mathrm{Trace}\ V \cdot \bar{\partial}_w \cdot V \cdot \partial_w, \qquad \Delta_2 = \mathrm{Trace}\ \bar{\partial}_w \cdot V \cdot \partial_w.$$

Then Δ_1 is the Laplace operator of the Bergman metric on H. A bounded function f satisfying $\Delta_1 f = 0$ is representable by a Poisson integral on the Furstenberg boundary $F(H)$. On the other hand, a bounded holomorphic function in H is representable by a Poisson integral on the Šilov boundary $S(H) = \{W; V = 0\}$. There is a natural fibering $F(H) \to S(H)$.

1.2.1. THEOREM. *Denote by f a bounded function satisfying $\Delta_1 f = 0$; then f is representable on $S(H)$ if and only if $\Delta_2 f = 0$.*

The proof is based on the study of the exit measure of a compound process obtained in mixing the process associated to Δ_1 and Δ_2. As a corollary a bounded solution $\Delta_1 f = \Delta_2 f = 0$ will satisfy the full Hua system.

2. Estimates on an open manifold.

2.1. *A comparison between geodesic balls* **[5]**. Denote by B a geodesic ball of center x_0 on a riemannian manifold M, k (resp. K) the lower (resp. the upper) bound of the sectional curvature of M on B, p_t^B (resp. p_t^k, p_t^K) the fundamental solution for the heat equation with Dirichlet boundary condition for B (resp. geodesic ball of the same radius for the spaces of constant curvature k, K).

2.1.1. THEOREM. *$p_t^k(x_0', x') \leqq p_t^B(x_0, x) \leqq p_t^K(x_0'', x'')$ where the points x, x', x'' correspond in the exponential map.*

2.1.2. COROLLARY. *If λ^B (resp. λ^k, λ^K) denotes the first eigenvalue of B, then $\lambda^K \leqq \lambda^B \leqq \lambda^k$.*

The method used is a study of the exit time T^B of B for the process starting from x_0, which proves

$$T^k \leqq T^B \leqq T^K. \tag{2.1.3}$$

2.2. *Estimates of the heat kernel on an open manifold* **[13]**. Denote by M a riemannian manifold which satisfies the hypothesis

($\mathscr{H}$) M has a global radius of injectivity; the curvature tensor and its covariant derivatives are uniformly bounded.

2.2.1. THEOREM. *There exist two constants c, γ depending only on the bounds appearing in ($\mathscr{H}$) and on the dimension of M, such that the fundamental solution of the heat equation satisfies*

$$p_t(x, y) < c \exp(-\gamma d^2(x, y)), \qquad 0 < t < 1, d(x, y) > 1,$$

where d is the geodesic distance.

This estimate is obtained starting with (2.1.3), replacing the diffusion by some Markov chain with holding time, for which the number of jumps can be estimated.

2.2.2. COROLLARY. *Let M be a hermitian manifold satisfying ($\mathscr{H}$), $\Box$ denote a complex Laplacian on (p, q) differential form, π_t a solution of the heat equation*

$$\partial \pi_t / \partial_t + \Box \pi_t = 0, \qquad \pi_0 = 0.$$

Suppose that, for $\gamma' < \gamma$,

$$\int_M \|\pi_t\| \exp(-\gamma' d^2(x_0, y))\, dy < c < +\infty.$$

Then $\pi_t \equiv 0$.

2.3. *Estimates for the Green operators* [12]. M will be a Stein manifold with a C^3 strictly plurisubharmonic function $p(z)$ which defines on M the Kählerian metric $\partial\bar{\partial}p$. Denote by G_0 the Green operator for the Laplacian on the functions of M. Then

2.3.1. THEOREM. *If $|\partial p| < c$ then G_0 operates from L^r to L^r (for every $1 < r < +\infty$). More generally if $|\partial p| < \varphi(p)$ then there exist two weight functions w_1, w_2 which can be computed from φ such that G_0 operates from $L^r(w_1)$ into $L^r(w_2)$.*

3. Singularity of parametrix.

3.1. *Explicit expression of the heat kernel on the Heisenberg group* [7]. On the Heisenberg group $H_1 = (z, t)$ $((z, u)(z', u') = (z + z', u + u' + 2\,\mathrm{Im}\, zz'))$ the heat kernel $p_t^{H_1}(z, u)$ is determined by

3.1.1. THEOREM.

$$p_t^{H_1}(z, u) = \frac{1}{s^2} \int_{-\infty}^{+\infty} \left(\frac{2\tau}{\sinh 2\tau}\right) \exp\left[\frac{1}{s}\left(i\tau t - \frac{2\tau(z)^2}{\tanh 2\tau}\right)\right] d\tau.$$

This formula is obtained by an exact description of the path of the diffusion associated; in the coordinate z it is a standard brownian path $b(t)$ on C, $u(t_0)$ being equal to the area swept out by the segment $[0, b(t)]$, $0 < t < t_0$. Formulae similar to 3.1.1 are obtained for the Heisenberg group H_n, where the C^n factor is an *arbitrary* hermitian metric. Then, integrating in t, we have an answer given to the problem raised [6, p. 520] of finding an explicit fundamental solution of the Laplacian on H_n with such a metric.

The expression 3.1.1 combined with stationary phase estimates leads to:

3.1.2. THEOREM. *When $t \to 0$, $\log p_t^{H_1}$ is equivalent to the classical action.*

3.2. REMARK. The singular integral method fully developed in [6] reconstructs from a local model the behaviour of the parametrix of a general hypoelliptic operator of the Heisenberg type. A first step for an alternative stochastic approach is made in [**10**].

4. Construction of semigroups on differential forms. The existence of semigroups associated with the heat equation on differential forms results classically from spectral decomposition.

The purpose is to reverse this approach; first to construct by a direct approach such a semigroup, and then from its behaviour for t large to deduce properties of the spectral decomposition.

4.1. *Compact Riemannian manifold* [**9**]. The semigroup is constructed in terms of a differential linear system along the path of a diffusion on a bundle of frames. Vanishing theorems are obtained where the positivity condition of S. Bochner is replaced by "positivity in mean" in a sufficiently strong sense.

4.2. *Compact manifold with boundary M with d-Neumann-Spencer condition* [**1**]. The approach consists of packing together by a functional integration the obvious solution when M is a flat half-space. A theorem of Privaloff type is also obtained for a system of n-conjugate functions.

5. Hull of holomorphy of some thin set [**4**]. Given a perfect set P on the sphere of C^2, saturated for the rotation of the z_2 coordinate, the holomorphy hull of this set is constructed. The fine potential theory on the projection of P on the z_1 plane plays an important role. The method of prolongation uses a heat equation where the time is identified to some "normal coordinate" to the sphere. (See [**15**] for an elementary description of this phenomenon. See [**16**] and [**17**] for other regularity theorems.)

6. Vanishing theorem in degree zero for line bundles ([18] and [19]).

BIBLIOGRAPHY

1. H. Airault, *Solution stochastique du d. Neumann Spencer*, C. R. Acad. Sci. Paris **280** (1975), 781; Séminaire Jean Leray January 1975, J. Math. Pures Appl. (to appear).

2. H. Bremermann, *On a generalized Dirichlet problem for pluri-subharmonic functions and pseudo-convex domains*, Trans. Amer. Math. Soc. **91** (1959), 246–276. MR **25** #227.

3. A. Debiard and B. Gaveau, *Sur une conjecture de Bremermann*, C. R. Acad. Sci. Paris **279** (1974), 407; Bull. Sci. Math. (1976), 18–33.

4. ———, *Algèbres de fonctions et potentiel fin*, C. R. Acad. Sci. Paris **278** (1974), 1025; J. Functional Analysis April 1976.

5. A. Debiard, B. Gaveau and E. Mazet, *Théorèmes de comparaison en géométrie*, C. R. Acad. Sci. Paris **278** (1974), 723–725; ibid. **281** (1975), 955; J. Res. Inst., Kyoto Univ. (to appear).

6. G. B. Folland and E. M. Stein, *Estimates for the $\bar{\partial}_b$ complex*, Comm. Pure Appl. Math. **27** (1974), 429–522.

7. B. Gaveau, *Le groupe de Heisenberg*, C. R. Acad. Sci. Paris **281** (1975), 571.

8. A. Korányi and P. Malliavin, *Poisson formula and compound diffusion on the Siegel upper half plane of rank two*, Acta Math. (1975), 181–206.

9. P. Malliavin, *Formules de la moyenne calcul de perturbations et théorèmes d'annalution pour les formes harmoniques*, J. Functional Analysis **17** (1974), 274–291; Bull. Sci. Math. (to appear).

10. P. Malliavin, *Parametrix trajectorielle*, C. R. Acad. Sci. Paris **281** (1975), 241.

11. H. Rossi, *Holomorphically convex sets in several complex variables*, Ann. of Math. (2) **74** (1961), 470–493. MR **24** #A3310.

12. J. Vauthier, *Estimées de la fonction de Green d'une variété de Stein*, Séminaire Jean Leray janvier 19th; C. R. Acad. Sci. Paris **279** (1974), 1539.

13. ———, *Unicité de la solution de l'équation de la chaleur*, C. R. Acad. Sci. Paris **281** (1975), 41.

14. P. Malliavin, *Diffusions et géométrie différentielle globale*, Centro Mathematico Estvo, Varenna, 1975.

15. ———, *Equation de la chaleur et valeur frontière des fonctions holomorphes*, C. R. Acad. Sci. Paris **281** (1975); Ann. Inst. Fourier **25** (1975), 447–466.

16. A. Debiard and B. Gaveau, *Potentiel fin et algébres de fonctions analytiques*. I, II. Functional Analysis **16** (1974), 289–304; ibid. **17** (1974), 296–310.

17. ———, *Différentiabilité des fonctions finement harmoniques*, Invent. Math. **29** (1975), 111–123.

18. ———, *Annulation en degré zéro form un fibré en droites negatif au dessus de certaines variétés kähleriennes non complètes*, Bull. Sci. Math. (to appear).

19. B. Gaveau and J. Vauthier, *Traces de la courbure et annulation form des fibrés non integrables*, Bull. Sci. Math. (to appear).

INSTITUT HENRI POINCARÉ, UNIVERSITÉ DE PARIS

Proceedings of Symposia in Pure Mathematics
Volume 30, 1977

HYPERFUNCTION BOUNDARY VALUES AND A GENERALIZED BOCHNER-HARTOGS THEOREM*

JOHN C. POLKING** AND R. O. WELLS, JR.†

1. Introduction. In 1943 Bochner proved the following generalization of Hartogs' theorem.

THEOREM (BOCHNER [**2**]). *Let D be a domain in C^n, $n \geqq 2$, with ∂D smooth and connected. Then, given a function $f \in C^1(\partial D)$ which satisfies the tangential Cauchy-Riemann equations on ∂D, there is a function $F \in C(\bar{D})$ which is holomorphic in D and is equal to f on ∂D.*

This theorem was the impetus for the study over the next thirty years of the tangential Cauchy-Riemann equations on a smooth real submanifold. In particular this theorem was localized by Hans Lewy in 1956 (Lewy [**13**]) assuming an additional local convexity condition expressed in terms of the Levi form. The concept of a $\bar{\partial}_b$-complex of (p,q)-forms on a real submanifold of C^n was introduced by Kohn and Rossi [**10**], who generalized Bochner's theorem by replacing $\bar{\partial}$-closed functions with $\bar{\partial}$-closed differential forms of degree (p, q). These papers were seminal for research efforts by numerous mathematicians in the 1960's and early 1970's (see the survey paper by Wells [**21**] and the monograph by Folland and Kohn [3]).

In particular, Weinstock [**20**] generalized Bochner's theorem by replacing C^1 functions satisfying the tangential Cauchy-Riemann equations with continuous functions which satisfy these equations weakly (in this case the condition on the boundary value is necessary and sufficient). Recently Harvey and Lawson [**5**] have

AMS (MOS) subject classifications (1970). Primary 32D15.

*Research supported by National Science Foundation Grant MPS-75-05270 at Rice University.

**Research supported by a Sloan Foundation Research and Instruction Grant while a Visiting Member at the Courant Institute.

†Alexander von Humboldt Awardee.

© 1977, American Mathematical Society

shown as a part of a generalization of Bochner's theorem that the function F in Bochner's theorem is actually in $C^1(\bar{D})$. The question arose: how general can boundary values of holomorphic functions be, do they satisfy the induced Cauchy-Riemann equations in some sense, and is there a version of the Bochner-Hartogs theorem for more general boundary values?

In another direction it is of interest to mention the classical theorem of Plemelj in one complex variable. Plemelj proved that if Ω is a domain in $\boldsymbol{C}$ which is divided into two pieces Ω^+ and Ω^- by a smooth curve γ, then for every Hölder continuous function f on γ there are holomorphic functions u^+ and u^- on Ω^+ and Ω^- which are Hölder continuous up to γ such that $f = u^+ - u^-$. Thus, roughly speaking, every function on γ is the jump or saltus of a holomorphic function defined on the complement of γ. A version of the Plemelj theorem in $\boldsymbol{C}^n$ is contained in the work of Andreotti and Hill [**1**]. They prove that if Ω is divided by a C^∞ hypersurface M, then every C^∞ function on M which satisfies the induced Cauchy-Riemann equations is the jump of a holomorphic function defined on $\Omega - M$ (provided that $H^1(\Omega, \mathcal{O}) = 0$).

It is our purpose here to present results similar to those of Bochner, Lewy, and Andreotti-Hill where more general boundary values are allowed. These results are proved in [**15**]. The analogs of the Bochner, Lewy, and Andreotti-Hill theorems are stated in § 2, and in § 3 we use these results to analyze the Hans Lewy example.

Our approach to the problems discussed above, using the Cauchy-Kowalewski theorem, was motivated by Komatsu [**11**] and Komatsu-Kawai [**12**]. This approach was also introduced independently by Schapira [**19**]. A different and more general approach to studying hyperfunction boundary values of general elliptic systems is announced in Kashiwara-Kawai [**8**]. The results of § 2 of this paper are essentially contained in [**8**]. The proof there depends on the fundamental work of Sato, Kawai, and Kashiwara [**17**], while the goal here is to describe an elementary proof of these facts given in [**15**].

Since the preparation of [**15**] we have learned the local generalization of H. Lewy's result (Theorem 8) has been proved also by Pallu de la Barriere [**14**] using different techniques. This work is closely related to the work of Kashiwara-Kawai, cited above. In addition we have recently learned that the results of § 2 of this paper have been obtained by Olle Stormark. Again the techniques are completely different.

2. Statement of results. The class of generalized functions which are admissible boundary values is the class of *hyperfunctions* of Sato and Martineau. Hyperfunctions contain the usual distributions and functions as a proper subset but are defined only on real analytic manifolds. Consequently we must restrict ourselves to real analytic manifolds. For an introduction to hyperfunctions see Schapira [**18**] or Sato-Kawai-Kashiwara [**17**].

Let X be a complex manifold of dimension n which is countable at infinity. Let $\mathscr{B}^{p,q}(X)$ denote the space of hyperfunction (p, q)-forms on X. Then we have the hyperfunction Dolbeault complex

$$\mathscr{B}^{p,0}(X) \xrightarrow{\bar\partial} \mathscr{B}^{p,1}(X) \xrightarrow{\bar\partial} \cdots \xrightarrow{\bar\partial} \mathscr{B}^{p,n}(X). \tag{1}$$

The cohomology of this complex is the standard Dolbeault cohomology and we will denote it by $H^{p,q}(X)$.

Suppose $M \subset X$ is a real analytic hypersurface. Let $\mathscr{B}^{p,q}(M)$ denote the hyperfunction (p, q)-forms on M. Then

(2) $$\mathscr{B}^{p,0}(M) \xrightarrow{\bar\partial_b} \mathscr{B}^{p,1}(M) \xrightarrow{\bar\partial_b} \cdots \xrightarrow{\bar\partial_b} \mathscr{B}^{p,n-1}(M)$$

is the hyperfunction $\bar\partial_b$ complex or the *boundary complex*. Let $H^{p,q}(M)$ denote the cohomology of the boundary complex.

The Dolbeault complex (1) and the boundary complex (2) are connected by the *complex push-forward* μ_*: $\mathscr{B}^{p,q}(M) \to \mathscr{B}^{p,q+1}_M(X)$, where $\mathscr{B}^{p,q+1}_M(X)$ denote the $(p, q+1)$ hyperfunction forms on X with support in M (for the definition of $\mathscr{B}^{p,q}(M)$ and μ_* see **[15]**). The main property of μ_* is the following.

PROPOSITION 1. $\mu_* \bar\partial_b = -\bar\partial\mu_*$.

In order to state our analogs of the Bochner, Lewy, and Andreotti-Hill theorems we must define the saltus or jump of a holomorphic function on $X - M$ across M. More generally we will define the saltus of a Dolbeault cohomology class in $X - M$ across M. Let the class $[u] \in H^{p,q}(X - M)$ be represented by $u \in \mathscr{B}^{p,q}(X - M)$. Then $\bar\partial u = 0$ in $X - M$. By the extension property of hyperfunctions, there is a $\tilde u \in \mathscr{B}^{p,q}(X)$ such that $\tilde u = u$ in $X - M$. There are of course many such extensions, but using the following lemma we can choose one with nice properties.

LEMMA 2. *Suppose* $g \in \mathscr{B}^{p,q}(M)$ *and* $v \in \mathscr{B}^{p,q}_M(X)$ *satisfy* $\mu_* g = \bar\partial v$. *Then there exist* $w \in \mathscr{B}^{p,q-1}_M(X)$ *and* $h \in \mathscr{B}^{p,q-1}(M)$ *such that* $v = \bar\partial w - \mu_* h$ *and* $g = \bar\partial_b h$.

REMARK. Lemma 2 is proved using the Cauchy-Kowalewski theorem. The prooj is contained in **[15]** and will not be given here. Lemma 2 provides the technical base upon which all of the following results depend.

If we apply Lemma 2 to $\bar\partial\tilde u \in \mathscr{B}^{p,q+1}_M(X)$ we get the following result.

PROPOSITION 3. *Let* $u \in \mathscr{B}^{p,q}(X - M)$ *satisfy* $\bar\partial u \equiv 0$. *Then there is an extension* $u' \in \mathscr{B}^{p,q}(X)$ *and* $h \in \mathscr{B}^{p,q}(M)$ *such that*

$$\bar\partial u' = -\mu_* h, \qquad \bar\partial_b h = 0.$$

Furthermore h is unique modulo $\bar\partial_b \mathscr{B}^{p,q-1}(M)$.

As a result we can make the following definition.

DEFINITION 4. For a class $[u] \in H^{p,q}(X - M)$ we define the *saltus* $b[u] \in H^{p,q}(M)$ by $b[u] = [h]$ where $h \in \mathscr{B}^{p,q-1}$ is as described in Proposition 3.

The most important case is when $p = q = 0$. If $u \in \mathscr{O}(X - M)$, then by Proposition 3 there is a unique extension $u' \in \mathscr{B}(X)$ of u and a unique hyperfunction $h \in \mathscr{B}(M)$ such that $\bar\partial u' = -\mu_* h$ and $\bar\partial_b h = 0$. Then the saltus of u across M is $bu = h \in H^{0,0}(M)$. We let $'\mathscr{O}(M) = H^{0,0}(M)$ and call an element $h \in {}'\mathscr{O}(M)$ a *CR hyperfunction.* Suppose M divides X into two connected components: $X - M = X^+ \cup X^-$, where $\int_{X^\pm} d\varphi = \pm \int_M d^*\varphi$ for compactly supported $(2n - 1)$-forms φ. If $u \in \mathscr{O}(X - M)$ is continuous on $\bar X^+$ and $\bar X^-$ with boundary values u^+ and u^- on M respectively then $bu = u^+ - u^-$. Consequently our definition of saltus generalizes the usual definition.

The hyperfunction version of the Andreotti-Hill theorem can now be stated.

THEOREM 5. *The sequence*

$$0 \longrightarrow \mathcal{O}(X) \longrightarrow \mathcal{O}(X - M) \xrightarrow{b} {}'\mathcal{O}(M) \xrightarrow{\mu_*} H^{0,1}(X) \longrightarrow$$

$$\cdots \longrightarrow H^{0,q}(X - M) \xrightarrow{b} H^{0,q}(M) \xrightarrow{\mu_*} H^{0,q+1}(X) \longrightarrow$$

$$\cdots H^{0,q+1}(X - M) \xrightarrow{b} H^{0,n}(X - M) \xrightarrow{\mu_*} H^{0,n}(X) \longrightarrow 0$$

is exact.

REMARK. The proof of Theorem 5 is quite easy using Lemma 2. Of course the analogous sequence for (p, q) cohomology is also exact.

A special case is worth mentioning.

COROLLARY 6. *Suppose $H^{0,1}(X) = 0$. Then*

$$0 \longrightarrow \mathcal{O}(X) \longrightarrow \mathcal{O}(X - M) \longrightarrow {}'\mathcal{O}(M) \longrightarrow 0$$

is exact.

In particular if $H^{0,1}(X) = 0$ (which is true for example if X is Stein), then the saltus map $b: \mathcal{O}(X - M) \to {}'\mathcal{O}(M)$ is subjective. Thus every CR hyperfunction on M is the saltus of a holomorphic function across M. Since every point of M has a neighborhood in X which is Stein, we see that every CR hyperfunction on M is locally the saltus of a holomorphic function. This is a fact of some significance in the study of CR functions (see **[7]**).

Next we consider the hyperfunction version of Bochner's theorem. In this case $M = \partial \Omega$, where Ω is a relatively compact open subset of X. For $u \in \mathcal{O}(\Omega)$ we will denote by bu the saltus of the extension of u by zero to $\mathcal{O}(X - M)$ and we will call bu the *boundary values of u*.

THEOREM 7. *Let X be a complex manifold of dimension greater than 1 which is countable at ∞ and suppose $H^{1,0}_*(X) = 0$. If $\Omega \subset X$ is a relatively compact open set with real analytic boundary M and with connected complement, then $b: \mathcal{O}(\Omega) \to {}'\mathcal{O}(M)$ is an isomorphism.*

Finally we present the hyperfunction version of Lewy's theorem.

THEOREM 8. *Let M be a real analytic hypersurface in $\mathbf{C}^n$ defined near $z_0 \in M$ by the equation $\rho(z) = 0$. Suppose there is a $\zeta \in \mathbf{C}^n$ such that*

$$\sum_{j=1}^{n} \zeta_j \frac{\partial \rho}{\partial z_j}(z_0) = 0 \quad \textit{and} \quad \sum_{j,k=1}^{n} \zeta_j \bar{\zeta}_k \frac{\partial^2 \rho}{\partial z_j \partial \bar{z}_k}(z_0) < 0.$$

Then for every neighborhood Ω of z_0 there is a neighborhood ω of z_0 with $\omega \subset \Omega$ such that the boundary value map

$$b: \mathcal{O}(\omega_+) \to {}'\mathcal{O}(\omega \cap M)$$

is an isomorphism, where $\omega_+ = \{z \in \omega \mid \rho(z) > 0\}$.

REMARK. For $\varphi \in \mathcal{O}(\omega_+)$ the boundary values $b\varphi$ are defined to be the saltus of the extension of φ by zero to all of ω.

The proof of Theorem 8 is in **[15]**.

REMARK. There are analogs of Theorems 5, 6, 7, and 8 for other categories. In particular there are results when we consider distribution forms on M and restrict the growth of the distribution forms on $X - M$. In this case we need only assume that M is a C^∞ manifold. For precise statements of the results and the proofs see **[15]**.

3. The Hans Lewy example. Using the results of the previous section we wish to analyze the question of when the inhomogeneous equation $\bar{\partial}_b f = g$ has a solution locally.

Since the problem is local, we may assume that X is Stein and that M divides X into two components. Thus $X - M = X^+ \cup X^-$. If $g \in \mathscr{B}^{p,q}(M)$ satisfies $\bar{\partial}_b g = 0$, then $\bar{\partial}\mu_* g = 0$ by Proposition 1, and since X is Stein there is a $v \in \mathscr{B}^{p,q}(X)$ with $\bar{\partial}v = \mu_* g$. Let $v^\pm = v\,|_{X^\pm}$.

PROPOSITION 9. *Suppose X is Stein. Let $g \in \mathscr{B}^{p,q}(M)$ and $v \in \mathscr{B}^{p,q}(X)$ satisfy $\bar{\partial}v = \mu_* g$. Let $z_0 \in M$. The following are equivalent:*

(a) *There is a neighborhood ω of z_0 in M and $f \in \mathscr{B}^{p,q-1}(\omega)$ such that $\bar{\partial}_b f = g$ in ω.*

(b) *There is a neighborhood Ω_1 of z_0 in X and $w^\pm \in \mathscr{B}^{p,q-1}(X^\pm \cap \Omega_1)$ such that $\bar{\partial}w^\pm = v^\pm$ in $X^\pm \cap \Omega_1$.*

(c) *There is a neighborhood Ω_2 of z_0 in X and $u^\pm \in \mathscr{B}^{p,q}(\Omega_2)$ such that $u^\pm = v^\pm$ in $X^\pm \cap \Omega_2$ and $\bar{\partial}u^\pm \equiv 0$ in Ω_2.*

PROOF. It follows from Definition 4 that

$$b([v^+] + [v^-]) = [g],$$

and since X is Stein, Theorem 5 shows that

$$b\colon H^{p,q}(X - M) \to H^{p,q}(M)$$

is an isomorphism for $q \geqq 1$. Thus the equivalence of (a) and (b) is immediate.

Assume (b) is true and let $W^\pm$ be any extensions of $w^\pm$ to a neighborhood of z_0. Then $u^\pm = \bar{\partial}W^\pm$ extends $v^\pm$ and is $\bar{\partial}$-closed. On the other hand if (c) is true choose $\Omega_1 \subset \Omega_2$ to be a Stein neighborhood of z_0. Choose $w^\pm$ to solve the equation $\bar{\partial}w^\pm = u^\pm$.

We will illustrate Proposition 9 by applying it to the Hans Lewy example. Let $X = \boldsymbol{C}^2$ with coordinates (z, w). Let $\rho(z, w) = |z|^2 - \operatorname{Im} w$ and let $M = \{(z, w) \mid \rho(z, w) = 0\}$. We parametrize M by $\boldsymbol{C} \times \boldsymbol{R}$ with the map $(z, t) \to (z, t + i|z|^2)$. Then $\bar{\partial}_b\colon \mathscr{B}^{0,0}(M) \to \mathscr{B}^{0,1}(M)$ is given by $\bar{\partial}_b f = L(f)d\bar{z}$, where $L(f) = \partial f/\partial\bar{z} + i\bar{z}\partial f/\partial t$ is the Hans Lewy operator. For nonnegative integers α, β let

$$g_{\alpha,\beta} = (\partial/\partial z)^\alpha(\partial/\partial t)^\beta\delta_0$$

where δ_0 is the delta measure with support at $0 \in M$.

PROPOSITION 10. *No finite linear combination of the distributions $\{g_{\alpha\beta}\}$ is in the range of the Hans Lewy operator.*

PROOF. It suffices to show that no linear combination of $g_{\alpha\beta}\,d\bar{z}$ is in the range of $\bar{\partial}_b$. Suppose $g = \sum c_{\alpha\beta}\,g_{\alpha\beta}\,d\bar{z}$ is in the range of $\bar{\partial}_b$ and let $v \in \mathscr{B}^{0,1}(\boldsymbol{C}^2)$ satisfy

$$\bar{\partial}v = \mu_* g = \sum c_{\alpha\beta}\,(\partial/\partial z)^\alpha\,(\partial/\partial t)^\beta\delta_0\,d\bar{z} \wedge d\bar{w}$$

$(w = t + is)$. Since g is in the range of $\bar{\partial}_b$, by Proposition 9 there is a (0, 1) hyperfunction u defined near 0 with $u = v$ in $X^+ = \{(z, w) \mid |z|^2 > \operatorname{Im} w\}$ and $\bar\partial u \equiv 0$. Since supp $(v - u) \subset \overline{X^-}$ there is a compactly supported hyperfunction h with supp $h \subset \overline{X^-}$ and $h \equiv v - u$ in $B(0, \eta)$ for some $\eta > 0$. Then $\bar\partial h = f + \mu_* g$ where supp $f \subset \overline{X^-} - B(0, \eta)$. Choose l and k such that $c_{lk} \neq 0$ and $c_{\alpha\beta} = 0$ for $\beta > k$. Let

$$\varphi_\varepsilon(z, w) = z^l w^k \exp\left(-\frac{1}{\varepsilon}\frac{w}{w + i}\right) dz \wedge dw.$$

Since the coefficient in φ_ε is holomorphic we have

$$0 = (\bar\partial h, \varphi_\varepsilon) = (f, \varphi_\varepsilon) + (\mu_* g, \varphi) = (f, \varphi_\varepsilon) + c_{lk} k!l!.$$

On the other hand $\varphi_\varepsilon \to 0$ uniformly on supp f so $(f, \varphi_\varepsilon) \to 0$ as $\varepsilon \to 0$ which gives us a contradiction.

Bibliography

1. A. Andreotti and C. D. Hill, *E. E. Levi convexity and the Hans Lewy problem*. I. *Reduction to vanishing theorems*, Ann. Scuola Norm. Sup. Pisa **26** (1972), 325–363.
2. S. Bochner, *Analytic and meromorphic continuation by means of Green's formula*, Ann. of Math. (2) **44** (1943), 652–673. MR **5**, 116.
3. G. B. Folland and J. J. Kohn, *The Neumann problem for the Cauchy-Riemann complex*, Ann. of Math. Studies, no. 75, Princeton Univ. Press, Princeton, N. J., 1972.
4. H. Grauert, *On Levi's problem and the imbedding of real-analytic manifolds*, Ann. of Math. (2) **68** (1958), 460–472. MR **20** #5299.
5. F. R. Harvey and H. B. Lawson, Jr., *Boundaries of complex analytic varieties*. I, Ann. of Math. **102** (1975), 233–290.
6. L. Hörmander, *An introduction to complex analysis in several variables*, Van Nostrand-Reinhold, Princeton, N. J., 1966. MR **34** #2933.
7. L. R. Hunt, J. Polking, and M. Strauss, *Unique continuation for solutions to the induced Cauchy-Riemann equations*, J. Differential Equations (to appear).
8. M. Kashiwara and T. Kawai, *On the boundary value problem for elliptic systems of linear differential equations*. I, Proc. Japan Acad. **48** (1972), 712–715.
9. J. J. Kohn, *Boundaries of complex manifolds*, Proc. Conf. Complex Analysis (Minneapolis, 1964), Springer-Verlag, Berlin and New York, 1965, pp. 81–94. MR **30** # 5334.
10. J. J. Kohn and H. Rossi, *On the extension of holomorphic functions from the boundary of a complex manifold*, Ann. of Math (2) **81** (1965), 451–472. MR **31** #1399.
11. H. Komatsu, *Boundary values for solutions of elliptic equations*, Proc. Internat. Conf. on Functional Analysis and Related Topics (Tokyo, 1969), Univ. of Tokyo Press, Tokyo, 1970, pp. 107–121. MR **42** #4879.
12. H. Komatsu and T. Kawai, *Boundary values of hyperfunction solutions of linear partial differential equations*, Publ. Res. Inst. Math. Sci. **7** (1971/72), 95–104. MR **46** #5817.
13. H. Lewy, *On the local character of the solutions of an atypical linear differential equation in three variables and a related theorem for regular functions of two complex variables*, Ann. of Math. (2) **64** (1956), 514–522. MR **18**, 473.
14. Ph. Pallu de la Barrière, *Existence et prolongement des solutions holomorphes des équations aux dérivées partielles*, Seminaire Goulaouic-Lions-Schwartz, 1974–75, Exposé, no. XII, École Polytechnique, Paris.
15. J. Polking and R. O. Wells, Jr., *Boundary values of Dolbeault cohomology classes and a generalized Bochner-Hartogs theorem*, Abh. Math. Sem. Univ. Hamburg (to appear).
16. H. Rossi, *A generalization of a theorem of Hans Lewy*, Proc. Amer. Math. Soc. **19** (1968), 436–440. MR **36** #5379.
17. M. Sato, T. Kawai and M. Kashiwara, *Microfunctions and pseudo-differential equations*, Proc. Conf. Katata, 1971, Springer-Verlag, New York, 1973.

18. P. Schapira, *Théorie des hyperfunctions*, Lecture Notes in Math., vol. 126, Springer-Verlag, New York, 1970.

19. ———, *Problème de Dirichlet et solutions hyperfunctions des équations elliptiques*, Boll. Un. Mat. Ital. (4) **2** (1969), 367–372. MR **42** #2323; erratum, **42** #2324.

20. B. M. Weinstock, *Continuous boundary values of analytic fonctions of several complex variables*, Proc. Amer. Math. Soc. **21** (1969), 463–466. MR **38** #6106.

21. R. O. Wells, Jr., *Function theory on differentiable submanifolds*, Contributions to Analysis, Academic Press, New York, 1974, pp. 407–441. MR **50** #10322.

22. ———, *Differential analysis on complex manifolds*, Prentice-Hall, Englewood Cliffs, N.J. 1973.

23. ———, *On the local holomorphic hull of a real submanifold in several complex variables*, Comm. Pure Appl. Math. **19** (1966), 145–165. MR **33** #5948.

RICE UNIVERSITY

Proceedings of Symposia in Pure Mathematics
Volume 30, 1977

PARAMETRICES FOR THE BOUNDARY LAPLACIAN AND RELATED HYPOELLIPTIC DIFFERENTIAL OPERATORS

LINDA PREISS ROTHSCHILD

1. Introduction. This is a discussion of some recent results on parametrices (approximate inverses) and estimates for a general class of second-order hypoelliptic partial differential operators. The operators to be considered are ones whose principal part is of the form $\sum a_{jk}X_jX_k$, where the X_j are real, smooth vector fields, and (a_{jk}) is a positive definite quadratic form. An example of such an operator which arises in several complex variables is the Laplacian $\Box_b$ of the boundary Cauchy-Riemann operator $\bar{\partial}_b$. Here the X_j are the real and imaginary parts of the tangential holomorphic vector fields. (See, e.g., Folland and Kohn [7] for an expository treatment of the Cauchy-Riemann complex and its boundary analogue.)

Kohn [**15**] proved that the operator $\Box_b$ is hypoelliptic under appropriate geometric conditions. The proof uses the observation that while the X_j themselves do not span the tangent space, the missing direction is obtained as a commutator $[X_j, X_k]$. A significant generalization of this regularity result was later obtained by Hörmander [**12**] (see also Kohn [**16**]). His theorem states that if $X_0, X_1, \cdots, X_n$ are real, smooth vector fields on a manifold M such that they and their commutators (up to some fixed length) span the tangent space, then the operator

$$\mathscr{L} = X_0 + \sum_{j=1}^{n} X_j^2$$

is hypoelliptic, i.e., $\mathscr{L}u = f, f \in C^\infty(U)$, implies $u \in C^\infty(U)$ for any open set U. The work to be discussed here involves the construction of a class of integral operators which contain parametrices for operators like $\Box_b$ or $\mathscr{L}$. The methods involve approximating the vector fields X_j by left-invariant vector fields on a nilpotent Lie

AMS (MOS) subject classifications (1970). Primary 35H05, 35N15, 35S99.

© 1977, American Mathematical Society

group. This technique was employed by Folland and Stein [8] to find a parametrix for $\Box_b$ and was then extended to operators like $\mathscr{L}$ by Stein and the author [18].

2. Theorems on parametrices and estimates. The results discussed here are proved in [18].

In order to obtain the best possible smoothness properties for solutions of $\mathscr{L}u = f$, it is desirable to define new Sobolev spaces which take into account the differences in the directions determined by $X_0, X_1, \cdots, X_n$ and their higher commutators. Thus X_j, $1 \leqq j \leqq n$, will have weight one, while X_0 and each $[X_j, X_k]$, $1 \leqq j, k \leqq n$, has weight two. (For $\Box_b$, $X_0 \equiv 0$ and X_j, $1 \leqq j \leqq 2l = n$, is defined by $Z_j = \frac{1}{2}(X_j - i X_{j+l})$, where the Z_j range over a spanning set of tangential holomorphic vector fields.) For the monomial $\mathscr{X} = X_{i_1} X_{i_2} \cdots X_{i_s}$ let $\sigma(\mathscr{X}) = s + t$, where t is the number of i_j's which are zero. Write r for the smallest integer such that $X_0, X_1, \cdots, X_n$ and $\{[X_{i_1}, [X_{i_2}, \cdots, X_{i_s}]\cdots]: \sigma(X_{i_1} X_{i_2} \cdots X_{i_s}) \leqq r\}$ span the tangent spaces at each point.

The new Sobolev spaces S_k^p, $1 < p < \infty$, $k = 0, 1, 2, \cdots$, are defined by

$$S_k^p = \{f \in L^p: \mathscr{X} f \in L^p \text{ for all } \mathscr{X} \text{ with } \sigma(\mathscr{X}) \leqq k\},$$

with norm given by

$$\| f \|_{S_k^p} = \sum_{\sigma(\mathscr{X}) \leqq k} \| \mathscr{X} f \|_{L^p} + \| f \|_{L^p}, \qquad \mathscr{X} = X_{i_1} X_{i_2} \cdots X_{i_s}.$$

Since all results will be local, we replace M by a relatively compact subset of itself and assume $M \subset \boldsymbol{R}^m$. The classical Sobolev spaces $L_\alpha^p(M)$ may then be defined relative to $\boldsymbol{R}^m$. An operator T is bounded from $L_{\alpha_1}^p(M)$ to $L_{\alpha_2}^p(M)$ if aTb is bounded from $L_{\alpha_1}^p(\boldsymbol{R}^m)$ to $L_{\alpha_2}^p(\boldsymbol{R}^m)$ for all $a, b \in C_0^\infty(\boldsymbol{R}^m)$. Finally, we let Λ_α denote the classical Lipschitz spaces (see, e.g., [20]).

The main results are then as follows. An operator T (initially defined on $C_0^\infty(M)$) will be said to be *smoothing of order* λ, λ a nonnegative integer, if T extends to a bounded operator from

$$\left.\begin{array}{lll} S_k^p(M) \text{ to } S_{k+\lambda}^p(M), & & \\ L_\alpha^p(M) \text{ to } L_{\alpha+\lambda/r}^p(M), & \alpha \geqq 0, & \\ \Lambda_\alpha(M) \cap L^p \text{ to } \Lambda_{\alpha+\lambda/r}(M) \cap L^p, & & \alpha > 0, \\ L^\infty(M) \cap L^p \text{ to } \Lambda_{\lambda/r}(M) \cap L^p, & & \end{array}\right\} \quad 1 < p < \infty.$$

THEOREM 1. *Let* $\mathscr{L} = \sum X_j^2 + X_0$. *For any* $a \in C_0^\infty(M)$, *there exist operators* P, S, *and* S' *smoothing of orders two, one and one, respectively, such that* $\mathscr{L}P = aI + S$ *and* $P\mathscr{L} = aI + S'$.

The operator P is then called a *parametrix* for $\mathscr{L}$.

As a consequence of this result we obtain regularity for solutions of the corresponding differential equation.

THEOREM 2. *Suppose* $f \in L^p(M)$, $1 < p < \infty$, *and* $\mathscr{L}f = g$. *Then the following hold for* $1 < p < \infty$.

(a) *If* $g \in L_\alpha^p(M)$, *then* $f \in L_{\alpha+2/r}^p(M)$, $\alpha > 0$.

(b) *If* $g \in \Lambda_\alpha(M)$, *then* $f \in \Lambda_{\alpha+2/r}(M)$, $\alpha > 0$.

(c) *If* $g \in L^\infty(M)$, *then* $f \in \Lambda_{2/r}(M)$.

(d) *If* $g \in S_k^p(M)$, *then* $af \in S_{k+2}^p(M)$, *for each* $a \in C_0^\infty(M)$, $k = 0, 1, \cdots$.

Analogous results hold for the operator $\square_b$ acting on q-forms of a partially complex (or CR) manifold M. Recall that $\square_b$ is defined to be the Laplacian $\square_b = \bar{\partial}_b\vartheta_b + \vartheta_b\bar{\partial}_b$, where ϑ_b is the adjoint of the boundary Cauchy-Riemann operator $\bar{\partial}_b$ with respect to some Hermitian metric on M. In order to obtain regularity a condition must be imposed on the Levi form ρ of M: For any point $\xi \in M$

(1) $\quad p_1 \geqq \max(q + 1, l + 1 - q) \quad$ or $\quad p_2 \geqq \min(q + 1, l + 1 - q),$

where p_1 is the larger of the number of eigenvalues of $\rho(\xi)$ of the same sign, and p_2 is the number of pairs of eigenvalues of opposite signs. In order to state our result, we use the same notation $S^p_k(M)$, $L^p_\alpha(\Lambda)$, etc., to denote spaces of q-forms in which each scalar component satisfies the appropriate condition.

THEOREM 3. *Suppose* (1) *is satisfied and* $\square_b f = g$, *where* f *and* g *are* q-*forms with* f *in* $L^p(M)$. *Then* (a)—(d) *of Theorem* 2 *hold for* f.

3. The construction of Folland and Stein. In **[8]**, a parametrix for $\square_b$ is constructed on a CR manifold with a definite Levi form using a nilpotent group as model. This is the basic approach to be used in the proofs of Theorems 1, 2 and 3.

We briefly outline some features of this construction. For simplicity, we shall assume that the Levi form is positive definite and that $\square_b$ is acting on 1-forms.

The "model" space will be $\boldsymbol{H}_l = \{(z, t): z \in \boldsymbol{C}^l, t \in \boldsymbol{R}\}$ with multiplication $(z, t) \cdot (z', t') = (z + z', t + t' + 2\,\mathrm{Im}\,(z \cdot z'))$, $z \cdot z' = \sum z_j \bar{z}_j$, a nilpotent Lie group. $\boldsymbol{H}_l$ may be given a CR structure by choosing the spanning holomorphic vector fields Z_j, $j = 1, 2, \cdots, l$, to be the left invariant vector fields agreeing with $\partial/\partial z_j$ at (0, 0). Then $Z_j = \partial/\partial z_j + i\bar{z}_j\,\partial/\partial t$. Writing $Z_j = \frac{1}{2}(Y_j - iY_{j+l})$ we obtain the important commutation relations

(2) $$[Y_j, Y_k] = -4\delta_{k,j+l}\,\partial/\partial t$$

if $j \leqq l$.

A calculation (see **[8]**) yields the following formula for $\square_b$ on 1-forms $\sum f_j \partial\bar{z}_j$ for the metric determined by the being an orthonormal basis $\{Z_j, \bar{Z}_j, \partial/\partial t\}$:

$$\square_b(\textstyle\sum f_j d\bar{z}_j) = -\sum(\mathscr{L}_l f_j) d\bar{z}_j,$$

where for any α, $\mathscr{L}_\alpha = \frac{1}{4}\sum Y_j^2 + i\alpha\,\partial/\partial t$.

To invert $\mathscr{L}_1$, and hence $\square_b$, it is necessary to consider homogeneity properties of $\boldsymbol{H}_l$. A crucial feature is the existence of a family of automorphisms $\{\delta_r\}$, given by

$$\delta_r(z, t) = (rz, r^2t), \qquad r \geqq 0,$$

which act as dilations. Homogeneous functions may be defined with respect to these dilations: f is *homogeneous of degree* λ if $f(\delta_r(z, t)) = r^\lambda f(z, t)$ for all $r \geqq 0$. For any such homogeneous f, $Y_j f$, $1 \leqq j \leqq 2l$, is homogeneous of degree $\lambda - 1$, while $\partial f/\partial t$ is homogeneous of degree $\lambda - 2$. We shall therefore say that Y_j (resp. $\partial/\partial t$) is a *homogenous differential operator* of degree 1 (resp. 2). In this sense, each $\mathscr{L}_\alpha$ is homogeneous of degree 2. Exploiting this homogeneity as well as the invariance of $\mathscr{L}_\alpha$ under unitary transformations among the z_j, one finds

$$\phi_\alpha = (|z|^2 - it)^{-(l+\alpha)/2}(|z|^2 + it)^{-(l-\alpha)/2}$$

as a candidate for a fundamental solution of $\mathscr{L}_\alpha$. A calculation shows that $\mathscr{L}_\alpha\phi_\alpha =$

$c_\alpha \delta_\alpha$ with $c_\alpha \neq 0$ if and only if $\alpha \neq \pm(l + 2k)$ for any nonnegative integer k. Since $\mathscr{L}_\alpha$ is left-invariant, $\mathscr{L}_\alpha(\phi_\alpha * f) = c_\alpha f$, $f \in C_0^\infty$, where $*$ denotes the group convolution.

Note that $\phi_\alpha \in C^\infty(\boldsymbol{H}_l - \{0\})$ and is homogeneous of degree $-(2l + 2) + 2$. The number $2l + 2$, called the *homogeneous dimension* of $\boldsymbol{H}_l$, arises from the transformation of volume elements $d(\delta_r(z, t)) = r^{2l+2}\, d(z, t)$. Thus one is led to study operators of the form $f \to f * k$, where $k \in C^\infty(\boldsymbol{H}_l - \{0\})$, and homogeneous of degree $-(2l + 2) + \lambda$ for some λ, $0 \leqq \lambda < 2l + 2$. For the critical case $\lambda = 0$ (in which case the convolution must be taken in the principal value sense) it has been proved (see Knapp-Stein [**14**]) that such an operator extends to a bounded operator on $L^2(\boldsymbol{H}_l)$. From this and extensions to L^p one obtains the following regularity result.

THEOREM 4 (FOLLAND-STEIN [**8**]). *Suppose* $\mathscr{L}_\alpha f = g$, $\alpha \neq \pm(l + 2k)$, k *a nonnegative integer. If* $g \in L^p$, *locally*, $1 < p < \infty$, *then* $Y_j Y_k f \in L^p$, *locally*, $1 < j, k \leqq 2l$, *and* $\partial f/\partial t \in L_p$, *locally.*

This result says that the solution f is two degrees smoother in the "good" directions, i.e., those given by the Y_j, and one degree smoother in the "bad" direction.

Now consider an arbitrary Cauchy-Riemann manifold with positive definite Levi form. It is shown that there exists a Hermitian metric on M with respect to which one can choose an orthonormal basis $Z_1, \cdots, Z_l, \bar{Z}_1, \cdots, \bar{Z}_l, T$ of holomorphic and antiholomorphic vector fields, respectively, with T real, satisfying

$$[Z_j, \bar{Z}_k] = -\tfrac{1}{2}\delta_{jk} + \text{linear terms in } Z_s, \bar{Z}_t.$$

Writing $Z_j = \frac{1}{2}(X_j - iX_{j+l})$, X_j real, we have

$$\Box_b\left(\textstyle\sum f_j\, d\bar{z}_j\right) = -\textstyle\sum (L_1 f_j) d\bar{z}_j + O(\bar{Z}_j f, Z_j f, f)$$

where $L_\alpha = \sum X_j^2 + i\alpha T$ and $O(\bar{Z}_j f, Z_j f, f)$ is an error term involving only f and first derivatives of f with respect to Z_j or $\bar{Z}_k$.

In order to develop a class of operators on M which approximate group convolution it is necessary to define a map on $M \times M$ which replaces the group map $(x, y) \to x^{-1}y$. For $\xi, \eta \in M$, η sufficiently close to ξ, let $\Theta(\xi, \eta) = \operatorname{Exp}(\sum u_j y_j + u_0\, \partial/\partial t) \in \boldsymbol{H}_l$ where $\eta = \exp(\sum u_j X_j + u_0 T) \cdot \xi$. Here Exp and exp are the exponential transformations on $\boldsymbol{H}_l$ and M respectively. The corresponding integral operators on M are then given essentially by

$$f \to \int f(\eta) K(\xi, \eta)\, d\eta, \tag{3}$$

where $K(\xi, \eta) = k(\Theta(\eta, \xi))$ with $k \in C^\infty(\boldsymbol{H}_l \sim \{0\})$ homogeneous of degree $-(2l + 2) + \lambda$. In particular, a parametrix for L_α, $\alpha \neq$ the exceptional values, is given essentially by the kernel $K(\xi, \eta) = \phi_\alpha(\Theta(\eta, \xi))$.

4. An outline of the program. A careful examination of the operators (3) shows that the approximation of $\Box_b$ on a CR manifold by the corresponding operator on $\boldsymbol{H}_l$ is successful because of the similarity of the commutation relations,

$$[X_j, X_k] = -4\delta_{k,j+l}\, T + \text{linear terms in } X,$$

and (2), on M and $\boldsymbol{H}_l$ respectively. Therefore, in analyzing an operator of the form

$\mathscr{L} = X_0 + \sum X_j^2$, and $\Box_b$ for nondefinite Levi form we set forth the following steps.

Step 1. Find an appropriate nilpotent Lie group N with dilations to serve as a model for M. In particular, the commutation relations of the left-invariant vector fields on N should approximate the commutation relations of the X_j in some sense.

Step 2. Find a homogeneous fundamental solution for the corresponding operator on N.

Step 3. Define a map $\Theta: M \times M \to N$, locally and use this to define a class of operators on M which contains an approximate inverse for $\mathscr{L}$ (or $\Box_b$).

It will be clear from examples that this program has to be modified. Consider the operator $\mathscr{L} = \partial^2/\partial x^2 + x^2\, \partial^2/\partial y^2$ on $M = \boldsymbol{R}^2$. Here $X_1 = \partial/\partial x$, $X_2 = x\,\partial/\partial y$, $X_0 \equiv 0$. X_1, X_2 and $[X_1, X_2] = \partial/\partial y$ are needed to span the tangent space. However, since there are no nonabelian nilpotent Lie algebras of dimension two, Step 1 already presents a difficulty. However, let s be a new variable, and write $\tilde{X}_1 = \partial/\partial x$, $\tilde{X}_2 = \partial/\partial s + x\,\partial/\partial y$ on $\boldsymbol{R}^3$. Then the Lie algebra spanned by $\tilde{X}_1$, $\tilde{X}_2$, and $[\tilde{X}_1, \tilde{X}_2] = \partial/\partial y$ is isomorphic to the three-dimensional Heisenberg algebra.

With this example in mind we modify the above program as follows. Replace Step 1 by

Step 1$'$. Lift the vector fields X_j to $\tilde{X}_j$ on a larger manifold $\tilde{M} = M \times \boldsymbol{R}^q$, so that the $\tilde{X}_j$ may be approximated by generators $\{Y_j\}$ of a nilpotent Lie algebra with dilations.

Now Step 2 is as before, but Step 3 must be carried out on the extended manifold $\tilde{M}$. In order to obtain results on M a new step is needed.

Step 4. Define extension and restriction operators to pass from functions on M to functions on $\tilde{M}$ and conversely. This must be done in such a way that the class of operators thus obtained on M has appropriate smoothing properties.

For simplicity, in what follows we shall always assume $X_0 \equiv 0$.

5. Associating a nilpotent group as model. We briefly describe here some of the ideas involved in Step 1$'$ above.

(a) Suppose $X_1, X_2, \cdots, X_n$ already span the tangent space at each point. Then the commutators $[X_j, X_k]$ are not needed for hypoellipticity. Take $\tilde{X}_j = X_j$, $j = 1, \cdots, n$, and associate the abelian nilpotent Lie algebra with generators Y_j, $j = 1, 2, \cdots, n$ satisfying $[Y_j, Y_k] = 0$, all j, k.

(b) Suppose $X_1, X_2, \cdots, X_n$ and $\{[X_j, X_k], 1 \leqq j < k \leqq n\}$ span the tangent space. (This is the case when M is a CR manifold with nowhere vanishing Levi form.) Then assign the 2-step nilpotent Lie algebra $\mathfrak{N}_{n,2}$ with generators $Y_1, Y_2, \cdots, Y_n$ such that $Y_1, Y_2, \cdots, Y_n$, and $\{[Y_j, Y_k], j < k\}$ form a linearly independent set. To construct $\tilde{X}_j$, locally, first fix a point ξ. Then find vector fields of the form $F_j = \sum_k \alpha_{jk}(\eta, t)\, \partial/\partial t_k$, where $t_1, t_2, \cdots, t_q$ are new variables and α_{jk} are smooth functions defined for η close to ξ and $t \in \boldsymbol{R}^q$ such that the $\tilde{X}_j = X_j + F$ satisfy the following. First, $\{\tilde{X}_j, [\tilde{X}_k, \tilde{X}_l], 1 \leqq j \leqq n, 1 \leqq k < l \leqq n\}$ should be linearly independent at $\tilde{\xi} = (\xi, 0) \in M \times \boldsymbol{R}^q$. Second, the $\tilde{X}_j$ together with their first commutators $[\tilde{X}_j, \tilde{X}_k]$, $1 \leqq j, k \leqq n$, should span the tangent space at ξ.

In general one finds an r-step nilpotent Lie algebra $\mathfrak{N}$ with generators $Y_1, Y_2, \cdots, Y_n$ such that the dilations $\delta_t(Y_j) = tY_j$ extend to automorphisms of $\mathfrak{N}$. These then give automorphisms of the group N corresponding to $\mathfrak{N}$. Furthermore the construction yields a one-to-one correspondence between a set of vector fields $\{\tilde{X}_{jk}\}$

spanning the tangent space at ξ and a basis $\{Y_{jk}\}$ of $\mathfrak{N}$. Here $\tilde{X}_{1k} = \tilde{X}_k$ and $Y_{1k} = Y_k$, $1 \leqq k \leqq n$, and more generally, $\tilde{X}_{jk}$ and Y_{jk} correspond to the same commutators of length j of the $\{\tilde{X}_k\}$ and $\{Y_k\}$ respectively.

As in §3, we may define the map $\Theta : \tilde{M} \times \tilde{M} \to N$, where N is the group corresponding to $\mathfrak{N}$, by

$$\Theta(\xi, \tilde{\eta}) = \operatorname{Exp}\left(\sum u_{jk}\tilde{X}_{jk}\right) \in N, \tag{4}$$

if $\tilde{\eta} = \exp(\sum u_{jk}\tilde{X}_{jk}) \cdot \xi$. For each fixed $\xi \in \tilde{M}$, the map $\tilde{\eta} \to \Theta(\xi, \tilde{\eta})$ identifies a neighborhood of ξ in $\tilde{M}$ with a neighborhood of 0 in the graded Lie group N. One of the main features of this identification is the following. In analogy with the terminology of §3, we say that a differential operator D on N is homogeneous of degree λ if $D\phi$ is a homogeneous function of degree $\alpha - \lambda$, for any homogeneous function ϕ of degree α.

THEOREM 5. *In the coordinate system around* $\xi \in \tilde{M}$ *given by* $\Theta(\xi, \tilde{\eta}) \leftrightarrow \tilde{\eta}$

$$\tilde{X}_j = Y_j + R_j, \qquad 1 \leqq j \leqq n,$$

where each term of the Taylor expansion of R_j *around* 0 *is a differential operator of degree* $\leqq 0$.

The proof of this Theorem is quite long and complicated. One of the main techniques is the use of Baker-Campbell-Hausdorff formula (see, e.g., **[19]**) to express the product of formal power series $e^x e^y$ as a power series $e^{h(x,y)}$.

6. Operators and parametrices on an extended space $\tilde{M} = M \times \boldsymbol{R}^q$**.** We shall assume that from the given vector fields $X_1, X_2, \cdots, X_n$, on the manifold M, we constructed the extended vector fields $\tilde{X}_1, \cdots, \tilde{X}_n$ on the manifold $\tilde{M} = M \times \boldsymbol{R}^q$. Since all constructions and results are local, assume $\tilde{M}$ is shrunk so that the map $\theta : \tilde{M} \times \tilde{M} \to N$ is defined everywhere.

Our first task is to define an appropriate class of integral operators of $\tilde{M}$, using homogeneous distributions on N. Recall that $\mathfrak{N}$ is spanned by $\{Y_{jk}\}$, where each Y_{jk} is a differential operator of degree j. The *homogeneous dimension* of N is defined to be $Q = \sum_j \dim V^j$, where V^j is the linear span of the Y_{jk}. As in §3 an appropriate class of operators on N are those given by

$$f \to f * k, \qquad f \in C_0^\infty(N),$$

where $k \in C^\infty(N - \{0\})$ and is homogeneous of degree $-Q + \lambda$, for some λ, $0 \leqq \lambda < Q$, and $*$ denotes group convolution on N. (For $\lambda = 0$ the integral defining the convolution must be taken in the principal value sense and k must satisfy a mean value zero condition.) Then the operators of type λ on $\tilde{M}$ are given essentially by

$$f \to \int K(\xi, \tilde{\eta}) f(\tilde{\eta})\, d\eta, \qquad f \in C_0^\infty(M),$$

where $K(\xi, \tilde{\eta}) = k(\Theta(\tilde{\eta}, \xi))$, with k as above. In the formal definition, the kernel $k(\Theta(\tilde{\eta}, \xi))$ is multiplied by cut-off functions in $\tilde{\eta}$ and ξ. Furthermore, k is not required to be homogeneous, but merely to have an expansion in terms of homogeneous functions of degrees $\geqq -Q + \lambda$. We omit these technicalities and refer the reader to **[8]** or **[18]**.

We now state the main properties of operators of type λ. First, using Theorem 5 one can prove that if T is an operator of type λ on $\tilde{M}$, then $\tilde{X}_k T$ and $T\tilde{X}_k$ are operators of type $\lambda - 1$. Next, we define the new Sobolev spaces $\tilde{S}^p_k(\tilde{M})$, $1 < p < \infty$, $k = 0, 1, 2, \cdots$, on $\tilde{M}$ as in §2, using $\tilde{X}_j$.

THEOREM 6. *An operator of type λ, $0 \leqq \lambda < Q$, is smoothing of order λ on $\tilde{M}$.*

In order to establish the analogue of Theorem 1 for the operator $\tilde{\mathscr{L}} = \sum \tilde{X}_j^2$, it will suffice to find an operator $\tilde{P}$ of type 2 such that $\tilde{\mathscr{L}}\tilde{P}$ and $\tilde{P}\tilde{\mathscr{L}}$ differ from the identity on compact sets by operators of type 1. For the construction of $\tilde{P}$, the following result on fundamental solutions of homogeneous differential operators on N is needed.

THEOREM 7. *Let D be a hypoelliptic differential operator on N which is left-invariant and homogeneous of degree λ, $0 < \lambda < Q$. Then there is a unique $k \in C^\infty(N - \{0\})$ which is homogeneous of degree $-Q + \lambda$ such that*

$$(Dk) * f(x) = D(k * f)(x) = f(x), \qquad f \in C_0^\infty(N).$$

For the proof of the above theorem see Folland [**6**]. This result was obtained independently by E. M. Stein and by R. Strichartz (unpublished).

We may now construct $\tilde{P}$. Suppose $a \in C_0^\infty(\tilde{M})$ is given. Let $k \in C^\infty(N - \{0\})$ define the fundamental solution of the homogeneous operator $D = \sum_{k=1}^{n} Y_{1k}^2$. Then let $\tilde{P}f(\tilde{\xi}) = \int \tilde{K}(\tilde{\xi}, \tilde{\eta}) f(\tilde{\eta})\, d\tilde{\eta}$, where $\tilde{K}(\tilde{\xi}, \tilde{\eta}) = a(\tilde{\xi}) k(\theta(\tilde{\eta}, \tilde{\xi})) b(\tilde{\eta})$, with $b \in C_0^\infty(\tilde{M})$ and $b \equiv 1$ on the support of a.

The analogous construction for the parametrix for $\tilde{\Box}_b$ on $\tilde{M}$ follows the same outline, but there are several significant differences. First, $\tilde{\Box}_b$ is defined on forms, not functions. Second, $\tilde{\Box}_b$ has an important lower order term which affects the conditions for hypoellipticity. Finally, the expression for $\tilde{\Box}_b$ involves varying coefficients, which means that the approximating differential operator on the group varies with $\tilde{\xi}$. Thus, the fundamental solutions involved vary with $\tilde{\xi} \in \tilde{M}$. The reader is referred to [**18**] for details.

7. Parametrix for $\sum_{j=1}^{n} X_j^2$. We have now lifted the original vector fields X_j to new vector fields $\tilde{X}_j$ on a higher dimensional space $\tilde{M} = M \times \boldsymbol{R}^q$, and constructed a parametrix for the lifted operator $\sum_{j=1}^{n} \tilde{X}_j^2$. The next step is to find an appropriate means of restricting operators on $\tilde{M}$ to M. For this purpose we define an extension mapping E taking functions on M to functions on $\tilde{M}$ by $(Ef)(\tilde{\xi}) = f(\xi)$, where $\tilde{\xi} = (\xi, t) \in \tilde{M}$, $\xi \in M$, $t \in \boldsymbol{R}^q$. Next, choose $\phi \in C_0^\infty(\boldsymbol{R}^q)$ with $\int_{\boldsymbol{R}^q} \phi(t)\, dt = 1$, and define the restriction map R by

$$(Rf)(\xi) = \int_{\boldsymbol{R}^q} f(\xi, t)\phi(t)\, dt$$

taking functions on $\tilde{M}$ to functions on M.

An operator T on M is said to be of *type* λ if $T = R\tilde{T}E$, for some operator $\tilde{T}$ of type λ, $0 \leqq \lambda < Q$, on $\tilde{M}$. Such operators are smoothing of order λ on M. To prove this, using Theorem 6 it suffices to show that E maps $S_k^p(M)$ to $S_k^p(\tilde{M})$, $L_\alpha^p(M)$ to $L_\alpha^p(\tilde{M})$ and $\Lambda_\alpha(M)$ to $\Lambda_\alpha(\tilde{M})$, and conversely for R. Finally, if $\tilde{P}$ is chosen to be a parametrix for $\tilde{\mathscr{L}} = \sum \tilde{X}_j^2$, then it can be shown that $P = R\tilde{P}E$ is a parametrix for

$\mathscr{L}$ in the sense of Theorem 1. This completes our brief outline of the proofs of the Theorems of §2.

8. Related results and problems. There is a considerable literature on hypoelliptic operators. We mention here a few results whose relationship to the present work is not yet understood, in the hope that some readers may be interested in pursuing this.

Oleinik and Radkevich [**17**] give sufficient conditions for hypoellipticity for second-order operators in a more general class. It would be interesting to construct parametrices for these operators. Grušin [**11**] gives necessary and sufficent conditions for hypoellipticity of certain homogeneous degenerate elliptic equations. By methods quite different from these, he constructs parametrices for the hypoelliptic operators. Beals [**1**] develops a general calculus of pseudodifferential operators which contains parametrices for certain hypoelliptic degenerate elliptic operators.

Boutet de Monvel [**2**] and Boutet de Monvel and Trèves [**3**], [**4**] study pseudodifferential operators which are formally similar to $\Box_b$. Their method of constructing parametrices involves using Fourier integral operators (see [**13**] and [**5**]) to transform the original operator into a canonical form. Trèves [**21**] gives new methods for establishing hypoellipticity.

Greiner and Stein [**9**] have used the parametrix for $\Box_b$ of [**8**] to obtain a parametrix for the Cauchy-Riemann operator $\bar{\partial}$ in the interior. Very recently, Greiner, Kohn, and Stein [**10**] used the explicit computation of [**8**] to give necessary and sufficient conditions for local solvability of the Lewy equation $(\partial/\partial z + i\bar{z}\,\partial/\partial t)u = f$.

References

1. R. Beals, *A general calculus of pseudodifferential operators*, Duke Math. J. **42** (1975), no. 1, 1–42.

2. L. Boutet de Monvel, *Hypoelliptic operators with double characteristics and related pseudodifferential operators*, Comm. Pure Appl. Math. **27** (1974), 585–639.

3. L. Boutet de Monvel and F. Trèves, *On a class of pseudodifferential operators with double characteristics*, Invent. Math. **24** (1974), 1–34. MR **50** #5550.

4. ———, *On a class of systems of pseudodifferential equations with double characteristics*, Comm. Pure Appl. Math. **27** (1974), 59–89. MR **50** #2725.

5. J. J. Duistermaat and L. Hörmander, *Fourier integral operators*. II, Acta Math. **128** (1972), 183–269.

6. G. B. Folland, *Subelliptic estimates and function spaces on nilpotent Lie groups*, Ark. Mat. **13** (1975), 161–207.

7. G. B. Folland and J. J. Kohn, *The Neumann problem for the Cauchy-Riemann complex*, Ann. of Math. Studies, no. 75, Princeton Univ. Press, Princeton, N. J., 1972.

8. G. B. Folland and E. M. Stein, *Estimates for the $\bar{\partial}_b$ complex and analysis on the Heisenberg group*, Comm. Pure Appl. Math. **27** (1974), 429–522.

9. P. C. Greiner and E. M. Stein, *A parametrix for the $\bar{\partial}$-Neumann problem* (to appear).

10. P. C. Greiner, J. J. Kohn and E. M. Stein, *Necessary and sufficient conditions for solvability of the Lewy equation*, Proc. Nat. Acad. Sci. **72** (1975), 3287–3289.

11. V. V. Grušin, *On a class of hypoelliptic operators*, Mat. Sb. (N.S.) **83 (125)** (1970), 456–473 = Math. USSR Sbornik **12** (1970), 458–476. MR **43** #5158.

12. L. Hörmander, *Hypoelliptic second order differential equations*, Acta Math. **119** (1967), 147–171. MR **36** #5526.

13. L. Hörmander, *Fourier integral operators*. I, Acta. Math. **127** (1971), 79–183.

14. A. W. Knapp and E. M. Stein, *Intertwining operators for semisimple groups*, Ann. Math. **93** (1971), no. 3, 489–578.

15. J. J. Kohn, *Boundaries of complex manifolds*, Proc. Conf. Complex Analysis (Minneapolis, 1964), Springer-Verlag, Berlin and New York, 1965, pp. 81–94. MR **30** #5534.

16. ———, *Pseudo-differential operators and hypoellipticity*, Proc. Sympos. Pure Math., vol. 23, Amer. Math. Soc., Providence, R. I., 1973, pp. 51–69. MR **49** #3356.

17. O. A. Oleĭnik and E. V. Radkevič, *Second order equations with non-negative characteristic form*, Plenum Press, New York, 1973.

18. L. P. Rothschild and E. M. Stein, *Hypoelliptic differential operators and nilpotent groups*, Acta Math. (to appear).

19. J.-P. Serre, *Lie algebras and Lie groups*, Benjamin, New York and Amsterdam, 1965. MR **36** #1582.

20. E. M. Stein, *Singular integrals and differentiability properties of functions*, Princeton Univ. Press, Princeton, N. J., 1970. MR **44** #7480

21. F. Trèves, *Concatenations of second-order evolution equations applied to local solvability and hypoellipticity*, Comm. Pure Appl. Math. **26** (1973), 201–250. MR **49** #5554.

UNIVERSITY OF WISCONSIN-MADISON

Proceedings of Symposia in Pure Mathematics
Volume 30, 1977

ON THE GLOBAL REAL ANALYTICITY OF SOLUTIONS TO $\Box_b$ ON COMPACT MANIFOLDS

DAVID S. TARTAKOFF

1. Introduction. On a compact, real analytic, $(2n - 1)$-dimensional Hermitian manifold X with a real analytic, formally integrable, CR structure we consider the analytic hypoellipticity of the Laplace-Beltrami operator $\Box_b$. If X is strictly pseudoconvex (i.e., if all eigenvalues of the Levi form are strictly positive) then $\Box_b$ is subelliptic with loss of $\frac{1}{2}$ derivative, hence hypoelliptic and even Gevrey-hypoelliptic, and the same is true under weaker convexity assumptions as long as subellipticity holds. In fact these results hold for a wide class of differential operators (see **[1]**, **[4]**, **[5]** and **[11]**). The question of real analytic hypoellipticity, i.e., the question "If $\Box_b u = f$ with f real analytic, is u real analytic?", however, has proved refractory even in the strictly pseudoconvex case. For the case of the Heisenberg group in C^n, Folland and Stein have recently proved local analytic hypoellipticity **[2]**, yet several subelliptic operators, even those which lose only $\frac{1}{2}$ derivative, are known not to be analytic hypoelliptic **[12]**.

Our results are of the following type: On a compact manifold, a large class of differential operators (namely, those satisfying an a priori estimate somewhat weaker, possibly, than the $\frac{1}{2}$ estimate, and stating roughly that the operator is elliptic in most directions and sufficiently positive overall) are globally analytic hypoelliptic provided the associated Levi form satisfies a nondegeneracy condition (which is automatic in the strictly pseudoconvex case for $\Box_b$).

In the present paper we sketch the proof for $\Box_b$; in addition, we should like to announce that recently Derridj and the author and, separately, G. Komatsu have proved the global analytic hypoellipticity for the $\bar{\partial}$-Neumann problem in many cases including the strictly pseudoconvex case.

One note on the formulation of our theorems is in order: We have proved that C^∞ forms u solving $\Box_b u = f$, f real analytic, are real analytic, but "analytic hypoel-

AMS (MOS) subject classifications (1970). Primary 32N15, 35H05.

© 1977, American Mathematical Society

liptic" should mean that distribution solutions are real analytic. If one assumes that a subelliptic estimate holds, then the work of Kohn [3], Kohn and Nirenberg [5] or Tartakoff [9] show that weak solutions with C^∞ data are C^∞, and Tartakoff's recent paper [10] shows the same for distribution solutions. If one assumes "arbitrary positivity" instead, the same result (globally) follows from arguments of Tartakoff [8], Kohn [4], and Hörmander. But the estimate we assume, which certainly holds in the important cases where $\frac{1}{2}$-subellipticity holds for $\bar\partial_b$, has not, to the author's knowledge, been treated. It would appear that an application of the techniques in [9] would suffice, but we leave this question open at the moment.

The author is indebted to J. J. Kohn, L. Nirenberg, and M. Kuranishi for encouragement and for several helpful discussions.

2. Statement of the problem and results. A concise definition of $\Box_b$ may be given as follows: We are given a (real analytic) subbundle S of complex dimension $n-1$ of $CT(X)$, the complexified tangent bundle of X, such that $S \cap \bar S = \{0\}$. Let $S \oplus \bar S \oplus F = CT(X)$. Then $\Lambda^{p,q}(X) = \Lambda^p(S \oplus F) \otimes \Lambda^q \bar S$ and sections of $\Lambda^{p,q}$ are called (p, q) forms. $\mathscr{D}^{p,q}(X)$ denotes $\{C_0^\infty$ sections of $\Lambda^{p,q}(X)\}$. If $\{\bar\zeta_1, \cdots, \bar\zeta_{n-1}\}$ is a local frame for $\bar S^*$, hence $\{\zeta_1, \cdots, \zeta_{n-1}\}$ for S^*, and we choose a dual basis (of vector fields) $\bar Y_1, \cdots, \bar Y_{n-1}$ of $\bar S$, then the operator $\bar\partial_b : \mathscr{D}^{p,q}(X) \to \mathscr{D}^{p,q+1}(X)$ is described locally on functions ((0, 0) forms) by

$$\bar\partial_b \varphi = (\bar Y_1 \varphi)\bar\zeta_1 + \cdots + (\bar Y_{n-1}\varphi)\bar\zeta_{n-1},$$

on (0, 1) forms $\varphi = \sum_{i=1}^{n-1} \varphi_i \bar\zeta_i$ by

$$\bar\partial_b \varphi = \sum_{i<j} \Big(\bar Y_i \varphi_j - \bar Y_j \varphi_i - \sum_k a_{ij}^k \varphi_k \Big) \bar\zeta_i \wedge \bar\zeta_j$$

where the fact that $[\bar Y_i, \bar Y_j] = \sum_{k=1}^{n-1} a_{ij}^k \bar Y_k$ is equivalent to the formal integrability of S, and on general (p, q) forms $\varphi = \sum_{|I|=p;\ |J|=q} \varphi^{IJ} \zeta^I \wedge \bar\zeta^J$ by

$$\bar\partial_b \varphi = \sum_{|I|=p;\ |J|=q} \sum_{k=1}^{n-1} (\bar Y_k \varphi^{IJ}) \zeta^I \wedge \bar\zeta^J \wedge \bar\zeta^k$$

modulo terms in which the φ^{IJ} are not differentiated, terms uniquely determined by the relation $\bar\partial_b \circ \bar\partial_b = 0$. Then using the Hermitian structure on the fibers of $CT(X)$ we may define the formal adjoint $\bar\partial_b^*$ at each level $\bar\partial_b^* : \mathscr{D}^{p,q}(X) \to \mathscr{D}^{p,q-1}(X)$. Next, one defines $\Box_b : \mathscr{D}^{p,q}(X) \to \mathscr{D}^{p,q}(X)$ by $\Box_b = \bar\partial_b \bar\partial_b^* + \bar\partial_b^* \bar\partial_b$. Since the $\{\bar\partial_b\}$ form a complex, given a $C^\infty(p, q)$ form g, $q > 0$, with $\bar\partial_b g = 0$, to find a $C^\infty(p, q-1)$ form f with $\bar\partial_b f = g$ we may instead find a $C^\infty(p, q)$ form h such that $\Box_b h = g$ since then $\bar\partial_b g = 0$ implies $(\bar\partial_b^* \bar\partial_b h, g) = 0$, and automatically $(\bar\partial_b^* \bar\partial_b h, -\bar\partial_b \bar\partial_b^* h) = 0$. Whence, adding, $(\bar\partial_b^* \bar\partial_b h, \bar\partial_b^* \bar\partial_b h) = 0$ so $\bar\partial_b^* \bar\partial_b h = 0$, $\Box_b h = \bar\partial_b \bar\partial_b^* h = g$ and we may choose $f = \bar\partial_b^* h$. This f has the property that it is orthogonal to ker $\bar\partial_b$, and of course any solution to $\bar\partial_b f = g$ with this property is unique.

The properties of $\Box_b$ which will interest us are:

(1) $\Box_b^* = \Box_b$.

(2) Locally, and if we consider the particular frame $\bar\zeta_1, \cdots, \bar\zeta_{n-1}, \zeta_1, \cdots, \zeta_{n-1}, \tau$ as generators of a C^∞ basis of $\mathscr{D}^{p,q}(X)$ locally, then in this basis we may write

$$\Box_b = \sum_{i,j} a_{ij} W_i W_j + \sum_i a_i W_i + a_0 \tag{2.1}$$

where each W_i is a $\bar{Y}_l$ or $\bar{Y}_l^*$ and the "coefficients" a_{ij}, a_i, and a_0 are real analytic square matrices. In fact the a_{ij} are constants, though we shall not need this fact. In particular, if we choose a vector field T dual to τ, T does not occur in this expression for $\square_b$ (although if one writes $\square_b$ differently, namely, as $\sum (\bar{Y}_j\bar{Y}_j^* + \bar{Y}_j^*\bar{Y}_j) + H$, T will occur in H; we shall never use this latter form).

Relative to the basis $\zeta_1, \cdots, \zeta_{n-1}, \bar{\zeta}_1, \cdots, \bar{\zeta}_{n-1}, \tau$, we may also define the Levi form:

$$\mathscr{L}(Y_i, \bar{Y}_j) = \text{the coefficient of } T \text{ in the expression } [Y_i, \bar{Y}_j] = c_{ij}\, T \qquad (\text{modulo } Y, \bar{Y}).$$

It is easy to see that if we choose τ (and thus T) to be purely imaginary, then $c_{ij} = \bar{c}_{ji}$, and it is well known that a different choice of basis yields the same number of nonzero eigenvalues and the same signature in absolute value of the Levi form. The manifold X is called (strictly) pseudoconvex if all eigenvalues are $\geqq 0$ (> 0).

We are now able to state our results.

THEOREM 1. *Let X be a compact, real analytic $(2n - 1)$-dimensional Hermitian manifold with a formally integrable CR structure. Assume that locally one may find a unique (up to a factor of ± 1) analytic vector field T, $T = -\bar{T}$, complementary to the Y's and $\bar{Y}$'s (cf. above) such that for all j, $[T, Y_j] \equiv 0$ (modulo Y's, $\bar{Y}$'s). Assume further that there exist a sufficiently large constant K and constants C, C_K such that for all $\varphi \in \mathscr{D}^{p,q}$ of small support*

$$(2.2) \qquad \sum_j \|\bar{Y}_j^*\varphi\|_{L^2}^2 + \sum_j \|\bar{Y}_j\varphi\|_{L^2}^2 + K\|\varphi\|_{L^2}^2 \leqq C(\square_b\varphi, \varphi) + \|\varphi\|_{-1}^2,$$

where $\| \ \|_{-1}$ is the usual Sobolev norm of order -1. Then if u is a C^∞ (p, q) form on X with $\square_b u = f$ with f real analytic on X, then u is real analytic on X.

REMARK 1. Proposition 1 below will show that a nondegeneracy condition on the Levi form will guarantee the existence of a T satisfying the hypotheses of the theorem.

REMARK 2. Both the a priori estimate and the hypotheses on the Levi form of Proposition 1 are satisfied in case X is strictly pseudoconvex and $n \geqq 3$. Thus we have

THEOREM 2. *Let X be a compact, real analytic, strictly pseudoconvex, $(2n - 1)$-dimensional Hermitian manifold, $n \geqq 3$, with a formally integrable, real analytic CR structure. Then $\square_b$ is globally real analytic hypoelliptic.*

COROLLARY. *Let the hypotheses of Theorem 1 be satisfied and let g be a real analytic $\bar{\partial}_b$-closed (p, q) form on X, $q > 0$. Then the unique, global $(p, q - 1)$ form f orthogonal to the kernel of $\bar{\partial}_b$ is real analytic.*

REMARK 3. One cannot in general expect *any* f with $\bar{\partial}_b f = g$, g given real analytic and $\bar{\partial}_b$ closed, to be real analytic; any other f' will differ from f by a $\bar{\partial}_b$ closed form, and these need not be at all smooth.

PROPOSITION 1. *Let T be a locally defined, real analytic vector field, with $\bar{T} = -T$ and $|T| = 1$, orthogonal to the Y's and $\bar{Y}$'s at each point. If the Levi form (c_{ij}) is invertible, then there exist, locally, unique a_j, real analytic, such that $T' =$*

$T + \sum_{j=1}^{n-1} a_j Y_j - \sum_{j=1}^{n-1} \bar{a}_j \bar{Y}_j$ satisfies

$$(2.3) \qquad [T', \overset{(-)}{Y_k}] \equiv 0 \ (\textit{modulo } Y_1, \cdots, Y_{n-1}, \bar{Y}_1, \cdots, \bar{Y}_{n-1}).$$

Such T' is the unique purely imaginary vector field, then, of the form $T + \gamma$, γ a real analytic section of $S \oplus \bar{S}$, such that $[T', \gamma'] \equiv 0$ (modulo $S \oplus \bar{S}$) for any section γ' of $S \oplus \bar{S}$.

PROOF. We solve for the a_j subject to (2.3):

$$\begin{aligned}[T', Y_k] &= [T, Y_k] + \sum_j [a_j, Y_k]\, Y_j \\ &\quad + \sum_j a_j[Y_k, Y_j] - \sum_j [\bar{a}_j, Y_k]\, \bar{Y}_j - \sum_j \bar{a}_j[Y_k, \bar{Y}_j] \\ &\equiv [T, Y_k] - \sum_j \bar{a}_j c_{k_j}\, T\end{aligned}$$

modulo $\{Y_k, \bar{Y}_k\}_k$. Now whatever the coefficient of T in $[T, Y_k]$, the invertibility of the Levi form allows us to solve uniquely for the $\bar{a}_j$. This finishes the proof since $[T', \bar{Y}_k] = -[T', Y_k]$.

REMARK. The proof of Proposition 1 shows that we need only assume that the c_{k_j} be sufficiently invertible to allow us to solve uniquely for the $\bar{a}_j$.

3. Sketch of proof of Theorem 1. We must show that, locally, there exists a constant R such that for all multi-indices α

$$(3.1) \qquad |(\partial/\partial x)^\alpha u| \leqq R R^{|\alpha|} |\alpha|!.$$

Equivalently (though one must prove this) there exists an R_1 such that for all k, locally,

$$(3.2) \qquad |\mathrm{Op}(k) u| \leqq R_1 R_1^k k!$$

where $\mathrm{Op}(k)$ is any kth order differential operator formed by k successive applications of the Y_j's, $\bar{Y}_j$'s, and T. And this will be the case provided, given $\omega_1 \subset\subset \omega_2 \subset\subset X$, and $\psi \in C_0^\infty(\omega_2)$, $\psi \equiv 1$ on ω_1 and real,

$$(3.3) \qquad \|\psi\, \mathrm{Op}(k) u\|_{L^2} \leqq R_2 R_2^k k!.$$

We denote by $\mathrm{Op}(l, k)$ any $\mathrm{Op}(k)$ with exactly l Y's or $\bar{Y}$'s and observe that if $l \geqq 1$, the estimate (2.2) looks like a coercive one with $\varphi = \psi\, \mathrm{Op}(l-1, k-1)$ once we write $\mathrm{Op}(l, k) = W\, \mathrm{Op}(l-1, k-1)$ modulo operators of the form $\mathrm{Op}(l, k-j)$, $j \geqq 1$ (as we may by Proposition 1; commuting a Y or $\bar{Y}$ past T always gives a Y or $\bar{Y}$).

Thus we first treat pure powers of T, $\varphi = \psi T^p u$, since these will be the hardest: (2.1) yields

$$\begin{aligned} I_{p,\psi}^2 &\equiv \sum_j \|\bar{Y}_j^* \psi T^p u\|_{L^2}^2 + \sum_j \|\bar{Y}_j \psi T^p u\|_{L^2}^2 + K\, \|\psi T^p u\|_{L^2}^2 \\ &\leqq C(\Box_b \psi T^p u, \psi T^p u) + C_K \|\psi T^p u\|_{-1}^2 \end{aligned}$$

and

$$\begin{aligned}(3.5) \quad \mathrm{Re}\,(\Box_b \psi T^p u, \psi T^p u) &= \mathrm{Re}(\psi[\Box_b, \psi]\, T^p u, T^p u) \\ &\quad + \mathrm{Re}([\Box_b, T^p]\, u, \psi^2 T^p u) + \mathrm{Re}(\psi T^p \Box_b u, \psi T^p u)\end{aligned}$$

where the last term is just $(\psi T^p f, \psi T^p u)$. Now since $\square_b = \square_b^*$,

$$\left|\operatorname{Re}(\psi[\square_b, \psi]\, v, v) = \operatorname{Re}([\psi, [\square_b, \psi]]\, v, v)\right| \leqq C\, \|\psi' v\|_{L^2}^2$$

while $[\square_b, T^p]$ is a sum of many terms, the "leading" ones of which are p in number and of the form $aW_1W_2T^{p-1}$, a some analytic function and each W_i a Y or a $\bar{Y}$ by virtue of Proposition 1 again. Thus using a weighted Schwarz inequality

$$(3.6) \qquad \left|\operatorname{Re}([\square_b, T^p]\, u, \psi^2 T^p u)\right| \leqq C_\varepsilon p^2 \,\|W\psi T^{p-1}u\|^2 + \varepsilon\, \|W\psi T^p u\|^2$$

modulo many more terms than this but with lower powers of T and no more W's. The last term (with ε) will be absorbed on the left side of (3.4) and, in all,

$$(3.7) \qquad (1 - \varepsilon\, C)\, I_{p,\psi}^2 \leqq C_\varepsilon \{\|\psi T^p f\|^2 + p^2 I_{p-1,\psi}^2\} + C\|\psi' T^p u\|^2,$$

again modulo less vital terms. We fix ε, say at $1/2\,C$, uniformly in p, and could begin induction except for the last term. Since there is no way to estimate $\|\psi' v\|$ by $\|\psi v\|$, we are led to choose ψ = some ψ_i where $\sum_{j=1}^N \psi_j = 1$ and estimate

$$(3.8) \qquad \|\psi' v\| \leqq \sup_i |\psi_i'| \sum_{j=1}^N \|\psi_j v\|,$$

sum (3.7) with $\psi = \psi_i$ from $i = 1$ to N, and assume $K > NC \sup_i |\psi_i'|$. This step is the (unfortunate) reason that our theorem is global and not local. But at any rate, we achieve

$$(3.9) \qquad \sum_{i=1}^N I_{p,\psi_i}^2 \leqq \tilde{C}_\varepsilon \sum_{i=1}^N \{\|\psi_i T^p f\|^2 + p^2 I_{p-1,\psi_i}^2\}$$

for which the induction hypothesis

$$(3.10) \qquad \sum_{i=1}^N I_{q,\psi_i} \leqq R_3 R_3^q q!, \qquad q < p,$$

is easily demonstrated also for $q = p$, provided R_3 is well chosen relative to $\tilde{C}_\varepsilon$ and constants associated with f. Notice that T, and hence T^p, is assumed globally defined so that (3.8) makes sense with $v = T^p u$. We conclude that

$$(3.11) \qquad \|T^p u\|_{L^2(X)} \leqq R_4 R_4^p p! \qquad \forall\, p.$$

Now the analysis of $\operatorname{Op}(l, p)u$ with $l \geqq 2$ is easier, for the estimate is in a sense coercive in the $\bar{Y}$ and Y directions. We discuss briefly the estimation of

$$\|\psi \operatorname{Op}(l, p)u\|, \qquad l \geqq 1,$$

noting that the estimate majorizes $\|\psi \operatorname{Op}(l, p)u\|^2$, written (principally) as

$$\|W\psi \operatorname{Op}(l-1, p-1)u\|^2, \qquad \text{by } (\square_b \psi \operatorname{Op}(l-1, p-1)u, \psi \operatorname{Op}(l-1, p-1)u),$$

whence as above by $\|\psi \operatorname{Op}(l-1, p-1)f\|^2$, $\|\psi' \operatorname{Op}(l-1, p-1)u\|^2$, and, most importantly, $|([\square_b, \operatorname{Op}(l-1, p-1)]u, \psi^2 \operatorname{Op}(l-1, p-1)u)|$. The leading terms in $[\square_b, \operatorname{Op}(l-1, p-1)]$ are proportional to p in number, of order p, and, thus,

$$\begin{aligned}&\left|([\Box_b, \operatorname{Op}(l-1, p-1)]\, u, \psi^2 \operatorname{Op}(l-1, p-1)u)\right| \\ (3.12) \qquad &\leqq p \text{ terms, each of the form } \left|(\psi \operatorname{Op}(p)\, u, \psi \operatorname{Op}(p-1)u)\right| \\ &\leqq \varepsilon \left\|\psi \operatorname{Op}(p)u\right\|^2 + C_\varepsilon p^2 \left\|\psi \operatorname{Op}(p-1)u\right\|^2.\end{aligned}$$

Now $\operatorname{Op}(p)$ is either of the form $\operatorname{Op}(l, p)$ with $l \geqq 1$, or T^p. The latter is estimated, while the former, for small ε (uniform in p), is absorbed in the estimate on the left. Thus modulo f and less harmful terms

$$\begin{aligned}(3.13) \qquad \sum_{l=1}^{p} \left\|\psi \operatorname{Op}(l, p)u\right\| &\leqq pC \sum_{l=1}^{p} \left\|\psi \operatorname{Op}(l, p-1)u\right\| \\ &\quad + C \sum_{l=1}^{p} \left\|\psi' \operatorname{Op}(l, p-1)u\right\| + CR_4R_4^p p!.\end{aligned}$$

This time we cannot choose ψ among a partition of unity, for the Y's which make up $\operatorname{Op}(l, p)$, $l \geqq 1$, are not global; but rather we use a special construction due to Ehrenpries and interesting in its own right.

LEMMA (EHRENPREIS). *Given $\omega_1 \subset\subset \omega_2$, open, there exists a constant S such that for any M there exists $\psi_M \in C_0^\infty(\omega_2)$, $\psi_M \equiv 1$ on ω_1 with*

$$(3.14) \qquad \left|D^\alpha \psi_M(X)\right| \leqq S(SM)^{|\alpha|} \qquad \forall\, \alpha \text{ if } |\alpha| \leqq M.$$

Thus up to order M, ψ_M acts like a real analytic, compactly supported function. And so (3.13) expresses the majorization of $\left\|\psi \operatorname{Op}(l, p)u\right\|$ by $\left\|\psi \operatorname{Op}(l, p-1)u\right\|$ (as with $T^p u$) and $\left\|\psi' \operatorname{Op}(l, p-1)u\right\|$, which, by (3.14) will contribute a factor of M ($=p$). Our inductive hypothesis is, denoting by $\psi^{(b)}$ some $D^\beta\psi$ (actually $\operatorname{Op}(|\beta|)\, \psi$) with $|\beta| \leqq b$,

$$(3.15) \qquad \sum_{l=1}^{q} \left\|\psi^{(b)} \operatorname{Op}(l, q)u\right\| \leqq S_0(S_0M)^b(S_1M)^q$$

assumed for $M = p \geqq b \geqq b_0 \geqq 0$, $b + q \leqq p$, and using (3.13) we demonstrate it also for $b = b_0$, provided S_0 and S_1 are well chosen relative to *each other* and to various other constants which have appeared. Finally, we show (3.15) is valid for $b = p$, $q = 0$, for suitable S_0 and S_1, uniformly in p. We conclude that

$$(3.16) \qquad \sum_{l=1}^{p} \left\|\psi \operatorname{Op}(l, p)u\right\| \leqq S_2(S_2 p)^p,$$

whence easily that

$$(3.17) \qquad \left\|\psi \operatorname{Op}(p)u\right\| \leqq R_5 R_5^p p! \quad \text{Q.E.D.}$$

BIBLIOGRAPHY

1. M. Derridj, *Gevrey regularity up to the boundary for the $\bar\partial$ Neumann problem*, Proc. Sympos. Pure Math., vol. 30, part 1, Amer. Math. Soc., Providence, R. I., 1977, pp. 123–126.

2. G. B. Folland and E. Stein, *Estimates for the $\bar\partial_b$ complex and analysis on the Heisenberg group*, Comm. Pure Appl. Math. **27** (1974), 429–522.

3. J. J. Kohn, *Harmonic integrals on strongly pseudo-convex manifolds*. I, II, Ann of Math (2) **78** (1963), 112–148; ibid. **79** (1964), 450–472. MR **27** #2999; **34** #8010.

4. ———, *Global regularity for $\bar\partial$ on weakly pseudoconvex manifolds*, Trans. Amer. Math. Soc. **181** (1973), 273–292. MR **49** #9442.

5. J. J. Kohn and L. Nirenberg, *Non-coercive boundary values problems*, Comm. Pure Appl. Math. **18** (1965), 443–492. MR **31** #6041.

6. C. B. Morrey and L. Nirenberg, *On the analyticity of the solutions of linear elliptic systems of partial differential equations*, Comm. Pure Appl. Math. **10** (1957), 271–290. MR **19**, 654.

7. M. Sato, T. Kawai and M. Kashiwara, *Microfunctions and pseudo-differential equations*, Lecture Notes in Math., vol. 287, Springer-Verlag, Berlin and New York, 1973.

8. D. S. Tartakoff, *Regularity of solutions to boundary value problems for first order systems*, Indiana Univ. Math. J. **21** (1972), 1113–1129.

9. ———, *On the regularity of non-unique solutions of degenerate elliptic-parabolic systems of partial differential equations*, Comm. Pure Appl. Math **24** (1971), 763–788. MR **45** #5543.

10. ———, *Remarks on the hypoellipticity of general subelliptic partial differential operators and quadratic forms* (to appear).

11. ———, *Gevrey hypoellipticity for subelliptic boundary value problems*, Comm. Pure Appl. Math. **26** (1973), 251–312. MR **49** #7586.

12. M. S. Baouendi and C. Goulaouic, *Analyticity for degenerate elliptic equations and applications*, Proc. Sympos. Pure Math., vol. 23, Amer. Math. Soc., Providence, R. I., 1971, pp. 79–84.

THE INSTITUTE FOR ADVANCED STUDY

PRINCIPAL LECTURE

J. J. KOHN

METHODS OF PARTIAL DIFFERENTIAL EQUATIONS IN COMPLEX ANALYSIS

Proceedings of Symposia in Pure Mathematics
Volume 30, 1977

METHODS OF PARTIAL DIFFERENTIAL EQUATIONS IN COMPLEX ANALYSIS

J. J. KOHN*

I. We will first consider the famous equation of Hans Lewy, show how it arises from the theory of several complex variables and explain the recent results obtained by P. Greiner, E. Stein and the author (see [1]). This topic will lead us directly to a discussion of the Hodge decomposition for the boundary Laplacian; which in turn will take us to our main subject, that is, the study of regularity via the $\bar{\partial}$ Neumann problem.

Let x, y, t denote coordinates in $\boldsymbol{R}^3$, set $z = x + iy$ and

$$\frac{\partial}{\partial z} = \frac{1}{2}\left(\frac{\partial}{\partial x} - i\frac{\partial}{\partial y}\right), \qquad \frac{\partial}{\partial \bar{z}} = \frac{1}{2}\left(\frac{\partial}{\partial x} + i\frac{\partial}{\partial y}\right).$$

The Lewy equation is then given by:

$$Lu = \partial u/\partial z + i\bar{z}\,\partial u/\partial t = f. \tag{1}$$

In [2] H. Lewy proved that for most f the above equation has no solutions. In [3] Sato gave necessary and sufficient conditions for microlocal solvability of (1). The result discussed here is given in [1]. It gives necessary and sufficient conditions for local solvability of (1) together with optimally smooth solutions.

Consider the domain $\Omega \subset \boldsymbol{C}^2$ given by

$$\Omega = \{(z_1, z_2) \in \boldsymbol{C}^2 \mid \operatorname{Im}(z_2) > |z_1|^2\}, \tag{2}$$

the boundary of Ω, denoted by $b\Omega$, is then given by

$$b\Omega = \{(z_1, z_2) \in \boldsymbol{C}^2 \mid \operatorname{Im}(z_2) = |z_1|^2\}. \tag{3}$$

Now we map $\boldsymbol{R}^3$ onto $b\Omega$ by

AMS (MOS) subject classifications (1970). Primary 32F15, 35N15.
*Supported by the National Science Foundation grant MPS75–01436.

© 1977, American Mathematical Society

(4) $$z_1 = z \quad \text{and} \quad z_2 = t + i|z|^2.$$

This map induces a map of tangent vectors on $\boldsymbol{R}^3$ to tangent vectors on $b\Omega$; the image of L then is

(5) $$L' = \partial/\partial z_1 + 2i\bar{z}_1 \partial/\partial z_2.$$

Observe that any vector of the form $a\partial/\partial z_1 + b\partial/\partial z_2$ tangent to $b\Omega$ is a multiple of L'.

Let $L_2(b\Omega)$ consist of the space of square-integrable functions with respect to the volume element $d\sigma = dxdydt$ and let $\mathscr{H}(b\Omega)$ denote the subspace of $L_2(b\Omega)$ consisting of "boundary values" of holomorphic functions on Ω. More precisely, we can take $\mathscr{H}(b\Omega)$ to be the closure of the boundary values of smooth holomorphic functions (see [4]). Now, let $H_b : L_2(b\Omega) \to \mathscr{H}(b\Omega)$ be the orthogonal projection and let $\tilde{H}_b : L_2(b\Omega) \to \mathscr{H}(\Omega)$, where $\mathscr{H}(\Omega)$ denotes the space of holomorphic functions on Ω, be the operator defined by requiring that $\tilde{H}_b f$ be the unique holomorphic function whose boundary values are $H_b f$. The operator $\tilde{H}_b$ can be expressed by the following (see [5]):

(6) $$\tilde{H}_b(f)(z_1, z_2) = \int_{b\Omega} S(z_1, z_2, x, y, t) f(x, y, t)\, dx\, dy\, dt$$

where

(7) $$S(z_1, z_2, x, y, t) = (i(\bar{w}_2 - z_2) - 2\bar{w}_1 z_1)^{-2}/\pi^2$$

with $w_1 = z$ and $w_2 = t + i|z|^2$. Observe that if $P \in \boldsymbol{R}^3$, then $\tilde{H}_b(f)$ is analytically continuable past P if and only if $H_b(f)$ is real analytic near P. Moreover, this property depends only on the behavior of f in a neighborhood of P.

THEOREM 1. *Given f, then the equation* (1) *has a solution in a neighborhood of a point $P \in \boldsymbol{R}^3$ if and only if $H_b(f)$ is real analytic in a neighborhood of P.*

We shall prove the necessity of the condition given in Theorem 1 in a more general setting (see [6]). Let $\Omega \subset \boldsymbol{C}^n$ be a domain with a smooth boundary $b\Omega$, i.e., there exists a real-valued function $r \in C^\infty(U)$, where U is a neighborhood of $b\Omega$, such that $dr \neq 0$, $r < 0$ in $\Omega \cap U$ and $r > 0$ in the complement of $\bar{\Omega}$. For simplicity we shall assume that $\bar{\Omega}$ is compact, although it will be clear how to apply the results for the noncompact type domain discussed above. Given a volume element $d\sigma$ on $b\Omega$ we define the spaces $L_2(b\Omega)$, $\mathscr{H}(b\Omega)$ and the operators H_b and $\tilde{H}_b$ as above and we then represent $\tilde{H}_b$ by a Cauchy-Szegö kernel $S(z, w)$ as follows:

(8) $$\tilde{H}_b f(z) = \int_{b\Omega} S(z, w) f(w)\, d\sigma_w$$

and S is a function on $\Omega \times b\Omega$ which is holomorphic for each fixed $w \in b\Omega$.

We will assume that S satisfies the following condition:

Analyticity hypothesis. If W is an open subset of $b\Omega$ then there exists an open set $\Omega' \supset \bar{\Omega} - \bar{W}$ such that the function $S(z, w)$ has a continuous extension to $\Omega' \times W$ which for each fixed w is holomorphic in z.

Unfortunately there are no general theorems known which would guarantee that S satisfy this hypothesis. Later we will see that it is natural to conjecture that S satisfies this hypothesis in the case that r is analytic and Ω is strongly pseudoconvex.

Of course for the case when S is given by an explicit formula (as in (7)) the hypothesis is easy to check. Conditions will be given which insure that S satisfies the following weaker hypothesis.

Differentiability hypothesis. If W is any open subset of $b\Omega$ then the function $S(z, w)$ has a continuous extension to $(\bar{\Omega} - \bar{W}) \times W$ which for each fixed $w \in W$ is in $C^\infty(\bar{\Omega} - \bar{W})$.

For $P \in b\Omega$ let A be a vector field defined in some neighborhood U of P by

$$A = \sum a_j \, \partial/\partial\bar{z}_j \quad \text{where } a_j \in C^\infty(U); \tag{9}$$

we assume that

$$A(r) = 0 \quad \text{when } r = 0. \tag{10}$$

The assumption (10) means that A can be restricted to functions on $U \cap b\Omega$; we denote this restriction by A_b and its formal adjoint by A_b^* (i.e., $(A_b^* v, w) = (v, A_b w)$ for all v and w in $C_0^\infty(U \cap b\Omega)$).

THEOREM 2. *If S satisfies the analyticity condition, if $f \in L_2(b\Omega)$ and if there exists a neighborhood U of P, a vector field A on U satisfying* (9) *and* (10) *and $u \in L_2(U \cap b\Omega)$ such that*

$$A_b^* u = f \quad \textit{on } U \cap b\Omega \tag{11}$$

then $\tilde{H}_b f$ has a holomorphic extension past P (i.e., to a domain which contains P in its interior).

PROOF. From the analyticity condition and (8) we have that if f is zero in a neighborhood of P, then $\tilde{H}_b f$ has a holomorphic extension past P. So that if $\zeta \in C_0^\infty(U \cap b\Omega)$ and $\zeta = 1$ in a neighborhood of P then $\tilde{H}_b(f - A_b^*(\zeta u))$ has a holomorphic extension past P. The proof is then concluded by observing that $\tilde{H}_b(f - A_b^*(\zeta u)) = \tilde{H}_b(f)$ since $A_b^*(\zeta u) \perp (b\Omega)$.

The following are immediate consequences of the theorem and its proof.

COROLLARY. *If f is a holomorphic function in Ω which cannot be holomorphically extended past P then* (11) *has no solution in a neighborhood of P.*

COROLLARY. *If r, f and the volume element on $b\Omega$ are analytic in a neighborhood of P then $\tilde{H}_b f$ has an analytic extension past P.*

PROOF. Set

$$A = \frac{\partial r}{\partial \bar{z}_2} \frac{\partial}{\partial \bar{z}_1} - \frac{\partial r}{\partial \bar{z}_1} \frac{\partial}{\partial \bar{z}_2};$$

then A_b^* has analytic coefficients and hence (11) has a solution by the Cauchy-Kowalevsky theorem.

COROLLARY. *For the domain given by* (2) *if we set $A = \partial/\partial\bar{z}_1 - 2iz_1 \, \partial/\partial\bar{z}_2$ then $A_b^* = L'$ and hence the theorem applies to the solvability of* (1).

If the Cauchy-Szegö kernel satisfies only the differentiability hypothesis then the above proof gives the following.

COROLLARY. *The solvability of* (11) *implies that $\tilde{H}_b(f)$ has a C^∞ extension to $\Omega \cap (U \cap b\Omega)$.*

It is clear from the above that the analyticity of $H_b(f)$ is a necessary condition for the solvability of (1). To prove its sufficiency we will give a short account of the theory of the induced Cauchy-Riemann equations in $b\Omega$.

Let $\mathscr{A}$ denote the space of complex-valued differential forms on $\bar{\Omega}$ (i.e., whose coefficients are in $C^\infty(\bar{\Omega})$). Then we have the usual decomposition

$$\mathscr{A} = \sum \mathscr{A}^{p,q}. \tag{12}$$

As in [7], let $\mathscr{C}$ denote the ideal which is generated by forms that near $b\Omega$ are equal to r or $\bar{\partial} r$ and $\tilde{\mathscr{C}}$ the ideal generated by r, ∂r, $\bar{\partial} r$ and $\partial\bar{\partial} r$. We set $\mathscr{C}^{p,q} = \mathscr{C} \cap \mathscr{A}^{p,q}$ and $\tilde{\mathscr{C}}^{p,q} = \tilde{\mathscr{C}} \cap \mathscr{A}^{p,q}$. Observe that

$$\bar{\partial}\mathscr{C}^{p,q} \subset \mathscr{C}^{p,q+1} \quad \text{and} \quad \bar{\partial}\tilde{\mathscr{C}}^{p,q} \subset \tilde{\mathscr{C}}^{p,q+1}. \tag{13}$$

Thus we have

$$\begin{array}{ccccccccc} 0 & \longrightarrow & \mathscr{C}^{p,q+1} & \longrightarrow & \mathscr{A}^{p,q+1} & \longrightarrow & \mathscr{B}^{p,q+1} & \longrightarrow & 0 \\ & & \uparrow \bar{\partial} & & \uparrow \bar{\partial} & & \uparrow \bar{\partial}_b & & \\ 0 & \longrightarrow & \mathscr{C}^{p,q} & \longrightarrow & \mathscr{A}^{p,q} & \longrightarrow & \mathscr{B}^{p,q} & \longrightarrow & 0 \end{array} \tag{14}$$

where $\mathscr{B}^{p,q}$ is the quotient of $\mathscr{A}^{p,q}$ by $\mathscr{C}^{p,q}$ and $\bar{\partial}_b$ is defined by virtue of (13). Similarly we can define $\bar{\partial}_b : \tilde{\mathscr{B}}^{p,q} \to \tilde{\mathscr{B}}^{p,q+1}$ and note that $\mathscr{B}^{0,q} = \tilde{\mathscr{B}}^{0,q}$ and $\bar{\partial}_b$ coincides there.

Let $T^{1,0}(b\Omega) = CT(b\Omega) \cap T^{1,0}$; then $A \in T^{1,0}_P(b\Omega)$, $P \in b\Omega$, if and only if

$$A = \sum a_j \frac{\partial}{z_j} \quad \text{and} \quad \sum a_j \left(\frac{\partial r}{\partial z_j}\right)_P = 0.$$

Similarly we set $T^{0,1}(b\Omega) = CT(b\Omega) \cap T^{0,1}$ and observe that $T^{1,0}(b\Omega)$ and $T^{0,1}(b\Omega)$ are dual to $\mathscr{B}^{1,0}$ and $\mathscr{B}^{0,1}$ respectively. The hermitian inner product on $CT(\Omega)$ induces an hermitian inner product on the spaces $T^{0,1}(b\Omega)$, $T^{1,0}(b\Omega)$ and $\mathscr{B}^{p,q}$. Given a volume element $d\sigma$ on $b\Omega$ we then have the L_2-inner product on $\mathscr{B}^{p,q}$ defined by

$$(\varphi, \psi) = \int_{b\Omega} \langle \varphi, \psi \rangle \, d\sigma, \tag{15}$$

where $\langle \varphi, \psi \rangle$ denotes the inner product of φ and ψ at each point of $b\Omega$. We then define the formal adjoint of $\bar{\partial}_b$ denoted by ϑ_b by

$$(\vartheta_b \varphi, \psi) = (\varphi, \bar{\partial}_b \psi); \tag{16}$$

then $\vartheta_b : \mathscr{B}^{p,q} \to \mathscr{B}^{p,q-1}$. We set

$$\Box_b = \bar{\partial}_b \vartheta_b + \vartheta_b \bar{\partial}_b, \tag{17}$$

$$\mathrm{Dom}(\Box_b^{p,q}) = \{\varphi \in L_2^{p,q}(b\Omega),\ \varphi \text{ in domain of } \Box_b\} \tag{18}$$

where $L_2^{p,q}(b\Omega)$ denotes the completion $\mathscr{B}^{p,q}$,

$$\mathscr{H}_b^{p,q} = \{\varphi \in \mathrm{Dom}(\Box_b^{p,q}) \mid \Box_b^{p,q}\varphi = 0\}. \tag{19}$$

It can be shown (with $b\Omega$ compact) that $\varphi \in \mathrm{Dom}(\Box_b)$ is equivalent to $\varphi \in \mathrm{Dom}(\bar{\partial}_b) \cap \mathrm{Dom}(\vartheta_b)$, $\bar{\partial}_b\varphi \in \mathrm{Dom}(\vartheta_b)$ and $\vartheta_b\varphi \in \mathrm{Dom}(\bar{\partial}_b)$. Further, denoting by ϑ_b the closure of the ϑ_b defined above we have

(20) $$(\bar{\partial}_b)^* = \vartheta_b,$$

where $\bar{\partial}_b^*$ denotes the L_2-adjoint of $\bar{\partial}_b$.

LEMMA.

$$\mathscr{H}_b^{p,q} = \{\varphi \in \mathrm{Dom}(\bar{\partial}_b) \cap \mathrm{Dom}(\vartheta_b) \cap L_2^{p,q}(b\Omega) \mid \bar{\partial}_b\varphi = \vartheta_b\varphi = 0\}.$$

PROOF. It follows from the above remarks that for $\varphi \in \mathrm{Dom}(\square_b^{p,q})$ we have

(21) $$(\square_b^{p,q}\varphi, \varphi) = \|\bar{\partial}_b\varphi\|^2 + \|\vartheta_b\varphi\|^2,$$

from which the result is easily deduced.

Observe that it follows from the lemma that $\mathscr{H}_b(\Omega) = \mathscr{H}_b^{0,0}$.

Observe that on functions $\square_b v = \vartheta_b\bar{\partial}_b v$, then arguing the same as in the proof of Theorem 2 we obtain the following result.

THEOREM 3. *If the Cauchy-Szegö kernel on Ω satisfies the analyticity hypothesis, if $f \in L_2(b\Omega)$ and if in a neighborhood U of $P \in b\Omega$ there exists $v \in L_2(U \cap b\Omega)$ such that*

(22) $$\square_b v = f$$

then $\tilde{H}_b f$ has holomorphic continuations past P.

Furthermore, if we can solve the equation (22) globally (for f orthogonal to $\mathscr{H}_b$) then we also obtain the converse of the above.

THEOREM 4. *If the range of the L_2-closure of $\square_b$ on functions is closed and if in addition the analyticity hypothesis holds, and further if r and $d\sigma$ are analytic in a neighborhood of $P \in b\Omega$, then the local solvability of* (22) *in a neighborhood of P is equivalent to the holomorphic extendibility of $\tilde{H}_b f$.*

PROOF. Denoting by $\square_b$ also the closure of $\square_b$ we see that the closed range of $\square_b$ implies that there exists a bounded selfadjoint operator $N_b : L_2(b\Omega) \ominus \mathscr{H}(b\Omega) \to \mathrm{Dom}(\square^{0,0}) \ominus \mathscr{H}(b\Omega)$ where $A \ominus B$ denotes the orthogonal complement of B in A; N_b is defined by

(23) $$\square_b N_b g = g.$$

We extend N_b to $L_2(b\Omega)$ by setting $N_b h = 0$ for all $h \in \mathscr{H}(b\Omega)$. We then have the orthogonal decomposition

(24) $$f = \square_b(N_b f) + H_b f.$$

Now if $H_b f$ is analytic in a neighborhood of P, then by the Cauchy-Kowalevsky theorem there exists a function w defined in a neighborhood of P such that $\square_b w = H_b f$, which completes the proof.

Returning to the Lewy equation, we have, in the case Ω is given by (2):

(25) $$\square_b = L\bar{L}.$$

In this case the operator N_b can be computed explicitly which then provides a solution of (22) and hence of (1) whenever $\tilde{H}_b f$ can be holomorphically extended past P.

The computation of N_b uses the fact that $b\Omega$ is a group (the Heisenberg group) under the operation

(26) $$(z, t) \cdot (z', t') = (z + z', t + t' + \mathrm{Im}(z \cdot \bar{z}')).$$

Setting

$$\Phi = \frac{1}{2\pi^2(|z|^2 - it)} \log\left(\frac{|z|^2 - it}{|z|^2 + it}\right) \tag{27}$$

we have

$$N_b f = f * \Phi \tag{28}$$

where $*$ denotes convolution on the group. Combining Theorems 2 and 4 with the above establishes Theorem 1. The regularity properties of homogeneous operators such as (28) are studied in **[8]** and this yields regularity results for our solution of (1).

II. One of our concerns will be the so-called $\bar\partial$ problem which can be described as follows. Let α be a (0, 1)-form with $\bar\partial\alpha = 0$. The problem is to find a function u such that

$$\bar\partial u = \alpha \tag{1}$$

and to study the dependence of u on α. Our main emphasis will be on regularity properties at the boundary. The philosophy here is that we are interested in proving the existence of holomorphic functions with certain properties. This is often accomplished with PDE methods by having good control of u in terms of α. To illustrate this consider the use of (1) to prove Hartog's theorem (see Serre **[9]** and Ehrenpreis **[10]**).

THEOREM 1. *If* $\alpha \in \Lambda^{1,0}(\boldsymbol{C}^n)$, $\bar\partial\alpha = 0$, *if* α *has compact support and if* $n \geqq 2$ *then there exists a solution of* (1) *with compact support.*

In fact u can be expressed by the classical one-variable formula

$$u(z_1, \cdots, z_n) = \frac{1}{2\pi i}\int\!\!\int_C \frac{\alpha_1(\tau, z_2, \cdots, z_n)}{\tau - z_1}\, d\tau \wedge d\bar\tau. \tag{2}$$

It is then easy to see that u has compact support since u is holomorphic outside of the support of α_1 and u is zero when $|z_2|$ is sufficiently large (see Hörmander **[11]** for a detailed discussion of the above theorem and of the following corollary).

COROLLARY (HARTOG'S THEOREM). *If* $\Omega \subset \boldsymbol{C}^n$, $n \geqq 2$, *and if* h *is a holomorphic function defined on* Ω *then* h *can be extended to a holomorphic function on the union of* Ω *and the bounded components of the complement of* Ω.

PROOF. Let U be a neighborhood of the closure of the bounded components of the complement of Ω such that U does not intersect the unbounded components of the complement of Ω; we further choose U so that its complement is unbounded and connected. Let $\rho \in C^\infty(\boldsymbol{C})$ such that $\rho = 0$ on bounded components of the complement of Ω and $\rho = 1$ outside of U. Let $\alpha = \bar\partial(\rho h) = h\bar\partial\rho$; then clearly α has compact support and hence there exists u with compact support satisfying (1). Now, since the function u is holomorphic in the complement of U, since it has compact support and since the complement of U is unbounded and connected we conclude that $u = 0$ on the complement of U. Hence the function $\rho h - u$, which is defined on the union of Ω with bounded components of the complement is holomorphic and equals h outside of U; thus it is the desired extension.

Hartog's theorem points up a fundamental difference between the theories of one and several variables. In several variables there may be points on the boundary of a domain such that all holomorphic functions extend past these points. These considerations lead to the following formulation of the Levi problem.

Local Levi problem. $\Omega \subset \boldsymbol{C}^n$, given $P \in b\Omega$, does there exist a neighborhood U of P and a holomorphic function h on $U \cap \Omega$ such that h does not have a holomorphic extension past P?

Global Levi problem. $\Omega \subset \boldsymbol{C}^n$, given $P \in b\Omega$, does there exist a holomorphic function on Ω which does not have a holomorphic extension past P?

Observe that if Ω is convex then the Levi problem has a simple solution, namely let w be a linear holomorphic function such that $\mathrm{Re}(w) = 0$ defines a supporting hyperplane through P; then $h = 1/w$ gives the solution. Similarly if there exists a neighborhood U of P such that $U \cap \Omega$ is convex then the local Levi problem has a solution.

Let $x_j = \mathrm{Re}(z_j)$ and $x_{j+m} = \mathrm{Im}(z_j)$ for $j = 1, \cdots, n$. Then consider the vector $X \in T_P(b\Omega)$ given by

$$X = \sum_1^{2n} a_k \frac{\partial}{\partial x_k}, \qquad a_k \in \boldsymbol{R}, \tag{3}$$

and

$$[X(r)]_P = \sum_1^{2n} a_k \left[\frac{\partial r}{\partial x_k}\right]_P = 0, \tag{4}$$

where r is, as usual, the function that defines $b\Omega$. The differential-geometric condition for convexity at P is then given by

$$[X^2(r)]_P = \sum a_k a_l \left[\frac{\partial^2 r}{\partial x_k \partial x}\right]_P \geqq 0, \tag{5}$$

for all X satisfying (3) and (4).

Condition (5) is not invariant under holomorphic transformations, however "part" of it is; to see this we express it in complex coordinates.

Then X can be written as:

$$X = \sum_1^n b_j \frac{\partial}{\partial z_j} + \sum_1^n \bar{b}_j \frac{\partial}{\partial \bar{z}_j} \tag{6}$$

where $b_j = a_j + i a_{j+n}, j = 1, \cdots, n$. Then (4) becomes:

$$\mathrm{Re}\left(\sum \left(\frac{\partial r}{\partial z_j}\right)_P\right) = 0 \tag{7}$$

and (5) becomes

$$\sum b_j \bar{b}_k \left[\frac{\partial^2 r}{\partial z_j \partial \bar{z}_k}\right]_P + \mathrm{Re}\left(\sum b_j b_k \left[\frac{\partial^2 r}{\partial z_j \partial z_k}\right]\right) \geqq 0. \tag{8}$$

Conditions (7) and (8) are then conditions on the n-tuple $(b_1, \cdots, b_n)$ or equivalently on the vectors $\sum b_j (\partial/\partial z_j) \in T_P^{1,0}$. Now the subset Q of $T_P^{1,0}$ defined by (7) is not invariant under holomorphic transformations. It is clear that the largest subset of Q which is invariant is the subspace given by the condition

$$\sum b_j \left(\frac{\partial r}{\partial z_j}\right)_P = 0; \tag{9}$$

this subspace is denoted, as in the previous section, by $T_P^{1,0}(b\Omega)$. Thus we wish to require that (8) hold for all vectors in $T_P^{1,0}(b\Omega)$. Observe that if a vector is multiplied by i then the first term in (8) is unchanged but the second changes sign. Hence requiring (8) to hold for all $\sum b_j(\partial/\partial z_j) \in T_P^{1,0}(b\Omega)$ is equivalent to the condition

$$\sum b_j \bar{b}_k \left[\frac{\partial^2 r}{\partial z_j \partial \bar{z}_k}\right]_P \geqq 0 \tag{10}$$

whenever (9) is satisfied. This motivates the following standard definitions.

DEFINITION. The Levi form on $T_P^{1,0}(\Omega)$ is the quadratic form defined by

$$\mathscr{L}(L, \bar{L}) = \langle \partial\bar{\partial} r, L \wedge \bar{L} \rangle_P, \tag{11}$$

where $\langle \ , \ \rangle_P$ denotes the contraction between covariant and contravariant tensors at P. If the form $\mathscr{L}$ is positive semidefinite at P we say that Ω is *pseudoconvex* at P and if it is positive definite we say that Ω is *strongly pseudoconvex* at P. We say that Ω is (strongly) pseudoconvex if (strong) pseudoconvexity holds for all $P \in b\Omega$.

If Ω is strongly pseudoconvex at P then the function R defined by

$$R = e^{\lambda r} - 1 \tag{12}$$

is strongly plurisubharmonic in a neighborhood of P if λ is sufficiently large. Strongly plurisubharmonic, as usual, means that the metric $R_{z_i \bar{z}_j}$ is positive definite.

The following lemma solves the local Levi problem at strongly pseudoconvex points.

LEMMA. *If Ω is strongly pseudoconvex at $P \in b\Omega$ then there exists a neighborhood U of P and a holomorphic function h on U such that $h(P) = 0$ and $h(Q) \neq 0$ for all $Q \in \bar{\Omega} \cap U - \{P\}$.*

PROOF. In fact h can be taken to be the second-order polynomial defined by expanding R in a Taylor series so that

$$\begin{aligned} h = &\sum R_{z_i}(P)(z_i - z_i(P)) \\ &+ \sum R_{z_i z_j}(P)(z_i - z_i(P))(z_j - z_j(P)). \end{aligned} \tag{13}$$

Thus we see that

$$R = \mathrm{Re}(h) + \sum R_{z_i \bar{z}_j}(P)(z_i - z_i(P))(\overline{z_j - z_j(P)}) + O(|z - z(P)|^3). \tag{14}$$

The desired result follows by observing that near P the function R is positive on the zeros of $\mathrm{Re}(h)$.

The above result is classical, dating back to E. E. Levi. It does not have a direct generalization to domains which are pseudoconvex but not strongly pseudoconvex as is shown by the following example.

Let r be a function in C^2 given by

$$r = \mathrm{Re}(w) + |z|^8 + (15/7)|z|^2 \mathrm{Re}(z^6) + |z|^2 |w|^2. \tag{15}$$

In **[12]** the following is proved.

THEOREM. *If h is a holomorphic function defined in a neighborhood U of $(0, 0) \in C^2$ such that $h(0, 0) = 0$ then there exist points $(z_1, w_1), (z_2, w_2) \in U$ such that $h(z_i, w_i) = 0, i = 1, 2; r(z_1, w_1) > 0$ and $r(z_2, w_2) < 0$ where r is given by* (15).

Turning now to the global Levi problem we have the following result.

THEOREM. *If $\Omega \subset C^n$ is strongly pseudoconvex at $P \in b\Omega$ and the range of the closure of $\bar\partial$ on $L_2(\Omega)$ is closed then there exists a holomorphic function g on Ω which cannot be continued past P (i.e., there is no holomorphic extension of g whose domain contains P).*

PROOF. Let h be the function defined by (13), let U be a small neighborhood of P such that $\mathrm{Re}(h) < 0$ on $U \cap \bar\Omega - \{P\}$ and let $\rho \in C_0^\infty(U)$ with $\rho = 1$ in a neighborhood of P. Define the (0,1)-form α by:

$$\alpha = \bar\partial\,(\rho/h^N) = \bar\partial\rho/h^N, \tag{16}$$

where N is such that h^{-N} restricted to $V \cap \bar\Omega$ is not square-integrable for any neighborhood V of P. Clearly α is in the closure of the range of $\bar\partial$ since $(h - \varepsilon)^{-N}$ approaches h^{-N} on the support of $\bar\partial\rho$ as $\varepsilon \to 0$. Since the range is closed there exists $u \in L_2(\Omega)$ such that $\bar\partial u = \alpha$; here we mean that u is in the domain of the closure of $\bar\partial$ and (as is common) we also denote by $\bar\partial$ the closure of $\bar\partial$. The required function g is then given by

$$g = \rho/h^N - u. \tag{17}$$

It is clear that g is holomorphic and that g is not square-integrable in any neighborhood of P.

The following proposition shows that, in general, there are no reasonable conditions on a (0, 1)-form to insure that there exists a u with $\bar\partial u = \alpha$.

PROPOSITION. *If $n > 1$ then there exist domains $\Omega \subset C^n$ on which $\bar\partial$ does not have a closed range.*

PROOF. Let $\Omega \subset C^n$ be a domain such that there exists $P \in b\Omega$ such that the Levi form is positive definite at P and such that P lies in a bounded component of the complement of Ω. Then if the range of $\bar\partial$ were closed we could find a holomorphic function on Ω which cannot be extended past P. However, Hartog's theorem gives a contradiction.

The following theorem gives a criterion for the closedness of the range of $\bar\partial$ which leads to the $\bar\partial$ Neumann problem.

THEOREM. *Given that A, B and C are Hilbert spaces and $T:A \to B$, $S:B \to C$ are densely defined closed operators such that $ST = 0$, let*

$$\begin{aligned} \mathscr{D} &= \mathrm{Dom}(T^*) \cap \mathrm{Dom}(S),\\ \mathscr{H} &= \mathscr{N}(T^*) \cap \mathscr{N}(S),\\ \mathscr{J} &= \{\varphi \in \mathscr{J} \,|\, \varphi \perp \mathscr{H}\}; \end{aligned} \tag{18}$$

where Dom(T) *and* $\mathscr{N}(T)$ *denote the domain and null space of T, and T^* denote the Hilbert space adjoint of T whose domain is defined (as usual) by:*

$$\text{Dom}(T^*) = \{\varphi \in B \mid \textit{there exists } C > 0 \textit{ and } |(\varphi, T_\psi)_B| \leqq C\|\psi\|_A \textit{ holds for all } \psi \in \text{Dom}(T)\}. \tag{19}$$

We assume that $\mathscr{D}$ is dense in B. Then the following statements are equivalent:

(i) *The ranges of T and S are closed.*

(ii) *The ranges of T^* and S are closed.*

(iii) *There exists a constant $C > 0$ such that*

$$\|\varphi\|_B^2 \leqq C(\|T^*\varphi\|_A^2 + \|S\varphi\|_C^2) \tag{20}$$

for all $\varphi \in \mathscr{J}$.

(iv) *Let L be the operator of B defined by*

$$L = TT^* + S^*S \tag{21}$$

with $\text{Dom}(L) = \{\varphi \in \mathscr{D} \mid T^*\varphi \in \text{Dom}(T)$ *and* $S\varphi \in \text{Dom}(S^*)\}$; *then L has a closed range.*

For a proof of this theorem see [**13**].

The approach of studying $\bar\partial$ via the operator defined by (21) is called the $\bar\partial$ Neumann problem. For domains in $\boldsymbol{C}^n$ the operator L is the Laplace operator and the Dom(L) is described by boundary conditions which will be discussed in the next section. We will conclude this section by mentioning some consequences of (iv).

First of all if the range of L is closed there exists a unique bounded selfadjoint operator $N:B \to B$ such that the range of N equals the domain of L and we have

$$LN = I - H \quad \text{and} \quad HN = NH = 0 \tag{22}$$

when $H:B \to B$ is the orthogonal projection of B onto $\mathscr{H}$.

It is clear that (22) implies uniqueness, and to define N we proceed in a manner entirely analogous to the definition of N_b in the first section. We observe that (22) is the orthogonal decomposition, i.e., for any $\varphi \in B$ we have

$$\varphi = TT^*N\varphi + S^*SN\varphi + H\varphi. \tag{23}$$

PROPOSITION. *If there exists a bounded selfadjoint operator $N:B \to B$ satisfying* (22), *then given $\alpha \in B$ the following conditions are equivalent:*

(i) *There exists $\psi \in \text{Dom}(T)$ with $T\psi = \alpha$.*

(ii) *$S\alpha = 0$ and α is orthogonal to $\mathscr{H}$.*

PROOF. Clearly (i) $\Rightarrow$ (ii). Assuming (α) satisfies (ii) we have

$$\alpha = TT^*N\alpha + S^*SN\alpha. \tag{24}$$

Applying S we obtain

$$SS^*SN\alpha = 0. \tag{25}$$

Taking inner products with $SN\alpha$ we get

$$(SS^*SN\alpha, SN\alpha) = \|S^*SN\alpha\|^2 = 0 \tag{26}$$

and hence

$$\alpha = T(T^*N\alpha) \tag{27}$$

so we can set $\psi = T^*N\alpha$; which is the unique solution of $T\psi = \alpha$ that is orthogonal to $\mathscr{H}$.

Observe that the above calculation shows that $SN\alpha = 0$ whenever $S\alpha = 0$; hence in particular we have

$$SNT = 0 \tag{28}$$

from which we will derive a formula for the orthogonal projection on the null space of T. Denoting by $H_T : A \to A$ the orthogonal projection on $\mathcal{N}(T)$ we have

PROPOSITION. *Assuming again* (iv) *of the above theorem we have*

$$H_T = I - T^*NT. \tag{29}$$

PROOF. Setting $R = I - T^*NT$ we see that for any $\psi \in \mathcal{N}(T)$ we have $R\psi = \psi$. Furthermore, R is selfadjoint; hence it suffices to show that $TR = 0$. By (23) we have

$$TR = T - TT^*NT = T - T + S^*SNT + HNT = 0;$$

the last two terms vanish by virtue of (28) and (22).

The solution given by (27) and the formula (29) will be the starting point of most of the applications of the theory. In case T is the closure of $\bar\partial$ on functions, (29) is a very useful formula when one has good control on N; in particular it gives information on the Bergman kernel function; see [**26**] and [**27**].

III. Our aim now is to study $\bar\partial$ on functions and on (0,1)-forms within the framework of the L_2-theory described in the previous section. Let $\mathscr{A}$ denote C^∞ complex-valued forms on $\bar\Omega$; $\mathscr{A}$ is bigraded as in (12) of § I. If $\varphi, \psi \in \mathscr{A}^{p,q}$ we define (φ, ψ), the L_2-inner product, as usual, by

$$(\varphi, \psi) = \sum \int_\Omega \varphi_{i_1\cdots i_p j_1\cdots j_q} \bar\psi_{i_1\cdots i_p j_1\cdots j_q}\, dV. \tag{1}$$

We will denote by $\tilde{\mathscr{A}}^{p,q}$ the Hilbert space obtained by completing $\mathscr{A}^{p,q}$. We will apply the previous discussion to the case when $A = \tilde{\mathscr{A}}^{0,0}$, $B = \tilde{\mathscr{A}}^{0,1}$, $C = \tilde{\mathscr{A}}^{0,2}$ and with T and S being the closures of $\bar\partial : \tilde{\mathscr{A}}^{0,0} \to \tilde{\mathscr{A}}^{0,1}$ and $\bar\partial : \tilde{\mathscr{A}}^{0,1} \to \tilde{\mathscr{A}}^{0,2}$, respectively. The first thing we will do is determine the smooth elements of $\mathscr{D}$, i.e., the elements in

$$\dot{\mathscr{D}} = \mathscr{D} \cap \mathscr{A}^{0,1} = \operatorname{Dom}(\bar\partial^*) \cap \mathscr{A}^{0,1} \tag{2}$$

where $\mathscr{D}$ was defined in the previous section (by (17)) and where we adopt the usual notation of using $\bar\partial$ and $\bar\partial^*$ to denote the closure and adjoint of $\bar\partial$, respectively. Suppose $\varphi \in \mathscr{A}^{0,1}$ and $u \in \mathscr{A}^{0,0}$; then, by integration by parts, we obtain

$$(\varphi, \bar\partial u) = \sum (\varphi_j, u_{\bar z_j}) = \Big(-\sum \varphi_{j z_j}, u\Big) + \int_{b\Omega} r_{z_j} \varphi_j \bar u\, dS. \tag{3}$$

Recall that $\varphi \in \operatorname{Dom}(\bar\partial^*)$ means that the functional sending u into $(\varphi, \bar\partial u)$ is bounded; then (3) implies that $\varphi \in \dot{\mathscr{D}}$ if and only if

$$\sum r_{z_j} \varphi_j = 0 \quad \text{on } b\Omega. \tag{4}$$

Furthermore, defining $\vartheta : \mathscr{A}^{0,1} \to \mathscr{A}^{0,0}$, the formal adjoint of $\bar\partial$, by

$$\vartheta\varphi = -\sum \varphi_{j z_j} \tag{5}$$

we see that on $\dot{\mathscr{D}}$ we have $\bar\partial^*\varphi = \vartheta\varphi$.

To study the $\bar\partial$ Neumann problem on pseudoconvex domains (which are not strongly pseudoconvex) we will use Carleman type weight functions as introduced by Hörmander in [**14**]. On the Hilbert space $\mathscr{A}^{p,q}$ we will introduce for each $t \geqq 0$ an inner product denoted by $(\ ,\ )_{(t)}$ and defined by

$$(\varphi, \psi)_{(t)} = (\varphi, e^{-t|z|^2}\psi) \tag{6}$$

where $|z|^2 = \sum_1^n |z_j|^2$. Setting

$$\vartheta_t\varphi = e^{t|z|^2}\vartheta(e^{-t|z|^2}\varphi) \tag{7}$$

we obtain the following generalization of (3):

$$(\varphi, \bar\partial u)_{(t)} = (\vartheta_t\varphi, u)_{(t)} + \int_{b\Omega} \sum r_{z_j}\varphi_j \bar u^{-t|z|^2}\, dS. \tag{8}$$

Thus $\mathscr{N}(\bar\partial_t^*) \cap \mathscr{A}^{0,1}$ is independent of t and given by (4).

The following proposition is proven in [**14**].

PROPOSITION. *If Ω is pseudoconvex then there exists a constant $C > 0$ such that*

$$tC\|\varphi\|_{(t)}^2 + \|\varphi_{i\bar z_j}\|_{(t)}^2 + \sum \int_{b\Omega} r_{z_i\bar z_j}\varphi_i\bar\varphi_j e^{-t|z|^2}\, dS \leqq \|\bar\partial\varphi\|_{(t)}^2 + \|\vartheta_t\varphi\|_{(t)}^2 \tag{9}$$

for all $\varphi \in \dot{\mathscr{D}}$ and all $t \geqq 0$.

In particular we see that the inequality (19) of the previous section holds for forms in $\dot{\mathscr{D}}$ whenever $t > 0$. It is, however, necessary to establish (19) for all elements $\mathscr{D}_t$ and not just the one in $\dot{\mathscr{D}}$. This is done by showing that $\dot{\mathscr{D}}$ is dense in $\mathscr{D}_t$ with respect to the norm defined by the right side of (9). Denseness is proved by use of smoothing operators; see [**14**]. Observe that (9) implies that $\mathscr{H}_t^{0,1} = 0$ (where $\mathscr{H}_t^{0,1}$ is defined by $\mathscr{H}_t^{0,1} = \mathscr{N}(\bar\partial) \cap \mathscr{N}(\bar\partial_t^*) \cap \mathscr{A}^{0,1}$). This implies in turn that the $\bar\partial$ L_2-cohomology is zero by the results of the previous section.

Now suppose that Ω is a bounded domain, with a smooth boundary, contained in a complex manifold X. In this case the $\bar\partial$-cohomology is not necessarily zero, however; under suitable assumptions, we may prove an estimate similar to (9), which then implies that the cohomology is finite dimensional. The additional assumption we make is that there exists a function which plays the role $|z|^2$.

PROPOSITION. *Suppose that $\Omega \subset X$ is a bounded pseudoconvex domain in the complex manifold X and that there exists a nonnegative function f on X which is strongly plurisubharmonic in a neighborhood U of $b\Omega$. Then, analogously to* (6), *we define for each $t \geqq 0$ an inner product by*

$$(\varphi, \psi)_{(t)} = (\varphi, e^{-tf}), \tag{10}$$

where $(\ ,\ )$ denotes the inner product induced by some fixed hermitian metric on Ω. Then, given $\zeta \in C_0^\infty(U)$, there exist positive constants C and C' independent of t such that

$$\begin{aligned} tC\|\zeta\varphi\|_{(t)}^2 + \sum \|\varphi_{i\bar z_j}\|_{(t)}^2 + \sum \int_{b\Omega} r_{z_i\bar z_j}\varphi_i\bar\varphi_j e^{-tf}\, dS \\ \leqq \|\bar\partial\varphi\|_{(t)}^2 + \|\bar\partial_t^*\varphi\|_{(t)}^2 + C'\|\varphi\|_{(t)}^2, \end{aligned} \tag{11}$$

for all $\varphi \in \dot{\mathscr{D}}$ and $t \geqq 0$. Furthermore, if f is strongly plurisubharmonic on $\bar\Omega$ then ζ can be chosen identically equal to 1 on Ω.

The proof of this inequality is quite analogous to the proof of (9) and is also contained in [**14**]. Observe that if t is large enough and $\zeta = 1$ near $b\Omega$ then we can have

$$\|\zeta\varphi\|^2_{(t)} \leqq \text{const}\,(\|\bar{\partial}\varphi\|^2_{(t)} + \|\bar{\partial}^*_t\varphi\|^2_{(t)} + \|K\varphi\|^2_{(t)}), \tag{12}$$

where K is a compact operator. This follows from the fact that the operator $(\bar{\partial}, \bar{\partial}^*_t)$ is elliptic in the interior and hence the L_2-norm of φ on a compact subset of Ω is dominated by the L_2-norm of the operator applied to φ plus any negative Sobolev norm on φ (which can be expressed by the norm of a compact operator). This shows then that $\mathscr{H}^{0,1}_t$ is finite dimensional and that the estimate (19) of the previous section holds. Of course, if f is strongly plurisubharmonic on all of $\bar{\Omega}$ then $\mathscr{H}^{0,1}_t = 0$ for t sufficiently large.

In [**11**] Hörmander uses weighting functions that go to infinity on $b\Omega$ and this greatly simplifies the proofs of the existence theorems. In our view, however, one of the most interesting problems is regularity behavior at the boundary; to study this we need smooth weighting functions. We will discuss regularity in the next two sections; we conclude this section by showing how the above existence theorems can be used to prove the Newlander-Nirenberg theorem.

DEFINITION. Let X be a differentiable manifold and denote by $\boldsymbol{C}T(X)$ the bundle of complex tangent vectors. An *almost-complex structure* on X is given by a subbundle of $\boldsymbol{C}T(X)$ denoted by $T^{1,0}(X)$ which satisfies the following two conditions.

(a) $T^{1,0}(X) \cap \overline{T^{1,0}(X)} = \{0\}$.

(b) $\boldsymbol{C}T(X) = T^{1,0}(X) + \overline{T^{1,0}(X)}$.

We say that the almost-complex structure given by $T^{1,0}(X)$ is *integrable* if whenever V and W are local vector fields with values in $T^{1,0}(X)$ then $[V, W] = VW - WV$ also has values in $T^{1,0}(X)$.

If X is a complex manifold the complex structure defines a unique *underlying almost-complex structure* by setting $T^{1,0}_P(X)$ to be those tangent vectors L, at $P \in X$, such that $L(\bar{h}) = 0$ for all h that are germs of holomorphic functions at P. It is clear that this structure is integrable.

THEOREM (NEWLANDER-NIRENBERG). *If X has an integrable almost-complex structure given by $T^{1,0}(X)$ then X has a complex structure whose underlying almost-complex structure is given by $T^{1,0}(X)$.*

OUTLINE OF PROOF. Set $T^{0,1}(X) = \overline{T^{1,0}(X)}$. It suffices to prove that every $P \in X$ has a neighborhood U and functions $z_1, \cdots, z_n$ on U such that $L(z_j) = 0$ for all $L \in T^{0,1}_Q(X)$, with $Q \in U$, and such that $(dz_1)_P, \cdots, (dt_n)_P$ are linearly independent.

The splitting in (b) induces direct sum decomposition of the complex-valued exterior forms on X denoted by:

$$\Lambda(X) = \sum \Lambda^{p,q}(X) \tag{13}$$

and we denote the corresponding projection by

$$\prod_{p,q};\ \Lambda(X) \to \Lambda^{p,q}(X). \tag{14}$$

We define the operator $\bar{\partial}: \Lambda^{p,q}(X) \to \Lambda^{p,q+1}(X)$ by

$$\bar{\partial} = \prod_{p,q+1} d. \tag{15}$$

It is easy to check that the almost-complex structure on X is integrable if and only if for every germ of a function f we have

$$\bar{\partial}^2 f = 0. \tag{16}$$

Then if Ω is a domain in X we can define the Levi form just as in the complex case (see (11) of the previous section). Furthermore, the estimate (12) holds under the same hypotheses as above.

Given $P \in X$ we can find n complex-valued functions $w_1, \cdots, w_n$ such that $\{\mathrm{Re}(w_j), \mathrm{Im}(w_j)\}$ are a coordinate system in a neighborhood of P and such that

$$w_j(P) = 0. \tag{17}$$

Then we have

$$\bar{\partial} f = \sum \left(a_k^j \frac{\partial f}{\partial w_j} + \tilde{a}_k^j \frac{\partial f}{\partial \bar{w}_j} \right) dw_k + \sum \left(b_k^j \frac{\partial f}{\partial w_j} + \tilde{b}_k^j \frac{\partial f}{\partial \bar{w}_j} \right) d\bar{w}_k. \tag{18}$$

For each small ε we define an almost complex structure on the ball $B = \{z \in \boldsymbol{C}^n : |z| < 1\}$ by

$$\begin{aligned} \bar{\partial}_\varepsilon f(z) &= \sum \left(a_k^j(\varepsilon z) \frac{\partial f}{\partial z_j} + \tilde{a}_k^j(\varepsilon z) \frac{\partial f}{\partial \bar{z}_j} \right) dz_k \\ &\quad + \sum \left(b_j^k(\varepsilon z) \frac{\partial f}{\partial z_j} + \tilde{b}_k^j(\varepsilon z) \frac{\partial f}{\partial \bar{z}_j} \right) d\bar{z}_k. \end{aligned} \tag{19}$$

Note that when $\varepsilon = 0$ the almost complex structure coincides with that of $\boldsymbol{C}^n$.

To complete the proof it now suffices to find $u_1, \cdots, u_n$ defined in a neighborhood of $0 \in \boldsymbol{C}^n$ with $du_1, \cdots, du_n$ independent such that $\bar{\partial}_\varepsilon u_j = 0$, $j = 1, \cdots, u$ for some fixed small ε. We choose ε sufficiently small so that $\partial_\varepsilon \bar{\partial}_\varepsilon(|z|^2)$ is positive definite. Applying (11), we observe that the constant in (11) is independent of ε if ε is small and so with t sufficiently large the operator $N_{t,\varepsilon}$ on $(0,1)_\varepsilon$-forms has trivial null space and depends smoothly on ε. We now define

$$u_{j,\varepsilon} = H_{t,\varepsilon}(z_j) = z_j - \mathscr{D}_\varepsilon N_{t,\varepsilon} \bar{\partial}_\varepsilon z_j. \tag{20}$$

Then $\partial_\varepsilon u_{j,\varepsilon} = 0$ and by interior ellipticity we have, for $\zeta \in C_0^\infty(B)$,

$$\|\zeta N_{t,\varepsilon} \alpha\|_{k+2} \leqq C_k \|\alpha\|_k. \tag{21}$$

We apply this to $\partial_\varepsilon z_j$ and note that we have $\|\bar{\partial}_\varepsilon z_j\|_k = O(\varepsilon^m)$ for all m. Hence the last term in (19) is small and its first derivatives are small, so that $du_{1,\varepsilon}, \cdots, du_{n,\varepsilon}$ are independent, which completes the proof.

IV. Consider the following two problems. Given a square-integrable (0, 1)-form α on a pseudoconvex domain $\Omega \subset X$, X is a complex manifold such that there exists $u \in L_2(\Omega)$ satisfying

$$\bar{\partial} u = \alpha. \tag{1}$$

Problem 1 (*global regularity*). Suppose α is smooth on $\bar{\Omega}$. Does there exist $u \in C^\infty(\bar{\Omega})$ satisfying (1)?

Problem 2 (*local regularity*). Suppose that α is smooth on $U \cap \bar{\Omega}$, where U is an open subset of $\boldsymbol{C}^n$. Does there exist a distribution u on Ω satisfying (1) and whose restriction to $U \cap \bar{\Omega}$ is smooth?

In Problem 2 if U does not intersect $b\Omega$ then every solution of (1) is smooth on

U, since $\bar{\partial}$ on functions is elliptic. However, if U intersects $b\Omega$ the problem is more delicate since, if u satisfies (1) then $u + h$ also satisfies (2) for any holomorphic function h. In fact (2) is false without additional hypotheses as is shown by $\bar{\Omega} \subset \boldsymbol{C}^2$ for which the boundary contains the origin and near the origin is given by $r = \operatorname{Re}(z_2)$. In this case, if we let ρ be a compactly supported function with $\rho \equiv 1$ near the origin and $\alpha = \bar{\partial}(\rho z_2^{-1})$, then the solution u cannot be smooth in both the sets where $\rho \equiv 1$ and the set when $\rho \equiv 0$, as is shown in **[16]**.

To investigate regularity and in particular the above problems it is useful to place the $\bar{\partial}$ Neumann in variational form. Returning to our Hilbert space set-up we have

$$A \underset{T^*}{\overset{T}{\rightleftarrows}} B \underset{S^*}{\overset{S}{\rightleftarrows}} C \tag{2}$$

with $\mathscr{D} = \operatorname{Dom}(T^*) \cap \operatorname{Dom}(S)$. We consider $\mathscr{D}$ the Hilbert space with the following inner product

$$Q(\varphi, \psi) = (T^*\varphi, T^*\psi) + (S\varphi, S\psi). \tag{3}$$

It then follows that if the inequality (19) of §II holds, then the operator N can be defined by setting $N\alpha$ as the unique element of $\mathscr{D}$ which is orthogonal to $\mathscr{H}$ such that

$$Q(N\alpha, \psi) = (\alpha, \psi) \tag{4}$$

for all $\psi \in \mathscr{D}$, where $\alpha \in B \ominus \mathscr{H}$.

The solution of Problem 1 is given in the following theorem.

THEOREM. *If $\Omega \subset X$, Ω pseudoconvex, and if there exists a strongly plurisubharmonic function in a neighborhood of $b\Omega$, then for the* (0, 1)*-form α which is in $C^\infty(\bar{\Omega})$ and for which there exists a square-integrable solution u of* (1) *there exists also a solution $u \in C^\infty(\bar{\Omega})$ of* (1).

OUTLINE OF PROOF. The first part of the proof is to show that for every m there exists $u_m \in C^m(\bar{\Omega})$ satisfying (1). We will indicate very briefly how this is done and refer to **[17]** for details.

Consider the inequality (11) of §III written in terms of Q_t:

$$tC\|\zeta\varphi\|^2_{(t)} \leqq Q_t(\varphi, \varphi) + C'\|\varphi\|^2_{(t)}. \tag{5}$$

For a point $P \in b\Omega$ we can choose coordinates $x_1, \cdots, x_{2n-1}, r$ with origin at P. Further, on a neighborhood U of P we can choose a local basis of (0, 1)-forms $\bar{\omega}_1, \cdots, \bar{\omega}_{n-1}, \bar{\omega}_n$ with $\bar{\omega}_n = \bar{\partial} r$. If $\varphi = \sum \varphi_j \bar{\omega}_j$ then the condition $\varphi \in \dot{\mathscr{D}}$ is expressed by $\varphi_n = 0$ on $U \cap b\Omega$. If $\rho \in C_0^\infty(U)$ we set $D^a(\rho\varphi) = \sum D^a(\rho\varphi_j)\bar{\omega}_j$, when $a = (a_1, \cdots, a_{2n})$,

$$D^a = (-i)^{|a|}(\partial/\partial x_1)^{a_1} \cdots (\partial/\partial x_{2n})^{a_{2n}}$$

and $|a| = \sum a_j$. Hence if a' is a multi-index such that $a'_{2n} = 0$ and if $\rho\varphi \in \dot{\mathscr{D}}$ then $D^{a'}(\rho\varphi) \in \dot{\mathscr{D}}$. Thus we can apply (5) to $D^{a'}(\rho\varphi)$, and use the estimate

$$\begin{aligned} Q_t(D^{a'}(\rho\varphi), D^{a'}(\rho\varphi)) &\leqq \text{const Re } Q_t(\varphi, D^{2a'}(\rho^2\varphi)) \\ &\quad + \text{const}\Big(\sum_{|b|=m} \|D^b\rho'\varphi\|^2_{(t)} + C_t \sum_{|b|<m} \|D^b\rho'\varphi\|^2_{(t)}\Big), \end{aligned} \tag{6}$$

where $|a'| = m$, $\rho' \in C_0^\infty(U)$ with $\rho' \equiv 1$ on the support of ρ and where the first two constants on the right are independent of t. The fact that $b\Omega$ is noncharacteristic yields the following estimate of arbitrary derivatives in terms of tangential derivatives.

$$\|D^a(\rho\varphi)\|^2_{(t)} \leqq \text{const} \sum_{|b'|=m} \|D^{b'}(\rho'\varphi)\|^2_{(t)} \tag{7}$$
$$+ \text{const}\Big(\sum_{|b|=m} \|D^b\rho'\alpha\|^2_{(t)} + \sum_{|b|<m} \|D^b(\rho'\varphi)\|^2_{(t)}\Big),$$

where $|a| = m$ and the first constant is independent of t. We choose a finite covering of $\bar{\Omega}$ by open sets such that $b\Omega$ is covered by open sets of the type described above. Let $\{\rho_i\}$ be a partition of unity that corresponds to this covering. We use (5) with φ replaced by $D^a(\rho_i\varphi)$ (where $\text{supp}(\rho_i) \cap b\Omega = \emptyset$), combine the resulting inequality with (7) and sum on i and on a with $|a| = m$. For ρ_i with $\text{supp}(\rho_i) \cap b\Omega = 0$ the desired estimate comes from standard elliptic theory. Finally, using (4) with $Q = Q_t$, $\varphi = N_t\alpha$ and $\psi = D^{2a'}(\rho^2\varphi)$ and choosing t sufficiently large we obtain:

$$\|N_t\alpha\|_s \leqq \text{const}\|\alpha\|_s \quad \text{when } s \leqq m; \tag{8}$$

here $\| \quad \|_s$ stands for the Sobolev s-norm. To estabiish (8) t has to be chosen sufficiently large to absorb the second constant in (6), since this constant depends on m; the size of t will depend on m. It is true, in general, that

$$\|\vartheta_t N_t\alpha\|_s \leqq \text{const}(\|\alpha\|_s + \|N_t\alpha\|_s) \tag{9}$$

and so from (8) we have that there exist t_m such that

$$\|\vartheta_{t_m} N_{t_m}\alpha\|_s \leqq \text{const}\|\alpha\|_s, \quad \text{when } s \leqq m. \tag{10}$$

Setting $u_m = \vartheta_{t_m} N_{t_m}\alpha$ we then obtain $\bar{\partial} u_m = 0$ and

$$\|u_m\|_s \leqq \text{const}\|\alpha\|_s \quad \text{if } s \leqq m. \tag{11}$$

The usual regularizing procedure then shows that $u_m \in H_m$, the Sobolev space of functions on Ω with square-integrable derivatives up to order m.

We now have to show that there is a solution $u \in C^\infty(\bar{\Omega})$. The proof that (11) implies the existence of $u \in C^\infty(\bar{\Omega})$ was communicated to me in a letter by L. Hörmander and (with his permission) it is given below.

It suffices to find a sequence of solutions $u_j \in H_j$ such that

$$\|u_{j+1} - u_j\|_j \leqq 2^{-j} \tag{12}$$

since we can then set $u = u_k + \sum_{j=1}^{\infty}(u_{k+j} - u_{k+j-1})$, clearly a solution in $C^\infty(\bar{\Omega})$. Suppose that $u_1, \cdots, u_{m-1}$ have already been chosen to satisfy (9) and we have a solution u_m satisfying (11); we wish to replace u_m by $u_m - V_\varepsilon$ with V_ε holomorphic, $V_\varepsilon \in H_{m-1}$ and

$$\|u_m - V_\varepsilon - n_{m-1}\|_{m-1} \leqq 2^{-m+1}. \tag{13}$$

Let $h = u_m - u_{m-1}$. For each $P \in b\Omega$ there is a coordinate neighborhood U with coordinates $z_1, \cdots, z_n$ with $z_j(0) = P$ and an n-tuple $(a_1, \cdots, a_n)$ such that for $\varepsilon > 0$ sufficiently small $\Phi_\varepsilon(z) = (z_1 + \varepsilon a_1, \cdots, z_n + \varepsilon a_n) \in \Omega$ whenever $(z_1, \cdots, z_n) \in U \cap \bar{\Omega}$. Let $\{U_j\}$, $j = 1, \cdots, \nu$, be a covering of $b\Omega$ by open coordinate neighborhoods of the above type and denote the corresponding transformations by Φ_ε^j. Let U_0 be an open subset of Ω such that $\{U_j\}$, $j = 0, \cdots, \nu$, is a covering and $\bar{U}_0 \cap b\Omega = \emptyset$. Let $\{\rho_j\}$ be a C^∞ partition of unity on $\bar{\Omega}$ subordinate to $\{U_j\}$. Then we set

(14) $$V_\varepsilon(z) = \sum \rho_j(z)h(\Phi_\varepsilon^j(z)) - w_\varepsilon(z).$$

We wish to find the appropriate w_ε; since we want V_ε holomorphic w_ε must satisfy

(15) $$\bar\partial w_\varepsilon = \sum \bar\partial\rho_j h_\varepsilon^j = \sum \bar\partial\rho_j(h_\varepsilon^j - h).$$

The first term on the right of (15) is $C^\infty(\bar\Omega)$; in particular we can choose $w_\varepsilon \in H_m$ such that w_ε satisfies (11). The second term on the right of (15) shows that $\lim_{\varepsilon\to 0}\|\bar\partial w_\varepsilon\|_{m-1} = 0$ and hence $\lim_{\varepsilon\to 0}\|w_\varepsilon\|_{m-1} = 0$. Hence by choosing ε sufficiently small we make $\|h - V_\varepsilon\|_{m-1}$ as small as we wish, which completes the proof.

We now turn to the question of local regularity. The method of establishing local regularity will be to prove subelliptic estimates which are proved below.

DEFINITION. The $\bar\partial$ Neumann problem for Ω is *subelliptic* at $P \in b\Omega$ if there exists a neighborhood U of P and an $\varepsilon > 0$ such that

(16) $$\|\varphi\|_{[\varepsilon]}^2 \leqq C(Q(\varphi, \varphi) + \|\varphi\|^2) \quad \text{for all } \varphi \in \dot{\mathscr{D}}$$

such that supp $\varphi \subset U \cap \bar\Omega$. To define $\|\varphi\|_{[\varepsilon]}$ we first express φ as $\varphi = \sum\varphi_j\bar\omega_j$ as in the preparation for (6), then we set $\|\varphi\|_{[\varepsilon]}^2 = \sum \|\varphi_j\|_{[\varepsilon]}^2$ where

(17) $$\|u\|_{[\varepsilon]}^2 = \|\Lambda^{[\varepsilon]}_u\|^2 = \int_{-\infty}^0 \int_{R^{2n-1}} \left(1 + \sum|\xi|^2\right)^\varepsilon |\tilde u(\xi, r)|^2\, d\xi\, dr$$

and $\tilde u(\xi, r)$ is the tangential Fourier transform given by

(18) $$\tilde u(\xi, r) = \int_{R^{2n-1}} e^{-ix'\cdot\xi}\, u(x, r)\, dx'$$

where $x' = (x_1, \cdots, x_{2n-1})$, $\xi = (\xi_1, \cdots, \xi_{2n-1})$, $x'\cdot\xi = \sum_1^{2n-1} x_j\xi_j$ and $dx' = dx_1 \cdots dx_{2n-1}$.

The following is a result of the theory of subelliptic estimates (see **[16]**).

THEOREM. *If Ω is pseudoconvex, if the $\bar\partial$ Neumann problem is subelliptic at $P \in b\Omega$ and if U is a neighborhood of P such that for $Q \in U \cap b\Omega$ the $\bar\partial$ Neumann problem is subelliptic at Q, then the operator N_t (defined above) is pseudolocal in $U \cap \bar\Omega$ for t sufficiently large. Furthermore, if the $\bar\partial$ Neumann problem is subelliptic at every $P \in b\Omega$ then N_t exists for all $t \geqq 0$ and is pseudolocal.*

Here by pseudolocal we mean (as usual) that the operator preserves smoothness on open subsets. Thus, in particular, the projection operator H_t on holomorphic functions is pseudolocal since it is given by the expression $H_t = I - \vartheta_k N_k \bar\partial$.

Inequality (16) never holds if $\varepsilon > 1$. If $\varepsilon = 1$ the inequality holds if and only if $n = 1$. Further, if (16) does not hold for $\varepsilon = 1/k$ then it cannot hold for $\varepsilon > 1/k + 1$. If Ω is pseudoconvex then the necessary and sufficient condition for (16) to hold at $P \in b\Omega$ with $\varepsilon = \frac{1}{2}$ is that the Levi form is positive definite at P. The problem is to characterize subellipticity. The basic idea is that analytic subsets of $b\Omega$ propagate singularities and thus it seems plausible that subellipticity holds at $P \in b\Omega$, if any analytic subset through P has a finite bounded order of contact with $b\Omega$ at P. The precise formulation of this condition should appear in the doctoral thesis of D'Angelo. We will now describe a special case; for this we need the following definitions:

DEFINITION. For $P \in b\Omega$ let $(T_P^{1,0}(b\Omega))^{\text{germ}}$ denote the germs of C^∞ vector fields at P with values in $T^{1,0}(b\Omega)$. If $L \in (T_P^{1,0}(b\Omega))^{\text{germ}}$ we define $(\mathscr{L}_P^0(L))^{\text{germ}}$ to be the subset of $(CT_P(b\Omega))^{\text{germ}}$ consisting of all germs of vector fields of the form $fL + g\bar{L}$ where $f, g \in (C_P^\infty(b\Omega))^{\text{germ}}$. We define $(\mathscr{L}_P^k(L))^{\text{germ}}$ inductively by

$$\mathscr{L}_P^k(L) = \mathscr{L}_P^{k-1}(L) + [\mathscr{L}_P^0(L), \mathscr{L}_P^{k-1}(L)]. \tag{19}$$

We say that $L \in T_P^{1,0}(b\Omega)$ is *finite at* P if for some k we have

$$\mathscr{L}_P^k(L) \not\subset T_P^{1,0}(b\Omega) + T_P^{0,1}(L). \tag{20}$$

We say k is the order of L at P if k is the least integer for which (20) holds.

Observe that if Ω is pseudoconvex at $P \in b\Omega$ then Ω is strongly pseudoconvex at P if and only if every $L \in (T_P^{1,0}(b\Omega))^{\text{germ}}$, such that $L_P \neq 0$, has order 1. Observe also that if $P \in b\Omega$ is a regular point of a germ of an analytic curve which lies in $b\Omega$ and if we take $L \in (T_P^{1,0}(b\Omega))^{\text{germ}}$ to be a vector field whose restriction to the curve is the holomorphic vector tangent to the curve, then (20) does not hold for any k.

DEFINITION. If U is a neighborhood of $P \in b\Omega$ we say that the *Levi form is diagonalizable on* $U \cap b\Omega$ if there exist local bases $L_1, \cdots, L_{n-1}$ of $T^{1,0}(U \cap b\Omega)$ such that $[L_i, \bar{L}_j]$ has values in $T^{1,0}(U \cap b\Omega) + T^{0,1}(U \cap b\Omega)$ if $i = j$. In other words, letting T be a vector field independent of $L_1, \cdots, L_{n-1}, \bar{L}_1, \cdots, \bar{L}_{n-1}$ we have

$$[L_i, \bar{L}_j] = \lambda_i \delta_{ij} T \ (\text{mod } L_1, \cdots, L_{n-1}, \bar{L}_1, \cdots, \bar{L}_{n-1}). \tag{21}$$

THEOREM. *If Ω is pseudoconvex and if $P \in b\Omega$ has a neighborhood U such that the Levi form is diagonalizable on $U \cap b\Omega$ then the $\bar{\partial}$ Neumann problem is subelliptic at P if and only if any local basis $L_1, \cdots, L_{n-1}$ of $(T_P^{1,0}(b\Omega))^{\text{germ}}$ which diagonalizes the Levi form in a neighborhood of P has the property that each L_j is of finite order at P.*

Here we will briefly outline the proof of the sufficiency in the above; the details can be found in **[16]** and **[19]**; the necessity is proved in **[18]**. Observe that the above theorem completely settles the question of subellipticity in the case of dimension $n = 2$, since then the Levi form is 1×1 and hence diagonalizable.

According to a theorem of Sweeney (see **[33]**) subellipticity of a Neumann problem is independent of the metric; hence we will choose a metric such that the vector fields $L_1, \cdots, L_{n-1}, L_n$ form an orthonormal basis of $T_Q^{1,0}(U)$ for each $Q \in U$ and U is some suitably small neighborhood of P. Now let $\omega^1, \cdots, \omega^n$ be the dual basis. Let $\varphi \in \dot{\mathscr{D}}$ with $\text{supp}(\varphi) \subset U \cap \bar{\Omega}$. Then $\varphi = \varphi_j \bar{\omega}_j$ and $\varphi_n = 0$ on $U \cap b\Omega$. We then have

$$\sum_{k,j} \|\bar{L}_k \varphi_j\|^2 + \sum_j \int_{b\Omega} \lambda_j |\varphi_j|^2 \, dS \leqq C(Q(\varphi, \varphi) + \|\varphi\|^2). \tag{22}$$

Now we have for $k \leqq n - 1$

$$\|L_k u\|^2 \leqq C\left(\|\bar{L}_k u\|^2 + \int_{b\Omega} \lambda_k |u|^2 \, dS + \|u\|^2\right). \tag{23}$$

Thus we see that (22) implies

$$\sum_{k,j} \|\bar{L}_k \varphi_j\|^2 + \sum_1^{n-1} \|L_j \varphi_j\|^2 + \sum_1^n \|L_j \varphi_n\|^2 \leqq C(Q(\varphi, \varphi) + \|\varphi\|^2) \tag{24}$$

and in case of strong pseudoconvexity (i.e., $\lambda_j > 0$):

(25) $$\sum_{k,j} \|\bar{L}_k\varphi_j\|^2 + \sum_1^{n-1} \|L_k\varphi_j\|^2 + \sum_1^{n} \|L_j\varphi_n\|^2 \leqq C(Q(\varphi, \varphi) + \|\varphi\|^2).$$

Subellipticity then follows from the following proposition.

PROPOSITION. *If $L_1, \cdots, L_{n-1}, \bar{L}_1, \cdots, \bar{L}_{n-1}, T$ are linearly independent vector fields in a neighborhood of the origin in $\boldsymbol{R}^{2n-1}$ and if* (21) *holds and further if L_1 is of finite order at the origin, then there exists a neighborhood U of the origin and $\varepsilon > 0$, $C > 0$ such that*

(26) $$\|u\|_\varepsilon^2 \leqq C\left(\sum_1^{n-1} \|\bar{L}_k u\|^2 + \|L_1 u\|^2 + \|u\|^2 \right),$$

for all $u \in C_0^\infty(U)$. Furthermore, the constant C depends only on the bounds of a finite number of derivatives of the L's. In case the λ_j in (21) *are positive we also have*

(27) $$\|u\|_{1/2}^2 \leqq C\left(\sum_1^{n-1} (\|\bar{L}_k u\|^2 + \|L_k u\|^2) + \|u\|^2 \right)$$

for all $u \in C_0^\infty(U)$.

To go from the proposition to the theorem we simply set $u = \varphi_1, j = 1$ and integrate on surfaces $r =$ const and then with respect to r, obtaining the estimate for φ_1, and proceed similarly for φ_j.

The estimate (26) is proved with $\varepsilon = 2^{-p}$, where p is the order of L_1, by the methods of [**20**]. In the case $n = 2$ it follows, from recent results of Rothschild and Stein (see [**21**]), that (26) holds with $\varepsilon = 1/(p + 1)$ and it is proved by Greiner in [**18**] that this is best possible. We should remark here the estimates (26) and (27) imply hypoellipticity of the operators $\sum_{k=1}^{n-1} L_k \bar{L}_k + \bar{L}_1 L_1$ and $\sum_1^{n-1}(L_k\bar{L}_k + \bar{L}_k L_k)$. The second is a simple example of the class of hypoelliptic operators introduced by Hörmander (see [**22**]). Operators having similar properties as the first have not as yet been systematically studied.

To conclude our discussion of subellipticity of the $\bar{\partial}$ Neumann problem we wish to mention the following result (see [**19**] and [**23**]).

THEOREM. *If Ω is pseudoconvex, $P \in b\Omega$, and if there is a neighborhood U of P such that the function r is analytic in U and the Levi form is positive definite in $U \cap b\Omega - \{P\}$ then the $\bar{\partial}$ Neumann problem is subelliptic at P.*

V. In this section we will return to the discussion of the induced Cauchy-Riemann equations.

DEFINITION. Let X be a real manifold of dimension m. A *CR structure* is given by a subbundle $T^{1,0}(X)$ of the complexified tangent bundle $\boldsymbol{C}T(X)$ that satisfies the condition that $T^{1,0}(X) \cap \overline{T^{1,0}(X)} = 0$. The *codimension* k of the CR structure is defined to be the codimension of $T^{1,0}(X) + T^{0,1}(X)$ in $\boldsymbol{C}T(X)$. We say that the CR structure is *integrable* if whenever $L, L' \in (T_P^{1,0}(X))^{\text{germ}}$ then $[L, L'] \in T_P^{1,0}(X)$.

We set $\Lambda^q(X) = \Lambda^q(\boldsymbol{C}T(X)^*)$, the complex-valued q-forms on X, and $\Lambda^{0,q}(X) = \Lambda^q(T^{0,1}(X)^*)$, where $T^{0,1}(X) = \overline{T^{1,0}(X)}$. Then we have the abstract version of (14) given in the first section:

(1) $$C^{0,q}(X) \xrightarrow{i_q} \Lambda^q(X) \xrightarrow{r_q} \Lambda^{0,q}(X) \to 0,$$

where r_q is induced by the dual of the injection mapping $T^{0,1}(X) \to CT(X)$ and $C^{0,q}(X)$ is the kernel of r_q. The integrability condition implies

$$r_{q+1} \circ d \circ i_q = 0 \tag{2}$$

and hence there exists a unique map $\bar{\partial}_b : (\Lambda^{0,q}(X))^{\text{germ}} \to (\Lambda^{0,q+1}(X))^{\text{germ}}$ such that

$$\bar{\partial}_b \circ r_q = r_{q+1} \circ d. \tag{3}$$

Observe that if the codimension of the CR structure is zero then it gives an almost-complex structure. We will restrict our attention to the case of integrable CR structures of codimension one. In this case we can define the Levi form as follows:

DEFINITION. If X is an integrable CR manifold of codimension 1, then the subbundle $D^{0,1}(X)$ has fiber dimension one and so for any $P \in X$ there is a neighborhood U of P and a nonvanishing one-form Υ defined on U; Υ is characterized by the fact that it annihilates $T^{1,0}(X) + T^{0,1}(X)$; we will choose Υ so that $\gamma = -\bar{\gamma}$. The *Levi form* at P is then the hermitian form on $T^{1,0}_P(X)$ defined by

$$\langle (d\Upsilon)_P, L_\wedge \bar{L} \rangle \tag{4}$$

for $L \in T^{1,0}_P(X)$.

It is clear then that the number of zero eigenvalues, the number of eigenvalues of one sign and the number of eigenvalues of the opposite sign at P are all invariants of the CR structure.

For each $P \in X$ we define a hermitian metric on $CT_P(X)$ such that $T^{1,0}_P(X)$ is orthogonal to $T^{0,1}_P(X)$ and whose dependence on P is C^∞. This metric then induces an L_2-inner product in $\Lambda^{0,q}(X)$ and we denote the formal adjoint of $\bar{\partial}_b : \Lambda^{0,q-1}(X) \to \Lambda^{0,q}(X)$ by $\vartheta_b : \Lambda^{0,q}(X) \to \Lambda^{0,q-1}(X)$. We define the Laplace operator $\Box_b$: $\Lambda^{0,q}(X) \to \Lambda^{0,q}(X)$ by

$$\Box_b = \bar{\partial}_b \vartheta_b + \vartheta_b \bar{\partial}_b \tag{5}$$

and set $\mathscr{H}^{0,q} = \{\varphi \in \Lambda^{0,q}(X) \mid \Box_b \varphi = 0\}$.

DEFINITION. If X is a CR manifold and $P \in X$ we say $\Box_b$ on q-forms is *subelliptic at P* if there exist constants $\varepsilon > 0$, $C > 0$ and a neighborhood U of P such that

$$\|\bar{\partial}_b \varphi\|^2 + \|\vartheta_b \varphi\|^2 + \|\varphi\|^2 \geqq C \|\varphi\|^2_\varepsilon \tag{6}$$

for all $\varphi \in \Lambda^{0,q}(X)$ with $\operatorname{supp}(\varphi) \subset U$.

The following characterization of subellipticity with $\varepsilon = \frac{1}{2}$ is proved in **[15]**.

THEOREM. (6) *holds with* $\varepsilon = \frac{1}{2}$ *if and only if the Levi form at* P *satisfies the following condition. Either the Levi form at* P *has at least* $\max(n - q, q + 1)$ *eigenvalues of the same sign or it has at least* $q + 1$ *eigenvalues of one sign and at least* $q + 1$ *eigenvalues of the opposite sign.*

The main consequences of this theorem are the existence and good control of the operator N_b which consequently yield solutions to the $\bar{\partial}_b$ equation and also control of the projection operators on the null space of $\bar{\partial}_b$ in both $\Lambda^{0,q}_b$ and $\Lambda^{0,q-1}_b$. In particular when $q = 1$ we obtain pseudolocality of the Cauchy-Szegö operator. In case $X = b\Omega$ with Ω a domain in an n-dimensional complex manifold and X compact, it can be shown that the estimate (6) with $\varepsilon = \frac{1}{2}$ holds for $\varphi \in \Lambda^{0,q}_b(X)$ and

$\varphi \perp \mathscr{H}_b^{0,q}$, provided that the Levi form satisfies the weaker conditions that at each point it has either at least $\max(n - q - 1, q)$ eigenvalues of the same sign or at least q eigenvalues of one sign and at least q eigenvalues of the opposite sign. This shows that the ranges of the closures of $\Box_b$ and $\bar{\partial}_b$ on $\Lambda_b^{0,q-1}$ and $\Lambda^{0,q}$ are closed. It appears that this result has also been proved in the general CR case by Sjöstrand and Boutet de Monvel.

Recently F. Trèves proved (see **[25]**) that the conditions given in the above theorem are sufficient for the local solvability of the equation $\bar{\partial}_b\varphi = \alpha$ with $\bar{\partial}_b = 0$; they are also necessary if the Levi form is nondegenerate. The important local problem is when can an integrable CR manifold of dimension $2n - 1$ and codimension 1 be locally imbedded in $\boldsymbol{C}^n$? This amounts to finding n functionally independent solutions of the equation $\bar{\partial}_b u = 0$. An example of L. Nirenberg (see **[29]**) shows that this is not possible in case $n = 2$. Recently Boutet de Monvel solved the problem (see **[28]**) in case X is compact with $n \geqq 3$, whose Levi form has all eigenvalues of the same sign.

THEOREM (BOUTET DE MONVEL). *If X is a compact CR manifold of codimension one and $\dim_{\boldsymbol{R}} X \geqq 5$ and if the Levi form has all eigenvalues of the same sign, then for each $P \in X$ there exist functions $h_1, \cdots, h_n \in C^\infty(X)$ such that $\bar{\partial}_b h_j = 0$, $h_j(P) = 0$ and $(dh_1)_P, \cdots, (dh_n)_P$ are linearly independent. Furthermore, there exist a finite number of C^∞ functions that separate points and satisfy $\bar{\partial}_b h = 0$.*

OUTLINE OF PROOF. First we show that there exists a function $g_P \in C^\infty(X)$ such that $(dg_P)_P \neq 0$, $\bar{\partial}_b g_P$ has a zero of infinite order and $\mathrm{Re}(g_P(Q)) \geqq Cd(P, Q)^2$, where $d(P, Q)$ denotes the distance between P and Q. In case $X \subset \boldsymbol{C}^n$ we can define g_P to be the Levi polynomial near P and extend it to be sufficiently large to satisfy the inequality. In the general case we can find functions $\psi_1, \cdots, \psi_n \in C^\infty(X)$ such that $\psi_j(P) = 0$, $(d\psi_1)_{P}, \cdots, (d\psi_n)_P$ are linearly independent and $\bar{\partial}_b\psi_j$ have zeros of infinite order at P. Then the image of X is strongly pseudoconvex at $(\psi_1(P), \cdots, \psi_n(P))$; hence we can take for g_P near P the pull-back of the corresponding Levi polynomial and extend it to X so it satisfies the inequality.

We define ω_λ by

$$\omega_\lambda = \bar{\partial}_b(e^{-\lambda g_P}) = -\lambda e^{-\lambda g_P}\bar{\partial}_b g_P. \tag{7}$$

Thus as $\lambda \to \infty$, $\omega_\lambda \to 0$ with all derivatives since

$$|(\omega_\lambda)_Q| \leqq C_N \lambda^{1-N}\{(\lambda d(P, Q)^2)^N e - (\lambda d(P, Q)^2)\} \tag{8}$$

and the term in $\{\ \}$ is bounded, similarly with higher derivatives of λ. Set $f_\lambda = H_b(e^{-\lambda g_P})$; thus we have

$$f_\lambda = e^{-\lambda g_P} - \vartheta_b N_b \omega_\lambda. \tag{9}$$

Hence $\bar{\partial}_b f_\lambda = 0$ and if $Q = P$ then $f_\lambda(P)$ and $f_\lambda(Q)$ can be made arbitrarily close to 1 and 0, respectively, by choosing λ sufficiently large (since N_B is continuous in the C^∞ topology). Thus we can separate points as required. The functions h_j are given by setting $h_j = H_b(\psi_j e^{-\lambda g_P})$ and arguing as above we see that the h_j satisfy the conditions in the theorem.

To conclude this paper we wish to make two observations. First, here we have concentrated on differentiability properties measured by the Sobolev norms. In recent years much more delicate estimates (involving Hölder norms) have been obtained in the strongly pseudoconvex case; see **[8]** and **[30]** for results in this direction and for further references. The outstanding unsolved problem is the question of analyticity. The global problem has been settled (see **[30]** and **[31]**), and the more important local problem is apparently on the verge of being solved by D. Tartakoff.

REFERENCES

1. P. C. Greiner, J. J. Kohn and E. M. Stein, *Necessary and sufficient conditions for the solvability of the Lewy equation*, Proc. Nat. Acad. Sci. U.S.A. **72** (1975), 3287–3289.
2. H. Lewy, *An example of a smooth linear partial differential equation without solution*, Ann. of Math (2) **66** (1957), 155–158. MR **19**, 551.
3. M. Sato, *Regularity of hyperfunction solutions of partial differential equations*, Proc. Internat. Congress Math. (Nice, 1970), vol. 2, Gauthier-Villars, Paris, 1971, pp. 785–794.
4. E. M. Stein, *Boundary behavior of holomorphic functions in several complex variables*, Math. Notes, Princeton Univ. Press, Princeton, N. J., 1972.
5. A. Korányi and S. Vági, *Singular integrals in homogeneous spaces and some problems of classical analysis*, Ann. Scuola Norm. Sup. Pisa **25** (1971), 575–648.
6. J. J. Kohn, *Homomorphic extensions of orthogonal projections into holomorphic functions*, Proc. Amer. Math. Soc. **52** (1975), 333–336.
7. J. J. Kohn and H. Rossi, *On the extension of holomorphic functions from the boundary of a complex manifold*, Ann. of Math. (2) **81** (1965), 451–472. MR **31** #1399.
8. G. B. Folland and E. M. Stein, *Estimates for the $\bar\partial_b$ complex and analysis on the Heisenberg group*, Comm. Pure Appl. Math. **27** (1974), 429–522.
9. J.-P. Serre, *Un théorème de dualité*, Comment. Math. Helv. **29** (1955), 9–26. MR **16**, 736.
10. L. Ehrenpreis, *A new proof and an extension of Hartog's theorem*, Bull. Amer. Math Soc. **67** (1961), 507–509. MR **24** #A1511.
11. L. Hörmander, *An introduction to complex analysis in several variables*, Van Nostrand, Princeton, N. J., 1966. MR **34** #2933.
12. J. J. Kohn and L. Nirenberg, *A pseudo-convex domain not admitting a holomorphic support function*, Math. Ann. **201** (1973), 265–268. MR **48** #8850.
13. J. J. Kohn, *Propagation of singularities for the Cauchy-Riemann equations*, C.I.M.E. Conf. Complex Analysis (1973), Edizioni Cremonese, Rome, 1974, pp. 179–280.
14. L. Hörmander, *Estimates and existence theorems for the $\bar\partial$ operator*, Acta Math. **113** (1965), 89–152. MR **31** #3691.
15. G. B. Folland and J. J. Kohn, *The Neumann problem for the Cauchy-Riemann complex*, Ann. of Math. Studies, no. 75, Princeton Univ. Press, Princeton, N. J., 1972.
16. J. J. Kohn, *Boundary behavior of $\bar\partial$ on weakly pseudo-convex manifolds of dimension two*, J. Differential Geometry **6** (1972), 523–542. MR **48** #727.
17. ———, *Global regularity for $\bar\partial$ on weakly pseudoconvex manifolds*, Trans. Amer. Math. Soc. **181** (1973), 273–292. MR **49** #9442.
18. P. Greiner, *Subelliptic estimates for the $\bar\partial$-Neumann problem in $\mathbf{C}^2$*, J. Differential Geometry **9** (1974), 239–250. MR **49** #9441.
19. J. J. Kohn, *Subellipticity of the $\bar\partial$-Neumann problem on weakly pseudo-convex domains*, Proc. Recoutre sur plusiers var. compl. et le prob. de Neumann, Montreal (1974) (to appear).
20. ———, *Complex hypoelliptic operators*, Symposia Matematica, vol. VII (Convegno sulle Equazioni Ipoellittiche e Spazi Funzionali, INDAM, Roma, Gennaro 1970), Academic Press, London, 1971, pp. 459–468. MR **50** #5857.
21. L. P. Rothschild and E. M. Stein, *Hypoelliptic differential operators and nilpotent groups*, Acta Math. (to appear).
22. L. Hörmander, *Hypoelliptic second order differential equations*, Acta. Math. **119** (1967), 147–171. MR **36** #5526.

23. J. J. Kohn, *Subellipticity on pseudo-convex domains with isolated degeneracies*, Proc. Nat. Acad. Sci. U.S.A. **71** (1974), 2912–2914. MR **50** #7840.

24. A. Andreotti and D. Hill, *E. E. Levi convexity and the Hans Lewy problem.* I, II, Ann. Scuola Norm. Sup. Pisa **26** (1974), 325–363, 747–806.

25. F. Trèves, *Study of a model in the theory of overdetermined systems in one unknown function* (to appear).

26. C. Fefferman, *The Bergmann kernel and biholomorphic equivalence of pseudo-convex domains*, Invent. Math. **26** (1974), 1–65. MR **50** #2562.

27. N. Kerzman, *The Bergmann kernel function: Differentiability at the boundary*, Math. Ann. **195** (1972), 149–158. MR **45** #3762.

28. L. Boutet de Monvel, *Intégration des equations de Cauchy-Riemann induites formelles*, Sèminaire Goulaouic-Lions-Schwartz 1974–1975, Exposé No. 9.

29. L. Nirenberg, *On a question of Hans Lewy*, Uspehi Mat. Nauk **29** (1974), 241–251.

30. P. C. Greiner and E. M. Stein, *A parametrix for the $\bar{\partial}$-Neumann problem* (to appear).

31. D. Tartakoff, *On the real analyticity of solutions to $\square_b$ on compact manifolds*, Proc. Sympos. Pure Math., vol. 30, part 1, Amer. Math. Soc., Providence, R. I., 1977, pp. 205–211.

32. M. Derridj and D. Tartakoff, *Sur la régularité analytique globale des solutions du problème de Neumann pour $\bar{\partial}$*, Séminaire Goulaouic-Schwartz Ecole Polytechnique, Nov. 1976.

33. W. J. Sweeney, *Coerciveness in the Neumann problem*, J. Differential Geometry **6** (1971/72), 375–393. MR **45** #7757.

PRINCETON UNIVERSITY

SEMINAR SERIES

COMPACT COMPLEX MANIFOLDS

Proceedings of Symposia in Pure Mathematics
Volume 30, 1977

COMPACTIFYING C^n

LAWRENCE BRENTON AND JAMES A. MORROW*

1. Introduction. Let X be a connected compact complex manifold. X is a *compactification of* C^n if there is a (proper, closed) analytic subset A of X such that $X - A = C^n$. We are interested in characterizing compactifications of C^n. One can easily prove that A is of pure codimension 1 and is connected. By resolution of singularities, given a compactification of C^n a suitable proper modification will produce another compactification of C^n in which the set A is a union of manifolds with normal crossings only. In the case $n \leqq 2$, such compactifications are completely classified in [**5**]. Here, we were interested in the case $n > 2$. Compactifications of C^n have been studied previously by Remmert and Van de Ven [**7**], and by Van de Ven [**8**].

2. The nonsingular case. If A is a manifold then we can compute its cohomology. In fact,

PROPOSITION 2.1. *Suppose* (X, A) *is a compactification of* C^n *with* A *nonsingular. Then for the integral cohomology,*

$$H^*(X) \simeq H^*(P^n) \quad \textit{(as rings), and}$$
$$H^*(A) \simeq H^*(P^{n-1}) \quad \textit{(as rings).}$$

PROOF. This proposition follows by applying the Thom-Gysin sequence to the manifold A and the circle bundle σ associated to its normal bundle. Since σ is homotopy equivalent to a sphere we find that cup product with the Euler class of σ defines an isomorphism: $H^k(A) \simeq H^{k+2}(A)$. This same sequence implies that $H^1(A) = 0$. Since $H^0(A) = Z$ (A is connected), we conclude that $H^*(A) \simeq H^*(P^{n-1})$. The cohomology ring of $H^*(X)$ is now easily computed by using the cohomology sequence of the pair (X, A) and Poincaré duality.

AMS (MOS) subject classifications (1970). Primary 32J05.

*Second author supported by NSF grant MPS 73–08658–A02.

© 1977, American Mathematical Society

In the case that $n \leqq 3$ we can actually prove much more. We have

THEOREM 2.2. *Suppose (X, A) is a compactification of $\boldsymbol{C}^n$ with A nonsingular and $n \leqq 3$. Then $A = \boldsymbol{P}^{n-1}$ and $X = \boldsymbol{P}^n$.*

PROOF. The theorem is trivial for $n = 1$. For $n = 2$ it is proved in [5]. For $n = 3$, we only sketch the proof, the details of which will appear elsewhere. A crucial point in the proof is to show that the hypotheses imply that X is projective algebraic. We first prove that A is projective. Let b^+ be the number of positive eigenvalues of the cup product as a quadratic form on $H^2(A; \boldsymbol{Z})$. Now using a theorem in Kodaira [3] we conclude that $b^+ = 1$. Using this and another theorem in Kodaira [3], we conclude that A is projective. Next we consider $[A]$, the bundle of the divisor $A \subset X$. By considering $H^2(A) \simeq \boldsymbol{Z}$ and the fact that A is projective we find that $[A]|_A$ is either positive, negative, or trivial. We easily check that $[A]|_A$ is not trivial. It is not negative, else Grauert's theorem on contractibility [2] would allow us to contract A to a point in a compact complex space $\tilde{X}$. Now Hartogs' theorem will allow us to construct a nonconstant holomorphic function on $\tilde{X}$. This contradiction forces us to conclude that $[A]|_A$ is positive. Now a theorem in [6] allows us to conclude that in fact $[A]$ is positive on all of X and hence X is projective.

Next one computes the first Chern class c_1 of X. This is done with the aid of the Riemann-Roch formula and the adjunction formula. We find that $c_1 = 4g$ where g is the positive generator of $H^2(X; \boldsymbol{Z})$. Applying a theorem of Kobayashi and Ochiai we conclude that $X = \boldsymbol{P}^3$ and $A = \boldsymbol{P}^2$.

One would naturally conjecture that Theorem 2.2 is true for all n. At present we cannot even prove that an X satisfying the hypothesis of the theorem is projective.

3. The rational singularity case. Let $p \in A$ be an isolated singular point. Let $\pi: \tilde{A} \to A$ be a resolution of the singularity at p. Then p is a *rational singularity* if $(R^1\pi_*\mathcal{O}_{\tilde{A}})_p = 0$. We now wish to consider the case that A has only rational singularities. We have the following result.

THEOREM 3.1. *Suppose (X, A) is a compactification of $\boldsymbol{C}^3$ where A has only rational singularities. Then X is projective.*

In fact we can probably prove much more. We believe that X is either $\boldsymbol{P}^3$, a quadric, or a third kind of surface (X, A) where the space A will be described later in this paper. In our search for a proof of this result we have discovered two theorems which are of independent interest. They are stated as follows.

THEOREM 3.2. *Let A be a normal compact two-dimensional analytic space with only rational singular points. If A is Moišeson, then A is projective algebraic.*

THEOREM 3.3. *Let A be a normal compact two-dimensional analytic space with only rational singularities. Suppose that $H^*(A; \boldsymbol{Z}) \simeq H^*(\boldsymbol{P}^2; \boldsymbol{Z})$ and that $H^2(A; \boldsymbol{Z})$ is generated by the Poincaré dual of an irreducible curve in A. Then the nonsingular model $\tilde{A}$ of A is rational (bimeromorphic to $\boldsymbol{P}^2$).*

REMARK. Theorem 3.2 was proved earlier by M. Artin [1]. We were originally unaware of this result and we feel that our proof (which differs from Artin's) is of independent interest.

SKETCH OF PROOF OF THEOREM 3.2. Let $\pi: \tilde{A} \to A$ resolve the singularities of A. Since A is Moišeson, so is $\tilde{A}$. Thus $\tilde{A}$ is a compact complex two-dimensional manifold with two algebraically independent meromorphic functions. Then by a theorem of Chow and Kodaira $\tilde{A}$ is projective. Let S be the singular locus of A. Then $\pi^{-1}(S) = \bigcup_{j=1}^{s} C_j$ where each C_j is an irreducible curve. (We may assume that each C_j is smooth, that all intersections are transverse, and that no three curves pass through the same point.) By relating the topology of $\tilde{A}$ to that of A we then can show that there is a positive line bundle F on $\tilde{A}$, and a line bundle G on A such that

$$(1) \qquad F = \pi^*(G) \otimes \left(\bigotimes_{j=1}^{s} [C_j]^{n_j} \right),$$

where $[C_j]$ is the bundle of the divisor C_j and the n_j are integers. By using the facts that F is positive and $(C_j \cdot C_k)$ is negative definite (where $C_j \cdot C_k$ is the intersection multiplicity) we show that each integer n_j is negative.

We want to show that G is positive. For each irreducible positive dimensional variety $V \subset A$ we produce a positive integer k such that $G^k|_V$ has a section which is not identically zero, but has some zeros on V. By a theorem of Grauert [2], this will prove that G is positive and hence A is projective. Using this same criterion on $F^k|\pi^{-1}(V)$ and using the fact that each $n_j < 0$ in (1), we can get such a section on V.

SKETCH OF PROOF OF THEOREM 3.3. Our hypothesis means that there is an irreducible curve $D \subset A$ so that $c([D]) = g \in H^2(A; \mathbf{Z}) \simeq \mathbf{Z}$ is a generator and $g^2 = 1$ via the canonical isomorphism $H^4(A; \mathbf{Z}) \simeq \mathbf{Z}$. Let $\pi: \tilde{A} \to A$ resolve the singularities of A as before. Let $\tilde{D} = \pi^{-1}(D)$. Then $(\tilde{D})^2 = 1$, so a theorem of Kodaira [3] implies that $\tilde{A}$ is projective. By Theorem 3.2, A is projective. To prove $\tilde{A}$ is rational we use Castelnuovo's criterion. Let $q = \dim H^1(\tilde{A}; \mathcal{O})$, and $P_2 = \dim H^0(\tilde{A}; K^2)$ where K is the canonical bundle of $\tilde{A}$. We claim $q = P_2 = 0$, which will complete the proof. Now one can show that the first Betti number $b_1(\tilde{A}) = 0$. Since $\tilde{A}$ is Kähler, $q = 0$. An argument involving the rationality of the singularities of A shows that $K = [\tilde{D}]^{-c_1}$. We claim that $c_1 > 0$. If not, K would have a section, but using Kodaira [3] we can show that $H^0(\tilde{A}, K) = 0$. So $c_1 > 0$ and $K^2 = [\tilde{D}]^{-2c_1}$. Hence $P_2 = 0$.

SKETCH OF PROOF OF THEOREM 3.1. We will prove that A is projective. It will then follow as in the proof of Theorem 2.2 that X is projective. If we show that $\tilde{A}$ is projective then, by Theorem 3.2, A is projective. We use a topological computation and some results of Kodaira [3] to prove that $b^+(\tilde{A}) = 1$ and $H^0(\tilde{A}; K) = 0$. Then another result in Kodaira [3] implies that $\tilde{A}$ is projective.

We conclude this section by describing the example mentioned in the paragraph following the statement of Theorem 3.1. We will construct a space Q which satisfies the hypotheses of Theorem 3.3. Let $\Gamma \subset \mathbf{P}^2$ be a nonsingular torus and let p_1 be an inflection point on Γ. Let L_1 be tangent to Γ at p_1. Let L_2 pass through p_1 and be tangent to Γ at p_2. The graph of these curves is as follows:

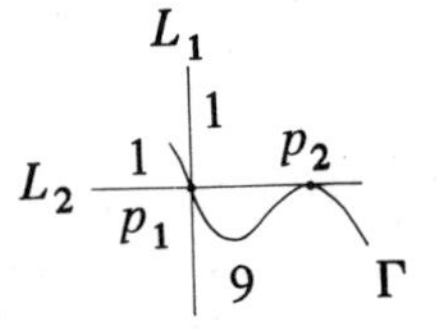

where the numbers indicate self-intersections. We perform the following sequence of blow-ups:

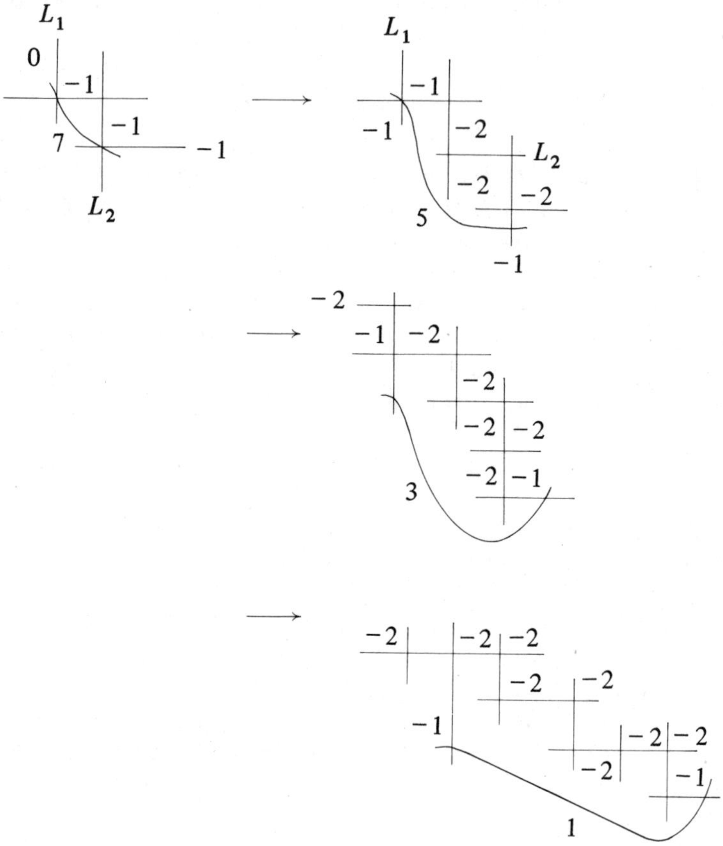

Now blow down all of the curves with self-intersection −2 whose graph is

to get a rational double point. The total space Q thus produced is bimeromorphic to P^2.

Question. Is there a compact complex manifold X with a subvariety $A \subset X$ such that $A \simeq Q$ and $X - A = C^3$?

4. Embeddings of P^n. Let X be a compact complex space of dimension $n+1$. We are interested in the extent to which an embedding of P^n in X determines X.

THEOREM 4.1. *Suppose X is a normal compact complex space of dimension $n+1 \geqq 4$ which is nonsingular in a neighborhood of A which is a submanifold isomorphic to P^n. Suppose that $[A]|_A$ is positive. Then X is obtained by point modification from H, where H is the projectivized normal bundle of A with the "∞" section collapsed to a point. These spaces H are in* 1-1 *correspondence with the positive integers.*

PROOF. We can show that there is a neighborhood of A which is isomorphic to a neighborhood of the zero section of $[A]|_A = N$. (Notice that N is just some positive power of the bundle of a hyperplane on P^n.) Since N is positive $X-A$ is holomorphically convex and we collapse each maximal compact subset of $X-A$ to a point obtaining a compact complex space $\tilde{X}$ such that $\tilde{X}-A$ is Stein. Now projectivize N by adding the ∞ section and then collapse this divisor to a point in a compact complex space H. Now $H-A$ and $\tilde{X}-A$ are Stein and biholomorphic outside of corresponding compact subsets. Then a version of Hartogs' theorem implies that we may extend this biholomorphism to see that $\tilde{X}$ is biholomorphic to H.

COROLLARY 4.2. *If $P^n = A$ is positively embedded in a compact complex manifold X of dimension $n+1 \geqq 4$ and if $X-A$ is Stein, then $X-A = C^{n+1}$ and $X = P^{n+1}$.*

The proof of this corollary uses a result of Moišeson **[4]** characterizing exceptional divisors of the first kind. If we do not assume that the embedding is positive we have

PROPOSITION 4.3. *Suppose $P^{n-1} \doteq A \subset X$ of dimension n with $X-A$ Stein and $H_c^k(X-A) = H_c^{k+1}(X-A) = 0$ for some k with $n < k < k+1 < 2n$. Then $X-A = C^n$ and $X = P^n$.*

We omit the proof of this proposition.

BIBLIOGRAPHY

1. M. Artin, *Some numerical criteria for contractibility of curves on algebraic surfaces*, Amer. J. Math. **84** (1962), 485–496. MR **26** #3704.
2. H. Grauert, *Über Modifikationen und exzeptionelle analytische Mengen*, Math. Ann. **146** (1962), 331–368. MR **25** #583.
3. K. Kodaira, *On the structure of compact complex analytic surfaces*. I, Amer. J. Math. **86** (1964), 751–798. MR **32** #4708.
4. B. G. Moĭšezon, *On n-dimensional compact complex varieties with n algebrically independent meromorphic functions*. I, II, III, Izv. Akad. Nauk SSSR Ser. Mat. **30** (1966), 133–174, 345–386, 621–656; English transl., Amer. Math. Soc. Transl. (2) **63** (1967), 51–177. MR **35** #7355a,b,c.
5. J. Morrow, *Minimal normal compactifications of C^2*, Rice Univ. Studies **59** (1973), no. 1, 97–112. MR **48** #11580.
6. J. Morrow and H. Rossi, *Some theorems of algebraicity for complex spaces*, J. Math. Soc. Japan **27** (1975). 167–183.

7. R. Remmert and T. Van de Ven, *Zwei Sätze über die komplex-projektive Ebene*, Nieuw. Arch. Wisk. (3) **8** (1960), 147–157. MR **24** #A2397.

8. A. J. H. M. Van de Ven, *Analytic compactifications of complex homology cells*, Math. Ann. **147** (1962), 189–204. MR **25** #3548.

Wayne State University

University of Washington

Proceedings of Symposia in Pure Mathematics
Volume 30, 1977

ON THE CANCELLATION PROBLEM FOR COMPACT COMPLEX-ANALYTIC MANIFOLDS

JÉRÔME BRUN

We give in this paper the following cancellation results:

THEOREM 1. *Let* $\boldsymbol{P}_n$ *be complex projective space. If* A *and* B *are two isomorphism classes of compact complex-analytic manifolds such that* $A \times \boldsymbol{P}_n = B \times \boldsymbol{P}_n$, *then* $A = B$.

THEOREM 2. *Let* Γ *be a compact Riemann surface of genus* $\geqq 2$. *If* A *and* B *are two isomorphism classes of compact complex-analytic manifolds such that* $A \times \Gamma = B \times \Gamma$, *then* $A = B$.

The proofs use holomorphic vector fields (or holomorphic 1-forms) and their decomposition on a product of compact manifolds. Note that these theorems hold as well in the algebraic category (see Serre [**1**]). Aron Simis in [**2**] has shown, in the algebraic case, that if $A \times \boldsymbol{P}_1 = B \times \boldsymbol{P}_1$ and if $\dim A = 1$ or $\dim A = 2$ and A is nonruled, then $A = B$.

I would like to thank Professor A. Van de Ven for having introduced me to this subject and for the helpful advice he gave to me during his stay in Nice.

Before giving the proofs, let us mention a few points about the cancellation problem:

(1) It is more reasonable to hope for positive theorems of cancellation in the compact case than in the noncompact case; for instance, examples are known of isomorphism classes A and B of complex-analytic manifolds such that $A \neq B$ and $A \times \boldsymbol{C} = B \times \boldsymbol{C}$.

(2) However, about tori, the lack of isolated zeros for holomorphic vector fields (or holomorphic 1-forms) limits our method to giving only this partial result: If T is a complex torus and A, B are two compact complex-analytic manifolds such that

AMS (*MOS*) *subject classifications* (1970). Primary 32C10; Secondary 14M15.

© 1977, American Mathematical Society

$A \times T$ and $B \times T$ are isomorphic, then either A and B are isomorphic, or A and B are bundles with tori isogenous to T as fibres. In particular, I would like to ask the following question:

"Let T_1, T_2, T be one-dimensional complex tori such that $T_1 \times T$ and $T_2 \times T$ are analytically equivalent. Does it follow that T_1 and T_2 are analytically equivalent?"

PROOF OF THEOREM 1. The result is trivial for $n = 0$. When $n \geqq 1$, we can assume by induction that the result is true up to $n - 1$.

Suppose now that $A \times \boldsymbol{P}_n = B \times \boldsymbol{P}_n$. We can assume, of course, that A and B are connected.

Note first that it is obvious that A and B have the same dimension; let $m = \dim A = \dim B$.

In showing that $A = B$, difficulties arise when A and B contain several copies of $\boldsymbol{P}_n$. So we shall face at once these difficulties and consider the following assertion, where $k \in \boldsymbol{N}$:

One of the following is true:
(i) $A = B$.
(S_k) (ii) There exist compact manifolds A_k and B_k of dimension $m-kn$ such that $A = A_k \times \boldsymbol{P}_n^k$ and $B = B_k \times \boldsymbol{P}_n^k$.

Since (ii) is impossible for k large enough, it is sufficient to prove (S_k) for all $k \in \boldsymbol{N}$, in order to achieve the proof of the theorem.

(S_0) is true. Assume by induction (S_k). To show (S_{k+1}), we can assume we are in case (ii) of (S_k). We deduce from (ii) and from $A \times \boldsymbol{P}_n = B \times \boldsymbol{P}_n$ that there exists an analytic isomorphism $f_k : A_k \times \boldsymbol{P}_n^{k+1} \to B_k \times \boldsymbol{P}_n^{k+1}$.

Let $H \subset \boldsymbol{P}_n$ be a hyperplane. Then there exists a holomorphic vector field on $\boldsymbol{P}_n$ vanishing only on H. Let ξ be this field lifted to $(A_k \times \boldsymbol{P}_n^k) \times \boldsymbol{P}_n$, so that $\{\xi = 0\} = A_k \times \boldsymbol{P}_n^k \times H$. Let $\eta = (fk)_*(\xi)$ be the image field of ξ on $B_k \times \boldsymbol{P}_n^{k+1}$, so that $\{\eta = 0\} = f_k(A_k \times \boldsymbol{P}_n^k \times H)$, a hypersurface of $B_k \times \boldsymbol{P}_n^{k+1}$. We shall use the following lemma:

LEMMA. *Let $X = X_1 \times \cdots \times X_p$ be a product manifold, each X_i being compact complex-analytic. Let η be a holomorphic vector field on X and $Y = \{\eta = 0\}$. Assume Y is a hypersurface of X. Then, there exists q, $1 \leqq q \leqq p$, and X_q', a hypersurface of X_q, such that $Y = X_1 \times \cdots \times X_{q-1} \times X_q' \times X_{q+1} \times \cdots \times X_p$. Moreover, X_q' is the zero set of a field on X_q.*

PROOF. Since the X_i are compact, $\eta = \sum_{i=1}^p \eta_i$, where η_i is a field on X_i. Moreover $Y = \{\eta = 0\} = \prod_{i=1}^p \{\eta_i = 0\}$, so that $\dim\{\eta = 0\} = \sum_{i=1}^p \dim\{\eta_i = 0\}$. Now $\dim\{\eta = 0\} = (\sum_{i=1}^p \dim X_i) - 1$ and $\dim\{\eta_i = 0\} \leqq \dim X_i$. Hence there exists q such that $\{\eta_q = 0\} = X_q'$, a hypersurface of X_q, and $\eta_i \equiv 0$ for $i \neq q$. Q.E.D.

A connected hypersurface of $\boldsymbol{P}_n$ which is the zero set of a field is a hyperplane. From this and from the lemma, we get two cases for $f_k(A_k \times \boldsymbol{P}_n^k \times H)$:

(α) $f_k(A_k \times \boldsymbol{P}_n^k \times H) = B_k \times \boldsymbol{P}_n^k \times H'$, where H' is a hyperplane; that is to say: $A_k \times \boldsymbol{P}_n^k \times \boldsymbol{P}_{n-1} = B_k \times \boldsymbol{P}_n^k \times \boldsymbol{P}_{n-1}$; thus $A_k \times \boldsymbol{P}_n^k = B_k \times \boldsymbol{P}_n^k$ by the induction hypothesis on n, and so $A = B$.

(β) $f_k(A_k \times \boldsymbol{P}_n^k \times H) = B_k' \times \boldsymbol{P}_n^{k+1}$, where B_k' is a hypersurface of B_k.

Now, we can assume we are in this case (β) for any hyperplane H (otherwise, the proof is over). Then, by taking intersections of such hyperplanes H, we get

$f_k(A_k \times \boldsymbol{P}_n^k \times p) = A_{k+1} \times \boldsymbol{P}_n^{k+1}$ where $p \in \boldsymbol{P}_n$ and A_{k+1} (intersection of the B_k''s) is a compact subspace of B_k of dimension $m - (k+1)n$. Thus $A_k \times \boldsymbol{P}_n^k = A = A_{k+1} \times \boldsymbol{P}_n^{k+1}$.

The same proof works if we start from the isomorphism $f_k^{-1} : B_k \times \boldsymbol{P}_n^{k+1} \to A_k \times \boldsymbol{P}_n^{k+1}$, and gives the conclusion $A = B$ or $B = B_{k+1} \times \boldsymbol{P}_n^{k+1}$, which finally gives the desired (S_{k+1}).

PROOF OF THEOREM 2. The method is exactly the same as above, taking holomorphic 1-forms instead of holomorphic vector fields; simplifications occur because of dimension one.

BIBLIOGRAPHY

1. J.-P. Serre, *Géométrie algébrique et géométrie analytique*, Ann. Inst. Fourier (Grenoble) **6** (1955/56), 1–42. MR **18**, 511.

2. A. Simis, *On the cancellation problem for projective varieties*, Comm. Algebra **2** (1974), 535–557.

UNIVERSITÉ DE NICE

Proceedings of Symposia in Pure Mathematics
Volume 30, 1977

VECTOR FIELDS, CHERN CLASSES, AND COHOMOLOGY

J. B. CARRELL* AND D. I. LIEBERMAN**

Given a holomorphic vector field V on a compact Kähler manifold X, one can say a lot about the topology of X from considering the zeroes of V as the subvariety of X defined by the sheaf of ideals $i(V)\Omega^1 \subset \mathcal{O}$. Denote this variety by Z and note that its structure sheaf $\mathcal{O}_Z = \mathcal{O}/i(V)\Omega^1$ is a coherent sheaf of rings on Z.

THEOREM 1. *If* $|p - q| > \dim_C Z \geqq 0$, *then* $H^p(X, \Omega^q) = 0$.

This theorem, which was proved in [**2**], generalizes theorems of Howard [**5**], Sommese [**7**], and Lusztig and Wright [**8**]. It is also related to work of Frankel [**4**] and Kobayashi [**6**, Theorem 11.2, p. 116]. After [**2**] appeared, we were told that Lusztig had originally conjectured this result.

COROLLARY. *When V has isolated zeroes, then X is projective and* $H^{2p}(X, \boldsymbol{C}) = H^p(X, \Omega^p)$.

From now on, Z is assumed to be finite and nonempty. Then the complex defined by contracting p-forms to $(p-1)$-forms via V, i.e., $i(V)\colon \Omega^p \to \Omega^{p-1}$, gives a free resolution of $\mathcal{O}_Z$; hence there exists a spectral sequence with $E_1^{-p,q} = H^q(X, \Omega^p)$ abutting to $H^0(X, \mathcal{O}_Z)$. Theorem 1 follows by the fact that this sequence degenerates at its E_1 term. Consequently the cohomology ring of X is completely determined by Z as follows.

THEOREM 2 [**3**]. *There exists a filtration*

$$(1) \qquad H^0(X, \mathcal{O}_Z) = F_{-n} \supset F_{-n+1} \supset \cdots \supset F_0 = C, \qquad n = \dim_C X,$$

AMS (MOS) subject classifications (1970). Primary 14F25; Secondary 32M99.
Key words and phrases. Vector fields, Chern classes, cohomology.
*Partially supported by an NRC grant.
**Partially supported by a Sloan Foundation fellowship and NSF grant GP-28323A3.

© 1977, American Mathematical Society

such that $F_iF_j \subset F_{i+j}$ *and having the property that*

$$\mathrm{Gr}_{F.}\ H^0(X, \mathcal{O}_Z) \cong H^{\cdot}(X, C).$$

The localization of the cohomology of X to Z may be described as follows: Fix a Leray cover of X, say $\mathcal{U}$, and consider the double Čech complex $C^{\cdot}(\mathcal{U}, \Omega^{\cdot})$ with differentials δ and $i(V)$. Consider the total complex

$$K^j = \bigoplus_{p-q=j} C^p(\mathcal{U}, \Omega^q)$$

with total differential which is given on $C^p(\mathcal{U}, \Omega^q)$ by $D = \delta + (-1)^p i(V)$. Then one has the fundamental isomorphism

$$H^0(K) \cong H^0(X, \mathcal{O}_Z)$$

induced by the natural map of complexes $K^p \to C^p(\mathcal{U}, \mathcal{O}_Z)$ obtained by first projecting K^p onto $C^p(\mathcal{U}, \mathcal{O})$ and following by the natural map onto $C^p(\mathcal{U}, \mathcal{O}_Z)$. In the remainder of this note, we apply these ideas to give new computations of the cohomology ring of $\boldsymbol{P}^n$ and the universal Chern classes of the Grassmannians.

EXAMPLE 1. *The cohomology ring of* $\boldsymbol{P}^n$. Let V be the vector field on $\boldsymbol{P}^n$ generated by the flow $\exp(tM)$ where M is the simple Jordan matrix

$$M = \begin{bmatrix} 0 & \\ \vdots & I_n \\ 0 & \cdots\ 0 \end{bmatrix}. \tag{2}$$

The only zero of V is $[1, 0, \cdots, 0]$, and in standard affine coordinates $w_1, \cdots, w_n$ at $[1, 0, \cdots, 0]$,

$$\begin{aligned} V = {} & (w_2 - w_1^2)\, \partial/\partial w_1 + (w_3 - w_1w_2)\, \partial/\partial w_2 + \cdots \\ & + (w_n - w_1w_{n-1})\, \partial/\partial w_{n-1} - w_1w_n\, \partial/\partial w_n. \end{aligned} \tag{3}$$

Hence $H^0(\boldsymbol{P}^n, \mathcal{O}_Z) = \boldsymbol{C}[w_1]/(w_1^{n+1})$. Note that this is already the cohomology ring of $\boldsymbol{P}^n$ with respect to the natural grading. In order to interpret w_1 as a class on $\boldsymbol{P}^n$, one first finds its image in $H^0(K)$. Recall that the hyperplane section bundle $\mathcal{O}(1)$ on $\boldsymbol{P}^n$ is given by patching data $f_{ij} = z_j/z_i$ on $U_i \cap U_j$, where $U_i = \{[z_0, \cdots, z_n]: z_i \neq 0\}$. We claim that

$$f_{ij}^{-1}\, df_{ij} + z_{i+1}/z_i \tag{4}$$

is a D cocycle of K^0. This can be seen by noting that

$$M = z_1\partial/\partial z_0 + \cdots + z_n\partial/\partial z_{n-1},$$

so

$$i(M) f_{ij}^{-1}\, df_{ij} = z_{j+1}/z_j - z_{i+1}/z_i. \tag{5}$$

The left-hand side of (5) is $i(V) f_{ij}^{-1}\, df_{ij}$ when viewed as a form on $\boldsymbol{P}^n$; hence the claim is established. Clearly the image of (4) in $H^0(\boldsymbol{P}^n, \mathcal{O}_Z)$ is w_1. Since $f_{ij}^{-1}\, df_{ij}$ is the cocycle on $\boldsymbol{P}^n$ representing the Chern class $c(\mathcal{O}(1))$ in $H^1(\boldsymbol{P}^n, \Omega^1)$, it follows that w_1 determines this class in the graded ring associated to the filtration (1). To complete the picture of $H^{\cdot}(\boldsymbol{P}^n, \boldsymbol{C})$ it is necessary to show that the filtration (1) is the natural filtration defined by degree in w_1. It is enough to show that there is no monic polynomial $p(w_1)$ of degree m such that $p(w_1) \in F_{-k}$ and $n \geqq m > k$. Given such p, then $w_1^{n-m} p(w_1) \in F_{-k+m-n} \subset F_{-n+1}$, so $p(w_1) = 0$ in $F_{-n}/F_{-n+1} \cong H^n(\boldsymbol{P}^n, \Omega^n)$ and this contradicts the fact that $c_1(\mathcal{O}(1))^n \neq 0$.

In the general case when X is projective and Z is finite, one may compute the Chern classes of V-equivariant vector bundles in a natural way in $H^0(X, \mathcal{O}_Z)$. One says that a locally free sheaf $\mathcal{E}$ of $\mathcal{O}$-modules is V-equivariant if there exists a $\boldsymbol{C}$-linear map $\tilde{V}: \mathcal{E} \to \mathcal{E}$ such that $\tilde{V}(fs) = V(f)s + f\,\tilde{V}(s)$ where f (resp. s) is a local section of $\mathcal{O}$ (resp. $\mathcal{E}$). Equivariance is obstructed precisely by $i(V)c(\mathcal{E}) \in H^1(X, \mathrm{Hom}(\mathcal{E}, \mathcal{E}))$ where $c(\mathcal{E}) \in H^1(X, \mathrm{Hom}(\mathcal{E}, \mathcal{E}) \otimes \Omega^1)$ is the Atiyah-Chern class of $\mathcal{E}$. Any V-equivariant $\mathcal{E}$ defines a total cocycle of degree 0 in the complex

$$K^\cdot = \bigoplus_{p-q=\cdot} C^p(\mathcal{U}, \Omega^q \otimes \mathrm{Hom}(\mathcal{E}, \mathcal{E}))$$

with differential $\delta \pm i(V)$, where $\mathcal{U}$ is a Leray cover of X. In effect, given local connections $D: \mathcal{E}\,|\,U \to \mathcal{E} \otimes \Omega^1\,|\,U$, representative of $c(\mathcal{E})$ is given by $k_{\alpha\beta} = D_\alpha - D_\beta$, and $i(V)c(\mathcal{E}) = 0$ if and only if there is an $L \in C^0(\mathcal{U}, \mathrm{Hom}(\mathcal{E}, \mathcal{E}))$ such that $i(V)$ $k_{\alpha\beta} = L_\beta - L_\alpha$. In this case $k_{\alpha\beta} + L_\alpha$ is a total cocycle of degree zero. Moreover, it also follows that L_α determines an element $L \in H^0(X, \mathrm{Hom}(\mathcal{E}_Z, \mathcal{E}_Z))$, where $\mathcal{E}_Z = \mathcal{E} \otimes_{\mathcal{O}} \mathcal{O}_Z$, which is unique up to an element of $H^0(X, \mathrm{Hom}(\mathcal{E}, \mathcal{E}))$. We view L as the localization of $c(\mathcal{E})$ to Z. It follows that the localization of $c_k(\mathcal{E}) \in H^k(X, \Omega^k)$ to Z is the kth elementary symmetric function $\sigma_k(L)$ of L viewed in the natural way as an element of $H^0(X, \mathcal{O}_Z)$. Note $c_k(\mathcal{E})$ is the Chern class of Atiyah [**1**].

EXAMPLE 2. *The universal Chern classes.* Let Grass (k, n) denote the Grassmann manifold of all k-planes in $\boldsymbol{C}^n$ and let $U_{k,n}$ denote the universal subbundle on Grass (k, n) consisting of all pairs (x, W) where $W \in$ Grass (k, n) and $x \in W$. If one views Grass (k, n) as the homogeneous space of left cosets G/H where $G = \mathrm{Gl}(n, C)$ and $H = \mathrm{Gl}\,(k, n-k; \boldsymbol{C})$ is the isotopy group of the k-plane $\boldsymbol{C}^k$ spanned by $e_1, \cdots, e_k$, then explicit patching data for $U_{k,n}$ are described as follows: For any sequence $I = (i_1, \cdots, i_k)$ with $1 \leqq i_1 < \cdots < i_k \leqq n$, and for any $\tau \in G$, let $W_I(\tau)$ be the $k \times k$ submatrix of τ obtained by selecting the first k columns and the rows I from τ. Set $U_I = \{\tau \in G: W_I(\tau)$ is nonsingular$\}$. The opens U_I in G are H-invariant so define an open cover $\{U_I/H = \tilde{U}_I\}$ of Grass (k, n). $U_{k,n}$ is trivial over $\tilde{U}_I$ and the patching data on $\tilde{U}_I \cap \tilde{U}_J$ are $f_{IJ} = W_I W_J^{-1}$. An explicit cocycle representing $c(U_{k,n})$ is $(f_{IJ}^{-1} df_{IJ})^*$ [**1**, p. 195].

Now any $M \in \mathrm{Hom}\,(\boldsymbol{C}^n, \boldsymbol{C}^n)$ defines by left multiplication a linear vector field on G which descends to a holomorphic vector field on Grass (k, n). An explicit computation shows that $U_{k,n}$ is equivariant with respect to any field V on Grass (k, n) and that the Atiyah-Chern class of $U_{k,n}$ in $H^0(\mathrm{Grass}, \mathrm{Hom}(U_{k,n}, U_{k,n}) \otimes \mathcal{O}_Z)$ is represented in $\tilde{U}_I$ by[1]

$$-\,(i(M)dW_I W_I^{-1})^*. \tag{6}$$

In particular, the nilpotent matrix M of §1 defines a vector field on Grass with unique zero the k-plane spanned by $e_1, \cdots, e_k$. Thus (6) defines an element of H^0 (Grass, $\mathrm{Hom}(U_{k,n}, U_{k,n}) \otimes \mathcal{O}_Z)$ which vanishes except on $\tilde{U}_{1\cdots k}$, and there

$$i(M)dW_I W_I^{-1} = \begin{bmatrix} 0 & \\ \vdots & I_{k-1} \\ 0 & \\ w_{11} & \cdots \quad w_{1k} \end{bmatrix} \tag{7}$$

[1]The authors thank E. Akyldiz for pointing out a sign error at this point in an earlier version.

where (w_{ij}) is the matrix of local coordinates on $\tilde{U}_I$ obtained by deleting the rows I from WW_I^{-1} $(I = (1, \cdots, k))$. The elementary symmetric functions of (6) are well known; in fact, $c_j(U_{k,n}) = -w_{1,k-j}$. Similar reasoning shows that the Chern classes of the universal quotient bundle, i.e., the special Schubert cycles, are $c_j(Q) = w_{j,k}$.

It remains to comment on the relations in $\mathcal{O}_Z$ for this example. Differentiating the flow e^{tM} in the local coordinates w_{ij} on $\tilde{U}_{1\cdots k}$, one gets

$$V = \text{last } n - k \text{ rows of } (d/dt)\,[(e^{tM}W)(e^{tM}W)_I^{-1}]_{t=0}.$$

Hence the coefficient of $\partial/\partial w_{ij}$ in V is the (i,j) entry of the matrix

$$\begin{bmatrix} w_{21} & \cdots & w_{2k} \\ \vdots & & \\ w_{p1} & \cdots & w_{pk} \\ 0 & \cdots & 0 \end{bmatrix} - \begin{bmatrix} w_{11} & \cdots & w_{1k} \\ \vdots & & \vdots \\ w_{p1} & \cdots & w_{pk} \end{bmatrix} \begin{bmatrix} 0 & & \\ \vdots & I_{k-1} & \\ 0 & & \\ w_{11} & \cdots & w_{1k} \end{bmatrix}, \qquad p = n - k.$$

Hence the defining relations for the algebra $\mathcal{O}_Z$ are

$$\begin{aligned} w_{i+1,j} - w_{i,j-1} - w_{1j}w_{ik} &= 0, && 1 \leqq i \leqq p-1,\ 2 \leqq j \leqq k, \\ w_{i+1,1} &= w_{ik}w_{11}, && 1 \leqq i \leqq p-1, \\ w_{pj} &= -w_{1,j+1}w_{pk}, && 1 \leqq j \leqq k-1, \\ w_{11}w_{pk} &= 0. \end{aligned}$$

These relations imply several relevant facts about $\mathcal{O}_Z$ which we include as a theorem.

THEOREM 3. *In the algebra* $\mathcal{O}_Z = H^0(\text{Grass}\,(k, n), \mathcal{O}_Z)$ *defined by the matrix* M, *the following statements are true*:

(1) *the universal Chern classes* $c_j(U) = -w_{1,k-j}$ *form a set of generators*;

(2) *the special Schubert cycles* $c_q(Q) = w_{qk}$ *also generate* $\mathcal{O}_Z$; *and*

(3) *the Whitney formula is valid in* $\mathcal{O}_Z$, *i.e.*,

$$(1 + c_1(U) + \cdots + c_k(U))\,(1 + c_1(Q) + \cdots + c_{n-k}(Q)) = 1.$$

REFERENCES

1. M. F. Atiyah, *Complex analytic connections in fibre bundles*, Trans. Amer. Math. Soc. **85** (1957), 181–207. MR **19**, 172.

2. J. Carrell and D. Lieberman, *Holomorphic vector fields and Kaehler manifolds*, Invent. Math. **21** (1973), 303–309. MR **48** #4356.

3. ———, *Vector fields and Chern numbers*, Math. Ann. (to appear).

4. T. Frankel, *Fixed points and torsion on Kähler manifolds*, Ann. of Math. (2) **70** (1959), 1–8. MR **24** #A1730.

5. A. Howard, *Holomorphic vector fields on algebraic manifolds*, Amer. J. Math. **94** (1972), 1282–1290. MR **46** #9384.

6. S. Kobayashi, *Transformation groups in differential geometry*, Ergebnisse der Math. und ihrer Grenzgebiete, Band 70, Springer-Verlag, Berlin and New York, 1972. MR **50** #8360.

7. A. Sommese, *Holomorphic vector fields on compact Kaehler manifolds*, Math. Ann. **210** (1974), 75–82. MR **50** #646.

8. E. T. Wright, Jr., *Killing vector fields and harmonic forms*, Trans. Amer. Math. Soc. **199** (1974), 199–202. MR **50** #3249.

THE UNIVERSITY OF BRITISH COLUMBIA

BRANDEIS UNIVERSITY

INSTITUT DES HAUTES ETUDES SCIENTIFIQUES

Proceedings of Symposia in Pure Mathematics
Volume 30, 1977

DIFFERENTIAL FORMS WITH SUBANALYTIC SINGULARITIES; INTEGRAL COHOMOLOGY; RESIDUES

PIERRE DOLBEAULT AND JEAN POLY

On a real analytic manifold, we first define subanalytic chains and integration along such chains and describe the homology of the manifold; this is a generalization of the semianalytic chains of Bloom and Herrera **[2]**, which behaves well with respect to proper maps, and of their properties.

Secondly, using a parametrix for d, we describe the integral cohomology of the manifold with pairs of differential forms having subanalytic singularities; this is analogous to a former theorem of Allendoerfer and Eells **[1]** in the differentiable case; moreover, this shows the relation between pairs and kernels associated to the singularities.

Let X be a real analytic, oriented, countable at infinity manifold of dimension n.

1. Subanalytic chains [11, Chapter IV].

1.1. A *subanalytic set* Y of X is, locally, a finite union of differences of images of proper morphisms f_i: $U_i \to X$; g_i: $V_i \to X$ where U_i and V_i are real analytic manifolds. If Y is closed, then it is the image of a proper morphism. Clearly, the image of a subanalytic set by a proper morphism is subanalytic (Hironaka **[7]**).

1.2. Let L be a principal ideal domain; a subanalytic p-chain with coefficients in L is an equivalence class of triples (Y', Y'', η) called p-prechains where Y', Y'' are closed subanalytic sets of X of dimension $\leqq p$ and $(p-1)$, resp.; $Y'' \subset Y'$; $\eta \in H_p(Y; L)$ where $Y = Y' \backslash Y''$. The equivalence relation is defined as follows: Let $y = (Y', Y'', \eta)$, $z = (Z', Z'', \zeta)$; $y < z$ means $Y' \subset Z'$; $Y'' \subset Z''$; $\zeta = \tilde{j}_{ZY}$ where $\tilde{j}_{ZY}$ is a convenient map $H_p(Y; L) \to H_p(Z; L)$ (it is induced by inclusion if $Y \subset Z$); y and z are equivalent if there exists a p-prechain t such that $y < t$ and $y < t$.

AMS (MOS) subject classifications (1970). Primary 32C05, 32C30, 55B30, 58A10, 58A25.

© 1977, American Mathematical Society

1.2. THEOREM. *The subanalytic chains of X define a presheaf* $\mathcal{S}_{.X}$ *of L-modules: for every paracompactifying family of supports* Φ *of X,* $\mathcal{S}_{.X}$ *is a* Φ*-soft sheaf.*

If $f\colon X \to X'$ *is a proper morphism of real analytic manifolds, then f defines a morphism* $f_*\colon \mathcal{S}_{.X} \to \mathcal{S}_{.X'}$.

1.3. For $L = \boldsymbol{R}$ or $\boldsymbol{C}$, starting with integration on a submanifold, one can define integration on subanalytic chains and prove

THEOREM. *The integration morphism* $I\colon \mathcal{S}_{.X} \to N^{\mathrm{loc}}_{.X}$ *(sheaf of locally normal currents) is an injection and is compatible with proper morphisms.*

2. Poincaré lemma for subanalytic chains and homology.

2.1. Let $\mathcal{Y}$ be a p-chain to which belongs the p-prechain $y = (Y', Y'', \eta)$. Then $b\mathcal{Y}$ is the equivalence class of the $(p-1)$-prechain $(Y'', \varnothing, b_{Y''Y'}\eta)$ where $b_{Y''Y'}$ is the canonical morphism $H_p(Y; L) \to H_{p-1}(Y''; L)$. Every subanalytic chain $\mathcal{Y}$ has a unique *faithful* representative $y = (\bar{Y}, bY, \eta) = (Y, \eta)$ which is the smallest element of $\mathcal{Y}$; the support of $\mathcal{Y}$ (supp $\mathcal{Y}$) is the support of the section of $\mathcal{S}_{.X}$ defined by $\mathcal{Y}$.

2.2. *For every* $x \in X$, *there exists a fundamental system of open neighborhoods of x such that, for any such neighborhood U, for every p-cycle* $\mathcal{Y}$ *(closed subanalytic p-chain) of U, we have:*

if $p \leqq n-1$, *there exists a* $(p+1)$*-subanalytic chain* $\mathcal{Z}$ *in U such that* $\mathcal{Y} = b\mathcal{Z}$;

if $p = n$, *then, if* $\mathcal{Y} \neq 0$, supp $\mathcal{Y}$ *is U and the group of n-subanalytic cycles of U is isomorphic to L.*

The proof uses the cone construction.

2.3. COMPACT POINCARÉ LEMMA. *For every point* $x \in X$, *there exists a fundamental system of open neighborhoods U of x such that, for any U and for every p-subanalytic compact cycle* $\mathcal{Y}$ *of U (i.e., such that* supp $\mathcal{Y}$ *is compact), we have:*

if $1 \leqq p \leqq n-1$, *there exists a compact* $(p+1)$*-subanalytic chain* $\mathcal{Z}$ *such that* $\mathcal{Y} = b\mathcal{Z}$;

if $p = n$, *then* $\mathcal{Y} = 0$;

if $p = 0$, *then the homology group of* 0*-compact chains is isomorphic to L.*

2.4. REMARK. From Hironaka's triangulation theorem of subanalytic sets **[8]**, for every integral subanalytic q-chain (with coefficients in $\boldsymbol{Z}$), the integration current $I(\mathcal{Y})$ is locally rectifiable; moreover, from [11, IV.4.1.4], for every locally closed subanalytic set Y of dimension $\leqq q$, we have $\mathcal{H}^r(Y) = 0$ for $r > q$. Then from the structure theorem of holomorphic chains **[14]**, we have: *On a complex analytic manifold X, the integration current* $I(\mathcal{Y})$ *of a* $2p$*-subanalytic chain* $\mathcal{Y}$ *is a holomorphic p-chain if and only if* $b\mathcal{Y} = 0$ *and* $I(\mathcal{Y})$ *is of bidimension* (p, p).

2.5. *Generalization.* Let M be a closed subanalytic set of a real analytic space X. We can define the subanalytic chains on X and consider the subanalytic chains which have their supports in M. Then using triangulation **[8]**, we get a compact Poincaré Lemma 2.3 for subanalytic chains on M as in **[2]**.

2.6. THEOREM. *Let* $\mathcal{F}$ *be a sheaf of L-modules on a real analytic manifold X; then, for any paracompactifying family* Φ, *there exists a canonical isomorphism* $H^{\Phi}_q(X; \mathcal{F}) \approxeq H_q(\Gamma_\Phi(\mathcal{S}_{.X} \otimes_L \mathcal{F})$.

This results from 1.2 and 2.3 as in Bloom and Herrera's proof for semianalytic chains [2].

2.6.1. REMARK. Using 2.5, Theorem 2.6 is valid for analytic spaces.

2.7. PROPOSITION. *Let $f: X \to X'$ be a proper morphism of real analytic manifolds; let Φ and Φ' be paracompactifying families of supports on X and X' such that $f(\Phi) \subset \Phi'$; then, for every sheaf of L-modules $\mathscr{F}'$ on X', the following diagram is commutative*:

$$\begin{array}{ccc} H_q^{\Phi}(X; f^* \mathscr{F}') & \approx & H_q(\Gamma_\Phi(\mathscr{S}_{.X} \otimes_L f^*\mathscr{F}')) \\ f_* \downarrow & & \downarrow \\ H_q^{\Phi'}(X, \mathscr{F}') & \approx & H_q(\Gamma_{\Phi'}(\mathscr{S}_{.X'} \otimes_L \mathscr{F}')) \end{array}$$

Use [3, V, 4], and 2.6.

2.8. REMARK. In [2], the sheaf of semianalytic chains is used as a tool; it can be replaced by the sheaf of subanalytic chains; then, from 1.2, 1.3 and 2.6.1, we get compatibility of the splittings of de Rham's theorems with proper morphisms.

3. Pairs of singular differential forms. We shall now give an interpretation of the isomorphisms of Theorem 2.6 in the case of integral coefficients.

3.1. LEMMA ([11, CHAPTER II], [10], CF. [9]). *Let X be a C^∞ differentiable manifold of dimension n; suppose that X is countable at infinity. Then, there exist continuous linear operators A and R on the space $\mathscr{D}'_{.}(X)$ of the currents on X such that*:

(α) *for every $T \in \mathscr{D}'_{.}(X)$, $T = dAT + AdT + RT$;*

(β) *for every $T \in \mathscr{D}'_{.}(X)$, RT is C^∞ and* sing supp $AT \subseteq$ sing supp T;

(γ) *if T is* 0-*continuous, then AT is locally integrable.*

3.2. *Pair defined by a subanalytic chain.* Let $\mathscr{Y}$ be an integral p-chain on X. We set $\omega = -AI(\mathscr{Y})$; $\theta = AdI(\mathscr{Y}) + RI(\mathscr{Y})$; then $I(\mathscr{Y}) = \theta - d\omega$ where ω and θ are locally integrable of respective degrees $(q-1)$ and q, with $q = n-p$. Since $dI(\mathscr{Y}) = \pm I(b\mathscr{Y})$, we have: sing supp $\theta \subset bY$; sing supp $\omega \subset \bar{Y}$, (Y, η) being the faithful representative of $\mathscr{Y}$.

Let $\mathscr{Z}$ be an integral, compact, subanalytic q-chain whose faithful representative is (Z, ζ) and suppose that $\mathscr{Y}$ and $\mathscr{Z}$ intersect *properly* (i.e., $\bar{Y} \cap \bar{Z}$ is a finite number of points and $\bar{Y} \cap bZ = \varnothing = \bar{Z} \cap bY$).

Then, from the definition and properties of the Kronecker index $\mathscr{K}(.,.)$ of two currents [12], $\mathscr{K}(I(\mathscr{Z}), \theta)$ and $\mathscr{K}(I(b\mathscr{Z}), \omega)$ exist and

$$R[(\theta, \omega), \mathscr{Z}] = \mathscr{K}(I(\mathscr{Z}), \theta) - \mathscr{K}(I(b\mathscr{Z}), \omega) = \mathscr{K}(I(\mathscr{Z}), I(\mathscr{Y})).$$

3.3. THEOREM. *Let $\mathscr{Y}$ and $\mathscr{Z}$ be two subanalytic integral chains of X, of dimensions p, q $(p+q = n)$, and (Y, η), (Z, ζ) their faithful representatives. Suppose that $K = \bar{Y} \cap \bar{Z}$ is compact and $\bar{Y} \cap bZ = \varnothing = bY \cap \bar{Z}$; then $\mathscr{K}(I(\mathscr{Y}), I(\mathscr{Z}))$ exists and is an integer.*

PROOF. (1) Let S and T be currents on X of dimensions p, q $(p+q = n)$, whose intersection of supports is a compact set K; let U be an open neighborhood of K. Then if $\mathscr{K}_X(S, T)$(Kronecker index with respect to X) exists, the Kronecker index $\mathscr{K}_U(S|U, T|U)$ exists and is equal to $\mathscr{K}_X(S, T)$.

As a consequence, if supp $S \cap$ supp $bT = \varnothing =$ supp $T \cap$ supp bS, to compute

$\mathscr{K}(S, T) = \mathscr{K}_X(S, T)$ we can suppose that S and T are closed currents; let Y and Z be closed sets of X which contain supp S and supp T, respectively, and such that $Y \cap Z = K$ compact; let $s \in H_p(Y; \boldsymbol{R})$ and $t \in H_q(Z; \boldsymbol{R})$ the homology classes of S and T, resp.

(2) Using cup product and Poincaré duality, we define the intersection product in homology

$$\begin{array}{ccccc} H_p(Y; \boldsymbol{R}) \otimes_{\boldsymbol{R}} & H_q(Z; \boldsymbol{R}) & \dashrightarrow & H_0(K; \boldsymbol{R}) \\ s & t & \mapsto & s.t \end{array}$$

The map of K into a point $*$ defines the evaluation map

$$\mathrm{ev}: H_0(K; \boldsymbol{R}) \to H_0(*; \boldsymbol{R}) = \boldsymbol{R};$$

then

LEMMA. *In the hypothesis of* (1), *$\mathscr{K}(S, T)$ exists and is equal to* ev$(s.t)$.

(3) Going back to the (Borel-Moore) definition of homology, it can be shown that the following diagram is commutative

$$\begin{array}{ccccccc} H_p(Y; \boldsymbol{Z}) \times H_q(Z; \boldsymbol{Z}) & \dashrightarrow & H_0(K; \boldsymbol{Z}) & \xrightarrow{\mathrm{ev}} & \boldsymbol{Z} \\ \downarrow & & \downarrow & & \downarrow \\ H_p(Y; \boldsymbol{R}) \times H_q(Z; \boldsymbol{R}) & \dashrightarrow & H_0(K; \boldsymbol{R}) & \xrightarrow{\mathrm{ev}} & \boldsymbol{R} \end{array}$$

where the vertical morphisms are defined by the inclusion $\boldsymbol{Z} \hookrightarrow \boldsymbol{R}$.

(4) Let (Y, η) be the faithful representative of an integral subanalytic cycle $\mathscr{Y}$ of X; then the image of η by the morphism $H_p(Y; \boldsymbol{Z}) \to H_p(Y; \boldsymbol{R})$ is the homology class s of the integration current $I(\mathscr{Y})$.

3.4. COROLLARY. *To every integral subanalytic p-chain $\mathscr{Y}$, whose faithful representative is (Y, η), is associated a pair of locally integrable differential forms (θ, ω) of respective degrees q, $q-1$ $(p+q = n)$, whose singular supports are contained in $\bar{Y}$ and bY, resp., and such that, for every integral compact subanalytic q-chain $\mathscr{Z}$ intersecting $\mathscr{Y}$ properly, $R[(\theta, \omega), \mathscr{Z}] \in \boldsymbol{Z}$.*

3.5. *Pairs of singular differential forms.* Consider the pairs of currents (θ, ω) such that:

(a) if $q \geqq 1$, the degrees of θ and ω are respectively q and $(q-1)$; if $q = 0$, then $\omega = 0$ and $d\theta = 0$;

(b) the currents θ and ω have their singular supports $e(\theta)$ and $e(\omega)$ contained in closed subanalytic sets Y'', Y' of codimension $(q+1)$ and q respectively and $e(\theta) \subset e(\omega)$;

(c) $\theta - d\omega$ is locally flat (this condition is satisfied, in particular, when θ and ω are locally integrable);

(d) for every compact subanalytic q-chain $\mathscr{Z}$ (of faithful representative (Z, ζ)) such that $\dim(\bar{Z} \cap Y') = 0$ and $bZ \cap Y' = \varnothing = \bar{Z} \cap Y''$, we have $\mathscr{K}((\theta - d\omega), I(\mathscr{Z})) \in \boldsymbol{Z}$; q is called the *degree* of the pair.

3.6. THEOREM. *For every pair of currents (θ, ω) satisfying conditions* (a), (b), (c), (d), *$\theta - d\omega$ is the integration current of an integral subanalytic p-chain $(p+q = n)$ having (Y', Y'', η) as a representative where η is the class of homology of $(\theta - d\omega)|_{X \setminus Y''}$.*

The proof uses a result of **[11**, Chapter IV] based on measure support theorem.

3.7. Theorem 3.6 and Corollary 3.4 show there is a surjection $(\theta, \omega) \mapsto \mathscr{Y}$ of the set of pairs (θ, ω) of degree q satisfying (a), (b), (c), (d), onto the set of integral subanalytic chains of dimension p $(p + q = n)$. In the set of pairs, we consider the equivalence relation

$$(\theta_1, \omega_1) \sim (\theta_2, \omega_2) \Leftrightarrow \theta_1 - d\omega_1 = \theta_2 - d\omega_2;$$

we denote by $[\theta, \omega]$ the equivalence class of (θ, ω) and by degree of $[\theta, \omega]$ the degree of a representative. A class of pairs $[\theta, \omega]$ will be called a *subanalytic pair*.

In the set of pairs, we define addition by $(\theta, \omega) + (\theta', \omega') = (\theta+\theta', \omega+\omega')$; then:

3.8. PROPOSITION. *The group $C^q(X; \mathbf{Z})$ of subanalytic pairs of degree q is isomorphic to the group of integral subanalytic chains $S_{n-q}(X; \mathbf{Z})$.*

3.9. Defining the differential of a subanalytic pair $[\theta, \omega]$ as $[0, -\theta]$, from 2.2, we get the following

POINCARÉ LEMMA. *Every point x of a real analytic manifold X has a fundamental system of open neighborhoods U such that every d-closed q-subanalytic pair $[\theta, \omega]$ of U, has the following properties*:

if $q \geqq 1$, there exists a $(q - 1)$-subanalytic pair $[\bar{\theta}, \bar{\omega}]$ such that $[\theta, \omega] = d[\bar{\theta}, \bar{\omega}]$;
if $q = 0$, then $\omega = 0$ and θ is a constant integer.

3.10. Using 3.8 and 3.9, we define the complex of sheaves of subanalytic pairs $\mathscr{C}^{\cdot}_X$ on X by identification with the complex of sheaves $\mathscr{S}_{\cdot}(X; \mathbf{Z})$ of integral subanalytic chains. We define the support of a subanalytic pair as the support of the associated subanalytic chain. Then, for every paracompactifying family Φ, $\mathscr{C}^{\cdot}_X$ is a Φ-soft sheaf, and from 3.9, it is a resolution of $\mathbf{Z}$, $0 \to \mathbf{Z} \to \mathscr{C}^{\cdot}_X$.

Hence we get a theorem of de Rham.

THEOREM. *There exists a canonical isomorphism $H^q\Gamma_\Phi(X, \mathscr{C}^{\cdot}_X) \approxeq H^q_\Phi(X; \mathbf{Z})$.*

4. Remarks on the complex case.

4.1. Suppose that X is a complex manifold. Let Y be a $2p$-closed subanalytic set and $\eta \in H_{2p}(Y; \mathbf{Z})$. Then $(Y, 0, \eta)$ is a representative of a closed subanalytic chain $\mathscr{Y}$; to $\mathscr{Y}$ is associated the pair (θ, ω) where $\omega = -AI(\mathscr{Y})$ is a locally integrable current and where $\theta = RI(\mathscr{Y})$ is a closed C^∞ differential form.

4.2. Suppose now that $(Y, 0, \eta)$ is a (complex analytic) divisor; if $\mathscr{Y}$ is locally defined by a meromorphic function f, then, locally, we can choose $AI(\mathscr{Y})$ as the current $(1/2\pi i)(df/f)$ up to a closed holomorphic form.

If X is a Stein manifold, it is known **[13]** that every element of $H^2(X; \mathbf{Z})$ is the cohomology class of a divisor D; moreover, if the class is 0, then D is the divisor of a meromorphic function.

Let $\mathscr{E}^2(X; \mathbf{Z})$ be the group of closed subanalytic pairs $[\theta, \omega]$ of degree 2, where ω is a closed meromorphic 1-form as above and θ is a closed holomorphic 2-form; let $\mathscr{B}^2(X; \mathbf{Z})$ be the subgroup of pairs $[0, \omega]$ where $\omega = (1/2\pi i)(d\varphi/\varphi) + du$ where φ and u are global meromorphic functions on X, then $H^2(X; \mathbf{Z})$ is isomorphic to $\mathscr{E}^2(X; \mathbf{Z})/\mathscr{B}^2(X; \mathbf{Z})$ **[5]**.

4.3. There are analogous results for $H^{1,1}(X; \mathbf{Z})$ where X is a complex projective manifold [**5**].

5. Residues.

5.1. For any closed subanalytic chain $\mathscr{Y}$ there exists a closed pair (θ, ω) of locally integrable currents such that $d\omega = -I(\mathscr{Y}) + \theta$; $I(\mathscr{Y})$ is the residue current of ω and the residue number is the constant function 1 supported by Y.

5.2. More generally, for any closed subanalytic chain $\mathscr{Y}$, one can consider the locally integrable current $\Phi = K \wedge \psi + \theta'$ where $K = AI(\mathscr{Y})$ and where ψ and θ' are C^∞ differential forms; moreover $dK = I(\mathscr{Y}) + L$ where L is C^∞; $\varphi = \Phi | X \backslash \bar{Y}$ is a C^∞ form.

Let now X be a complex manifold and Y be a closed, complex analytic set of X; differential forms as above are called *K-simple* [**10**], [**11**], and for such forms, Poly [**10**], [**11**] proved a theorem of Leray whose simplest form is the following:

Let Y be a complex submanifold of X of any dimension; then every complex cohomology class α of $X \backslash Y$ contains a closed K-simple form. When Y has singularities, a condition on the residue class of α has to be satisfied. $\psi | Y$ is called the residue form of φ.

5.3. *Problem.* Given a complex analytic subset Y of a complex manifold X, to study locally integrable currents Φ on X such that $\varphi = \Phi | X \backslash Y$ is C^∞; $d\varphi = 0$.

It may be possible to define locally flat currents with supports on Y playing the part of residue forms, but having a nonempty singular support contained in the singular set of Y and for which the theorem of Leray would be true.

A result in this direction is obtained by G. Gordon [**6**] for $\dim_C X = 2$; $\dim_C Y = 1$; forms of degree 2; cf. also [**4**, IV,D] for the singularities of the residue.

REFERENCES

1. C. B. Allendoerfer and J. Eells, *On the cohomology of smooth manifolds*, Comment. Math. Helv. **32** (1958), 165–179. MR **21** #868.

2. T. Bloom and M. Herrera, *de Rham cohomology of an analytic space*, Invent. Math. **7** (1969), 275–296. MR **40** #1601.

3. G. Bredon, *Sheaf theory*, McGraw-Hill, New York, 1967. MR **36** #4552.

4. P. Dolbeault, *Formes différentielles et cohomologie sur une variété analytique complexe*. I, II, Ann. of Math. (2) **64** (1956), 83–130; ibid. (2) **65** (1957), 282–330. MR **18**, 670; **19**, 171.

5. ———, *Sur le groupe de cohomologie entière de dimension deux d'une variété analytique complexe*, Rend. Mat. e Appl. **21** (1962), 219–239. MR **31** #5222.

6. G. Gordon, *Les formes semi-méromorphes sur une variété analytique complexe*, 1975 (preprint).

7. H. Hironaka, *Subanalytic sets*, Number theory, Alg. Geom. and Comm. Alg., in honour of Y. Akizuki, Kinokuniya, Tokyo, 1973, pp. 453–493.

8. ———, *Triangulations of algebraic sets*, Algebraic Geometry, Arcata, 1974, Proc. Sympos. Pure Math. vol. 29, Amer. Math. Soc., Providence, R.I., 1975, pp. 165–185.

9. J. King, *Global residues and intersections on a complex manifold*, Trans. Amer. Math. Soc. **192** (1974), 163–199. MR **49** #3198.

10. J. Poly, *Sur un théorème de J. Leray en théorie des résidus*, C. R. Acad. Sci. Paris Sér. A **274** (1972), 171–174. MR **44** #6994.

Formes-résidus (*en codimension quelconque*), Journées complexes de Metz, février 1972, Publi. I. R. M. A. Strasbourg, 1973.

11. ———, *Formule des résidus et intersection des chaînes sous-analytiones*, Thèse, Poitiers, 1974.

12. G. de Rham, *Variétés différentiables. Formes, courants, formes harmoniques*, Actualités Sci. Indust., no. 1222, Hermann, Paris, 1953. MR **16**, 957.

13. J.-P. Serre, *Quelques problèmes globaux relatifs aux variétés de Stein*, Colloq. sur les fonctions de plusieurs variables (Bruxelles, 1953), Georges Thone, Liège; Masson, Paris, 1953, pp. 57–68. MR **16**, 235.

14. R. Harvey and B. Shiffman, *A characterization of holomorphic chains*, Ann. of Math. (2) **99** (1974), 553–587. MR **50** #7572.

Université de Paris VI

Université de Bordeaux I

Proceedings of Symposia in Pure Mathematics
Volume 30, 1977

ON THE HOMOTOPY TYPE OF WEIGHTED HOMOGENEOUS NORMAL COMPLEX SURFACES

LUDGER KAUP AND GOTTFRIED BARTHEL

0. Introduction. Though the underlying topological structure of compact complex curves has been understood for a long time, the situation for general dimension n seems to be fairly complicated. In the case of complex dimension two, the following result is available (cf. [**BK**]): The homotopy type of a simply connected irreducible compact complex surface X is completely determined by the integral homology $H_*(X, \boldsymbol{Z})$ together with the second Poincaré homomorphisms $P_2(X, \boldsymbol{Z}/(m))$ for certain integers m, depending on the torsion subgroup T of $H_2(X, \boldsymbol{Z})$; in case of two-torsion in T, moreover, certain Pontrjagin squares have to be considered. This result may be looked at as a generalisation of a theorem of Milnor [**Mi**]. Via the Poincaré homomorphisms, there is an interesting interplay between the global homotopy invariants of X and the local homology in the singularities of X, which in fact is not invariant under homotopy. One of the aims of the present article is to turn this qualitative relation into a more quantitative description by calculating explicitly invariants for a class of normal surfaces in $\boldsymbol{P}_3$, which belong to the "weighted homogeneous" surfaces.

1. The homotopy classification theorem HCT. In [**BK**] an HCT has been proved for simply connected compact locally complex surfaces X. There a "locally complex surface" is a connected oriented topological space, which is locally homeomorphic to a (possibly singular) complex surface. There is a generalisation for a finite number of compact irreducible components, which also includes noncompact irreducible components with finitely generated homology (the map $P_2\pi^2$ comes from the Poincaré homomorphism; cf. §2):

HOMOTOPY CLASSIFICATION THEOREM (HCT). *Let X denote a simply connected*

AMS (MOS) subject classifications (1970). Primary 32J15, 55D15; Secondary 14J25, 32C40.

© 1977, American Mathematical Society

locally complex surface with finitely generated homology $H_*^c(X, \mathbf{Z})$. *Then the homotopy type of X is uniquely determined by*

(i) *the Betti numbers* b_3, b_4, *and* $H_2^c(X, \mathbf{Z})$;

(ii) *the maps* $(i_\rho)_2\ P_2(X_\rho, \mathbf{Z}/(m))\pi^2 i_\rho^2$: $H^2(X, \mathbf{Z}/(m)) \to H_2^{cld}(X, \mathbf{Z}/(m))$, *where* $i_\rho: X_\rho \hookrightarrow X$ *denotes for* $1 \leqq \rho \leqq b_4$ *the natural inclusion of the compact irreducible component* X_ρ *of X, for* $m = 0$, $m = p_i^{n_i}$;

(iii) *the Pontrjagin squares*

$\mathfrak{p}_{p_i}^{n_i}$: $H^2(X, \mathbf{Z}/(p_i^{n_i})) \to H^4(X, \mathbf{Z}/(p_i^{n_i+1}))$, *for* $p_i = 2$, *where* $p_i^{n_i}$ *denotes prime powers which occur as the order in the natural decomposition of*

$$S := \mathrm{im}\Big(\bigoplus_{\rho=1}^{b_4} H_2(X_\rho, \mathbf{Z}) \to H_2^c(X, \mathbf{Z}) \twoheadrightarrow \mathrm{Tors}\, H_2^c(X, \mathbf{Z})\Big)$$

into a direct sum of subgroups, which are cyclic and whose order is a prime power.

The proof of this result uses the fact that the simply connected space X is of the homotopy type of a finite CW complex, since it has finitely generated homology [**W**, Proposition 4.1]. Thus Whitehead's classification [**Wh**] of simply connected finite (real) four-dimensional CW complexes can be applied. This gives the following consequence:

Let X and Y denote two simply connected locally complex surfaces, where $H_*^c(X, \mathbf{Z})$ is a finitely generated abelian group. Then X and Y are of the same oriented homotopy type, if the following hold:

(i) $b_3(X) = b_3(Y)$ and X and Y have the same number of compact irreducible components $X_1, \cdots, X_r$ and $Y_1, \cdots, Y_r$.

(ii) There is an isomorphism of groups θ_2: $H_2^c(X, \mathbf{Z}) \to H_2^c(Y, \mathbf{Z})$ and after an appropriate enumeration of the irreducible components there are commutative diagrams with homomorphisms of abelian groups

$$\begin{array}{ccc} H^2(Y, \mathbf{Z}/(m)) & \xrightarrow{j_\rho^2} & T_\rho(m) \\ & {\scriptstyle i_\rho^2\theta^2(m)}\searrow \quad \swarrow{\scriptstyle \gamma_\rho} & \\ & H^2(X_\rho, \mathbf{Z}/(m)) & \end{array}$$

and

$$\begin{array}{ccc} T_\rho(m) & \xrightarrow{P_2(Y_\rho, \mathbf{Z}/(m))} & H_2^c(Y_\rho, \mathbf{Z}/(m)) \\ {\scriptstyle \gamma_\rho(m)}\downarrow & & \downarrow{\scriptstyle \theta_2(m)} \\ H^2(X_\rho, \mathbf{Z}/(m)) & \xrightarrow{P_2(X_\rho, \mathbf{Z}/(m))} & H_2^c(X_\rho, \mathbf{Z}/(m)) \end{array}$$

(iii) The Pontrjagin squares are compatible:

$$\mathfrak{p}_m(X)\,\theta^2 = \theta^4 \mathfrak{p}_m(Y), \qquad m = 2^{n_i}.$$

There $m = 0$ or $m =$ prime power, which arises in the decomposition of $H_2^c(X, \mathbf{Z})$ as in HCT, and

$$T_\rho(m) := \mathrm{im}(H^2(Y, \mathbf{Z}/(m)) \to H^2(Y_\rho, \mathbf{Z}/(m)).$$

2. Singular Poincaré duality. The locally analytic surface X has a topological normalisation $\pi: \tilde{X} \to X$, that is, $\tilde{X}$ is a locally irreducible locally analytic surface,

π is a finite continuous surjection, which is homeomorphic outside a closed subset $S \subset X$, where S is locally analytic and of complex dimension at most one. If $A \subset X$ is locally closed, ϕ a family of supports on X and $\mathscr{F}$ a locally constant sheaf of finitely generated L-modules on X (arbitrary sheaf, if A is closed and ϕ is paracompactifying) for a fixed principal ideal domain L, then there are natural *Poincaré homomorphisms*,

$$P_j^{\phi}(X, A; \mathscr{F}): H_{\phi}^j(\tilde{X}, \tilde{A}; \tilde{\mathscr{F}}) \to H_{4-j}^{\phi|X\backslash A}(X\backslash A, \mathscr{F}),$$
$$Q_j^{\phi}(A, \mathscr{F}): H_{\phi \cap \tilde{A}}^j(\tilde{A}, \tilde{\mathscr{F}}) \to H_{4-j}^{\phi}(X, X\backslash A; \mathscr{F}),$$

where $\sim$ always indicates the inverse image with respect to π. If X is paracompact and ϕ paracompactifying, then $P_j^{\phi}(X, \mathscr{F})\, \pi^j: H_{\phi}^j(X, \mathscr{F}) \to H_{4-j}^{\phi}(X, \mathscr{F})$ is given by the cap product with the fundamental class.

For our purposes the case of at most isolated reducible points is of particular interest. Then there exists an exact sequence (where $\mathscr{H}_*(X, \mathscr{F})$ denotes the local homology sheaf)

$$0 \to H_{\phi}^1(\tilde{X}, \tilde{A}; \tilde{\mathscr{F}}) \xrightarrow{P_1} H_3^{\phi|X\backslash A}(X\backslash A, \mathscr{F}) \to H_{\phi}^0(X\backslash A, \mathscr{H}_3(X, \mathscr{F})) \to H_{\phi}^2(X, A; \mathscr{F})$$
$$\xrightarrow{P_2} H_2^{\phi|X\backslash A}(\tilde{X}\backslash \tilde{A}, \mathscr{F}) \to \cdots \to H_{\phi}^4(X, A; \mathscr{F}) \xrightarrow{P_4} H_0^{\phi|X\backslash A}(X\backslash A, \mathscr{F}) \to 0.$$

An analogous result is true for $Q_j^{\phi}(A, \mathscr{F})$, and there is a big commutative diagram combining $P_*^{\phi}(X, A; \mathscr{F})$, $P_*^{\phi}(X, \mathscr{F})$ and $Q_*(A, \mathscr{F})$. For more details, cf. [**K**$_1$,] [**K**$_2$] and [**BK**].

3. The equivalence type of the "symmetric bilinear form" $P_2(X, L)$**.** Let us assume that X is moreover compact, connected and locally irreducible. Then there is a useful *rank formula* for coefficients in L:

(3.1) $$\operatorname{rank}_L P_2(X, A) = b_2(X, A) + b_3(X\backslash A) - \sum_{y \in X\backslash A} b_3^y(X),$$

where b_j denotes the appropriate global and b_j^y for $y \in X$ the local Betti number. For a given basis of the free L-module $\bar{H}_2(X, L) = H_2(X, L)/\operatorname{Tors}_L H_2(X, L)$ we choose in $H^2(X, L)$ the dual basis. Then the induced homomorphism $P_2^+(X, L)$: $H^2(X, L) \to \bar{H}_2(X, L)$ is given by a symmetric bilinear matrix σ. If we denote by $I(L)$ the isotropy subspace of the symmetric bilinear form σ in $H^2(X, L)$, then the "nontrivial part" of $P_2(X, L)$ is given by

$$\bar{P}_2(X, L): H^2(X, L)/I(L) \to I^{\perp}/\operatorname{Tors}_L H_2(X, L),$$

where $I^{\perp} \subset H_2(X, L)$ denotes the Kronecker orthogonal complement of $I(R)$. It is a consequence of §1 that for the homotopy classification it is sufficient to determine $b_2(X, L)$ and the equivalence type of the "symmetric bilinear form" $\bar{P}_2(X, L)$ rather than $P_2(X, L)$ itself (in fact we identify the symmetric bilinear form and its correlation). Evidently

$$\operatorname{rank}_L P_2(X, L) = \operatorname{rank}_L \bar{P}_2(X, L).$$

In general, $\bar{P}_2(X, L)$ is not nonsingular, but it is always nondegenerate by construction. One obtains nevertheless

3.1. PROPOSITION. (a) *If* $\mathscr{H}_2(X, L)$ *has no L-torsion, then* $\bar{P}_2(X, L)$ *is an isomorphism.*

(b) *If* $H_2(X, L)$ *has no L-torsion, then the determinant of* $\bar{P}_2(X, L)$ *equals* $\pm$ *order* $\mathrm{Tors}_L H^0(X, \mathscr{H}_2(X, L))$, *if this expression is finite.*

(c) *If X is a Q(L)-homology manifold (Q(L) field of quotients of L), then* $P_2(X, L)$ *is nondegenerate and*

$$\det P_2(X, L) = \det \bar{P}_2(X, L) = \pm \frac{\mathit{order}\ \mathrm{Tors}_L H^0(X, \mathscr{H}_2(X, L))}{[\mathit{order}\ \mathrm{Tors}_L H_2(X, L)]^2},$$

if this expression is finite.

A case of particular interest is that of $L = \boldsymbol{Z}$. *We obtain*

(d) *If a prime number p divides* $\det \bar{P}_2(X, \boldsymbol{Z})$, *then* $\mathscr{H}_2(X, \boldsymbol{Z})$ *has p-torsion.*

(e) *If* $H_2(X, \boldsymbol{Z})$ *has no p-torsion, then there is an integer m such that* $(m, p) = 1$ *and*

$$\det \bar{P}_2(X, \boldsymbol{Z}) = m \cdot \mathit{order}(p\text{-}\mathrm{Tors}\ \mathscr{H}^0(X, H_2(X, \boldsymbol{Z}))).$$

As in general $H_2(X, \boldsymbol{Z})$ has direct factors $\boldsymbol{Z}/(p^n)$ for $n > 1$, it is not always sufficient to consider coefficients in an integral domain L. One has to generalise some of the relations for coefficients in $\boldsymbol{Z}/(p^n)$, which at least is a finite local ring. Instead of considering Betti numbers one counts elements. For simplicity let us assume that X is moreover *projective algebraic*; for a set D put $|D| :=$ order D. Then for any locally closed $A \subset X$ such that $H_2(X \backslash A, \boldsymbol{Z}/(p)) = 0$ one has

$$p^n \leqq |\mathrm{im}\ P_2(X, \boldsymbol{Z}/(p^n))| = \frac{p^{n(b_2+b_3)}|T \otimes \boldsymbol{Z}/(p^n)|^2}{\prod_{y \in X} |T_y \otimes \boldsymbol{Z}/(p^n)| p^{nb_2^y}} \leqq p^{nb_2(A)}$$

where $T = \mathrm{Tors}\ H_2(X, \boldsymbol{Z})$, $T_y = \mathrm{Tors}\ \mathscr{H}_2(X, \boldsymbol{Z})_y$, $b_j = b_j(X, \boldsymbol{Z})$, etc. (The first inequality holds only for $(p, \deg X) = 1$.)

If A is a connected curve such that $X \backslash A$ is contractible, then

(3.2) $$b_2(X) - b_3(X) = b_2(A) - b_1(A)$$

and hence

(3.3) $$1 \leqq 2b_3(X) - \sum_{y \in X} b_2^y + b_2(A) - b_1(A) \leqq b_2(A),$$

(3.4) $$2b_3(X) \leqq b_1(A) + \sum_{y \in X} b_2^y.$$

4. Homogeneous polynomials. If $p \in \boldsymbol{C}[z_1, z_2, z_3]$ is a polynomial, let us denote by $V(p) := \{z \in \boldsymbol{C}^3 : p(z) = 0\}$ and by $X(p)$ the closure of $V(p)$ in $\boldsymbol{P}_3$.

4.1. PROPOSITION. *The homotopy type of a normal surface* $X(p)$, *where p is homogeneous of degree a, is given by*

$$H_2(X, \boldsymbol{Z}) = \boldsymbol{Z}, \quad H_3(X, \boldsymbol{Z}) = (a-1)(a-2)\boldsymbol{Z}, \quad P_2(X, \boldsymbol{Z}) = (a);$$

hence it is completely determined by the degree of p.

PROOF. We remark that $X(p)$ is normal if and only if $V(p)$ has an isolated singularity. Let $m_i \in \boldsymbol{C}[z_1, z_2, z_3]$, $1 \leqq i \leqq r$, denote the monic monomials of degree a and let U denote the open connected submanifold of $\boldsymbol{C}^r$ such that $p_\alpha = \sum \alpha_i m_i$ has an isolated singularity for $\alpha = (\alpha_1, \cdots, \alpha_r) \in U$. Then

$$Y := \{(z, \alpha) \in \boldsymbol{P}_3 \times U : p_\alpha(z_1, z_2, z_3) = 0\}$$

has singular set $S(Y) = \{[1:0:0:0]\} \times U$. The projection map $Y \to U$ is proper and transversal to the strata of the Whitney stratification $\{Y \backslash S(Y), S(Y)\}$ of Y, hence is topologically locally trivial [**Ve**]. The homotopy type of its fibers $X(p_\alpha)$ is thus given by the data for $X(z_1^q + z_2^q + z_3^q)$, calculated in [**BK**, 3.10].

5. Weighted homogeneous polynomials. This bigger class is not as easy to handle as the homogeneous case. By definition, $p \in C[z_1, z_2, z_3]$ is *weighted homogeneous*, if there are suitable nonzero natural numbers d, q_1, q_2, q_3 with $p(t^{q_1}z_1, t^{q_2}z_2, t^{q_3}z_3) = t^d p(z_1, z_2, z_3)$ for all $t \in C$ or, equivalently, if there are positive rational numbers w_1, w_2, w_3 (the *weights*) such that $\sum_{i=1}^3 a_i/w_i = 1$ holds for any monomial $z_1^{a_1} z_2^{a_2} z_3^{a_3}$ occurring in p.

In particular $V(p)$ is contractible; hence (3.2)—(3.4) can be applied with $A := X(p)\backslash V(p)$. We call $X(p)$ a *weighted homogeneous surface* if $V(p)$ is normal. For this case, it has been proved in [**OW**, Proposition 3.1.2] and in [**A**, Proposition 11.1] that $p = f + g$, where f and g are weighted homogeneous with the same weights, f and g have no monomial in common, $V(p)\backslash\{0\}$ and $V(f)\backslash\{0\}$ are diffeomorphic, and f belongs (with a suitable numeration of the variables, multiplied by nonzero complex numbers) to one of the following seven classes:

Class	Equation	Weights
I	$Z_1^{a_1} + Z_2^{a_2} + Z_3^{a_3}$,	$w_1 = a_1$, $w_2 = a_2$, $w_3 = a_3$;
II	$Z_1^{a_1} + Z_2^{a_2} + Z_2 Z_3^{a_3}$, $a_2 \geqq 2$,	$w_1 = a_1$, $w_2 = a_2$, $w_3 = a_2 a_3/(a_2 - 1)$;
III	$Z_1^{a_1} + Z_2^{a_2} Z_3 + Z_2 Z_3^{a_3}$, $a_2, a_3 \geqq 2$,	$w_1 = a_1$, $w_2 = (a_2 a_3 - 1)/(a_3 - 1)$, $w_3 = (a_2 a_3 - 1)/(a_2 - 1)$;
IV	$Z_1^{a_1} + Z_1 Z_2^{a_2} + Z_2 Z_3^{a_3}$, $a_1 \geqq 2$,	$w_1 = a_1$, $w_2 = a_1 a_2/(a_1 - 1)$, $w_3 = a_1 a_2 a_3/(a_1 a_2 - a_1 + 1)$;
V	$Z_1^{a_1} Z_2 + Z_2^{a_2} Z_3 + Z_1 Z_3^{a_3}$,	$w_1 = (a_1 a_2 a_3 + 1)/(a_2 a_3 - a_3 + 1)$, $w_2 = (a_1 a_2 a_3 + 1)/(a_3 a_1 - a_1 + 1)$, $w_3 = (a_1 a_2 a_3 + 1)/(a_1 a_2 - a_2 + 1)$;
VI	$Z_1^{a_1} + Z_1 Z_2^{a_2} + Z_1 Z_3^{a_3} + Z_2^{c_2} Z_3^{c_3}$, $a_1 \geqq 2$, $c_2/w_2 + c_3/w_3 = 1$,	$w_1 = a_1$, $w_2 = a_1 a_2/(a_1 - 1)$, $w_3 = a_1 a_3/(a_1 - 1)$;
VII	$Z_1^{a_1} Z_2 + Z_1 Z_2^{a_2} + Z_1 Z_3^{a_3} + Z_2^{c_2} Z_3^{c_3}$, $a_1, a_2 \geqq 2$, $c_2/w_2 + c_3/w_3 = 1$,	$w_1 = (a_1 a_2 - 1)/(a_2 - 1)$, $w_2 = (a_1 a_2 - 1)/(a_1 - 1)$, $w_3 = a_3(a_1 a_2 - 1)/a_2(a_1 - 1)$.

Weighted homogeneous polynomials of Class VI exist (cf. [**A**, Proposition 11.2]) if and only if $(a_1 - 1) \mid \operatorname{lcm}(a_2, a_3)$ holds, those of Class VII if and only if $(a_1 - 1)\,(a_2, a_3) \mid (a_2 - 1)a_3$ holds.

If p is the sum of monomials $\sum_{i=1}^r m_i$, ordered by $\deg m_i =: c_i \geqq c_{i+1} \geqq 1$, then $X(p)$ is given by the homogeneous polynomial

$$\hat{p}(z_0, z_1, z_2, z_3) := \sum_{i=1}^{r} z_0^{c_1-c_i}\, m_i(z_1, z_2, z_3).$$

Corresponding to the behaviour of the "infinite curve" $X_0 = X(p) \cap \{z_0 = 0\}$ we distinguish the cases

A: $c_1 > c_2$,

B: $c_1 = c_2 > c_r$.

The case A is subdivided into the subcases A_i: "m_1 has i different irreducible factors". We are in case B if $c_1 = c_2 > c_3$ (except for the polynomials of Classes VI and VII with $c_1 = c_3 > c_4$—see Proposition 8.1 for these; the remaining polynomials of Classes VII and VI are homogeneous).

By the homotopy Lefschetz theorem on hypersurface sections all surfaces $X(p)$ under consideration are simply connected.

6. The case A_1. This follows from a more general result, which holds without any normalcy assumption.

6.1. PROPOSITION. *Let Y denote a compact locally complex surface, $A \subset Y$ a locally closed subset and $\pi: \tilde{Y} \to Y$ the normalisation map. Then*

(i) *If $H_2(\tilde{Y}\backslash\tilde{A}, G) = 0$ for some abelian group G, then*

$$P_2(X, G)\, \pi^2 \,|\, r^2 H^2(X, A; G) = 0, \quad \textit{where } r:(X, \varnothing) \hookrightarrow (X, A).$$

(ii) *If furthermore X is projective algebraic and $i^2 j^2$: $H^2(\boldsymbol{P}_n, G) \to H^2(A, G)$ is an isomorphism (where $A \hookrightarrow^i X \hookrightarrow^j \boldsymbol{P}_n$), then*

$$H^2(X, G) = r^2 H^2(X, A; G) \oplus j^2 H^2(\boldsymbol{P}_n, G).$$

(iii) *If in addition $G = L/(m)$, A homologous to $\boldsymbol{P}_1$, $H_1(\tilde{X}\backslash\tilde{A}, G) = 0$, then the form $P_2(X, G)\pi^2$ on $H^2(X, G)$ splits into the zero form and the unary form* (deg X).

The proof uses the properties of the big diagram of Poincaré homomorphisms as well as the properties of the canonical generator ω of $H^2(\boldsymbol{P}_n, G)$.

In case A_1, the infinite part X_0 of $X(p)$ is a curve, which is in fact topologically a projective line; $X\backslash X_0$ is contractible. Thus Proposition 6.1 applies, if X has an isolated singularity in the affine point 0. Now, it is easy to see that $H^2(X, X_0; G) \to^{\pi^2} H^2(\tilde{X}, \tilde{X}_0; G)$ is an isomorphism; consequently by singular Poincaré duality we have $r^2 H^2(X, X_0; G) \cong H^2(X, X_0; G) = \mathscr{H}_3(X, G)_0$. Finally, we have to apply (3.2) to obtain for all $m \geqq 0$:

$$H_3(X, \boldsymbol{Z}/(m)) \cong \mathscr{H}_3(X, \boldsymbol{Z}/(m))_0,$$
$$H_2(X, \boldsymbol{Z}/(m)) \cong \mathscr{H}_2(X, \boldsymbol{Z}/(m))_0 \oplus \boldsymbol{Z}/(m) j^* \omega;$$

$$P_2(X, \boldsymbol{Z}/(m))\, \pi^2 \,|\, \mathscr{H}_2(X, \boldsymbol{Z}/(m))_0 = 0,$$
$$P_2(X, \boldsymbol{Z}/(m))\, \pi^2\, (j^* \omega) = (\deg p) i_*[X_0].$$

7. The case A_2. Now, we will consider only the normal case; again the coefficients are taken in an integral domain L. Then we have

7.1. PROPOSITION. *If there is a connected analytic subset A with $\chi(A) \leqq 3$ in the locally irreducible projective algebraic surface X, such that* (i) *$X\backslash A$ is contractible,* (ii) *$b_{2,0} :=$ rank $H^0(X\backslash A, \mathscr{H}_2(X)) \leqq$ rank $H^0(A, \mathscr{H}_2(X)) =: b_{2,\infty}$, then*

$$b_3(X) = b_{2,0}; \qquad b_2(X) = b_3(X) + \chi(A) - 1;$$
$$0 \leqq b_{2,\infty} - b_{2,0} \leqq \chi(A) - 1 - \mathit{rank}\ P_2 \leqq 1.$$

While this result will cover the free part of $H_*(X, \mathbf{Z})$, for the torsion there is another result.

COROLLARY 1. *If moreover* $b_1(A) = 0$, $b_2(A) = 2$ *and* $b_{2,0} = b_{2,\infty}$, *then the torsion subgroups* $T := \text{Tors } H_2(X)$, $T_{2,0} := \text{Tors } H^0(X\backslash A, \mathscr{H}_2(X))$ *and* $T_{2,\infty} :=$ $\text{Tors } H^0(A, H_2(X))$ *are related by*

$$T_{2,\infty} \cong T \oplus L/(l_1) \oplus L/(l_2)$$

and an exact sequence

$$0 \to L/(s_1) \oplus L/(s_2) \to T_{2,0} \to T \to 0$$

for suitable $l_1, l_2, s_1, s_2 \in L\backslash\{0\}$.

COROLLARY 2. *There is a unit* $\varepsilon \in L$ *such that*

$$\det \bar{P}_2(X) = \varepsilon\, l_1\, l_2\, s_1\, s_2.$$

COROLLARY 3. *If the assumption of Corollary* 1 *holds for* $L = \mathbf{Z}$, *then*

$$\det \bar{P}_2(X, \mathbf{Z}) = \pm |T_{2,0}|\,|T_{2,\infty}|/|T|^2.$$

Now, it is a matter of calculation that, for all normal surfaces X defined by weighted homogeneous polynomials of case A_2 for $A := X \cap \{z_0 = 0\}$, in fact $b_{2,0} = b_{2,\infty}$. Since A is the union of two projective lines, which intersect in a point, $\chi(A) = 3$, $b_1(A) = 0$; hence all the results can be applied. In fact, the proof of Corollary 1 shows that one can actually choose $s_2 = 1 \in L$. Thus for the $\mathbf{Z}$-Betti numbers we have:

$$b_3(X) = b_{2,0}(X), \quad b_2(X) = b_{2,0}(X) + 2, \quad \text{rank } P_2(X) = 2.$$

If $T_{2,0} \otimes_{\mathbf{Z}} T_{2,\infty} = 0$, then $H(X, \mathbf{Z})$ is free and $\det \bar{P}_2(X, \mathbf{Z}) = \pm |T_{2,0}| \cdot |T_{2,\infty}|$.

Moreover, as J.M.H. Steenbrinck pointed out to us, the *signature* of $\bar{P}_2(X, \mathbf{Z})$ is zero.

8. The case B. The following polynomials of Classes II—V, case B, define surfaces which, in a different local chart $z_i = 1$, are given by polynomials of Classes I — V, case A (described here shortly by class and exponents a_1, a_2, a_3 in suitable coordinates, $c \leqq a - 1, d \leqq a - 2$).

II $(a, c, a-1)$: z_2: I $(a, a-1, a-c)$ (A_1);
III $(a, a-1, d)$: z_3: II $(a, a-1, a-1-d)$ (A_1);
IV $(a, d, a-1)$: z_2: II $(a-1, a, a-1-d)$ (A_1);
IV $(c, a-1, a-1)$: z_1: II $(a-c, a-1, a-1)$ (A_2);
V $(a-1, a-1, d)$: z_3: IV $(a-1, a-1, a-1-d)$ (A_2).

The remaining polynomials of Classes I — V, case B, and $z_1^a + z_1z_2^{a-1} + z_1z_3^c + z_2^a$ and $z_1^a + z_1z_2^{a-1} + z_1z_3^c + z_2z_3^c$ of Class VI as well as $z_1^{a-1}z_2 + z_1z_2^{a-1} + {}_1z_3^c +$ z_2^a and $z_1^{a-1}z_2 + z_1z_2^{a-1} + z_1z_3^c + \lambda z_2z_3^c$ $(0 \neq \lambda \neq 1)$ of Class VII $(c \leqq a - 2)$ are, up to a change of variables, of the form $F(z_1, z_2, z_3) = z_1^b + p(z_2, z_3)$, resp.

$G(z_1, z_2, z_3) = z_1^c z_2 + p(z_2, z_3)$, where p is homogeneous of degree a and b, $c + 1 \leqq a - 1$. For these, there is the following result:

8.1. PROPOSITION. *The homotopy type of a normal surface $X(F)$, resp. $X(G)$, depends only on the numbers a, b, c. For integral coefficients, the homology groups are free, and we have*:

(i) *For $X(F)$*: $b_3 = ((a, b) - 1)(a - 2)$, $b_2 = b_3 + a$,

$$\textit{rank}\ P_2 = a, \quad \det \bar{P}_2 = \pm (a, b)^2 m^{a-1},$$
$$\text{coker}\ P_2 = b_3 \boldsymbol{Z} \oplus \boldsymbol{Z}/(b) \oplus \boldsymbol{Z}/(a-b) \oplus (a-2)\boldsymbol{Z}/(m),$$

where $m := b(a-b)/(a, b)^2$.

(ii) *for $X(G)$*: $b_3 = ((a-1, c) - 1)(a-1)$, $b_2 = b_3 + a$,

$$\textit{rank}\ P_2 = a, \quad \det p_2 = \pm\ a^2 n^{a-1},$$
$$\text{coker}\ P_2 = b_3 \boldsymbol{Z} \oplus \boldsymbol{Z}/(ac/r) \oplus \boldsymbol{Z}/(a(a-1-c)/r) \oplus (a-2)\ \boldsymbol{Z}/(n),$$

where $r = (a-1, c)$ *and* $n = c(a-1-c)/r$.

PROOF. We denote by p_α the homogeneous polynomial $\sum_{i=0}^a \alpha_i z_2^i z_3^{a-i}$ for $\alpha = (\alpha_0, \cdots, \alpha_a) \in \boldsymbol{C}^{a+1}$. The set U (resp. W) of all α such that $X(F_\alpha)$ (resp. $X(G_\alpha)$) is normal is an open connected submanifold of $\boldsymbol{C}^{a+1}$. As in the proof of 4.1, the natural projection of $Y(F) = \{(z, \alpha) \in \boldsymbol{P}_3 \times U: z_0^{a-b} z_1^b + p_\alpha(z_2, z_3) = 0\}$ onto U (resp. $Y(G) = \{(z, \alpha) \in \boldsymbol{P}_3 \times W: z_0^{a-c-1} z_1^c z_2 + p_\alpha(z_2, z_3) = 0\}$ onto W) is topologically locally trivial, so all fibers are homeomorphic. For $F = z_1^b + z_2^a + z_3^a$, the homotopy data of (i) have been calculated in [**BK**, 3.13]. For $X(z_1^c z_2 + z_2^a + z_3^a)$, we have $X_2 = X \cap \{z_2 = 0\} \cong \boldsymbol{P}_1$ and $V_2 = X \backslash X_2$ nonsingular. The groups $H_i(V_2, \boldsymbol{Z})$ are free with Betti numbers $b_1 = (a - 1)((a - 1, c) - 1)$, $b_2 = b_1 + a - 1$ (cf. [**Ok**, §6]), and (ii) follows by application of the diagram (P) of [**BK**, 2.1] to X and $A = X_2$.

For some normal surfaces defined by polynomials of Classes VI and VII, case B, we get a defining polynomial in a different local chart $z_i = 1$ which is weighted homogeneous, case A, and with isolated affine singularity given by the indicated subpolynomial in suitable coordinates.

VI $(a_1, a-1, a_3; a-c, c)$: z_1:	$c=a-1$: II $(a-a_1, a-1, a-1)$	(A_2)
	$a_3=a-2$: II $(a-1, a-a_1, a-2)$	(A_2)
		$(a_1 \leqq a-2)$
VII $(a-1, a_2, a_3; a-c, c)$: z_2:	$1 \leqq c \leqq a-1$: II $(c, a-1, a-1-a_2)$	(A_3)
VII $(a_1, a-1, a_3; a-c, c)$: z_1:	$a_3=a-2$: III $(a-1, a-1-a_1, a-2)$	(A_2)
VII $(a_1, a_2, a-1; a-c, c)$: z_1:	$c=0$: II $(a-1, a, a-1-a_1)$	(A_1)
	$c=1$: IV $(a-1, a-1, a-1-a_1)$	(A_2)
	$a_2=a-2$: III $(a-1, a-2, a-1-a_1)$	(A_2)

9. Local homology groups. In the preceding sections, the homotopy classification has been reduced to the calculation of the local homology groups in the singular points. For isolated singularities defined by weighted homogeneous polynomials, these groups can be computed explicitly in terms of the weights (see [**Or**] for details). With the notation $\mathscr{H}_{2,0} = b_{2,0}\boldsymbol{Z} \oplus T_{2,0}$, we obtain the following list for Classes I – VII:

Class I. $b_{2,0} = c^2 c_1 c_2 c_3 - c(c_1 + c_2 + c_3) + 2$, $c = (a_1, a_2, a_3)$, $c_i = (a_j, a_k)/c$.

(α) All $(a_i, a_j) > 1$: $T_{2,0} = \boldsymbol{Z}/(c c_{12} c_{13} c_{23}) \oplus \sum (c c_k - 2)\boldsymbol{Z}/(c_{ij})$.

(β) $(a_i,a_j) = 1$: $T_{2,0}=(cc_j-1)\,\boldsymbol{Z}/(c_{ik}) \oplus (cc_i-1)\,\boldsymbol{Z}/(c_{jk})$, $c_{ij} = a_k/cc_ic_j$.

Class II. $b_{2,0} = (a_2-1, a_3)((a_1, a_2m)-1)-(a_1, a_2)+1$, $m = a_3/(a_2-1, a_3)$.

(α) $c := (a_1, a_2) = 1 : T_{2,0} = l_{23}\ \boldsymbol{Z}/(c_{23})$, $l_{23} := (a_2-1, a_3)$, $c_{23} := a_1/(a_1, a_2 m)$

(β) $c > 1 : T_{2,0} = \boldsymbol{Z}/(cc_{12}c_{23}) \oplus (c-2)\boldsymbol{Z}/(c_{12}) \oplus (l_{23}-1)\boldsymbol{Z}/(c_{23})$, $c_{12} := cm/(a_1, a_2m)$.

Class III. $b_{2,0} = d(c-1)$, $d = (a_2-1, a_3-1)$, $c = (a_1, (a_2a_3-1)/d)$, $T_{2,0} = \boldsymbol{Z}/(a_1) \oplus d\boldsymbol{Z}/(a_1/c)$.

Class IV. $b_{2,0}=d-e$, $d=(a_1a_2-a_1+1, a_2a_3)$, $e = (a_1-1, a_2)$, $T_{2,0} = \boldsymbol{Z}/(a_1a_3e/d) \oplus (e-1)\boldsymbol{Z}/(a_3e/d)$.

Class V. $b_{2,0} = (a_1a_2 - a_2+1, a_2a_3 - a_3+1) - 1$, $T_{2,0} = \boldsymbol{Z}/(c)$, $c = (a_1a_2a_3+1)/(b_{2,0}+1)$.

Class VI. $b_{2,0} = a_1l_{12}\,l_{13}/(a_1-1)t - l_{12} - l_{13}$, $t:= (a_2, a_3)/(a_1, -1, a_2, a_3)$, $l_{12} = (a_1-1, a_2)$, $l_{13} :=(a_1-1, a_3)$, $T_{2,0}=\boldsymbol{Z}/(a_1c_{12}c_{13}) \oplus (l_{12}-1)\boldsymbol{Z}/(c_{12}) \oplus (l_{13}-1)\boldsymbol{Z}/(c_{13})$, $c_{12} := a_3/tl_{13}$, $c_{13} := a_2/tl_{12}$.

Class VII. $b_{2,0} = (a_2, a_3)a_1-d-1$, $d = (a_1-1, a_2-1)$, $T_{2,0} = \boldsymbol{Z}/(c_{12}(a_1a_2-1)/d) \oplus d\boldsymbol{Z}/(c_{12})$, $c_{12} = a_3d/(a_1-1)(a_2, a_3)$.

ADDED IN PROOF. In the case of a weighted homogeneous surface X with simply connected curve at infinity, the Poincaré homomorphism $P_2\,(X, \boldsymbol{Z}/(m))$ is completely determined by $P_2\,(X, \boldsymbol{Z})$, and moreover we have

$$\det \bar{P}_2(X, \boldsymbol{Z}) = \pm \frac{|\operatorname{Tors} H^0(X, \mathscr{H}_2(X, \boldsymbol{Z}))|}{|\operatorname{Tors} H_2(X, \boldsymbol{Z})|^2}.$$

REFERENCES

[A] V. I. Arnol'd, *Normal forms of functions in neighbourhoods of degenerate critical points*, Uspehi Mat. Nauk **29** (1974), no. 2, 11–49 = Russian Math. Surveys **29** (1974), no. 2, 10–50.

[BK] G. Barthel and L. Kaup, *Homotopieklassifikation einfach zusammenhängender normaler kompakter komplexer Flächen*, Math. Ann. **212** (1974), 113–144.

[Kt] M. Kato, *Partial Poincaré duality for k-regular spaces and complex algebraic sets*, Sonderforschungsbereich 40 Theoretische Mathematik, Bonn, 1974 (preprint).

[K$_1$] L. Kaup, *Poincaré-Dualitäte für Räume mit Normalisierung*, Ann. Scuola Norm. Sup. Pisa **26** (1972), 1–31.

[K$_2$] ———, *Zur Homologie projektiv algebraischer Varietäten*, Ann. Scuola Norm. Sup. Pisa **26** (1972), 479–513.

[Mi] J. Milnor, *On simply connected* 4-*manifolds*, Internat. Sympos. Algebraic Topology, Universidad Nacional Autónoma de Mexico; UNESCO, Mexico City, 1958, pp. 122–128. MR **21** #2240.

[Ok] M. Oka, *On the homotopy types of hypersurfaces defined by weighted homogeneous polynomials*, Topology **12** (1973), 19–31.

[Or] P. Orlik, *On the homology groups of weighted homogeneous manifolds*, Proc. Second Conf. Compact Transformation Groups, part I, Lecture Notes in Math., vol. 298, Springer-Verlag, Berlin and New York, 1972, pp. 260–269.

[OW] P. Orlik and P. Wagreich, *Isolated singularities of algebraic surfaces with C*-action*, Ann. of Math. (2) **93** (1971), 205–228. MR **44** #1662.

[Ve] J.-L. Verdier, *Constructability*, Proceedings of the Summer Institute, 1975.

[W] C. T. C. Wall, *Finiteness conditions for* CW-*complexes*, Ann. of Math. (2) **81** (1965), 56–69. MR **30** #1515.

[Wh] J. H. C. Whitehead, *On simply connected* 4-*dimensional polyhedra*, Comment. Math. Helv. **22** (1949), 48–92. MR **10**, 559.

UNIVERSITÄT KONSTANZ

Proceedings of Symposia in Pure Mathematics
Volume 30, 1977

HOLOMORPHIC VECTOR FIELDS ON PROJECTIVE VARIETIES

DAVID LIEBERMAN*

Given a holomorphic vector field V on a compact Kähler manifold X much of the global geometric information about X is determinable on the zero set Z of V. For example, the cohomology of X and Chern invariants of X are determined on Z as noted in several works. In particular we recall that if Z is nonempty then the plurigenera of X vanish [9] and that $H^q(X, \Omega^p) = 0$ if $|p - q| > \dim(Z)$ [6]. From this latter result it follows that a Kähler manifold having a holomorphic V with isolated zeroes is necessarily algebraic. Employing such numerical information and the classification of surfaces, Carrell-Howard-Kosniowski [5] yields a nearly complete classification of vector fields on compact complex surfaces, showing in particular that the only surfaces having vector fields with isolated zeroes are rational, i.e., obtained from $\boldsymbol{P}^2$ by a sequence of blowing up and blowing down with isolated zeroes as centers.

Carrell has conjectured that isolated zeroes imply rationality in arbitrary dimension. In studying this problem we have obtained the birational classification of vector fields on algebraic varieties indicated below. Our initial proof of this theorem was based upon an equivariant projection argument inspired by A. Howard's work [8]. The present argument, employing algebraic groups and the fundamental work [3], was suggested to us by P. Deligne and D. Mumford. We gratefully acknowledge several stimulating conversations with R. Risch who brought to our attention the prior work of R. Hall (cf. [7], [15]) which gives the birational classification by employing an inductive argument in which the work of Painlevé [13] on surfaces plays the key role. By employing the theory of algebraic

AMS (*MOS*) *subject classifications* (1970). Primary 32M05, 14E05.
Key words and phrases. Birational classification, Zariski dense orbit, quasi-homogeneous manifold.
*Partially supported by NSF grant MPS 71–02759 AO4 and Sloan Foundation grant BR 1391.

© 1977, American Mathematical Society

groups one is led to a unified, simple treatment and to sharper, more extensive results, including the generalized Poincaré-Bendixson Theorem 1(d), and the fact that X is rational if V has nondegenerate rational zeroes, 1(e). We should remark that Hall's work has been ignored though it implies many later results (cf. **[8]**, **[9]**, **[11]**).

The classification theorem. Given an arbitrary manifold M, a torus $T = C^m/\Gamma$ and a projective space $\boldsymbol{P}^r$ one obtains vector fields on $M \times T \times \boldsymbol{P}^r$ of the form $W = (0, \sum_i c_i \, \partial/\partial z_i, \sum c_{ij} X_j \, \partial/\partial X_i)$ where the c_i, c_{ij} are constants, z_i are the coordinates on C^m and X_i are the homogeneous coordinates on $\boldsymbol{P}^r$. Furthermore, if Λ is any finite group of translations on T and an action $\Lambda \times M \to M$ is given then W induces a vector field W on the manifold $S = (M \times T \times \boldsymbol{P}^r)/\Lambda$, obtained by identifying (m, t, p) and $(\lambda m, t + \lambda, p)$. Finally, if $Y \subseteq S$ is an invariant submanifold (i.e., the ideal sheaf $\mathscr{I}_Y$ satisfies $W(\mathscr{I}_Y) \subseteq \mathscr{I}_Y$) then W lifts to the monoidal transform of S along Y. We write $(\tilde{X}, \tilde{V}) \approx (X, V)$ if there exists an $(\tilde{\tilde{X}}, \tilde{\tilde{V}})$ obtained from (X, V) by a sequence of such equivariant monoidal transforms and holomorphic modification $f\colon \tilde{\tilde{X}} \to \tilde{X}$ with $f_*(\tilde{\tilde{V}}) = \tilde{V}$.[1] We denote by Z, the zeroes of V, the subvariety of X defined by the ideal sheaf $\mathscr{I}_Z$ generated by $V(\mathcal{O}_X) \subseteq \mathcal{O}_X$ and let $\mathcal{O}_Z = \mathcal{O}_X/\mathscr{I}_Z$. Clearly $V(\mathscr{I}^i_Z) \subseteq \mathscr{I}^i_Z$ and we denote by $\mathscr{L}_V : \mathscr{I}_Z/\mathscr{I}^2_Z \to \mathscr{I}_Z/\mathscr{I}^2_Z$ the $\mathcal{O}_Z$ linear map induced by V. A point $z \in Z$ is said to be nondegenerate if z is a nonsingular point on Z and X and $\det(\mathscr{L}_V)_Z \neq 0$. We say V has *nondegenerate* zeroes if every $z \in Z$ is nondegenerate (X necessarily nonsingular). The zeroes of V are called *rational* if every component of Z is birational to a projective space (e.g., if $\dim Z = 0$).

THEOREM 1. *Given a holomorphic vector field V on a projective algebraic variety X then*

	PROPERTIES OF V	PROPERTIES OF X
(a)	*arbitrary*	$\approx S$ (*as above*)
(b)	$Z \neq \emptyset$	$\approx M \times \boldsymbol{P}^r$ (*ruled*)
(c)	*Zariski dense orbit*	$\approx T \times \boldsymbol{P}^r$
(d)	*dense real orbit*	$= T$
(e)	*nondegenerate rational zeroes*	$\approx \boldsymbol{P}^n$

REMARKS. The general classification (a) is given in [7] and one readily deduces (b), (c). Severi had earlier obtained case (c), the "quasi-abelian varieties". An analog of case (c) for Kähler manifolds has been studied in **[1]**, **[12]**, **[14]**. The result (d), which asserts that the only holomorphic flows with a dense real orbit are the skew translations on tori, generalizes the classical Poincaré-Bendixson theory. This result is false in the compact complex category since every compact nilmanifold **[2]** admits such vector fields. The analogous question for Kähler manifolds remains open. It is not known whether such a manifold is even quasi-homogeneous.

PROOF. Recalling that Aut(X) is an algebraic group we denote by G the algebraic subgroup obtained as Zariski closure of the 1-parameter group generated by V. Clearly G is a connected abelian group, hence of the form

[1]The factorization of an arbitrary equivariant birational map by a sequence of equivariant monoidal transforms has been established by Hironaka (to appear).

$$(1.1) \qquad 0 \to L \to G \to T \to 0,$$

where T is an abelian variety and L is a product of $\boldsymbol{C}$'s and $\boldsymbol{C}^*$'s. (In fact G is the quotient of its Lie algebra $\mathscr{G}$ by a discrete subgroup Γ. Note that if the real 1-parameter group $\exp(tV)$ is dense in G then necessarily Γ is maximal rank, i.e., $G = T$. Similarly, one concludes in general that at most one $\boldsymbol{C}$ factor appears in L, by decomposing $\mathscr{G} = \mathscr{G}_1 \oplus \boldsymbol{C}\cdot\Gamma$ and noting that the projection of $\exp(tV)$ in $\mathscr{G}_1 = \exp(\mathscr{G}_1)$ must be Zariski dense (since G is the closure of $\exp(tV)$) whence $\dim \mathscr{G}_1 \leqq 1$.)

(c) Assuming $G\cdot x$ dense in X for some x, then $G \to Gx$ is an isomorphism of G with an *open* subset of X (the isotropy group G_x is trivial since G is abelian and $G\cdot x$ dense; moreover, $G\cdot x$ is constructible (Chevalley) whence open). Thus X is an equivariant completion of G, and since (1.1) splits birationally [**15**, §5] we have $X \approx \boldsymbol{P}^r \times T$.

(d) The real orbit $\exp(tV)\cdot x$ is assumed dense in X, thus $\exp(tV)$ is dense in G since the topology on $G = G\cdot x$ is *induced* by X. Necessarily $G = T$ (see above) and is hence compact, whence $T = \overline{G\cdot x} = X$.

(a) Let M' be a nonsingular projective model for the (rational) quotient X/G [**15**, Theorem 2] and let N be a nonsingular projective model for X/L on which T acts, so that $M' = N/T$. Since L is solvable, $X \approx N \times \boldsymbol{P}^r$ by virtue of Theorem 10 in [**15**] (cf. [**10**]). (In fact one may show more precisely $X \approx X/L \times L$ by arguing that, for $x \in U$ Zariski open in X, the stabilizer L_x is trivial.) If the map $N \to M' = N/T$ admits a (rational) section, then $N \approx T \times M'$. In general no rational section exists, although such sections will exist if M' is replaced by a suitable variety M whose field of meromorphic functions $k(M)$ is finite Galois over $k(M')$ with group Λ and one may assume Λ acts biregularly on M with $M' = M/\Lambda$. Taking $N' = N \times_{M'} M$ one has $X \approx N' \times \boldsymbol{P}^r/\Lambda$ with $N' \approx T \times M$, as described in (a). The description of $N = T \times M/\Lambda$ may also be obtained as in [**4**] by studying the Albanese map.

(b) Fixing a zero $z \in Z$, note that the linear representation of G on the m-jets at z is faithful if $m \gg 0$. Hence $T = 0$ [**15**, §5].

(e) We proceed by induction on $\dim(X) = n$. Since $Z \neq \emptyset$ we know $G = L$ is a product of $\boldsymbol{C}^*$'s and at most one $\boldsymbol{C}$. The hypothesis that V have nondegenerate zeroes is equivalent to G acting without eigenvectors of weight 1 in the conormal bundle. Note therefore $G \neq \boldsymbol{C}$ since algebraic representations of $\boldsymbol{C}$ are unipotent. Thus $G = G_1 \times \boldsymbol{C}^*$. Note that $X_1 = X^{\boldsymbol{C}^*}$ is nonsingular and that $Z = X_1^{G_1}$ and G_1 must act nondegenerately on the conormal of Z in X_1. By induction, all components of X_1 are rational. However one has $X \approx X' \times \boldsymbol{P}^r$ for X' a suitable component of X_1, in view of [**3**]. Hence X is rational.

BIBLIOGRAPHY

1. K. Akao, *On prehomogeneous compact Kähler manifolds*, Proc. Japan Acad. **49** (1973), 483–485. MR **48** #11584.

2. L. Auslander, L. Green and F. Hahn, *Flows on homogeneous spaces*, Ann. of Math. Studies, no. 53, Princeton Univ. Press, Princeton, N. J., 1963, p. 157. MR **29** #4841.

3. A. Bialynicki-Birula, *Some theorems on actions of algebraic groups*, Ann. of Math. (2) **98** (1973), 480–497.

4. J. Carrell, *Holomorphically injective complex toral actions*, Proc. Conf. on Compact Trans-

formation Groups, Lecture Notes in Math., vol. 299, Springer-Verlag, Berlin and New York, 1971, pp. 205–236.

5. J. Carrell, A. Howard and C. Kosniowski, *Holomorphic vector fields on complex surfaces*, Math. Ann. **204** (1973), 73–82.

6. J. Carrell and D. Lieberman, *Holomorphic vector fields and Kähler manifolds*, Invent. Math. **21** (1973), 303–309. MR **48** #4356.

7. R. Hall, *On algebraic varieties which possess finite continuous commutative groups of birational self-transformations*, J. London Math. Soc. **30** (1955), 507–511. MR **17**, 411.

8. A. Howard, *Holomorphic vector fields on algebraic manifolds*, Amer. J. Math. **94** (1972), 1282–1290. MR **46** #9384.

9. S. Kobayashi, *Transformation groups in differential geometry*, Ergebnisse der Math. und ihrer Grenzgebiete, Band 70, Springer-Verlag, New York and Heidelberg, 1972. MR **50** #8360.

10. H. Matsumura, *On algebraic groups of birational transformations*, Atti Accad. Naz. Lincei Rend. Cl. Sci. Fis. Mat. Natur. (8) **34** (1963), 151–155. MR **28** #3041.

11. Y. Matsushima, *Holomorphic vector fields and the first Chern class of Hodge manifold*, J. Differential Geometry **3** (1969), 477–480. MR **42** #8431.

12. E. Oeljeklaus, *Fasthomogne Kählermannigfaltigkeiten mit verschwindender erster Bettizahl*, Manuscripta Math. **7** (1972), 175–183. MR **46** #6256.

13. P. Painlevé, *Leçons sur la théorie analytique des équations différentielles*, Paris, 1897, pp. 257–266.

14. J. Potters, *On almost homogeneous compact complex analytic surfaces*, Invent. Math. **8** (1969), 244–266. MR **41** #3808.

15. M. Rosenlicht, *Some basic theorems on algebraic groups*, Amer. J. Math. **78** (1956), 401–443. MR **18**, 514.

16. L. Roth, *Algebraic threefolds*, Ergebnisse der Math. und ihrer Grenzgebiete, Band 6, Springer-Verlag, Berlin, 1955. MR **17**, 897.

17. F. Severi, *Funzioni quasi abeliane*, Pontificiae Academiae Scientiarum Scripta Varia, vol. 4, Vatican City, 1947. MR **9**, 587.

BRANDEIS UNIVERSITY

Proceedings of Symposia in Pure Mathematics
Volume 30, 1977

THE TOPOLOGICAL TYPE OF ALGEBRAIC SURFACES: HYPERSURFACES IN $P^3(C)$

R. MANDELBAUM AND B. MOISHEZON

1. Introduction. It is well known (Pontryagin, Whitehead, Milnor) that the homotopy type of compact simply connected 4-manifolds is determined by the congruence class of their intersection form. Wall has extended this result to show that two simply connected such manifolds are h-cobordant iff they have congruent intersection forms. If the h-cobordism theorem were known to be true in dimension 4 then we would also know the diffeomorphy type of our manifolds; however, its truth is not known and, again by Wall, the most we can assert about h-cobordant simply connected 4-manifolds V_1, V_2 is that there exists some integer k such that $V_1 \# k(S^2 \times S^2) = V_2 \# k(S^2 \times S^2)$ (where $\#$ is the connected sum operation in the sense of Kervaire).

Now Wall's result is a pure existence proof and offers no clue on how to estimate k. It is then a question of some interest to find estimates on k. We can slightly rephrase this question as follows. It is well known that $(S^2 \times S^2) \# P = 2P \# Q$ (where $P = P^2(C)$ with its usual orientation and Q is $P^2(C)$ with opposite orientation), and so Wall's result implies the existence of integers k_1, k_2 such that

$$(*) \qquad V_1 \# k_1 P \# k_2 Q = V_2 \# k_1 P \# k_2 Q.$$

We can now ask about the minimum k_1, k_2 necessary for equality to hold in $(*)$.

A promising source of simply connected 4-manifolds are the nonsingular hypersurfaces in $P^3(C)$. Since all such hypersurfaces of the same degree are diffeomorphic we can consider explicit models V_n with affine equation $x_1^n + x_2^n + x_3^n = 1$ (or, in projective coordinates, $w_0^n + w_1^n + w_2^n + w_3^n = 0$).

Now it is clear that $V_1 = P$, $V_2 = S^2 \times S^2$, and it is a classical result that V_3 is simply P blown up at 6 points lying on a cubic, or in the language of differential topology, $V_3 = P \# 6Q$. V_4 is the diffeomorphic model of the $K3$-surface and essentially nothing is known about its topology.

AMS (MOS) subject classifications (1970). Primary 57D55, 57A15, 14J99.

© 1977, American Mathematical Society

We can now show that for the V_n, $n \geqq 4$, we can take $k_1 = 1$, $k_2 = 0$ in (2) to obtain a topological decomposition. More specifically we can show that

$$V_n \# P = k_n P \# l_n Q$$

where $k_n = (n/3)(n^2 - 6n + 11)$, $l_n = ((n-1)/3)(2n^2 - 4n + 3)$. Thus, for example, the $K3$-surface V_4 can be topologically decomposed as $V_4 \# P = 4P \# 19Q$.

We first indicate the basic analytic idea behind our proof for V_4 and then show what modifications are necessary to obtain the whole proof.

2. A homology basis for V_4. Let $V^4(\tau, \lambda)$ be the projective algebraic variety with affine part $x_1^3 + x_2^3 + x_3^3 + \tau(x_1^4 + x_2^4 + x_3^4) = \lambda$. Fixing τ, λ such that $0 < |\tau|^2 + |\lambda|^2 < 1/4$, we can insure that $V^4(\tau, \lambda)$ is nonsingular and of degree 4. Let $Z^3(\lambda)$ be the projective algebraic variety given (affinely) by $Z^3(\lambda) = V^4(0, \lambda) = \{Z_1^3 + Z_2^3 + Z_3^3 = \lambda\}$ and let $W^4(\tau) = V^4(\tau, 0)$.

Let B_ε be a closed ball of radius ε about 0 in $C^3 \subset P^3(C)$ and let B'_ε be the closure of its exterior in $P^3(C)$. Choosing $\lambda, \tau, \varepsilon$ sufficiently small we can verify that $B_\varepsilon \cap V^4(\tau, \lambda)$ is diffeomorphic to $B_\varepsilon \cap Z^3(\lambda)$. But as in Milnor [**2**], $B_\varepsilon \cap Z^3(\lambda)$ is diffeomorphic to $V_3 - \{$tubular neighborhood of hyperplane section at $\infty\}$. Now it is easy to see that $B'_\varepsilon \cap V^4(\tau, \lambda)$ is diffeomorphic to $B'_\varepsilon \cap W^4(\tau)$. Blowing up $W^4(\tau)$ at the origin to obtain $\tilde{W}^4$ we find that $\tilde{W}^4$ is simply P blown up at 12 points realizable as the points of intersection of a generic C_3 and C_4 in P and $B'_\varepsilon \cap W^4(\tau) = \tilde{W}^4 - T$, where T is a tubular neighborhood of the strict image of C_3 in $\tilde{W}_4$.

Denoting P blown up at k points by $\sigma_k(P)$ we have found (using our previously mentioned result on V_3) that

$$B_\varepsilon \cap V^4(\tau, \lambda) = \overline{V_3 - T_1} = \overline{\sigma_6(P) - T_1},$$
$$B'_\varepsilon \cap V^4(\tau, \lambda) = \overline{\tilde{W}_4 - T_2} = \overline{\sigma_{12}(P) - T_2},$$

where T_1, T_2 are both tubular neighborhoods of elliptic curves whose boundary circle bundles have Euler number $+3$ and -3 respectively.

In particular $\partial T_1 \approx_f \partial T_2$ by means of an orientation-reversing diffeomorphism and we obtain

$$V_4 = \overline{\sigma_6(P) - T_1} \cup_f \overline{\sigma_{12}(P) - T_2}.$$

Let us analyze this decomposition a bit more closely. We have the following picture

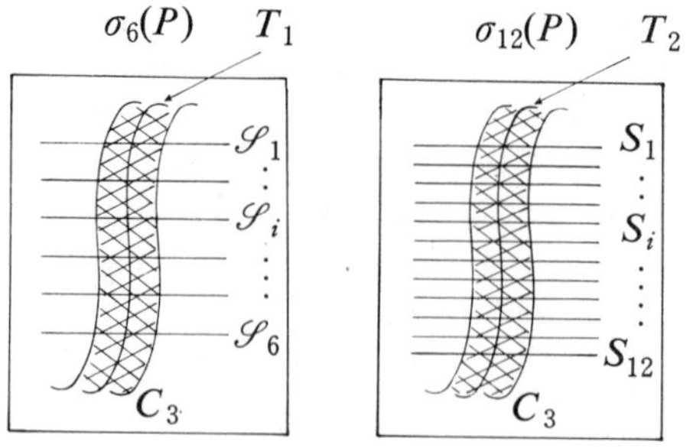

FIGURE 1. Note that $\mathscr{S}_i^2 = S_i^2 = -1$; all other intersections are zero.

Performing the indicated identification we can construct the following independent 2-cycles on V_4:

$$\begin{aligned} \tau_i &= [\mathscr{S}_i - (\mathscr{S}_i \cap T_1)] - [\mathscr{S}_{i+1} - (\mathscr{S}_1 \cap T_1)], & i &= 1, \cdots, 5, \\ \mathscr{T}_i &= [S_i - (S_2 \cap T_2)] - [S_{i+1} - (S_i \cap T_1)], & i &= 1, \cdots, 11, \\ \Sigma &= S_1 - (S_1 \cap T_2) + \mathscr{S}_1 - (\mathscr{S}_1 \cap T_1), \end{aligned}$$

which will have the following intersection matrix

$$\begin{array}{c|cccccccc} & \tau_5 & & & & & & & \mathscr{T}_{11} \\ \hline \tau_5 & -2 & 1 & & & & & & \\ & 1 & -2 & & & & & & \\ & & 1 & -2 & 1 & & & & \\ & & & 1 & -2 & 1 & & & \\ & & & & 1 & -2 & 1 & & \\ & & & & & & \ddots & & \\ \mathscr{T}_{11} & & & & & & & 1 & -2 \end{array}$$

FIGURE 2

We obtain an additional cycle l_1 by taking the strict image in $\sigma_{12}(P)$ of a projective line l in P interesting C_3 at 3 points e_1, e_2, e_3 (not the centers of any of our σ-processes) and setting

$$L_1 = \overline{l_1 - (l_1 \cap T_2)} + \bigcup_{i=4}^{i=6} \overline{(\mathscr{S}_i - (\mathscr{S}_i \cap T_1))}.$$

L_1 is then an additional 2-cycle in V_4 with intersection characteristics as in Figure 3.

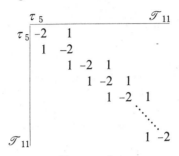

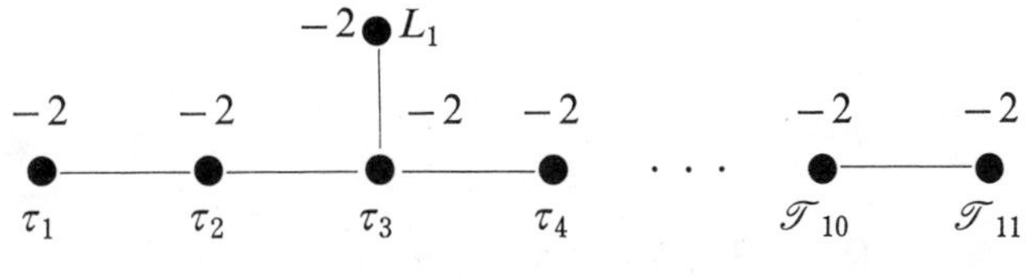

FIGURE 3

The last 4 2-cycles of V_4 arise out of a consideration of the tubular neighborhood T of C_3. C_3 is topologically a torus and we can take a primitive basis α, β of 1-cycles on it, with intersection matrix $\left(\begin{smallmatrix} 0 & 1 \\ 1 & 0 \end{smallmatrix}\right)$. Then in $\partial T \subset V_4$, α, β will have as preimages tori A_1, A_2 which are then 2-cycles in V_4. Furthermore we can suppose C_3 is a general member of a pencil of elliptic curves in P and suppose that α, β are then Lefschetz vanishing cycles bounding Lefschetz relative cycles D, E of this pencil. Then in $\sigma_6(P)$ and $\sigma_{12}(P)$ we can find 2-discs D_1, E_1 and D_2, E_2, respectively, whose boundaries are the strict images of α, β in $\sigma_6(P)$ and $\sigma_{12}(P)$, respectively. We set

$$\Delta_1 = \overline{(D_1 - (D_1 \cap T_1))} + \overline{(D_2 - (D_2 \cap T_2))},$$
$$\Delta_2 = (E_1 - (E_1 \cap T_1)) + (E_2 - (E_2 \cap T_2)),$$

to get an additional two 2-cycles in V_4.

The resulting intersection configuration has intersection matrix

	A_1	A_2	Δ_1	Δ_2
A_1	0	0	0	1
A_2	0	0	1	0
Δ_1	0	1	-2	0
Δ_2	1	0	0	-2

FIGURE 4

We have thus obtained all 22 2-cycles in $H_2(V_4)$. We note that topologically the classical σ-process at a point P of an algebraic surface M is simply the operation $M \# Q$. Similarly we can interpret $M \# P$ simply as the performing of a σ-process at p, but in conjugate local coordinates. (Note that this operation destroys complex structure.) We call such a process a $\bar{\sigma}$-process. We then denote M with $l\bar{\sigma}$-process by $\sigma^l(M)$. Then $V_4 \# P$ can be constructed by performing a $\bar{\sigma}$-process at an arbitrary point of V_4. We perform this process at some point

$$\zeta \in S_{12} - (\overline{S_{12} - T_2}) \subset \sigma_{12}(P) - T_2 \subset V_4.$$

We call the strict image of S_{12} under this process S_{12}. S_{12} now has self-intersection 0 and can be used to perform ambient surgery on the hitherto obtained tori to replace them by homologically equivalent spheres. Note also that T_{11}^2 now becomes -1. We get a basis for $V_4 \# P$ as follows: Let l_2 be the strict image in $\sigma_6(P)$ of the projective line l in P. Set $\tilde{L}_2 = l_1 - (l_1 \cap T_2) + l_2 - (l_2 \cap T_1)$. Perform ambient surgery on $\tilde{L}_2$ to get a homologically equivalent sphere L_2. Then $L_2^2 = +2$, $L_2 \cdot L_1 = +1$ and L_2 has empty intersection with all the previously listed cycles of V_4. We thus obtain the following configuration of 23 2-cycles in $V_4 \# P$

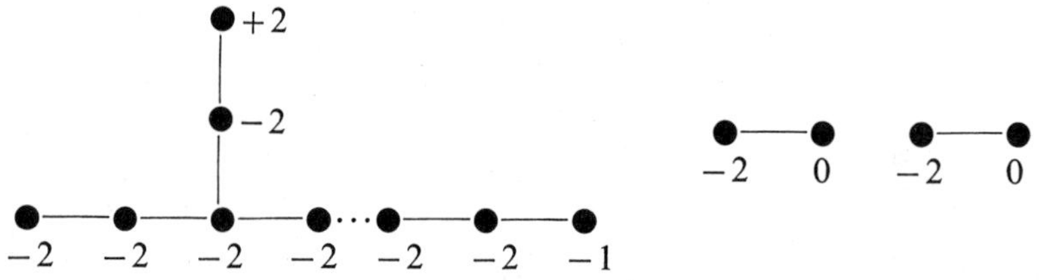

FIGURE 5

Now the long configuration above can be blown down to the configuration

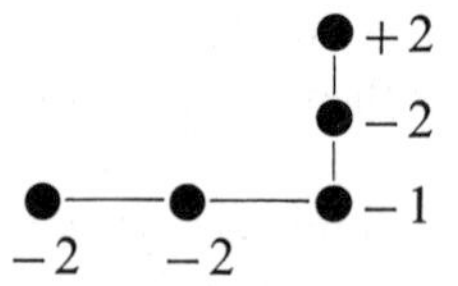

which can be successively blown down as follows

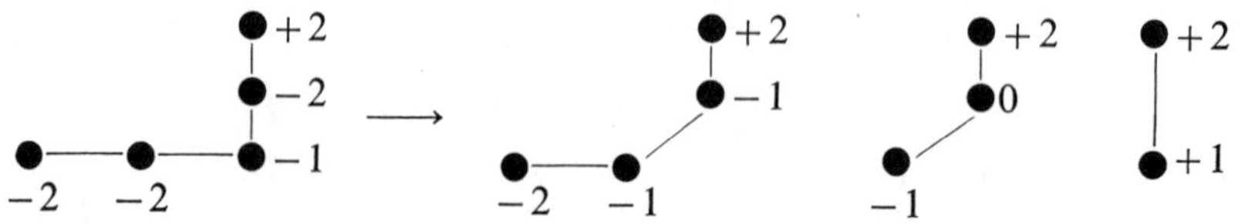

FIGURE 6

But

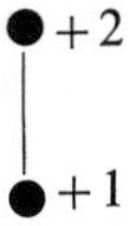

is simply two successive $\bar{\sigma}$-processes so our sequence of blowing downs gives us

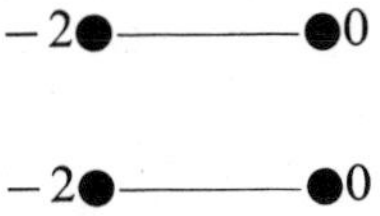

$\# \, 2P \, \# \, 17Q$

However

can be split off as an $S^2 \times S^2$ and so we can essentially obtain $V_4 \# P = M^4 \# 2P \# 17Q \# 2(S^2 \times S^2) = M^4 \# 4P \# 19Q$ where M^4 is a homotopy 4-sphere. The question which is left is then whether $M^4 \approx S^4$. To prove this requires redoing the above in a more subtle manner keeping tighter control of the blowing down processes, especially the last one. We sketch this tight control as well as the proof of the theorem in the general case as follows:

3. Topological structure of V_n.

THEOREM. *Let V_n be a nonsingular hypersurface of degree n in $P^3(C)$. Let $k_n = (n/3)(n^2 - 6n + 11)$, $l_n = (1/3)(n-1)(2n^2 - 4n + 3)$. Then*

(1) $$V_n \# P = k_n P \# l_n Q.$$

SKETCH OF PROOF. We proceed by induction. The cases $n = 1, 2, 3$ are well known and we assume the theorem is true for all integers $\leqq n$ $(n \geqq 3)$. We now represent V_{n+1} as the closure in $P^3(C)$ of the affine variety

(2) $$X^n + Y^n + Z^n + \tau(X^{n+1} + Y^{n+1} + Z^{n+1}) = \varepsilon.$$

Then just as in the case of $n = 4$ we can in general decompose V_{n+1} as

(3) $$V_{n+1} = \overline{V_n - T(E_n)} \cup \overline{\sigma_{n(n+1)}(P) - T(C_n)}$$

(where C_n is the strict image of a curve of degree n in P under $n(n+1)$ σ-processes with centers on it and E_n is a hyperplane section of V_n).

Note that $E_n^2 = +n$, $C_n^2 = -n$ so that the boundaries above are identifiable. We now perform a $\bar{\sigma}$-process on $\overline{V_n - T(E_n)}$. Then using induction we obtain

(4) $$(V_n - T(E_n)) \# P = [\sigma'_{n(n-1)}(P) - T(C'_n)] \# M_n$$

with

$$M_n = (1/3)(n^3 - 6n^2 + 11n - 6)P \# (1/3)(n-1)(2n^2 - 5n + 3)Q.$$

$\sigma'_{n(n-1)}(P) = \sigma^1(\sigma_{n(n-1)}(P))$ and C'_n is the strict image in $\sigma'_{n(n-1)}(P)$ of a curve of degree n in P under $n^2 - n$ σ-processes with centers on it and a $\bar{\sigma}$-process with center off it. We thus have that

(5) $$V_{n+1} \# P = M_n \# [(\sigma'_{n(n-1)}(P) - T(C'_n)] \cup [\sigma_{n(n+1)}(P) - T(C_n)].$$

The remaining topological problem is then the following:

Let C be a nonsingular algebraic curve of degree n in P ($g = \frac{1}{2}(n-1)(n-2)$).

Let $p_1, \cdots, p_k, q_1, \cdots, q_l$ be distinct points in C with $k + l = 2n^2$ and $k \geq 1$. Let Σ_1 be P blown up by σ-processes at p_i, Σ_2 be P blown up by σ-processes at q_j, C_1, C_2 the strict image of C in Σ_1, Σ_2, respectively, with tubular neighborhoods T_1, T_2. Let S be the total image of p_1 in Σ_1 and let Σ_1' be Σ_1 blown up by a $\bar{\sigma}$-process of some point $\xi \in S - (S \cap T_1)$ and denote the strict images of C_1, T_1 in Σ_1' again by C_1, T_1. Then $\partial(\Sigma_1' - T_1) \approx \partial(\Sigma_2 - T_2)$ and if

$$V = \overline{\Sigma_1' - T_1} \cup \overline{\Sigma_2 - T_2} = V_1' \cup V_2,$$

then we must show

$$(6) \qquad V = (2g+2)P \# (2g + 2n^2 - 1)Q \quad \text{where } g = \tfrac{1}{2}(n-1)(n-2).$$

(Actually in the case $n + 1 = 4$ this is all that is to be shown.)

We break this up into three parts. First of all we note that the $\bar{\sigma}$-process on S gives us a trivial 2-sphere bundle ξ with base a 2-disc M in C' around p_1' ('means strict image in Σ_1') such that $B = \xi \cap V_1'$ is a 4-handle $\approx D^2 \times D^2$ in V_1'. Then by a careful analysis of our blowing up and blowing downs we can verify that:

$$(7) \qquad \overline{V_1' - B} = \overline{S^2 \times D^2 - T^*} \# P \# (k-1)Q$$

where T^* is the image of $T_1|_{\overline{C-M}}$ under a smooth embedding;

$$(8) \qquad V_2 \cup B \approx \overline{S^4 - T^\wedge} \# P \# lQ$$

where $T^\wedge$ is the image of $T_2|_{\overline{C-N}}$ (N a 2-disc in C) in S^4 under a smooth embedding. Thus we can conclude that $V = V_1' \cup V_2 = \overline{(V_1' - B)} \cup (V_2 \cup B)$ so that

$$(9) \qquad V = 2P \# (2n^2 - 1)Q \# [\overline{S^2 \times D^2 - T^*} \cup \overline{S^4 - T^\wedge}].$$

Thus our topological problem is reduced to showing that

$$\overline{S^2 \times D^2 - T^*} \cup \overline{S^4 - T^\wedge} = W$$

where $W = 2g(S^2 \times S^2)$ or $W = 2gP \# 2gQ$.

This is done by the use of surgery in the following way.

Step 1. We can add $2g$ 4-handles $D_i = (D^2 \times D^2)_i$ to T ($T \approx T_2|_{\overline{C-M}}$) such that

(1) $Z = (S^4 - T^\wedge) \cup (\bigcup D_i)$ is diffeomorphic to $W - D^4$ with W as above.

(2) $X = T \cup \bigcup D_i$ is diffeomorphic to D^4.

Suppose the $2g$ 4-handles D_i we used in (1) can be found already in $\overline{S^2 \times D^2 - T^*}$. In this case we can again by a careful combinatorial argument show that

$$(10) \qquad \overline{S^2 \times D^2 - T^*} - \bigcup D_i \approx S^4 - D^4 \approx D^4.$$

But then we would have

$$\begin{aligned}\overline{S^4 - T^\wedge} \cup \overline{S^2 \times D^2 - T^*} &= (\overline{S^4 - T^\wedge} \cup \bigcup D_i) \cup (\overline{S^2 \times D^2 - T^*} - \bigcup D_i) \\ &= W - D^4 \cup S^4 - D^4 = W \# S^4 = W.\end{aligned}$$

Our topological problem is then reduced to the question of whether *ambient* surgery can be performed on T^* in $\overline{S^2 \times D^2}$.

This problem however can be reduced to the following algebraic-geometric theorem due to Wajnryb.

THEOREM (WAJNRYB). *Let $\tilde{C}$ be a nonsingular curve of genus g in P. Then it is possible to choose a pencil of curves in P containing $\tilde{C}$ and Lefschetz relative cycles $\Delta_1, \cdots, \Delta_{2g}$ whose boundaries $e_1, \cdots, e_{2g}$ are Lefschetz vanishing cycles forming a bouquet of* 1*-spheres at some point $a \in \tilde{C}$ such that $\tilde{C} - \bigcup_{i=1} e_i$ is an open* 2*-disc.*

4-handles D_i usable to kill $T|_{\overline{C-M}}$ in some $\sigma(P)$ are found by taking tubular neighborhoods $\tilde{D}_i$ of the Lefschetz relative cycles Δ_i above and setting $D_i = \tilde{D}_i \cap \overline{\sigma(P) - T}$. Then 4-handles in $S^2 \times D^2 - T^*$ can be obtained by noting that $S^2 \times D^2$ can be obtained from $\sigma_1(P) = P \# Q$ (which is the unique S^2 bundle over S^2) by subtracting an $S^2 \times D^2$ (with D^2 arbitrarily small). As the constructions of the D_i above can all be done within some ball or radius r in $C^2 \subset P$, subtracting $S^2 \times D^2$ from $\sigma_1(P)$ can be done without affecting the D_i. This essentially summarizes how the ambient 4-handles are obtained thus concluding our treatment.

Full details and complete proofs will appear in Topology **15** (1976), 23–40.

REFERENCES

1. M. A. Kervaire and J. Milnor, *Groups of homotopy spheres*, Ann. of Math. (2) **77** (1963), 504–537. MR **26** #5584.
2. J. Milnor, *Singular points of complex hypersurfaces*, Ann. of Math. Studies, no. 61, Princeton Univ. Press, Princeton, N. J.; Univ. of Tokyo Press, Tokyo, 1968. MR **39** #969.
3. J. Milnor and D. Husemoller, *Symmetric bilinear forms*, Ergebnisse der Math. und ihrer Grenzgebiete, Band 73, Springer-Verlag, Berlin and New York, 1973.
4. L. S. Pontryagin, *On the classification of four-dimensional manifolds*, Uspehi Mat. Nauk. **4** (1949), no. 4 (32), 157–158. (Russian) MR **11**, 194.
5. C. T. C. Wall, *Diffeomorphisms of 4-manifolds*, J. London Math. Soc. **39** (1964), 131–140; *On simply connected 4-manifolds*, J. London Math. Soc. **39** (1964), 141–149. MR **29** #626, #627.
6. B. Wajnryb, *The Lefschetz vanishing cycles on a projective non-singular plane curve*, Preprint, Hebrew University of Jerusalem (to appear).
7. J. H. C. Whitehead, *On simply connected 4-dimensional polyhedra*, Comment. Math. Helv. **22** (1949), 48–92. MR **10**, 559.

WEIZMANN INSTITUTE

Proceedings of Symposia in Pure Mathematics
Volume 30, 1977

L-DIMENSION AND DEFORMATIONS

EDOARDO SERNESI

The results described in this report are contained in a joint paper of David Lieberman and myself [**5**]. Preliminary announcement was given in [**4**].

Consider a compact complex space X and an invertible sheaf L on X (all spaces are assumed to be connected and not necessarily reduced). Whenever it is defined, denote by $\phi_i : X \dashrightarrow \boldsymbol{P}^N$ the rational map given by the global sections of $L^{\otimes i}$, $N + 1 = \dim_C(H^0(X, L^{\otimes i}))$, and by $I(\phi_i)$ the indeterminacy locus of ϕ_i. By $\phi_i(X)$ we mean (with an abuse of language) the projective image of X, i.e., the closure in $\boldsymbol{P}^N$ of $\phi_i(X \backslash I(\phi_i))$.

The *L-dimension of X* is defined as

$$k(X, L) = \sup_{i>0} \dim(\phi_i(X)) \quad \text{if } \phi_i \text{ is defined for some } i > 0,$$
$$= -\infty \quad \text{otherwise.}$$

This notion has been introduced and studied by Iitaka in [**3**].

If X is a compact complex manifold and $\mathscr{K}_X$ is the canonical invertible sheaf (of holomorphic differential forms of highest degree), $k(X, \mathscr{K}_X)$ is easily seen to be a bimeromorphic invariant of X (see [**9**]) known as the *canonical dimension* or the *Kodaira dimension* of X and denoted k-dim(X). It is the fundamental invariant in the Enriques-Kodaira classification of projective algebraic and compact analytic surfaces. For Riemann surfaces the possible values $-\infty$, 0, 1 of the canonical dimension correspond to the genus being 0, 1, $\geqq 2$ respectively.

Now let a proper and flat morphism $\pi : \mathscr{X} \to S$ of complex spaces and an invertible sheaf $\mathscr{L}$ on $\mathscr{X}$ be given. We shall say that these data define a family of deformations of a compact space X together with an invertible sheaf L on it, briefly of (X, L), if there is a point $0 \in S$ and an isomorphism $\mathscr{X}_0 := \pi^{-1}(0) \simeq X$ such that $\mathscr{L}_0 := \mathscr{L}|_{\mathscr{X}_0} \simeq L$. In particular if X is a compact manifold, any proper and smooth family $\pi : \mathscr{X} \to S$ of deformations of X is automatically a family of deformations of $(X, \mathscr{K}_X)$ (take $\mathscr{L} = \mathscr{K}_{\mathscr{X}}$).

AMS (MOS) subject classifications (1970). Primary 32G05.

© 1977, American Mathematical Society

The problem of investigating the behaviour of L-dimension and of canonical dimension under deformation naturally arises.

Canonical dimension of a compact Riemann surface depends only on the genus and therefore is invariant under arbitrary deformations. The same result holds for the canonical dimension of compact complex surfaces; this has been proved by Iitaka in **[2]**. It is known that canonical dimension of threefolds need not be constant under deformation **[7]**.

One of the conjectures on the subject says that canonical dimension is upper semicontinuous under deformation (see **[9]**). It has also been conjectured **[6]** that arbitrary (or small) deformations of a manifold of general type are again of general type; recall that a compact manifold X is called of general type if $k\text{-dim}(X) = \dim(X)$. Note that both conjectures are true for manifolds of complex dimension 1 or 2.

Our contribution to these problems consists of the following two theorems.

THEOREM 1. *Let $\mathscr{L}$ be an invertible sheaf on a complex space $\mathscr{X}$ and $\pi : \mathscr{X} \to S$ a proper and flat morphism onto an irreducible complex space S. There is a constant k (possibly $-\infty$) and a set $W \subseteq S$, which is the complement of the union of a countable number of proper closed subvarieties, such that*

$$k(\mathscr{X}_s, \mathscr{L}_s) = k, \quad \text{if } s \in W,$$
$$k(\mathscr{X}_s, \mathscr{L}_s) > k, \quad \text{if } s \in S \backslash W.$$

THEOREM 2. *Let L be an invertible sheaf on a compact, reduced, irreducible complex space X such that the following condition is satisfied for some $n > 0$:*

(α) *$k(X, L) \geqq 0$ and there is a polynomial $Q(T)$ with rational coefficients, of degree strictly less than $k(X, L)$, such that $\dim_C H^1(X, L^{\otimes in}) \leqq Q(i)$ for all $i \gg 0$.*

Then, for every family $\pi : \mathscr{X} \to S$, $\mathscr{L}$, of deformations of $(X, L) \simeq (\mathscr{X}_0, \mathscr{L}_0)$, $0 \in S$, *over an irreducible base space S, we have $k(\mathscr{X}_s, \mathscr{L}_s) \geqq k(\mathscr{X}_0, \mathscr{L}_0)$ for all $s \in S$.*

I will outline the proof of Theorem 2 (the reader is referred to **[5]** for complete proofs), but first let me make some observations.

Note that in Theorem 1 both W and its complement might be dense in S. Perhaps the easiest example is the following. Let X be a compact Riemann surface of positive genus, S the Jacobian variety of X, and take $\mathscr{X} = X \times S$, $\pi : \mathscr{X} \to S$ the projection, $\mathscr{L}$ the Poincaré sheaf on $\mathscr{X}$. Here W and $S \backslash W$ are the sets of points of the group S which are of infinite order and of finite order respectively. It is well known that both W and $S \backslash W$ are dense in S.

Note also that the conclusion of Theorem 2 is *global* with respect to S; in other words it is asserted that, under condition (α), $k(\mathscr{X}_s, \mathscr{L}_s) \geqq k(\mathscr{X}_0, \mathscr{L}_0)$ for *every* point $s \in S$.

Condition (α) is satisfied in the important case when $k(X, L) = \dim(X)$ and $L^{\otimes n}$ is generated by its global sections for some $n > 0$ (e.g., L is ample). In particular we have the following.

COROLLARY. *Let X be a compact manifold of general type such that $\mathscr{K}_X^{\otimes n}$ is generated by its global sections for some $n > 0$ (e.g., X is canonically polarized). Then every deformation of X (contained in a proper smooth family of deformations of X over irreducible base space) is again a manifold of general type.*

OUTLINE OF THE PROOF OF THEOREM 2. One easily reduces to the case when S is a nonsingular curve.

Let m be a positive integer such that $\pi_*\mathscr{L}^{\otimes m} \neq 0$. Notice that since π is proper $\pi_*\mathscr{L}^{\otimes m}$ is a coherent sheaf on S [**1**], and in fact it is locally free of finite rank because S is a nonsingular curve [**8**]. We get the following diagram

$$\begin{array}{ccc} \mathscr{X} & \overset{f_m}{\dashrightarrow} & \mathscr{Y}_m \subseteq \boldsymbol{P}(\pi_*\mathscr{L}^{\otimes m}) \\ & {\scriptstyle\pi}\searrow \quad \swarrow{\scriptstyle\sigma} & \\ & S & \end{array}$$

$\boldsymbol{P}(\pi_*\mathscr{L}^{\otimes m})$ is the projective bundle associated to $\pi_*\mathscr{L}^{\otimes m}$, f_m is the rational S-map defined by the canonical homomorphism $\pi^*\pi_*\mathscr{L}^{\otimes m} \to \mathscr{L}^{\otimes m}$ of sheaves on $\mathscr{X}$, $\mathscr{Y}_m$ is the projective image of $\mathscr{X}$ under f_m, and σ is the projection of $\mathscr{Y}_m$ onto S.

Since S is a curve, all fibres $\mathscr{Y}_{m,s} := \sigma^{-1}(s)$ have the same dimension, and

$$\dim(\mathscr{Y}_{m,s}) \leqq k(\mathscr{X}_s, \mathscr{L}_s)$$

essentially because the base change maps

$$t_s^m : (\pi_*\mathscr{L}^{\otimes m})_s \otimes_{\mathscr{O}_{S,s}} C \to H^0(\mathscr{X}_s, \mathscr{L}_s^{\otimes m})$$

are injective. Therefore we see that the theorem is reduced to proving that $\dim(\mathscr{Y}_{m,0}) = k(\mathscr{X}_0, \mathscr{L}_0)$ for some $m > 0$.

It is not restrictive to assume that $k(\mathscr{X}_0, \mathscr{L}_0) = \dim(\phi(\mathscr{X}_0))$, where $\phi : \mathscr{X}_0 \dashrightarrow \boldsymbol{P}^N$ is the rational map given by the sections of $\mathscr{L}_0$, and that condition (α) is satisfied with $n = 1$.

Well-known properties of projective varieties (see [**10**]) imply that, when m grows, $\dim_C(H^0(\mathscr{X}_0, \mathscr{L}_0^{\otimes m}))$ grows at least like a polynomial of degree $k(\mathscr{X}_0, \mathscr{L}_0)$. The maximum of the dimensions of the $\mathscr{Y}_{m,0}$'s, when m varies, can also be deduced from the order of growth of $\dim_C((\pi_*\mathscr{L}^{\otimes m})_0 \otimes C)$. Hypothesis ($\alpha$) and standard semicontinuity arguments show that

$$\dim_C((\pi_*\mathscr{L}^{\otimes m})_0 \otimes C) \geqq \dim_C(H^0(\mathscr{X}_0, \mathscr{L}_0^{\otimes m})) - Q(m)$$

when $m \gg 0$ and this ensures that $\dim_C((\pi_*\mathscr{L}^{\otimes m})_0 \otimes C)$ grows at least like a polynomial of degree $k(\mathscr{X}_0, \mathscr{L}_0)$.

This implies that, for some $m \gg 0$, $\dim(\mathscr{Y}_{m,0}) = k(\mathscr{X}_0, \mathscr{L}_0)$ and proves Theorem 2.

BIBLIOGRAPHY

1. H. Grauert, *Ein Theorem der analytischen Garbentheorie und die Modulräume komplexer Strukturen*, Inst. Hautes Études Sci. Publ. Math. No. 5 (1960). MR **22** #12544.

2. S. Iitaka, *Deformations of compact complex surfaces*. I, II, III, Global Analysis (Papers in Honor of K. Kodaira), Univ. of Tokyo Press, Tokyo, 1969, pp. 267–272; J. Math. Soc. Japan **22** (1970), 247–261; ibid. **23** (1971), 692–705. MR **40** #8086; **41** #6252; **44** #7598.

3. ———, *On D-dimensions of algebraic varieties*, J. Math. Soc. Japan **23** (1971), 356–373. MR **44** #2749.

4. D. Lieberman and E. Sernesi, *Semicontinuity of Kodaira dimension*, Bull. Amer. Math. Soc. **81** (1975), 459–460.

5. ———, *Semicontinuity of L-dimension*, Math. Ann. (to appear).

6. B. G. Moĭšezon, *Algebraic varieties and compact complex spaces*, Proc. Internat. Congress Math. (Nice, 1970), vol. 2, Gauthier-Villars, Paris, 1971, pp. 643–648.

7. I. Nakamura, *Complex parallelisable manifolds and their small deformations*, J. Differential Geometry **10** (1975), 85–112.

8. O. Riemenschneider, *Über die Anwendung algebraischer Methoden in der Deformationstheorie komplexer Räume*, Math. Ann. **187** (1970), 40–55. MR **41** #5659.

9. K. Ueno, *Classification theory of algebraic varieties and compact complex spaces*, Lecture Notes in Math., vol. 439, Springer-Verlag, Berlin and New York, 1975.

10. O. Zariski and P. Samuel, *Commutative algebra*. Vol. II, Univ. Ser. in Higher Math., Van Nostrand, Princeton, N. J., 1960. MR **22** #11006.

Università degli Studi di Ferrara

Proceedings of Symposia in Pure Mathematics
Volume 30, 1977

ON AMPLE DIVISORS

ANDREW JOHN SOMMESE

It is well known that for a projective manifold A to be an ample divisor [cf. §I] in a given projective manifold X puts severe restrictions on A. The purpose of this note is to show that the existence of any such X puts restrictions on A.

Full details and proofs are in **[6]**.

The philosophy behind the results below is that a projective manifold X is more special than any of its ample divisors and if some projective manifold A has some unusual property (especially one involving some form of flatness), then if any X can exist with A an ample divisor in X, it must just about be unique.

I hope the below amply illustrates this principle.

I. Let X be a projective manifold. A codimension one submanifold A is said to be an *ample divisor* in X if the sections of some power of the line bundle associated to A give a projective embedding of X. More geometrically, a codimension one submanifold A of X is ample if there exists an embedding of X into some $\boldsymbol{P}^N$ such that A is the set-theoretic (not necessarily transversal) intersection in $\boldsymbol{P}^N$ of X with a hypersurface of $\boldsymbol{P}^N$.

The main facts about ample divisors are given in the Lefschetz theorem.

LEFSCHETZ THEOREM. *Let A and X be as above. Then*:

$$\begin{array}{llcll}
\text{(I-A)} & \pi_i(A) & \approx & \pi_i(X), & i < \dim_C A,\\
\text{(I-B)} & \pi_i(A) & \to & \pi_i(X) \to 0, & i = \dim_C A,\\
\text{(II-A)} & H_i(A, \boldsymbol{Z}) & \approx & H_i(X, \boldsymbol{Z}), & i < \dim_C A,\\
\text{(II-B)} & H_i(A, \boldsymbol{Z}) & \to & H_i(X, \boldsymbol{Z}), & i = \dim_C A,\\
\text{(III-A)} & H^p(A, \Omega^q_A) & \approx & H^p(X, \Omega^q_X), & p + q < \dim_C A,\\
\text{(III-B)} & H^p(A, \Omega^q_A) & \leftarrow & H^p(X, \Omega^q_X) \leftarrow 0, & p + q = \dim_C A,\\
\text{(IV-A)} & \mathrm{Pic}(X) & \approx & \mathrm{Pic}(A), & \dim_C A > 2,\\
\text{(IV-B)} & 0 & \to & \mathrm{Pic}(X) \to \mathrm{Pic}(A), & \dim_C A = 2,
\end{array}$$

AMS (MOS) subject classifications (1970). Primary 32J25, 32L10, 14F05, 14M15.
Key words and phrases. Ample divisor, hyperplane section, Lefschetz theorem.

© 1977, American Mathematical Society

where $\pi_i(\ )$ *denotes the ith homotopy group with basepoints suppressed,* $H_i(\ ,\mathbf{Z})$ *denotes the ith integral singular homology group,* Ω^q_A *and* Ω^q_X *denote the sheaves of germs of holomorphic exterior q-forms, and where* Pic( ) *denotes the Picard group.*

One can find (I-A) and (I-B) in [**2**]; the rest are standard consequences of these (cf. [**4**] for a discussion).

LEMMA I-A. *Let A and X be as above. If* $\dim_{\mathbf{C}} A \geqq 2$, *then there exists no holomorphic map* $p: X \to A$ *that restricts to the identity on A.*

PROOF. Assume that such a p existed. Then the set S where p is not of maximal rank would be finite since S cannot intersect A. Thus $\pi_i(X) \approx \pi_i(X - p^{-1}(p(S)))$ and $\pi_i(A) \approx \pi_i(A - p(S))$ for $i \leqq 2$ by dimension considerations where basepoints are suppressed. Thus $\pi_i(X) \approx \pi_i(A) \oplus \pi_i(F)$ for $i \leqq 2$ where $F = p^{-1}(y)$ for some $y \in A - p(S)$. By the Lefschetz theorem one concludes that $\pi_i(F) = 0$ for $i \leqq 2$ which is absurd since F is a compact Riemann surface. Q.E.D.

PROPOSITION I. *Let A be a submanifold of an abelian variety. If the holomorphic tangent bundle of A has a nontrivial direct sum decomposition, then A cannot be an ample divisor in any projective manifold X. In particular if A is an ample divisor in some X, then A is of general type, i. e., the holomorphic sections of* K_A, *the canonical bundle of A, raised to some power give a birational embedding of A into projective space.*

The proof follows from an analysis of the Albanese maps of A and X combined with Lemma I-A and the Lefschetz theorem.

COROLLARY. *If A is an abelian variety of dimension* $\geqq 2$ *or if A is a nontrivial product of Riemann surfaces of genus* $\geqq 1$, *then A cannot be an ample divisor in any projective manifold X.*

REMARK. Define a codimension q submanifold A of a projective manifold X to be positive if there is a holomorphic vector bundle F of rank q that is positive (in the sense of [3]), that possesses a holomorphic section that vanishes only on A, and precisely to the first order on A. Then it follows from [3, pp. 204 ff.] that the analogue to the Lefschetz theorem holds. One can generalize Lemma I-A to read:

LEMMA I-A′. *If A is a positive codimension q submanifold of X and if* $q < \dim_{\mathbf{C}} A$, *then there can exist no holomorphic map* $p: X \to A$, *that is the identity on A.*

This lets one prove Theorem I′ where one has the same conclusion but instead of 'A being an ample divisor in X' one substitutes 'A is a positive q-dimensional submanifold of X with $q < \dim_{\mathbf{C}} A$'.

PROPOSITION II. *Let A and X be projective manifolds with* $\dim_{\mathbf{C}} A \geqq 2$ *and A an ample divisor in X. If the canonical bundle of A is trivial, then* $H^i(A, \mathcal{O}_A) = 0$ *for* $0 < i < \dim_{\mathbf{C}} A$ *and A is the anticanonical divisor of X.*

II.

PROPOSITION III. *Let A and X be projective manifolds with A an ample divisor of X. If* $p: A \to Y$ *is a holomorphic surjection with Y a projective analytic space where* $\dim_{\mathbf{C}} Y + 2 \leqq \dim_{\mathbf{C}} A$, *then p extends to a holomorphic surjection* $\tilde{p}: X \to Y$.

The proof proceeds by first reducing to the case when p has connected fibres and Y is normal.

Now one chooses a holomorphic line bundle L on Y whose holomorphic sections give a projective embedding of Y. p^*L extends to a line bundle $\tilde{L}$ on X by the Lefschetz theorem. Now it is not too hard to show one would be done if all the sections of p^*L extended to sections of $\tilde{L}$. To show this it would suffice to show $H^1(X, \tilde{L} \otimes [-A]) = 0$ as one sees from the sequence:

$$0 \to \tilde{L} \otimes [-A] \to L \to p^*L \to 0.$$

Now one can show this will be the case if $H^1(A, p^*L \otimes [-nA]|_A) = 0$ for all $n > 0$. Using Serre duality and the Leray spectral sequence one reduces to the following relative form of Kodaira's vanishing theorem:

LEMMA. *Let $p: X \to Y$ be a proper holomorphic map where X is a complex manifold and Y is an analytic space. Let E be a holomorphic line bundle on X and assume, for each $y \in p(X)$, there exists a neighborhood $U(y)$ such that $L|_{p^{-1}(U(y))}$ has a Hermitian norm with positive curvature. If $\mathscr{S}$ is any coherent sheaf on Y, then $R^i_{p*}(X, p^*\mathscr{S} \otimes_{\mathcal{O}_X} \mathcal{O}(L \otimes K_X)) = 0$ for all $i > 0$ where K_X is the canonical bundle of X.*

This lemma is proved by use of the L^2-estimates for the $\bar{\partial}$ operator [**1**].

COROLLARY. *If A is a nontrivial product of projective manifolds and if A is an ample divisor in a projective manifold X, then A has two factors and one is a curve.*

I conjecture that $\boldsymbol{P}^1 \times \boldsymbol{P}^1$ is the only product that is an ample divisor.

The above raises the question of whether there are any ample divisors A that are fibre bundles. The answer is yes.

EXAMPLE. Let $T^*_{\boldsymbol{P}^n}$ be the holomorphic cotangent bundle of $\boldsymbol{P}^n$. One has the sequence:

$$0 \to T^*_{\boldsymbol{P}^n} \to \bigoplus_{n+1 \text{ copies}} H^{-1} \to \mathcal{O}_{\boldsymbol{P}^n} \to 0$$

where H^{-1} is the dual of the hyperplane section bundle. One can, using the above sequence, show that $P(T^*_{\boldsymbol{P}^n})$ is an ample divisor in $\boldsymbol{P}^n \times \boldsymbol{P}^n$ where $P(E)$ for a vector bundle E is E minus its zero section quotiented by the natural C^* action. By pulling back this situation by finite maps $q: Y \to \boldsymbol{P}^n$ where Y is a projective manifold one constructs a wide variety of examples.

I know no other examples. One should not though:

PROPOSITION IV. *Let A be an ample divisor in a projective manifold X. Let $p: A \to Y$ be a holomorphic surjection of maximal rank onto a projective manifold Y. If* $\dim_C A \neq 1 + \dim_C Y$, *then* $\dim_C A \geqq 2 \dim_C Y - 1$. *If* $\dim_C A = 2 \dim_C Y - 1$ *or* $2 \dim_C Y$, *then the fibre of p has even Betti numbers* 1 *and odd Betti numbers* 0.

The proof follows from the Lefschetz theorem and the technique of Lemma I-A but is much more involved.

III. In this section I will mention some partial results.

PROPOSITION V. *If $\boldsymbol{P}^n$ is an ample divisor in X and $n \geqq 3$, then X is $\boldsymbol{P}^{n+1}$. If $\mathscr{Q}^n$ is an ample divisor in X and $n \geqq 3$, then X is either $\boldsymbol{P}^{n+1}$ or $\mathscr{Q}^{n+1}$ where $\mathscr{Q}^m$ is the nonsingular quadric of dimension m.*

This follows from the adjunction formula, the Lefschetz theorem, and the following consequence of Kodaira's vanishing theorem.

LEMMA. *Let A and X be projective manifolds with A an ample divisor in X. Assume* $H^2(A, \mathbf{Z}) \approx \mathbf{Z}$ *and* $\dim_C A \geqq 3$. *Then* $[A]|_A$ *is the ample generator of* $\operatorname{Pic}(A)$ *raised to the kth power where* $0 < k \leqq n + r + 2$ *and where* $n = \dim_C A$ *and the canonical bundle of A is the ample generator of* $\operatorname{Pic}(A)$ *raised to the rth power.*

Finally one shows:

PROPOSITION VI. *Let A be a projective manifold that is an ample divisor in a projective manifold X. Let* $p: A \to B$ *be a holomorphic surjection onto a complex manifold B that expresses A as B with a submanifold Y blown up. Let* $\tilde{Y} = p^{-1}(Y)$. *If* $\dim_C \tilde{Y} - \dim_C Y \geqq 2$, *then there exists a unique pure codimension one analytic subspace Z of X such that* $Z|_A = \tilde{Y}$.

One proves this by the same technique as in Proposition III. Z is the candidate to be blown down if p could be extended to a holomorphic map $\tilde{p}: X \to W$ where W is an analytic space that contains B as an analytic subspace.

ADDED IN PROOF. Proposition III has now been proved if A has slight singularities, e.g., if A admits a desingularization $q: \tilde{A} \to A$ such that $q_* \mathcal{O}_{\tilde{A}} = \mathcal{O}_A$. Counterexamples have been constructed showing that the fibre dimension of $p: A \to Y$ must be at least two if Proposition III is to be true. It has also been shown that $\mathcal{R} \times \mathbf{C}P^N$ can be an ample divisor for any N where $\mathcal{R}$ is any compact connected Riemann surface. The proof of Proposition V can be simplified by using a result of S. Kobayashi and T. Ochiai in *Characterizations of complex projective spaces and hyperquadrics*, J. Math. Kyoto Univ. **13** (1972), 31–47.

BIBLIOGRAPHY

1. A. Andreotti and E. Vesentini, *Carleman estimates for the Laplace-Beltrami equation on complex manifolds*, Inst. Hautes Études Sci. Publ. Math. No. **25** (1965), 81–130. MR **30** #5333; erratum, **32** #465.

2. R. Bott, *On a theorem of Lefschetz*, Michigan Math. J. **6** (1959), 211–216. MR **35** #6164.

3. P. A. Griffiths, *Hermitian differential geometry, chern classes, and positive vector bundles*, Global Analysis (Paper in honor of K. Kodaira), Princeton Univ. Press, Princeton, N. J.; Univ. of Tokyo Press, Tokyo, 1969, pp. 185–251. MR **41** #2717.

4. R. Hartshorne, *Ample subvarieties of algebraic varieties*, Lecture Notes in Math., vol. 156, Springer-Verlag, New York, 1970.

5. ———, *Ample vector bundles*, Inst. Hautes Études Sci. Publ. Math. No. **29** (1966), 63–94. MR **33** #1313.

6. A. J. Sommese, *On manifolds that cannot be ample divisors*, Math. Ann. **221** (1976), 55–72.

INSTITUTE FOR ADVANCED STUDY

Proceedings of Symposia in Pure Mathematics
Volume 30, 1977

COMPACT QUOTIENTS OF C^3 BY AFFINE TRANSFORMATION GROUPS

TATSUO SUWA*

In [4], the author classified the compact quotients of C^2 by affine transformation groups. This is a preliminary report on the three-dimensional case. Detailed proofs will appear elsewhere.

1. A fundamental theorem. The problem is formulated as follows:

Let G denote a group of affine transformations of the three-dimensional complex vector space C^3. Assume that the action of G is (i) properly discontinuous, i.e., for any pair (K_1, K_2) of compact sets in C^3, the set $\{g \in G \mid gK_1 \cap K_2 \neq \emptyset\}$ is finite, and (ii) free, i.e., for all $g \in G$, $g \neq 1$, g has no fixed points on C^3. Thus the quotient C^3/G is a complex manifold of complex dimension 3. Also assume that (iii) C^3/G is compact. The problem is to classify the compact complex threefolds of the form C^3/G.

Each element g in G is represented by a 4×4 matrix

$$g = \begin{pmatrix} A(g) & b(g) \\ 0 & 1 \end{pmatrix},$$

where $A(g)$ (the holonomy part of g) is in $GL(3, C)$ and $b(g)$ (the translation part of g) is in C^3. The action of g on $C^3 = \{z \mid z = {}^t(z_1, z_2, z_3)\}$ is given by

$$\begin{pmatrix} z \\ 1 \end{pmatrix} \mapsto \begin{pmatrix} A(g) & b(g) \\ 0 & 1 \end{pmatrix} \begin{pmatrix} z \\ 1 \end{pmatrix}.$$

From the assumption (ii), we have

$$\det(A(g) - I) = 0. \tag{1}$$

In particular, one of the eigenvalues of $A(g)$ must be equal to 1.

AMS (MOS) subject classifications (1970). Primary 32C10, 32L05; Secondary 14J15, 53C55.

*Supported in part by NSF grant GP43614X.

© 1977, American Mathematical Society

THEOREM 1. *Suppose G contains no elements whose holonomy parts have three different eigenvalues. Then G contains a solvable subgroup S of finite index. Moreover, S contains a nilpotent subgroup N of finite index.*

The proof of the theorem is computational. Using equation (1), it is shown that certain entries of the holonomy parts must be simultaneously zero for all elements in G. From the proper discontinuity of G, it is also shown that all elements can be simultaneously represented by upper triangular matrices and that the eigenvalues of the holonomy part of each element must be roots of unity.

Thus if G satisfies the condition in the theorem, the quotient $\boldsymbol{C}^3/G$ is finitely covered by $\boldsymbol{C}^3/N$, where N is nilpotent. Moreover the proof shows that every element g in N is represented by an upper triangular matrix whose diagonal entries are all 1;

$$g = \begin{pmatrix} 1 & a_{12}(g) & a_{13}(g) & b_1(g) \\ 0 & 1 & a_{23}(g) & b_2(g) \\ 0 & 0 & 1 & b_3(g) \\ 0 & 0 & 0 & 1 \end{pmatrix} \quad \text{for all } g \in N. \tag{2}$$

2. Classification of the quotients $\boldsymbol{C}^3/N$. From now on we take the representation (2). Let $N^{(1)}$ denote the commutator subgroup of N. Note that $N^{(1)}$ is commutative. For functions $a_i : N \to \boldsymbol{C}$, $1 \leqq i \leqq r$, we define $[a_1, \cdots, a_r] : N^r = N \times \cdots \times N \to \boldsymbol{C}$ (N occurs r times) by

$$[a_1, \cdots, a_r](g_1, \cdots, g_r) = \det(a_i(g_j)).$$

LEMMA 1. *By a suitable coordinate transformation (possibly nonlinear) of $\boldsymbol{C}^3$, every case is reduced to one of the following:*

(I) $N^{(1)} = 0$,

(II) $N^{(1)} \neq 0$, $a_{23} = 0$, *and* $[b_2, b_3] \neq 0$,

(III) $a_{12} = 0$ *and* $[a_{13}, a_{23}, b_3] \neq 0$,

(IV) $a_{12} \neq 0$ *and* $[a_{23}, b_3] \neq 0$.

LEMMA 2. *In each case, we can choose the following set as a basis of the closed holomorphic 1-forms on $\boldsymbol{C}^3/N$ (= the N-invariant closed holomorphic 1-forms on $\boldsymbol{C}^3$).*

(I) dz_1, dz_2, dz_3,

(II) dz_2, dz_3,

(III), (IV) dz_3.

Let d and $h^{p,0}$ denote, respectively, the numbers of linearly independent closed holomorphic 1-forms and holomorphic p-forms on $\boldsymbol{C}^3/N$. We define the Albanese variety of $\boldsymbol{C}^3/N$ and the Albanese map according to Blanchard [**1**]. The structure of the quotient $\boldsymbol{C}^3/N$ is determined by analyzing the Albanese map.

THEOREM 2. *The quotients $\boldsymbol{C}^3/N$ are classified as follows:*

	d	$h^{1,0}$	$h^{2,0}$	$H_1(\boldsymbol{C}^3/N, \boldsymbol{Z})$	*structure (Albanese map)*
I	3	3	3	$\boldsymbol{Z}^6$	(i) T^3
II	2	2	2	$\boldsymbol{Z}^5 \oplus \boldsymbol{Z}_m$	(ii) T^2-*bundle over* T^1
			1, 2	$\boldsymbol{Z}^5 \oplus \boldsymbol{Z}_m$	(iii) T^1-*bundle over* T^2
		2, 3	1, 2, 3	$\boldsymbol{Z}^4 \oplus \boldsymbol{Z}_{m_1} \oplus \boldsymbol{Z}_{m_2}$	(iv) T^1-*bundle over* T^2

III	1	1	2	$Z^3 \oplus Z_{m_1} \oplus Z_{m_2} \oplus Z_{m_3}$	(v) Alb = 0
					(vi) *regular fiber space of complex 2-tori over* T^1
IV	1	1, 2	1, 2	$Z^4 \oplus Z_{m_1} \oplus Z_{m_2}$ or $Z^3 \oplus Z_{m_1} \oplus Z_{m_2} \oplus Z_{m_3}$	(vii) *regular fiber space of complex 2-tori over* T^1

REMARKS. (1) T^k denotes a complex k-torus.

(2) The manifolds in the class (v) are differentiably real 3-torus bundles over the real 3-torus.

(3) By a regular fiber space, we mean a complex manifold X together with a proper holomorphic map $f: X \to Y$ onto another complex manifold Y such that the Jacobian matrix of f is everywhere of maximal rank.

(4) The manifolds in the classes (ii) — (vii) are non-Kähler (see Sakane **[3]**).

(5) It is not difficult to construct all manifolds in each of the classes (i) — (vi).

EXAMPLES. (1) The Iwasawa manifold (see, e.g., **[2]**) is an example of (iv).

(2) Take complex numbers ω_i, $i = 1, 2, 3, 4$, such that $(1, 0)$, $(0, 1)$, (ω_1, ω_2), (ω_3, ω_4) are linearly independent over the reals. Let N be the group generated by the following five elements:

$$g_1 = \begin{pmatrix} 1 & 0 & 1 & 0 \\ 0 & 1 & 0 & 0 \\ 0 & 0 & 1 & i \\ 0 & 0 & 0 & 1 \end{pmatrix}, \quad g_2 = \begin{pmatrix} 1 & 0 & 0 & 0 \\ 0 & 1 & 0 & 0 \\ 0 & 0 & 1 & 1 \\ 0 & 0 & 0 & 1 \end{pmatrix}, \quad g_3 = \begin{pmatrix} 1 & 0 & 0 & 0 \\ 0 & 1 & 0 & 1 \\ 0 & 0 & 1 & 0 \\ 0 & 0 & 0 & 1 \end{pmatrix},$$

$$g_4 = \begin{pmatrix} 1 & 0 & 0 & \omega_1 \\ 0 & 1 & 0 & \omega_2 \\ 0 & 0 & 1 & 0 \\ 0 & 0 & 0 & 1 \end{pmatrix}, \quad g_5 = \begin{pmatrix} 1 & 0 & 0 & \omega_3 \\ 0 & 1 & 0 & \omega_4 \\ 0 & 0 & 1 & 0 \\ 0 & 0 & 0 & 1 \end{pmatrix}, \quad i = \sqrt{-1}.$$

Then the quotient C^3/N is an example of (ii).

(3) Let N be the group generated by the following three elements:

$$g_1 = \begin{pmatrix} 1 & 0 & 0 & 0 \\ 0 & 1 & 0 & 0 \\ 0 & 0 & 1 & 1 \\ 0 & 0 & 0 & 1 \end{pmatrix}, \quad g_2 = \begin{pmatrix} 1 & 0 & 0 & i \\ 0 & 1 & 1 & 0 \\ 0 & 0 & 1 & \sqrt{2i} \\ 0 & 0 & 0 & 1 \end{pmatrix}, \quad g_3 = \begin{pmatrix} 1 & 0 & 1 & 0 \\ 0 & 1 & 0 & 0 \\ 0 & 0 & 1 & i \\ 0 & 0 & 0 & 1 \end{pmatrix}.$$

Then the quotient C^3/N is an example of (v).

(4) For an example of G containing an element whose holonomy part has three different eigenvalues, see **[2]**.

REFERENCES

1. A. Blanchard, *Sur les variétés analytiques complexes*, Ann. Sci. École Norm. Sup. (3) **73** (1956), 157–202. MR **19**, 316.

2. I. Nakamura, *Complex parallelisable manifolds and their small deformations*, J. Differential Geometry **10** (1975), 85–112.

3. Y. Sakane, *On compact complex affine manifolds* (to appear).

4. T. Suwa, *Compact quotient spaces of* C^2 *by affine transformation groups*, J. Differential Geometry **10** (1975), 239–252.

UNIVERSITY OF MICHIGAN

Proceedings of Symposia in Pure Mathematics
Volume 30, 1977

DEFORMATIONS AND ANALYTIC EQUATIONS

JOHN J. WAVRIK

The original goal of deformation theory was to illuminate an aspect of the study of all possible complex structures on a compact differentiable manifold. The initial problem was one of classification: Find a family containing all the complex structures (or at least all those "near" a given one) and count the dimension of the parameter space (number of moduli). This problem dates back to Riemann **[17]** who discussed the case of compact one-dimensional manifolds. The first results in higher dimension were obtained by Max Noether **[15]**; however the development of a comprehensive theory of deformations must be credited to Kodaira and Spencer. In their monumental work **[9]** they formulated the notion of a family of complex structures and began the investigation of problems of deformation theory.

In the classical theory, many deformation problems are handled by a somewhat typical approach: There is first a step-by-step process analogous (and often equivalent) to the process of determining a power series coefficient-by-coefficient to provide a sort of "formal" solution to the problem; and second there is the task of developing machinery to guarantee that the formal process converges or can be made to converge. This approach can be seen in **[9]**, **[10]**, **[12]**, **[8]**, **[11]**, **[7]**, **[4]**, **[5]**. Almost inevitably the formal part of the problem is relatively straightforward and the difficulty lies in the convergence step. In the classical theory, convergence machinery was developed ad hoc for each particular problem. It appeared to be one of the features of deformation theory that problems which could be solved formally could also be solved actually. D. C. Spencer posed in **[22]** the problem of understanding why this is so. It is the purpose of this paper to outline an approach to deal with this problem.

The approach will be to make use of theorems on solutions of analytic equations.

THEOREM **[1]**. *Let* $f_i(x; y) \in \mathbf{C}\{x; y\}$, $i = 1, \cdots, m$, *with* $x = (x_1, \cdots, x_n)$, $y =$

AMS (*MOS*) *subject classifications* (1970). Primary 32G05.

© 1977, American Mathematical Society

$(y_1, \cdots, y_p)$ *and let c be a positive integer. If* $\bar{y}_\nu(x) \in \boldsymbol{C}[[x]]$, $\bar{y}_\nu(0) = 0$ *satisfy* $f(x; \bar{y}(x)) = 0$ *then there are* $y_\nu(x) \in \boldsymbol{C}\{x\}$ *satisfying*

(1) $f(x; y(x)) = 0$,

(2) $y_\nu(x) \equiv \bar{y}_\nu(x) \bmod (x)^c$.

More generally,

THEOREM [**26**]. *If* $f_i(x; y; z) \in C\{x; y\}[z]$ *and* α *is a positive integer, then there is a* $\beta = \beta(f, \alpha)$ *such that if* $\bar{y}_\nu(x)$, $\bar{z}_\mu(x) \in \boldsymbol{C}[[x]]$, $\bar{y}_\nu(0) = 0$ *satisfy* $f(x; \bar{y}(x); \bar{z}(x)) \equiv 0 \bmod (x)^\beta$ *then there are* $y_\nu(x)$, $z_\mu(x) \in C\{x\}$ *satisfying*

(1) $f(x; y(x); z(x)) = 0$,

(2) $y_\nu(x) \equiv \bar{y}_\nu(x)$, $z_\mu(x) \equiv \bar{z}_\mu(x) \bmod (x)^\alpha$.

These theorems have consequences in the study of morphisms of analytic algebras or, dually, germs of analytic spaces. They imply that formal (or even sufficiently good approximate) morphisms can be approximated by actual morphisms. The following are examples of these consequences.

PROPOSITION [**1**]. *Given the diagram*

$$\begin{array}{ccc} X & & Y \\ & \searrow \; \swarrow & \\ & Z & \end{array}$$

of germs of analytic spaces and a positive integer c, any formal morphism $\hat{X} \to \hat{Y}$ *making the analogous diagram of formal neighborhoods commute can be approximated to order c by a morphism making the original diagram commute.*

PROPOSITION [**26**]. *Given germs of analytic spaces X, Y, there is an integer* $n_0 = n_0(X, Y)$ *so that a necessary and sufficient condition for a morphism* $f: X \to Y$ *to have a left* (*resp. right*) *inverse is that* $f^{(n_0)}: X^{(n_0)} \to Y^{(n_0)}$ *have a left* (*resp. right*) *inverse.*

($X^{(n)}$ is the nth infinitesimal neighborhood of the distinguished point in X, $f^{(n)}$ is the restriction of f to $X^{(n)}$.)

One uses the families constructed by Kuranishi [**13**], [**14**], and others [**2**], [**3**], [**6**], [**16**] somewhat in the fashion of classifying spaces to convert deformation theoretic problems into problems about morphisms of germs of analytic spaces. In this way the original Artin theorem is seen to assure that certain problems which can be solved formally can also be solved actually. The generalized theorem is the genesis of results which assert the existence of only finitely many obstructions to deformation. Two difficulties arise in carrying out this program. First, families (like Kuranishi's) are not in general universal; they are not really classifying spaces: different morphisms can induce the same family. Second, the Artin theorem does not assert that formal solutions are convergent, but only that they may be approximated by convergent solutions. Thus one often has at hand a formal morphism having certain desirable properties and can only, ab initio, assert the existence of actual morphisms which have the desirable properties up to a certain order (the order to which the actual morphism approximates the formal one). The way to handle these difficulties is to use further information so that the desirable properties are expressed as commutativity of some appropriate diagrams. We illustrate this approach with a discussion of extension problems.

Notation. X_0 is a compact analytic space.

X/T is a family of deformations of $X_0 \cong X_{t_0}$.

D_0 is some sort of analytic object attached to X_0 (e.g., a vector bundle, map to a fixed space, cohomology class, etc.).

Problem. To show (under some conditions) that if D_0 can be extended formally to a family D/T of objects D_t attached to the fibres X_t, then it can be extended actually.

Let F be the deformation functor of X_0 and G the deformation functor for the pair (X_0, D_0). F and G are contravariant from the category of germs of analytic spaces to the category of sets.

THEOREM. *If F, G are representable and D_0 is formally extendible, then D_0 is actually extendible.*

PROOF. If $\mathscr{F}$, $\mathscr{G}$ represent F, G and $f\colon T \to \mathscr{F}$ is the morphism inducing X/T while $\pi\colon \mathscr{G} \to \mathscr{F}$ is the obvious map, the problem is equivalent to showing that if f can be formally lifted to a morphism to $\mathscr{G}$ then it can be actually lifted. This is a consequence of Artin's theorem.

As we have pointed out, however, deformation functors are not in general representable. The theorem we have just proved has overly restrictive hypotheses. Using additional (and available) information about F, we obtain a more realistic theorem.

THEOREM. *If a complete family exists for G and D_0 is formally extendible, then D_0 is actually extendible.*

PROOF. Let $\mathscr{G}$ parametrize a complete family for G. Let $\mathscr{F}$ parametrize the versal family for F [3], [6]. We have a map $\pi\colon \mathscr{G} \to \mathscr{F}$. Let $p_i\colon \mathscr{F} \times \mathscr{F} \to \mathscr{F}$ $(i=1,2)$ be the projection and X_i the family over $\mathscr{F} \times \mathscr{F}$ induced by p_i from the versal family over $\mathscr{F}$. Let $\mathscr{I} \to \mathscr{F} \times \mathscr{F}$ represent the functor $\mathrm{Iso}_{\mathscr{F}\times\mathscr{F}}(X_1, X_2)$ [16], [20]. We consider $\mathscr{I}$ a space over $\mathscr{F}$ by the composition $\mathscr{I} \to \mathscr{F} \times \mathscr{F} \to^{p_2} \mathscr{F}$ and form $\mathscr{I} \times_{\mathscr{F}} \mathscr{G}$. Let $\sigma_1\colon \mathscr{I} \times_{\mathscr{F}} \mathscr{G} \to \mathscr{F}$ be the composition $\mathscr{I} \times_{\mathscr{F}} \mathscr{G} \to \mathscr{I} \to \mathscr{F} \times \mathscr{F} \to^{p_1} \mathscr{F}$. The family X/T gives us a morphism $f\colon T \to \mathscr{F}$ and our hypothesis that the extension problem is formally solvable provides, for each n, a morphism $g_n : T^{(n)} \to \mathscr{F}$ such that the families induced by $\pi \circ g_n$ and $f^{(n)}$ are isomorphic. From $f^{(n)}$ and $\pi \circ g_n$ we obtain $h_n : T^{(n)} \to (\mathscr{I} \times_{\mathscr{F}} \mathscr{G})^{(n)}$ making the diagram

$$\begin{array}{ccc} T^{(n)} & \longrightarrow & (\mathscr{I} \times_{\mathscr{F}} \mathscr{G})^{(n)} \\ & {\scriptstyle f^{(n)}} \searrow \quad \swarrow {\scriptstyle \sigma_1^{(n)}} & \\ & \mathscr{F}^{(n)} & \end{array}$$

commute. By Artin's theorem, we obtain $h : T \to \mathscr{I} \times_{\mathscr{F}} \mathscr{G}$ so that $f = \sigma_1 \circ h$. The map $g : T \to \mathscr{I} \times_{\mathscr{F}} \mathscr{G} \to \mathscr{G}$ provides us with an extension of D_0 to an actual family D/T.

COROLLARY. *If X/T is a family of deformations of the compact analytic space $X_0 \cong X_{t_0}$, Z is a fixed analytic space, and $f_0\colon X_0 \to Z$ a morphism, then if f_0 can be extended formally to a family of maps $f_t : X_t \to Z$, it can be extended actually.*

PROOF. The construction of a complete family of deformations of a holomorphic map is found in [25].

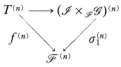

In the above proof we have used the existence of a versal family of deformations of X_0 and the representability of the Iso functor. It should be recognized that convergence problems arise in establishing these results. The theorems on analytic equations do not handle all of the convergence problems of deformation theory. Instead, the approach we have outlined shows that many convergence assertions follow from certain basic ones. Thus, for example, the Kodaira-Nirenberg-Spencer theorem [**12**], the theorem of completeness [**10**], and the rigidity theorem [**9**], all of which were originally proved with ad hoc techniques to establish convergence, can be seen to follow from the Kuranishi theorem (cf. [**23**]). Moreover the particular details of Kuranishi's construction are not involved, only the fact that the construction provides a versal family. Occasionally even the existence of a formally versal family (as provided by the work of Schlessinger [**18**]) will suffice (cf. [**23**]). Some further illustrations of the approach we have outlined will be found in the papers of the author [**24**], [**26**] and Schuster [**19**], [**21**].

References

1. M. Artin, *On the solutions of analytic equations*, Invent. Math. **5** (1968), 277–291. MR **38** #344.

2. A. Douady, *Le problème des modules pour les sous-éspaces analytiques compacts d'un éspace analytique donné*, Ann. Inst. Fourier (Grenoble) **16** (1966), fasc. 1, 1–95. MR **34** #2940.

3. ———, *Le problème des modules locaux pour les éspaces C-analytiques compacts*, Ann. Sci. Ecole Norm. Sup. **7** (1974), 569–602.

4. O. Forster and K. Knorr, *Über die Deformationen von Vektorraumbündeln auf kompakten komplexen Raümen*, Math. Ann. **209** (1974), 291–346.

5. ———, *Ein neuer Beweis des Satzes von Kodaira-Nirenberg-Spencer* (to appear).

6. H. Grauert, *Der Satz von Kuranishi für kompakte komplexer Räume*, Invent. Math. **25** (1974), 107–142. MR **49** #10920.

7. P. A. Griffiths, *The extension problem for compact submanifolds of complex manifolds*. I. *The case of a trivial normal bundle*, Proc. Conf. Complex Analysis (Minneapolis, 1964), Springer, Berlin, 1965, pp. 113–142. MR **32** #8362.

8. K. Kodaira, *A theorem of completeness of characteristic systems for analytic families of compact submanifolds of complex manifolds*, Ann. of Math. (2) **75** (1962), 146–162. MR **24** #A 3665b.

9. K. Kodaira and D. C. Spencer, *On deformations of complex analytic structures*. I, II, Ann. of Math. (2) **67** (1958), 328–466. MR **22** #3009.

10. ———, *A theorem of completeness for complex analytic fibre spaces*, Acta Math. **100** (1958), 281–294. MR **22** #3010.

11. ———, *A theorem of completeness of characteristic systems of complete continuous systems*, Amer. J. Math. **81** (1959), 477–500. MR **22** #3011.

12. K. Kodaira and L. Nirenberg, *On the existence of deformations of complex analytic structures*, Ann. of Math. (2) **68** (1958), 450–459. MR **22** #3012.

13. M. Kuranishi, *On the locally complete families of complex analytic structures*, Ann. of Math. (2) **75** (1962), 536–577. MR **25** #4550.

14. ———, *New proof for the existence of locally complete families of complex structures*, Proc. Conf. Complex Analysis (Minneapolis, 1964), Springer, Berlin, 1965, pp. 142–154. MR **31** #768.

15. M. Noether, *Anzahl der Moduln einer Classe algebraischer Flachen*, Sitzungsber Koniglich Preuss. Akad. Wiss. Berlin, **1888**, 123–127.

16. G. Pourcin, *Théorème de Douady au-dessus de S*, Ann. Scuola Norm. Sup. Pisa (3) **23** (1969), 451–459. MR **41** #2053.

17. B. Riemann, *Theorie der abel'schen Functionen*, 1857.

18. M. Schlessinger, *Functors of Artin rings*, Trans. Amer Math. Soc. **130** (1968), 208–222. MR **36** #184.

19. H. W. Schuster, *Über die Starrheit kompakter komplexer Räume*, Manuscripta Math. **1** (1969), 125–137. MR **40** #7478.

20. ———, *Zur Theorie der Deformationen kompakter komplexer Räume*, Invent. Math. **9** (1969/70), 284–294. MR **42** #3818.

21. ———, *Formale Deformationstheorie*, Habilitationsschrift, Munchen, 1971.

22. D. C. Spencer, *Some remarks on homological analysis and structures*, Proc. Sympos. Pure Math., vol. 3, Amer. Math. Soc., Providence, R. I., 1961, pp. 56–86. MR **23** #A3585.

23. J. J. Wavrik, *A theorem of completeness for families of compact analytic spaces*, Trans. Amer. Math. Soc. **163** (1972), 147–155. MR **45** #3770.

24. ———, *First order completeness theorems*, Math. Ann. **206** (1973), 249–264. MR **48** #11573.

25. ———, *A note on deformations of holomorphic maps* (unpublished).

26. ———, *A theorem on solutions of analytic equations with applications to deformations of complex structures*, Math. Ann. **216** (1975), 127–142.

UNIVERSITY OF CALIFORNIA, SAN DIEGO

Proceedings of Symposia in Pure Mathematics
Volume 30, 1977

INTERMEDIATE JACOBIANS AND THE HODGE CONJECTURE FOR CUBIC FOURFOLDS

STEVEN ZUCKER

This article is written to provide a summary and announcement of results on intermediate Jacobians which lead to a proof of the Hodge conjecture for cubic fourfolds.[1] Details will appear in [6] and [7].

Let $f: Y \to S$ be a surjective holomorphic mapping between the nonsingular complex projective varieties Y and S, with $\dim S = 1$. The basic idea is to study the Hodge conjecture for Y in terms of the mapping f. For the record, the present form of this conjecture is:

(H) Every cohomology class in $H^{2p}(Y, \mathbf{Q})$ of Hodge type (p, p) is the fundamental class of an algebraic (= analytic) cycle of codimension p with rational coefficients.

By a reduction, we may assume that $\dim Y = 2m$, and consider (H) for $p = m$. (In the subsequent discussion, integral cohomology is meant modulo torsion.)

Let Y_s denote the fiber $f^{-1}(s)$ for $s \in S$. For smooth fibers, we have an associated complex torus, the *mth intermediate Jacobian* of Y_s (in the sense of Griffiths), defined by

$$J(Y_s) = (F^m H^{2m-1}(Y_s))^\vee / H_{2m-1}(Y_s, \mathbf{Z}) \qquad ({}^\vee = \text{dual space}),$$

where $F^m H^{2m-1}(Y_s) = H^{2m-1,0}(Y_s) \oplus \cdots \oplus H^{m,m-1}(Y_s)$ is the mth Hodge filtration level. As s varies, $J(Y_s)$ varies holomorphically, in the sense that there is an analytic fiber space of tori $\mathbf{J}^*$ defined over S^*, the complement of finitely many points in S (the *singular points*), with fibers $J(Y_s)$.

Let $\Theta(Y_s)$ denote the group of codimension m algebraic cycles on Y_s which are

AMS (MOS) subject classifications (1970). Primary 14C30, 14F25, 32J25, 32L05; Secondary 14D05.

[1]The proof is based on an outline presented by Phillip Griffiths.

© 1977, American Mathematical Society

homologous to zero. There is an *Abel-Jacobi homomorphism*, defined by integrating over chains bounded by the cycles

$$\phi_s : \Theta(Y_s) \to J(Y_s),$$

generalizing the classical case where Y_s is a compact Riemann surface. For families of algebraic cycles parametrized by S^*, the Abel-Jacobi mappings along the fibers give holomorphic cross-sections of $\boldsymbol{J}^*$ [3]. Furthermore, these sections satisfy a certain differential equation, the *horizontality* condition of Griffiths [3].

We make assumptions on $f : Y \to S$ as they are needed, but in the end we are still left with some generality. I should remark here that given any manifold X embedded in some projective space, a fibering satisfying all restrictions can be constructed from X by taking a sufficiently general pencil of hyperplane sections. What results is a blown-up model Y of X, mapping onto $S = \boldsymbol{P}^1$, in which we can still keep hold of the cohomology of X.

1. Generalized intermediate Jacobians [6]. It is useful to partially compactify $\boldsymbol{J}^*$ in order to keep track of things at the singular points. As $\boldsymbol{J}^*$ is the quotient of a vector bundle $\boldsymbol{E}^*$ by a family of lattices, we extend both of these and then take the quotient $\boldsymbol{J}$. Specifically, under the assumption

(A1) The local monodromy transformations on $H^{2m-1}(Y_s)$ around the singular points are *unipotent*,

$\boldsymbol{E}^*$ has a canonical extension $\boldsymbol{E}$, and we take the local invariant cohomology to extend the lattices. It can be shown, using [4], that $\boldsymbol{J}$ is a complex manifold, fibered over S, with (in general) noncompact quotient groups over the singular points.

A global holomorphic cross-section of $\boldsymbol{J}$ will be called a *normal function*, following the classical terminology of Poincaré.

2. The cohomology class of a normal function. Let j denote the inclusion of S^* into S, $\tilde{f}$ the restriction of f over S^*, and $\mathscr{E}$ and $\mathscr{J}$ the respective sheaves of germs of holomorphic sections of $\boldsymbol{E}$ and $\boldsymbol{J}$. We have, by definition, the exact sequence

$$0 \to j_* R^{2m-1} \tilde{f}_* \boldsymbol{Z} \to \mathscr{E} \to \mathscr{J} \to 0.$$

Given $\nu \in \Gamma(S, \mathscr{J})$, its image $\delta(\nu) \in H^1(S, j_* R^{2m-1} \tilde{f}_* \boldsymbol{Z})$ under the connecting homomorphism is its *cohomology class*. Its significance is clearer if we make further restrictions on f [6] so that

(A2) $K = H^1(S, j_* R^{2m-1} \tilde{f}_* \boldsymbol{C})$ is a quotient of the "primitive" cohomology $\ker[H^{2m}(Y) \to H^{2m}(Y_s)]$ $(s \in S^*)$.

If V is an algebraic cycle whose intersections $V \cdot Y_s$ belong to $\Theta(Y_s)$, its fundamental class $[V]$ projects into K, and

PROPOSITION [6]. *If $\nu \in \Gamma(S, \mathscr{J})$ is determined by V via the Abel-Jacobi mapping, then $\delta(\nu) = [V]$ in K.*

(It is a hypothesis in the above statement that the Abel-Jacobi mapping for V extend across the singular points.)

Moreover, K inherits a Hodge structure from $H^{2m}(Y)$, and if one is trying to prove (H) inductively (i.e., if one knows (H) for the fibers), it suffices to prove the corresponding statement for K.

3. The theorem on normal functions. Consider the statement

(N) The cohomology classes of horizontal normal functions are precisely the integral classes in K of type (m, m).

Note that (N) is a necessary condition for (H), provided there are no "extraneous" normal functions. The main result of **[6]** is:

THEOREM 1. *If Y is the variety obtained from a general pencil of hyperplane sections of a variety X,* (N) *is true when restricted to $H^{2m}(X)$.* (*Recall the last paragraph of the introduction.*)

The proof of Theorem 1 uses cohomological methods, and a comparison of the bundle $\boldsymbol{E}$ to more geometrically defined bundles arising from de Rham theory via differential forms with prescribed singularities. The proof is complicated by the difficulty in reading the Hodge filtration on K. When X is a hypersurface, there is less trouble, and we may drop the restriction to the cohomology of X.

REMARK. (N) is also true when there are no singular fibers.

4. Inverting a normal function. Let ν be a normal function, and suppose that $\nu(s)$ lies in the image of the Abel-Jacobi mapping ϕ_s for all $s \in S^*$. We would like to construct an algebraic cycle from ν. From the theory of Hilbert schemes, and the resolution of singularities, we know there exists a nonsingular parameter space W for cycles over S, with a diagram of analytic mappings

$$\begin{array}{ccc} W^* & \xrightarrow{\Phi} & \boldsymbol{J}^* \\ \downarrow & & \downarrow \\ S^* & = & S^* \end{array} \qquad \text{(Φ is the Abel-Jacobi mapping defined for cycles over S^*)}$$

such that $\nu(S^*) \subset \Phi(W^*)$. We want to say that Φ is meromorphic on W, and it suffices to know that

(1) Φ extends across a Zariski-open subset of the fibers of W over the singular points as a mapping into $\boldsymbol{J}$.

(2) $\boldsymbol{J}$ embeds analytically in a manifold satisfying an extension theorem for meromorphic mappings (see **[5]**).

When these conditions are met, the closure of the graph Γ of Φ is a Moišezon space, so we can find algebraic curves in Γ, and this provides cycles on Y which give integer multiples of ν. We obtain

THEOREM 2. *If Φ is meromorphic, and ν an invertible normal function* (*as above*), *then for some positive integer k, $k\nu$ is the normal function of an algebraic cycle.*

COROLLARY. *If η is the cohomology class of an invertible normal function, η is the fundamental class of an algebraic cycle with rational coefficients.*

5. Application to cubic fourfolds. We recall the relevant facts about cubic threefolds (cubic hypersurfaces in $\boldsymbol{P}^4$) **[1]**:

(a) There is a nonsingular *Fano surface* parametrizing lines on a nonsingular cubic threefold.

(b) The intermediate Jacobians are principally polarized abelian varieties.

(c) The image of the Abel-Jacobi mapping for the Fano surface generates the entire intermediate Jacobian.

As the hyperplane sections of a cubic fourfold are cubic threefolds, this information is used in conjunction with Theorems 1 and 2 to conclude:

THEOREM 3. *If Y is a smooth cubic fourfold, then the Hodge conjecture* (H) *is true for Y.*

To apply Theorem 2, one must check that the conditions (1) and (2) for meromorphic extension are met. (1) is a consequence of (4.58) of **[6]**, and (2) follows from

THEOREM 4 **[8]**. *Let* $\pi: A \to \Delta = \{z \in \boldsymbol{C} : |z| < 1\}$ *be a family of principally-polarized abelian varieties which degenerates at the origin, with unipotent monodromy, to a noncompact group. Then A embeds in* $\boldsymbol{P}^N \times \Delta$ *via suitably chosen theta functions, with* $\pi^{-1}(0)$ *Zariski-open in* $\boldsymbol{P}^N \times \{0\}$.

REMARK. By the remark following Theorem 1, we can easily deduce the Hodge conjecture for smooth fibrations by cubic threefolds (e.g., products with a curve), for we may bypass Theorem 2.

REFERENCES

1. C. H. Clemens and P. A. Griffiths, *The intermediate Jacobian of the cubic threefold*, Ann. of Math. (2) **95** (1972), 281–356. MR **46** #1796.

2. P. A. Griffiths, *On the periods of certain rational integrals*. I, II, Ann. of Math. (2) **90** (1969), 460–541. MR **41** #5357.

3. ———, *Periods of integrals on algebraic manifolds*. III. *Some global differential-geometric properties of the period mapping*, Inst. Hautes Études Sci. Publ. Math. No. **38** (1970), 125–180. MR **44** #224.

4. W Schmid, *Variation of Hodge structure*: *The singularities of the period mapping*, Invent. Math. **22** (1973), 211–319.

5. Y.-T. Siu, *Extension of meromorphic maps into Kähler manifolds*, Ann. of Math. (2) **102** (1975), 421–462.

6. S. Zucker, *Generalized intermediate Jacobians and the theorem on normal functions*, Invent. Math. (to appear).

7. ———, *The Hodge conjecture for cubic fourfolds* (to appear).

8. ———, *Theta functions for degenerating abelian varieties* (manuscript).

RUTGERS UNIVERSITY

PRINCIPAL LECTURE

REESE HARVEY

HOLOMORPHIC CHAINS AND THEIR BOUNDARIES

Proceedings of Symposia in Pure Mathematics
Volume 30, 1977

HOLOMORPHIC CHAINS AND THEIR BOUNDARIES

REESE HARVEY*

Table of Contents

AMS (MOS) subject classifications (1970). Primary 32C25, 53C99, 49F20.
*Partially supported by NSF grant MPS75-05270 and a Sloan Fellowship.

© 1977, American Mathematical Society

Introduction. This paper could have been subtitled *Some applications of real variable techniques to several complex variables.* The real variable theory includes potential theory and large portions of geometric measure theory. However, the notes are written keeping in mind that some of the readers are, at best, casually familiar with geometric measure theory. On the other hand, to make the notes self-contained would have been a hindrance to one of the objectives, namely, to encourage the reader to become better acquainted with the very useful spaces of currents developed in Federer's book **[10]**. Consequently, on occasion, we make use of his results even when a more traditional development is possible.

In the first section several topics are chosen which, although essentially classical, can be better understood by making use of the point of view provided by currents. In addition, most of these topics are necessary for the remaining sections. The Wirtinger inequality and the Poincaré-Lelong formula are both extremely fruitful results which are central to this paper, as well as to many topics not discussed here.

In the second section the structure theorem of Harvey-Shiffman **[21]** is discussed. This result characterizes the holomorphic p-chains on a complex manifold as the currents which are d-closed, of bidegree k, k and locally rectifiable. The important positive case is due to King **[26]** (a new proof of King's theorem has been slipped into §1.5).

In the third section we consider the problem of which odd-dimensional real submanifolds M of $\boldsymbol{C}^n$ can occur as boundaries of complex submanifolds (or subvarieties) of $\boldsymbol{C}^n - M$. This problem is solved in Theorem 3.3 (Harvey-Lawson [**23**]). The necessary and sufficient condition is that M have the "maximal" amount of complex structure in its tangent space (for $\dim_{\boldsymbol{R}} M > 1$). The problem for a real curve M is a little different, and was treated by Wermer [**56**], Stolzenberg [**50**] and many others. Actually the problem they pursued was to find complex structure in the maximal ideal space of the subalgebra of $C(M)$ generated by the polynomials. The presentation in §3 differs from that in [**23**] in several respects. In particular, the two cases, where M is smooth and where M is a real-analytic subvariety, are combined into one case by only requiring M to be smooth outside of a scar set S (corresponding to the singular locus if M is a real-analytic subvariety). In addition, the proof presented here is valid in the geometrically natural situation where smooth only means class C^1. The new material in this §3 is joint work with Blaine Lawson.

In the fourth section we present several applications of the structure theorem for holomorphic p-chains given in §2. We discuss the Remmert-Stein-Shiffman and the Bishop extension theorems, uniqueness in the Plateau problem, and an application of Lawson and Simons [**30**] to the study of stable currents on $\boldsymbol{P}^n(\boldsymbol{C})$. Moreover, a relationship is established between the Hodge conjecture and the (homology) Plateau problem on a projective manifold. In addition, the results on boundaries in §3 enable us to generalize the classical Bochner extension theorem, for CR functions, in the last §4.7.

Besides the results discussed in these notes, another area of complex analysis with an equally close affinity for currents, potential theory, and geometric measure theory is value distribution theory. Many of the classical one variable results have now been generalized to several variables in the important work of Griffths and others. Unfortunately, space limitations only allowed us to brush against this beautiful subject in §1.11.

Perhaps these notes have provided some additional evidence that, for problems of a geometric nature which primarily involve first-order phenomena, geometric measure theory constitutes an elegant and powerful vehicle for development.

It is my pleasure to thank Robin Graham for his careful reading of the manuscript. I am also indebted to Wanna King for her proficiency in typing the manuscript.

1. Holomorphic chains and positive currents. Suppose V is a pure p-dimensional complex subvariety of a complex manifold X, and let Sing V denote the singular locus of V. Recall that the irreducible components $\{V_j\}$ of V are obtained by taking the connected components $\{W_j\}$ of the manifold $V - \text{Sing } V$ and setting $V_j = \bar{W}_j$. As a preliminary definition of a *holomorphic p-chain on X*, consider all formal sums of the form $\sum n_j V_j$ where each n_j is an integer called the multiplicity of V_j. That is, a holomorphic p-chain on X is just a subvariety V with (integer) multiplicities attached to each component of V.

1.1. *Wirtinger's inequality.* Let $\langle\ ,\ \rangle$ denote the usual Hilbert inner product on r-forms $\bigwedge^r \boldsymbol{R}^m$ and r-vectors $\bigwedge_r \boldsymbol{R}^m$, and let $|\ |$ denote the associated norm. An r-vector $\zeta \in \bigwedge_r \boldsymbol{R}^m$ is *decomposable* if it is of the form $\zeta = e_1 \wedge \cdots \wedge e_r$ with each $e_j \in \boldsymbol{R}^m$; and in this case ζ is said to *represent* the oriented r-plane spanned by $e_1, \cdots,$

e_r. (We will identify $G_{\boldsymbol{R}}(r, m)$, the grassmannian of oriented r-planes in $\boldsymbol{R}^m$ with $\{\zeta \in \bigwedge_r \boldsymbol{R}^m : \zeta$ is decomposable and $|\zeta| = 1\}$, by letting $\zeta = e_1 \wedge \cdots \wedge e_r$ correspond to the oriented r-plane spanned by an orthonormal basis $e_1, \cdots, e_r$.)

Norms other than $|\ |$ are of basic geometric importance. If $\varphi \in \bigwedge^r \boldsymbol{R}^m$ is an r-form, the *comass* of φ, denoted $\|\varphi\|^*$, is defined to be

$$\sup\{\varphi(\zeta) : \zeta \in G(r, m)\}.$$

Let K denote the convex hull of the subset $G(r, m)$ of $\bigwedge_r \boldsymbol{R}^m$. The *mass norm* on $\bigwedge_r \boldsymbol{R}^m$, denoted $\|\ \|$, is by definition the norm associated with the convex set K. Obviously, the comass and mass norms are dual to each other, and $K = \{\zeta \in \bigwedge_r \boldsymbol{R}^m : \|\zeta\| \leqq 1\}$ by definition. The mass norm can also be described as $\|\zeta\| = \inf\{\sum |\zeta_j| : \zeta = \sum \zeta_j$ and each ζ_j is decomposable$\}$ (but this description will not be used here and the reader may wish to skip the proof). Obviously, $\|\zeta\| \leqq$ the infimum above. To show equality is obtained we may assume $\|\zeta\| = 1$, or equivalently $\zeta \in \partial K$. Now ζ can be expressed as a convex combination of elements of $G(r, m)$, say $\zeta = \sum \lambda_j \eta_j$ with $0 \leqq \lambda_j \leqq 1$, $\sum \lambda_j = 1$, and each $\eta_j \in G(r, m)$. Let ζ_j denote $\lambda_j \eta_j$ so that $\zeta = \sum \zeta_j$ and $\|\zeta_j\| = \lambda_j$. Then $\|\zeta\| = 1 = \sum \|\zeta_j\|$ as desired.

Let

$$\omega = \sum_{j=1}^{n} \frac{i}{2} dz_j \wedge d\bar{z}_j = \sum_{j=1}^{n} dx_j \wedge dy_j \tag{1.1}$$

denote the standard Kähler form on $\boldsymbol{C}^n$ (we will sometimes use the notation $\omega_n = \omega$). If $H = \sum_{j=1}^n dz_j \otimes d\bar{z}_j$ denotes the standard hermitian form on $\boldsymbol{C}^n$, and $S = \sum_{j=1}^n (dx_j \otimes dx_j + dy_j \otimes dy_j)$ denotes the standard Riemannian metric on $\boldsymbol{C}^n \cong \boldsymbol{R}^{2n}$ then S is the real part and $-\omega$ the imaginary part of H. That is,

$$H = S - i\omega. \tag{1.2}$$

Consider the set of unit decomposable $2p$-vectors $G_{\boldsymbol{C}}(p, n) = \{\zeta \in \bigwedge_{2p} \boldsymbol{C}^n : \zeta$ represents a complex p-plane in $\boldsymbol{C}^n$ and $|\zeta| = 1\}$. If $\zeta \in G_{\boldsymbol{C}}(p, n)$ we can choose an oriented orthonormal basis for ζ of the form $e_1, Je_1, \cdots, e_p, Je_p$, and hence $\zeta = e_1 \wedge Je_1 \wedge \cdots \wedge e_p \wedge Je_p$. ($J : \boldsymbol{C}^n \to \boldsymbol{C}^n$ denotes multiplication by $\sqrt{-1}$.) The convex cone generated by $G_{\boldsymbol{C}}(p, n)$ will be denoted $\mathrm{P}_{p,p}$ (or sometimes $S\mathrm{P}_{p,p}$) and an element $\zeta \in \mathrm{P}_{p,p}$ will be referred to as a (*strongly*) *positive* (p, p)-*vector*. That is, $\mathrm{P}_{p,p}$ consists of all linear combinations $\sum \lambda_j \zeta_j$, where $\lambda_j \geqq 0$ and $\zeta_j \in G_{\boldsymbol{C}}(p, n)$, $j = 1, \cdots, N$. (See §1.6 for the definition of (*weakly*) *positive*.)

The next result is fundamental.

THEOREM 1.1 (WIRTINGER'S INEQUALITY).

$$\frac{1}{p!}\,\omega^p(\zeta) \leqq \|\zeta\|, \quad \textit{for all } 2p\textit{-vectors } \zeta \in \textstyle\bigwedge_{2p} \boldsymbol{C}^n. \tag{1.3}$$

Moreover, equality holds if and only if ζ is positive ($\zeta \in \mathrm{P}_{p,p}$).

PROOF. We (essentially) give the proof of Federer [**10**]. Obviously, it suffices to prove $\omega^p(\zeta) \leqq p!$ for all $\zeta \in K$ (i.e., $\|\zeta\| \leqq 1$). Since K is the convex hull of the unit decomposable $2p$-vectors $G_{\boldsymbol{R}}(2p, 2n) \subset \bigwedge_{2p} \boldsymbol{C}^n$, we need only prove the inequality for ζ decomposable and $|\zeta| = 1$.

First consider the case $p = 1$, so that ζ is of the form $\zeta = u \wedge v$ where u and

v are orthonormal. Since $H(Ju, v) = iH(u, v)$, taking real parts yields $S(Ju, v) = \omega(u, v)$. Therefore $\omega(\zeta) = \omega(u, v) = S(Ju, v) \leqq |Ju|\,|v| = |u|\,|v|$, by the Schwartz inequality, with equality if and only if $v = Ju$. This proves Theorem 1.1 if $p = 1$ and ζ is decomposable.

Next consider the general case $p \geqq 1$ with ζ decomposable (and $|\zeta| = 1$). Let $\boldsymbol{R}^{2p}$ denote the $2p$-plane represented by ζ and let $i: \boldsymbol{R}^{2p} \to \boldsymbol{C}^n$ denote the natural inclusion. Since $i^*(\omega)$ is a 2-form on $\boldsymbol{R}^{2p}$, it can be put in canonical form

$$i^*(\omega) = \sum_1^p \lambda_j e_{2j-1}^* \wedge e_{2j}^*,$$

where each $\lambda_j \geqq 0$, $j = 1, \cdots, p$, and $e_1^*, \cdots, e_{2p}^*$ is an orthonormal basis for $(\boldsymbol{R}^{2p})^*$. Now we can compute the pth power,

$$i^*(\omega)^p = p!\,\lambda_1 \cdots \lambda_p e_1^* \wedge \cdots \wedge e_{2p}^*.$$

Note that $\zeta = \varepsilon e_1 \wedge \cdots \wedge e_{2p}$ where $\varepsilon = \pm 1$, and that

$$\lambda_j = \langle i^*(\omega), e_{2j-1} \wedge e_{2j} \rangle = \omega(e_{2j-1}, e_{2j}).$$

Therefore $\omega^p(\zeta) = i^*(\omega)^p(\zeta) = \varepsilon\, p!\,\lambda_1 \cdots \lambda_p$, which is $\leqq p!$ with equality if and only if $\varepsilon = 1$ and $e_{2j} = Je_{2j-1}$ for $j = 1, \cdots, p$. This proves Theorem 1.1 if ζ is decomposable.

Consider the *face*, or *supporting set*, $F = \{\zeta \in K: \omega^p(\zeta)/p! = 1\}$ determined by $\omega^p/p!$. This face is convex and hence generated by its extreme points. To prove the equality part of Theorem 1.1 we will show that the set of extreme points of F is the set $G_C(p, n)$ of positive decomposable ζ with $|\zeta| = 1$. Each point of $G_C(p, n)$ is extreme in F since it is extreme in the ball $\{\zeta \in \bigwedge_{2p}\boldsymbol{C}^n: |\zeta| \leqq 1\}$ which contains F. Conversely, suppose ζ is extreme in F. If ζ is not extreme in K, then ζ is the interior point of a line segment $I \subset K$. But, since $\omega^p/p!$ attains a maximum on I at ζ, $\omega^p/p!$ must be identically one on I, or $I \subset F$. This contradicts ζ being extreme in F. Therefore, ζ must be extreme in K, i.e., $\zeta \in G_{\boldsymbol{R}}(2p, 2n)$. We have already proved (since $\zeta \in F$) that this implies $\zeta \in G_C(p, n)$.

Now suppose that V is a smooth oriented $2p$-dimensional real submanifold of a complex manifold X and that X has a hermitian metric H (let H_z denote the metric on the tangent space $T_z(X)$ at z). As in (1.2), define ω to be minus the imaginary part of H.

COROLLARY 1.2. $\int_V \omega^p/p! \leqq \mathrm{vol}_{2p}(V)$ *with equality if and only if V is a complex submanifold.*

PROOF. Let dv denote the volume form on V and let $\zeta(z) \in \bigwedge_{2p}T_z(X)$ denote the original tangent space to V at z. Then

$$\int_V \frac{\omega^p}{p!} = \int_V \left\langle \frac{\omega^p(z)}{p!}, \zeta(z) \right\rangle dv \leqq \int_V 1 \cdot dv = \mathrm{vol}(V)$$

with equality holding if and only if $\zeta(z)$ is complex linear. Graphing V locally, the implicit function theorem implies V is a complex submanifold if and only if $\zeta(z)$ is complex linear.

1.2. *Elementary consequences of Wirtinger's equality.* In this subsection we give two applications of the equality portion of Corollary 1.2. See §§1.8 and 4.4, 4.5 for applications of the full corollary.

LEMMA 1.3. *Suppose V is a pure p-dimensional closed subvariety of an open subset Ω in C^n. Then the manifold* Reg V *has locally finite volume and the singular locus has locally finite Hausdorff* $2p-2$ *measure (denoted $\mathscr{H}_{2p-2}$) in Ω.*

Before giving the proof we establish some useful notation. Given a complex p-plane $\pi \in G_C(p, n)$ in C^n we will sometimes identify π with C^p and then (with an abuse of notation) let $\pi : C^n \to C^p$ denote orthogonal projection onto the p-plane $\pi \cong C^p$. If coordinates $z = (z_1, \cdots, z_n)$ are prescribed, and $I = (i_1, \cdots, i_p)$ with $i_1 < \cdots < i_p$ then let $\pi_I : C^n \to C^p$ denote orthogonal projection onto the *axis* p-plane spanned by $z_{i_1}, \cdots, z_{i_p}$. Let $\sum'$ denote summation of strictly increasing multi-indices. The following formula for $\omega^p = \omega \wedge \cdots \wedge \omega$ (p times) is very useful. Let $\tau_p = (i/2)dw_1 \wedge d\bar{w}_1 \wedge \cdots \wedge (i/2)dw_p \wedge d\bar{w}_p$ denote the volume form on C^p. Let $(dz \wedge d\bar{z})^I$ denote $\pi_I^*(\tau_p) = (i/2)dz_{i_1} \wedge d\bar{z}_{i_1} \wedge \cdots \wedge (i/2)dz_{i_p} \wedge d\bar{z}_{i_p}$. Expanding out ω^p one obtains

(1.4) $$\omega^p = p! \sum_{|I|=p}' (dz \wedge d\bar{z})^I.$$

Now we can sketch the proof of Lemma 1.3.

PROOF. Choose a point $z \in$ Sing V which we may assume to be the origin, $z = 0$. Choose coordinates on C^n so that, for each axis p-plane π_I, the origin is an isolated point of the intersection of V with the fibre $\pi_I^\perp = \pi_I^{-1}(0)$. Suppose B is a small ball about the origin. Then

$$\mathrm{vol}(\mathrm{Reg}\, V \cap B) = \int_{\mathrm{Reg}\, V\cap B} \frac{1}{p!}\,\omega^p = \sum_{|I|=p}' \int_{\mathrm{Reg}\, V\cap B} \pi_I^*(\tau_p)$$

so it suffices to estimate each term $\int_{\mathrm{Reg}\, V\cap B} \pi_I^*(\tau_p)$. We may choose a connected product neighborhood $\Delta_I = \Delta_I' \times \Delta_I''$ of 0 with $\Delta_I' \subset \pi_I$ and $\Delta_I'' \subset \pi_I^\perp$ so that π_I considered on Δ_I is a proper map of $V \cap \Delta_I$ onto Δ_I' with generic sheeting number m_I. Obviously

$$\int_{\mathrm{Reg}\, V\cap \Delta_I} \pi_I^*(\tau_p) = m_I \,\mathrm{vol}(\Delta_I').$$

Therefore, if B is chosen to be contained in Δ_I for all I, then

$$\mathrm{vol}(B \cap \mathrm{Reg}\, V) \leqq \sum_{|I|=p}' m_I \,\mathrm{vol}(\Delta_I') < +\infty.$$

This proves that Reg V has locally finite volume.

Now Sing V is a subvariety of dimension $p-1$ or less. Therefore, Sing V is the disjoint union of a $(p-1)$-dimensional manifold and a subvariety of dimension $p-2$ or less. The fact that $\mathscr{H}_{2p-2}$(Sing V) is locally finite follows by induction.

A current T defined on an open subset Ω of R^m which acts on test forms of degree r is said to be a current of *dimension* r or *degree* $n-r$. The next result was first proved by Lelong **[31]**.

COROLLARY 1.4. *Suppose V is a pure p-dimensional subvariety of an open subset Ω of C^n. Then*

(1.5) $$[V](\varphi) \equiv \int_{\mathrm{Reg}\, V} \varphi, \quad \textit{for all test forms } \varphi \textit{ on } \Omega,$$

defines a $2p$*-dimensional current* $[V]$ *on* Ω *(with measure coefficients).*

PROOF. Suppose K is a compact subset of Ω. For each smooth $2p$-form φ on Ω, let $\|\varphi\|_K^*$ denote the supremum of $\|\varphi(z)\|^*$ over all points $z \in K$. Let dv denote the volume form on Reg V, and $\zeta(z) \in \bigwedge_{2p} \mathbf{C}^n$ the tangent space to Reg V at z. Then

(1.6) $$\int_{\mathrm{Reg}\, V} \varphi = \int_{\mathrm{Reg}\, V} \langle \zeta(z), \varphi(z) \rangle \, dv \leqq \|\varphi\|_K^* \operatorname{vol}(K \cap \mathrm{Reg}\, V)$$

if supp $\varphi \subset K$. This proves that $[V]$ is a current on Ω and, by the Riesz representation theorem, that each coefficient of $[V]$ (considering the current as a differential form with distribution coefficients) is a regular Borel measure (locally finite).

A slight refinement of the argument used in Lemma 1.3 will be needed in Corollary 3.23.

LEMMA 1.5. *Suppose that* M *is a* $(2p-1)$*-dimensional submanifold of an open set* Ω *of* $\mathbf{C}^n$. *If* V *is a pure* p*-dimensional subvariety of* $\Omega - M$ *then the manifold* Reg V *has locally finite volume across* M *in* Ω.

PROOF. Suppose $0 \in M \cap \bar{V}$, and adopt the notation in the proof of Lemma 1.3. Since $\mathscr{H}_{2p+1}(V \cup M) = 0$, it follows from a lemma of Shiffman [**46**] that coordinates and product neighborhoods $\Delta_I = \Delta_I' \times \Delta_I''$ can be chosen as in Lemma 1.3 so that, for each I, $\pi_I \colon (M \cup \bar{V}) \cap \Delta_I \to \Delta_I'$ is proper (see Lemmas 2.3 and 2.4 for a further discussion). Consequently, the map $\pi \colon V \cap \Delta_I \to \Delta_I'$ is a local biholomorphism outside a proper subvariety Σ of V. In addition, one can insure that, for each I, $\pi_I \colon M \cap \Delta_I \to \Delta_I'$ is an embedding. Let $\mathfrak{M}_I$ denote the image, and let $S_I = \pi_I^{-1}(\mathfrak{M}_I)$ denote the shadow of M in Δ_I. Then the set $(V - \Sigma) \cap S_I$ is a regular submanifold of dimension $2p - 1$ and Σ has locally finite $\mathscr{H}_{2p-2}$ measure. Therefore $\mathscr{H}_{2p}(V \cap S_I) = 0$. Consequently, the estimate for $\operatorname{vol}(V \cap B)$ where B is a small ball about 0 contained in each Δ_I proceeds as before to yield

$$\operatorname{vol}(V \cap B) \leqq \sum_{|I|=p}{}' \int_{\Delta'_I - \mathfrak{M}_I} m_I$$

except now the sheeting number m_I need only be locally constant on $\Delta_I' - \mathfrak{M}_I$. However, by choosing Δ_I' small we can insure that $\Delta_I' - \mathfrak{M}_I$ has precisely two components since $\mathfrak{M}_I$ is a submanifold.

1.3. *Holomorphic chains.*

DEFINITION 1.6. A current T defined on a complex n-dimensional manifold X is called a *holomorphic* p*-chain* if T is of the form $T = \sum n_j[V_j]$, where $\{V_j\}$ denotes the irreducible components of a pure p-dimensional subvariety V and the *multiplicity* n_j of each component V_j is an integer. A holomorphic $(n-1)$-chain is also called a *divisor*.

In this section we present two results about holomorphic chains which illustrate the applicability of geometric measure theory to complex analysis. Both results are concerned with the concept of a locally flat current (Federer [**10**]). These currents provide a very useful generalization of locally integrable functions to currents of higher degree. However, contrary to the obvious guess, a locally flat current is not just a current with locally integrable coefficients, but the exterior derivative of such a current is also allowed. A current T defined on an open subset Ω of $\boldsymbol{R}^m$ is called *locally flat* if every cut-off φT by a function $\varphi \in C_0^\infty(\Omega)$ can be expressed in the form $\varphi T = R + dS$ where R and S have locally integrable functions as coefficients. If T is locally flat then dT is also locally flat, because $\varphi dT = d(\varphi T) - d\varphi \wedge T$.

In order to verify that a given current (say $T = [X]$ where X is an oriented submanifold) is locally flat it is necessary to have an alternate definition. First, we refer to a current T such that both T and dT have measure coefficients as a *locally normal current*. Second, consider the *flat norm* $F_K(T) = \sup\{|T(\varphi)| : \|\varphi\|_K^* \leqq 1 \text{ and } \|d\varphi\|_K^* \leqq 1\}$ for each compact set K. A *flat current on* K is a limit in the norm F_K of locally normal currents supported in K. A current T is *locally flat* if each cut-off φT is a flat current on some compact set containing supp φ. The fact that the two definitions of locally flat agree is a nontrivial result of Federer [**10**, 4.1.18] (as remarked in [**21**] an alternate proof can be based on the Newtonian kernel).

Let $F_p^{\text{loc}}(\Omega)$ denote the space of locally flat currents on Ω of dimension p with the topology induced by the seminorms F_K.

EXAMPLE. Suppose X is an oriented locally closed submanifold of an open subset Ω of $\boldsymbol{R}^n$. If X has locally finite volume in Ω then $[X](\varphi) \equiv \int_X \varphi$ defines a current $[X]$ on Ω *corresponding to integration* (*of forms*) *over* X. The current $[X]$ is locally flat. To prove this suppose A is a closed subset of Ω such that X is a closed subset of $\Omega - A$. Choose an exhaustive sequence of functions $\chi_j \in C_0^\infty(\Omega - A)$ for $\Omega - A$. That is, each $0 \leqq \chi_j \leqq 1$, and for each compact subset K of $\Omega - A$, $\chi_j \equiv 1$ if j is large. Each $\chi_j[X]$ is normal (with compact support in $\Omega - A$), since $d(\chi_j[X]) = d\chi_j \wedge [X]$. Also for each cut-off function $\psi \in C_0^\infty(\Omega)$ with $0 \leqq \psi \leqq 1$ and supp $\psi = K$, $\psi[X]$ is approximated in the norm F_K by $\psi\chi_j[X]$, since

$$\begin{aligned} |\psi[X](\varphi) - \psi\chi_j[X](\varphi)| &\leqq \left| \int_{X \cap K} (1 - \chi_j)\, \varphi \right| \\ &\leqq \operatorname{vol}(X \cap K \cap (\operatorname{supp}(1 - \chi_j))) \|\varphi\|_K^*, \end{aligned}$$

which converges to zero.

In particular, each holomorphic chain is locally flat.

The next theorem explains the value of locally flat currents.

THEOREM 1.7 (THE SUPPORT THEOREM). *Suppose T is a locally flat current of dimension r defined on an open subset Ω of $\boldsymbol{R}^m$, and supported in a closed subset A of Ω.*

(a) *If $\mathscr{H}_r(A) = 0$ then $T = 0$.*

(b) *If A is an r-dimensional oriented submanifold of Ω then T is of the form $T = f[A]$ where f is a locally integrable function on A.*

See Federer [**10**, 4.1.15 and 4.1.20] for the proof.

We illustrate how this theorem may be put to use in the next two results concerning holomorphic chains. Several other applications are given later.

LEMMA 1.8. *Every holomorphic p-chain $T = \sum n_j[V_j]$ on Ω is d-closed.*

PROOF. Let Sing V denote the singular locus of $V = \operatorname{supp} T$. Of course, $dT = 0$ on $\Omega -$ Sing V. By Lemma 1.3, $\mathscr{H}_{2p-1}(\text{Sing } V) = 0$. The holomorphic p-chain T is locally flat, and hence dT is locally flat. Therefore dT is a locally flat current of dimension $2p - 1$ supported in a set Sing V which is too small. Theorem 1.7(a) implies $dT = 0$.

LEMMA 1.9. *Suppose T is a d-closed locally flat current of dimension $2p$, and that T is supported in a pure p-dimensional subvariety V. Then T is of the form $T = \sum c_j[V_j]$ where $\{V_j\}$ is the set of irreducible components of V and each $c_j \in \boldsymbol{R}$.*

PROOF. First we compute T on Ω − Sing V. Since Reg V is a submanifold of Ω − Sing V, Theorem 1.7 (b) implies that on Ω − Sing V the current T is of the form $T = f[\text{Reg } V]$ where f is locally integrable. Since $dT = 0$, f must be locally constant, say $f \equiv c_j$ on the component W_j of Reg V, where $V_j = \bar{W}_j$. Therefore $T - \sum c_j[V_j]$ is supported in Sing V. Since this current is of dimension $2p$ and $\mathscr{H}_{2p}(\text{Sing } V) = 0$, the support is too small and hence, by Theorem 1.7(a), $T - \sum c_j[V_j]$ must vanish.

1.4. *The Poincaré-Lelong equation.* This is a differential equation relating a meromorphic function f to its zeros and poles. It (along with generalizations to higher codimension) may be considered the most important formula in complex analysis. In combination with the Stokes theorem, one obtains the Cauchy integral theorem and the Poisson-Jensen equation. Moreover, it provides the basis for discussing the zeros of a holomorphic function in value distribution theory.

First we review some facts about plurisubharmonic functions.

DEFINITION 1.10. An upper semicontinuous $[-\infty, \infty)$-valued function φ defined on an open subset Ω of $\boldsymbol{C}^n$ (and not identically $-\infty$ on any component of Ω) is said to be *plurisubharmonic* on Ω if, for each point $z \in \Omega$ and each $\sigma \in \boldsymbol{C}^n$ (complex line), the function $\varphi(z + \lambda\sigma)$ of the one variable λ is either subharmonic, or identically $-\infty$, on each component of its domain.

EXAMPLE. If f is holomorphic then $\log|f|$ is plurisubharmonic.

The only facts that we will need about plurisubharmonic functions in discussing the Poincaré-Lelong formula are:

(1.7) Each plurisubharmonic function is subharmonic in the $2n$-real variables.

(1.8) If φ is subharmonic on an open subset Ω of $\boldsymbol{R}^m$ then φ and $\partial\varphi/\partial x_j$, $j = 1, \cdots, m$, are locally integrable functions on Ω.

Consequently, if f is holomorphic on an open subset Ω of $\boldsymbol{C}^n$ then $\log|f|$ is locally integrable and, in particular, defines a distribution on Ω so that it is meaningful to discuss derivatives of $\log|f|$.

Suppose f is a meromorphic function defined on a complex manifold X. Since f is locally of the form g/h with g and h holomorphic, and any two representations g/h differ (multiplicatively) by a never vanishing holomorphic function, we can associate a zero set Z and a pole set P which are both hypersurfaces in X. Let $\{Z_i\}$ and $\{P_i\}$ denote the irreducible components of Z and P respectively. Suppose z is a manifold point of $Z \cup P$. Then we can associate a unique integer $m(z)$ to z as follows. We may assume that $z = 0$ and that $Z \cup P$ is defined by $w = 0$ near 0, where w is a coordinate on $\boldsymbol{C}^n$. Consequently f must be of the form $f = w^m h$ near 0, where h is a never vanishing holomorphic function. The integer m is obviously unique and locally constant on the manifold points of $Z \cup P$. Consequently we have associated to each component Z_i a positive integer n_i called the *multiplicity of f on Z_i*. Similarly, for each component P_i of P there exists a (unique) negative integer m_i called the *multiplicity of f on P_i*.

Finally note that since f is locally of the form g/h, the function $\log|f|$ is locally integrable, because of (1.7) and (1.8) above. The holomorphic $(n-1)$-chain

$$D_f = \sum n_j[Z_j] + \sum m_j[P_j]$$

is called the *divisor of the meromorphic function f.*

THEOREM 1.11 (POINCARÉ-LELONG). *Suppose f is a meromorphic function on a complex manifold X. Then:*

$$(i/\pi)\,\partial\bar\partial \log|f| = D_f \quad \text{on } X. \tag{1.9}$$

First we prove the more familiar case where $f(w) = w$ on $\mathbf{C}$. Let $\delta_0 = [0]$ denote the holomorphic zero chain otherwise known as the delta function at the origin.

LEMMA 1.12.

$$(i/\pi)\,\partial\bar\partial \log|w| = \delta_0 \quad \text{on } \mathbf{C}. \tag{1.10}$$

PROOF. First we calculate that for the smooth function $\varphi(w) \equiv (1/2\pi)\log(1 + |w|^2)$,

$$i\partial\bar\partial\varphi = \psi \frac{i}{2}\, dw \wedge d\bar{w}, \tag{1.11}$$

where $\psi(w) = (1/\pi)(1 + |w|^2)^{-2}$. Consider multiplication by $1/\varepsilon$ as a map of $\mathbf{C}$ onto $\mathbf{C}$, and pull back (1.11), i.e., apply $(1/\varepsilon)^*$, to obtain (using the fact that ∂ and $\bar\partial$ commute with pull-backs):

$$i\partial\bar\partial\varphi\left(\frac{w}{\varepsilon}\right) = \frac{1}{\varepsilon^2}\,\psi\left(\frac{w}{\varepsilon}\right)\frac{i}{2}\, dw \wedge d\bar{w}. \tag{1.12}$$

Since $\varphi(w/\varepsilon) = (1/2\pi)\log(|w|^2 + |\varepsilon|^2) - (1/2\pi)\log|\varepsilon|^2$, on the left of (1.12)

$$\lim_{\varepsilon\to 0} i\partial\bar\partial\varphi\left(\frac{w}{\varepsilon}\right) = \frac{i}{\pi}\partial\bar\partial \log|w|.$$

Since $\int \psi(w) = \int_0^\infty 2r\, dr/(1 + r^2)^2 = \int_1^\infty dt/t^2 = 1$, $\psi_\varepsilon(w) \equiv (1/\varepsilon^2)\,\psi\,(w/\varepsilon)$ is an approximate identity. Therefore, the limit on the right of (1.12) is δ_0.

REMARK. The classical proof is exactly the same except one uses the function $\varphi(w) \equiv (1/\pi)\,\chi(w)\log|w|$ (where χ denotes the characteristic function of $\mathbf{C} - \Delta$ with Δ the unit disk) instead of the above φ. In this case ψ is the unit measure $(1/2\pi)d\theta$ supported on $\partial\Delta$, and

$$i\partial\bar\partial\varphi = \psi\frac{i}{2}\, dw \wedge d\bar{w} \tag{1.11$'$}$$

is more difficult involving Green's theorem. On the other hand, the approximate identity argument is easier in this case reducing to the fact that

$$\lim_{\varepsilon\to 0} (1/2\pi)\int \alpha(\varepsilon e^{i\theta})\, d\theta = \alpha(0)$$

for all continuous functions α. For more details see the proof of Theorem 1.33.

PROOF OF THEOREM 1.11. First we establish the formula (1.9) on $X - \operatorname{Sing}(Z \cup P)$. Since the equation is local it suffices to prove it near a point $z \notin \operatorname{Sing}(Z \cup P)$ which we take to be the origin $z = 0$. Making use of a biholomorphic coordinate change taking 0 to 0, we may assume that f is of the form $f = w^m h$ where w is a coordinate on $\mathbf{C}^n$ and h is a never vanishing holomorphic function near 0. Here m is the multiplicity of f at $z = 0$ (which is zero if $z \notin Z \cup P$). Note $\log|f|^2 = m\log|w|^2 + \log|h|^2$. Since near zero $\log|h| = \operatorname{Re}\log h$ where $\log h$ is a single-valued branch of the logarithm,

$$i\partial\bar{\partial}\log|h| = i\partial\bar{\partial}\,\mathrm{Re}\log h = \mathrm{Re}[i\partial\bar{\partial}\log h] = 0.$$

Therefore $(i/\pi)\,\partial\bar{\partial}\log|f| = m(i/\pi)\,\partial\bar{\partial}\log|w|$ which by Lemma 1.12 must equal $m[V]$, where V denotes the hyperplane $w = 0$. This proves the Poincaré-Lelong formula on $X - \mathrm{Sing}(Z \cup P)$. Or, said differently $T \equiv (i/\pi)\,\partial\bar{\partial}\log|f| - D_f$ has support in $\mathrm{Sing}(Z \cup P)$. In order to show $T \equiv 0$ we next apply Theorem 1.7(a). By Lemma 1.3, $\mathscr{H}_{2p}(\mathrm{Sing}(Z \cup P)) = 0$. As discussed above, every holomorphic p-chain is locally flat; hence it remains to show that $i\partial\bar{\partial}\log|f|$ is locally flat. Now $d^c\log|f|$ has locally integrable coefficients and hence is locally flat, because of (1.8) above and the fact that $d^c = i(\bar{\partial} - \partial)$ only involves first partial derivatives. Consequently $d[d^c\log|f|]$ is also locally flat. Since $dd^c = 2i\partial\bar{\partial}$ the proof is complete.

1.5. *A current approach to the Weierstrass problem.* The Weierstrass, or Cousin II, problem can be stated as follows. Suppose that $T = \sum n_j[V_j]$ is a divisor on a complex n-dimensional manifold X (i.e., T is a holomorphic $(n-1)$-chain on X). Can one find a meromorphic function f on X such that D_f, the divisor of f on X, equals T?

In this section we give sufficient conditions for this problem to be solvable. The result (Theorem 1.13) is standard; see, for example, Gunning and Rossi [**14**]. However, the proof presented here is different from the usual sheaf theory development, with several advantages (which are discussed in the remarks following the proof).

Lelong [**32**a] first considered the relationship between solving $(i/2\pi)\partial\bar{\partial}\varphi = T$ and solving the Cousin II problem for a prescribed divisor T in the case $X = \boldsymbol{C}^n$.

In stating the theorem it is convenient to utilize the notion of a locally rectifiable current. (See the appendix to this chapter for a brief discussion of locally rectifiable currents.) Let $H^k(X, \boldsymbol{Z})$, or $H^k_{\mathrm{loc\ rect}}(X, \boldsymbol{Z})$, denote the quotient $\{T \in \mathscr{R}^k_{\mathrm{loc}}(X) : dT = 0\}/d\mathscr{R}^{k-1}_{\mathrm{loc}}(X)$. Actually, the reader unfamiliar with locally rectifiable currents may replace $\mathscr{R}^k_{\mathrm{loc}}(X)$ by the group of (locally finite) smooth singular $(2n-k)$-chains on X throughout this section. Federer has essentially shown that $H^k(X, \boldsymbol{Z})$ as defined above is isomorphic to the kth singular cohomology group of X (cf. §4.5).

Now, because of Lemmas 1.3 and 1.8 (and one of the definitions of a locally rectifiable current given in the appendix) each holomorphic $(n-k)$-chain $T = \sum n_j[V_j]$ is a locally rectifiable current of degree $2k$ (i.e., $T \in \mathscr{R}^{2k}_{\mathrm{loc}}(X)$ and $dT = 0$). Consequently, T determines a cohomology class $\bar{T} \in H^{2k}(X, \boldsymbol{Z})$.

THEOREM 1.13. *Suppose $T = \sum n_j[V_j]$ is a divisor on a complex manifold X, and that $H^1(X, \mathscr{O}) = 0$. There exists a meromorphic function f on X with D_f, the divisor of f, equal to T if and only if the class of T in $H^2(X, \boldsymbol{Z})$ is zero.*

REMARK. Suppose $H^2(X, \boldsymbol{Z}) = 0$. Then Theorem 1.13 for the restricted class of positive divisors implies the general Theorem 1.13. Just let $T = T^+ - T^-$, where T^+ and T^- are the positive and negative parts of T respectively. Since $H^2(X, \boldsymbol{Z}) = 0$ we can apply Theorem 1.13 to both of the positive divisors T^+ and T^-. However, if $H^2(X, \boldsymbol{Z}) \neq 0$, it is interesting to note that Theorem 1.13 is applicable to a general divisor $T = T^+ - T^-$ which vanishes in $H^2(X, \boldsymbol{Z})$ even though T^+ and T^- will not, in general, vanish in $H^2(X, \boldsymbol{Z})$.

PROOF. First, we assume that $\bar{T} \in H^2(X, \boldsymbol{Z})$ is zero and find f on X. We have axiomatized the first two of the three steps in the method of proof.

Step 1. Solve the equation

(1.13) $$dS = T \quad \text{on } X,$$

with $S \in \mathcal{R}^1_{\text{loc}}(X)$. (Actually, the remainder of the proof will show that one need only solve $dS = T$ with S a real current of the form $R + d\psi$, where $R \in \mathcal{R}^1_{\text{loc}}(X)$ but ψ is allowed to be any generalized function. This greater flexibility might prove important in the applications of this current method to solving Cousin II with growth.)

Step 2. Since $H^1(X, \mathcal{O}) = 0$ we can solve

(1.14) $$(i/4\pi)\,\bar{\partial}\Phi = S^{0,1} \quad \text{on } X,$$

where $S \equiv S^{1,0} + S^{0,1}$ with $S^{i,j}$ of bidegree i, j. Then. $T = \partial S^{0,1} + \bar{\partial} S^{1,0} = 2 \operatorname{Re} \partial S^{0,1} = \operatorname{Re}\,(i/2\pi)\,\partial\bar{\partial}\Phi$; or since $i\partial\bar{\partial}$ is a real operator

(1.15) $$(i/2\pi)\,\partial\bar{\partial}\varphi = T \quad \text{on } X,$$

where $\Phi \equiv \varphi + i\psi$ with φ and ψ real.

Step 3. Because of the Poincaré-Lelong formula (Theorem 1.11), it remains to find a meromorphic function f on X with

(1.16) $$\varphi = \log|f|^2.$$

First we concentrate on $\Omega = X - (\operatorname{supp} T)$. Suppose we could find f with $\varphi = \log|f|^2$. Then $\partial\varphi = \partial f/f = d \log f$, where $\log f$ denotes the multi-valued logarithm. Consequently, $f(z) = f(a) \exp \int_a^z \partial\varphi$, where the integration is along any smooth path in Ω with initial point at a fixed point a and with terminal point z.

Therefore, it is very reasonable to define

(1.17) $$f(z) \equiv \exp \int_a^z \partial\varphi \quad \text{on } X - (\operatorname{supp} T)$$

where $a \in X - (\operatorname{supp} T)$ is fixed and the integration is along a path in $X - (\operatorname{supp} T)$ from a to z (however, $f(z)$ may a priori be multi-valued). Note that $\operatorname{Re} \partial\varphi = \frac{1}{2} d\varphi$ so that $\operatorname{Re} \int_a^z \partial\varphi = \frac{1}{2}\int_a^z d\varphi = \frac{1}{2}\varphi(z) - \frac{1}{2}\varphi(a)$. It is convenient to redefine φ by subtracting a real constant so that $\varphi(a) = 0$. Then

(1.18) $$\operatorname{Re} \int_a^z \partial\varphi = \tfrac{1}{2}\varphi(z)$$

or, equivalently, (1.16) is valid on $X - (\operatorname{supp} T)$. In particular (1.18) says that $\operatorname{Re} \int_a^z \partial\varphi$ is single-valued on $X - (\operatorname{supp} T)$.

Let $d^c = i(\bar{\partial} - \partial)$. Then $\operatorname{Im} \partial\varphi = \frac{1}{2} d^c\varphi$, since φ is real. Consequently, $f(z)$ is single-valued on $X - (\operatorname{supp} T)$ if and only if

(1.19) $$\frac{1}{2}\int_\tau d^c\,\varphi \in 2\pi \mathbf{Z}$$

for every closed path τ in $X - (\operatorname{supp} T)$.

Before verifying that (1.19) is true, note that by equation (1.14) above and the fact that S is real, $S = 2 \operatorname{Re} S^{1,0} = (i/4\pi)\bar{\partial}\Phi - (i/4\pi)\partial\bar{\Phi} = (1/4\pi)d^c\varphi - (1/4\pi)d\psi$. That is,

(1.20) $$(1/4\pi)d^c\varphi = S + (1/4\pi)d\psi \quad \text{on } X.$$

Intuitively, the proof of (1.19) can be completed as follows. Since $(1/4\pi)d^c\varphi$ and S differ by d of something, $(1/4\pi)\int_\tau d^c\varphi$ is the same as the integral of S over τ which is the intersection number of S with τ and hence an integer. There are several ways this argument can be made rigorous. We present a Čech approach to the intersection pairing of S and τ as follows.

Choose a covering $\mathscr{U} = \{U_j\}$ of $X -$ (supp T) by open contractible sets U_j. Since $dS = 0$ on U_j we may find $\chi_j \in \mathscr{R}^0_{\text{loc}}(U_j)$ such that $d\chi_j = S$ on U_j. Let c_{ij} denote $\chi_i - \chi_j$ on $U_{ij} = U_i \cap U_j$. Then $dc_{ij} = 0$ and $c_{ij} \in \mathscr{R}^0_{\text{loc}}(U_{ij})$. Therefore, c_{ij} is locally constant and integer-valued. Now express τ as a finite sum $\tau = \sum \tau_j$ where each τ_j is a 1-chain with compact support in U_j (or τ_j is a rectifiable current of dimension 1 on U_j). Then $(1/4\pi)\int_\tau d^c\varphi = \sum(1/4\pi)\int_{\tau_j} d^c\varphi$. On U_j, $(1/4\pi)d^c\varphi$ has the primitive $X_j + (1/4\pi)\psi$ by (1.20). Therefore $(1/4\pi)\int_\tau d^c\varphi = \sum (X_j + (1/4\pi)\,\psi)\big|_{\partial\tau_j}$. Suppose $+b$ is a boundary point of τ_j. Since τ is a cycle, $-b$ must be a boundary point of τ_i for some $i \neq j$. That is, the terms of the above sum can be grouped into pairs of the form $(X_j + (1/4\pi)\,\psi)(b) - (X_i + (1/4\pi)\psi)(b) = c_{ij}(b)$. Since each $c_{ij}(b) \in \boldsymbol{Z}$, this proves (1.19). In summary, we have shown that the function $f(z)$ defined by (1.17), on $X -$ (supp T), is a single-valued holomorphic function with $\varphi = \log|f|^2$.

If we can show that f has a meromorphic extension $\tilde{f}$ to all of X then the proof of the first half of Theorem 1.13 is complete. This is because both φ and $\log|\tilde{f}|^2$ are locally integrable on X and agree on $X -$ (supp T), thus verifying (1.16) as desired.

The divisor $T = \sum n_j[V_j]$ can be expressed as $T = T^+ - T^-$ where $T^+ = \sum_{n_j>0} n_j[V_j]$ and $T^- = -\sum_{n_j<0} n_j[V_j]$. Let Z denote supp T^+, P denote supp T^-, and I denote the indeterminancy set $Z \cap P$. First we extend f holomorphically from $X - (Z \cup P)$ to $X - P$ across $Z - P$. Since $\log|f|^2 = \varphi$ on $X - (Z \cup P)$ and φ is plurisubharmonic on $X - P$, it follows that $\log|f|^2$ is locally bounded above across $Z - P$. Therefore, f is locally bounded across $Z - P$ and hence by the Riemann removable singularity theorem f extends holomorphically across $Z - P$. Similarly $1/f$ extends holomorphically across $P - Z$. Consequently, f extends meromorphically to $X - I$. Because the intersection $I = Z \cap P$ has codimension at least two a standard result implies that f must extend meromorphically across I to all of X. This completes our development of the Cousin II problem.

We will only sketch the current proof of the converse. If $T = D_f$, the divisor of a global meromorphic function f on X, then $\bar{T} = 0$ in $H^2(X, \boldsymbol{Z})$ (of course, there is the standard proof where one shows that the line bundle of D_f, and hence the chern class of the line bundle, is trivial). Consider the special case where $X' = \boldsymbol{P}^1 = \boldsymbol{C} \cup \{\infty\}$ and $T' = [0] - [\infty]$. Let S' denote the current corresponding to integration of 1-forms from ∞ to 0 along the negative real axis. Then $dS' = T'$.

Suppose f is meromorphic on a complex manifold X. We omit the proofs but one can show (by considering the graph of f in $X \times \boldsymbol{P}^1$ and "slicing") that $f^*([0]) - f^*([\infty])$, the pull-back of $[0] - [\infty]$, is meaningful and equal to $T = D_f$. Similarly, $f^*(S')$ is a well-defined locally rectifiable current of degree 1 on X. Let $S \equiv f^*(S')$. Then $dS = df^*(S') = f^*(dS') = f^*([0] - [\infty]) = T$. This completes the proof of Theorem 1.13.

Some additional insight into the above proof of Theorem 1.13 is gained by examining the special case $X' = \boldsymbol{P}^1$ and $T' = [0] - [\infty]$ a little further. One has that $S' = X^-(x)\delta_0(y)dy$ where χ^- denotes the characteristic function of the negative real numbers. Let arg z denote the multi-valued argument of z and let Arg z denote the single-valued L^∞ function defined on C by requiring $-\pi \leqq \arg z < \pi$. Let $\operatorname{Log} z \equiv \log|z| + i \operatorname{Arg} z$. An elementary calculation shows that $\Phi' = 2 \operatorname{Log} z$ (i.e., $\varphi' + i\psi' = 2 \log|z| + i2 \operatorname{Arg} z$) satisfies (1.14). If f exists then $\Phi = f^*(\operatorname{Log} z) = \operatorname{Log} f$ denotes a single-valued branch of $\log f$ on $X - \operatorname{supp} S$.

REMARK 1. The above construction of f provides a new proof of the structure theorem of King [**26**]. See the Remark 1 after Lemma 2.2 in §2 for a full discussion of this proof of King's theorem.

REMARK 2. The beautiful work of Henkin [**58**] and Skoda [**59**] generalizing the Blaschke theorem to several variables can be made compatible with the above point of view of the Weierstrass problem.

1.6. *Positive forms.* In this subsection we restrict our attention to the grassmann algebra $\bigwedge^* C^n$ and its complexification $C \otimes_R \bigwedge^* C^n$ (which will be denoted $\bigwedge_C^* C^n$), although the results obviously extend to smooth differential forms on a complex manifold by considering the forms pointwise. Any complexification, such as $\bigwedge_C^* C^n$, comes equipped with a natural conjugation (and $\{\alpha \in \bigwedge_C^* C^n : \alpha = \bar\alpha\} = \bigwedge^* C^n$) related to the real and imaginary parts in the usual way. The reader familiar with the exterior algebra $\bigwedge_C^* C^n$ will probably wish to skip over the following review to Definition 1.14 below.

The real linear operator $J : C^n \to C^n$ is defined to be multiplication by $\sqrt{-1}$. In terms of a standard basis $\partial/\partial x_1, \partial/\partial y_1, \cdots, \partial/\partial x_n, \partial/\partial y_n$, $J(\partial/\partial x_j) = \partial/\partial y_j$ and $J(\partial/\partial y_j) = -\partial/\partial x_j$, $j = 1, \cdots, n$. The operator J has an adjoint, also denoted J on $\bigwedge^1 C^n = (C^n)^*$, and in terms of a standard basis $J(dx_j) = -dy_j$, $J(dy_j) = dx_j$, $j = 1, \cdots, n$. The operator J extends (complex linearly) to the complexification $\bigwedge_C^1 C^n$ of $(C^n)^*$ and $J^2 = -1$. Therefore, J has eigenvalues $\pm i$. Let $\bigwedge^{1,0} C^n$ and $\bigwedge^{0,1} C^n$ denote the corresponding eigenspaces. Let $dz_j \equiv dx_j + i dy_j$ $(= dx_j - iJdx_j)$ and $d\bar z_j \equiv dx_j - i dy_j$ $(= dx_j + iJdx_j)$. Then $dz_1, \cdots, dz_n$ is a basis for $\bigwedge^{1,0} C^n$ and $d\bar z_1, \cdots, d\bar z_n$ is a basis for $\bigwedge^{0,1} C^n$. Let $\bigwedge^{p,q} C^n$ denote the span of $\{dz^I \wedge d\bar z^J : |I| = p, |J| = q$ both increasing$\}$ (here we employ the usual multi-index notation). Then one has the standard direct sum decomposition $\bigwedge_C^r C^n = \sum_{p+q=r} \bigwedge^{p,q} C^n$. The operator J on $\bigwedge_C^1 C^n$ can be extended as a derivation to $\bigwedge_C^* C^n$, i.e.,

$$J(e_1 \wedge \cdots \wedge e_r) \equiv \sum e_1 \wedge \cdots \wedge Je_j \wedge \cdots \wedge e_r.$$

Then $\bigwedge^{p,q} C^n$ may be (alternately) defined as the subset of $\bigwedge_C^r C^n$ consisting of all α with $J\alpha = i(p-q)\alpha$, since $J(dz^I \wedge d\bar z^J) = i(p-q) dz^I \wedge d\bar z^J$. Let $\bigwedge_R^{p,p} C^n$ denote the space of $\alpha \in \bigwedge^{p,p} C^n$ with $\alpha = \bar\alpha$. Then, in particular, $\bigwedge_R^{p,p} C^n = \{\alpha \in \bigwedge^{2p} C^n : J\alpha = 0\}$ (cf. Lawson and Simons [**30**]).

Let $G_C(p, n)$ denote the set of orthogonal projections π of C^n onto a complex p-plane, say $\pi : C^n \to C^p$ with $\pi = (\pi_1, \cdots, \pi_p)$. We associate the $2p$-form

$$(i/2)\, \pi_1 \wedge \bar\pi_1 \wedge \cdots \wedge (i/2)\, \pi_p \wedge \bar\pi_p \tag{1.21}$$

to π and denote this $2p$-form by π as well. Each $2p$-form arising this way is easily seen to be real and of bidegree p, p. Thus we consider $G_C(p, n)$ as a submanifold of $\bigwedge_R^{p,p} C^n$.

Let τ_n denote the volume form on C^n, i.e.,

$$\tau_n = (i/2)\, dz_1 \wedge d\bar{z}_1 \wedge \cdots \wedge (i/2)\, dz_n \wedge d\bar{z}_n$$

in terms of coordinates. A $2n$-form $\alpha \in \bigwedge_C^{2n} C^n$ is said to be *positive* if when expressed in terms of τ_n, say $\alpha = \lambda\tau_n$ with $\lambda \in C$, the coefficient $\lambda \geqq 0$.

DEFINITION 1.14. (a) (*Strongly positive.*) A real (p, p)-form α is said to be (*strongly*) *positive* if α belongs to the convex cone generated by $G_C(p, n)$, that is, if α is of the form $\alpha = \sum \lambda_j \pi^j$ with each $\lambda_j \geqq 0$ and each π^j of the form (1.21). Let $\mathrm{SP}^{p,p}$ denote this cone in $\bigwedge_R^{p,p} C^n$.

(b) (*Weakly positive.*) A real (k, k)-form β is said to be (*weakly*) *positive* if the $2n$-form $\alpha \wedge \beta$ is positive for each $\beta \in \mathrm{SP}^{p,p}$, where $p + k = n$. Let $\mathrm{WP}^{k,k}$ denote this dual cone in $\bigwedge_R^{k,k} C^n$.

REMARK. Obviously, $\beta \in \bigwedge_R^{k,k} C^n$ is weakly positive if and only if

(1.22) $\qquad \beta \wedge \pi$ is a positive $2n$-form for each $\pi \in G_C(p, n)$

(expressed as in (1.21) above). Also note that,

$$\mathrm{SP}^{k,k} \subset \mathrm{WP}^{k,k}. \tag{1.23}$$

LEMMA 1.15. *The notions of weakly positive and strongly positive (p, p)-vectors agree if $p = 1$ or $p = n - 1$, and disagree for $2 \leqq p \leqq n - 2$.*

PROOF. First consider $p = 1$ (this covers the case $p = n - 1$ by duality). We must show that $\mathrm{WP}^{1,1} \subset \mathrm{SP}^{1,1}$. Now $\alpha \in \bigwedge_R^{1,1} C^n$ is of the form $\alpha = \sum \alpha_{ij}(i/2)\, dz_i \wedge d\bar{z}_j$; and (α_{ij}) is a hermitian matrix since $\alpha = \bar{\alpha}$. Consequently, we may put α in the canonical form

$$\alpha = \sum_{j=1}^{n} \lambda_j e_j \wedge Je_j \tag{1.24}$$

with each λ_j real and $e_1, Je_1, \cdots, e_n, Je_n$ an orthonormal basis for $\bigwedge^1 C^n$. Now $\alpha \wedge \pi_j = \lambda_j \tau_n$ if $\pi_j \equiv e_1 \wedge Je_1 \wedge \cdots \wedge \hat{e}_j \wedge \widehat{Je_j} \wedge \cdots \wedge e_n \wedge Je_n$. Therefore, if $\alpha \in \mathrm{WP}^{1,1}$ then each λ_j must be $\geqq 0$, which proves that $\alpha \in \mathrm{SP}^{1,1}$.

For examples of weakly positive (p, p)-forms which are not strongly positive, when $2 \leqq p \leqq n - 2$, see Harvey and Knapp [**20**].

REMARK. Obviously SP^* is closed under the exterior product. However, WP^* is not closed under the exterior product (see [**20**]).

LEMMA 1.16. *There exists a basis for $\bigwedge_R^{p,p} C^n$ contained in the cone $\mathrm{SP}^{p,p}$, or equivalently $\mathrm{SP}^{p,p}$ has nonempty interior in $\bigwedge_R^{p,p} C^n$.*

PROOF. First assume that $p = 1$. The fact that $\mathrm{WP}^{1,1}$ has nonempty interior can be easily established. Alternatively, a basis for $\bigwedge_R^{1,1} C^n$, consisting of vectors in the cone $\mathrm{SP}^{1,1}$, can be given explicitly in terms of the standard basis for $\bigwedge^1 C^n$. In fact, both the real and imaginary parts of $dz_i \wedge d\bar{z}_j$ can be expressed as real linear combinations of the vectors $dx_i \wedge dy_i$, $i = 1, \cdots, n$, $(dx_i + dx_j) \wedge (dy_i + dy_j)$, $i < j$, and $(dx_i + dy_j) \wedge (dy_i - dx_j)$, $i < j$; which all belong to $G_C(1, n) \subset \mathrm{SP}^{1,1}$. The general case reduces to this case $p = 1$ since $\bigwedge_R^{p,p} C^n$ is generated by elements of $\bigwedge_R^{1,1} C^n$.

REMARK 1. The dual cone $\mathrm{WP}^{k,k}$ $(k + p = n)$ contains no lines. That is, if $\sum \alpha_j = 0$ with each $\alpha_j \in \mathrm{WP}^{k,k}$ then each $\alpha_j = 0$.

REMARK 2. Another interpretation of Lemma 1.16 will be useful in discussing positive currents. Namely, there exists a basis $\mathscr{B}$ for $\bigwedge^{k,k}\mathbf{C}^n$ such that the dual basis $\mathscr{B}^*$ for $\bigwedge^{p,p}\mathbf{C}^n$ consists of $\pi \in G_C(p, n)$.

COROLLARY 1.17. *The* (p, p)*-form* $\omega^p = \omega \wedge \cdots \wedge \omega$ (p *times*), *where* ω *is the standard Kähler form, belongs to the interior of the cone* $\mathrm{SP}^{p,p}$ (*i.e.*, ω^p *is* (*strongly*) *positive definite*).

PROOF. It suffices to show that $\omega^p \wedge \beta$ is a strictly positive $2n$-form, for each $\beta \in \mathrm{WP}^{k,k}$. Recall $\omega^p = p! \sum'_{|I|=p}(dz \wedge d\bar{z})^I$ in (1.4). The (k, k)-form β can be expressed as

$$\beta = \sum_{|I|=k}{}' \beta_I (dz \wedge d\bar{z})^I + \sum_{I \neq J} \beta_{IJ} dz^I \wedge d\bar{z}^J.$$

Therefore $\omega^p \wedge \beta = p!(\sum'_{|I|=k} \beta_I)\tau_n$. In addition, if J denotes the multi-index complementary to I, then $\beta \wedge (dz \wedge d\bar{z})^J = \beta_I \tau_n$ so that $\beta_I \geqq 0$. If $\omega^p \wedge \beta = 0$ then $\sum' \beta_I = 0$, and hence each $\beta_I = 0$.

Now choose a basis $\mathscr{B}$ for $\bigwedge^{p,p}_{\mathbf{R}}\mathbf{C}^n$ consisting of elements in $G_C(p, n)$. By choosing new unitary coordinates we may assume that a particular $\pi \in \mathscr{B}$ is of the form $\pi = (dz \wedge d\bar{z})^J$ for some J. Consequently, we have proved above that if $\omega^p \wedge \beta = 0$ with $\beta \in \mathrm{WP}^{k,k}$ then $\pi \wedge \beta = 0$ for each $\pi \in \mathscr{B}$, and hence that β must vanish.

It is useful to have an alternate definition of weakly positive in analyzing plurisubharmonic functions, and in slicing positive currents. Suppose $\pi : \mathbf{C}^n \to \mathbf{C}^p$ is an orthogonal projection onto a p-plane π and let $\pi = (i/2)\,\pi_1 \wedge \bar{\pi}_1 \wedge \cdots \wedge (i/2)\,\pi_p \wedge \bar{\pi}_p$ as above. Choose an orthonormal basis $\sigma^1, \cdots, \sigma^{n-p}$ for the $(n-p)$-plane $\pi^\perp$ orthogonal to π and let $\sigma : \mathbf{C}^{n-p} \to \mathbf{C}^n$ denote the natural inclusion $\sigma(w) = \sum_{j=1}^{n-p} w_j \sigma^j$. Let $\tau_{n-p} \equiv (i/2)\, dw_1 \wedge d\bar{w}_1 \wedge \cdots \wedge (i/2)\, dw_{n-p} \wedge d\bar{w}_{n-p}$ denote the volume form on $\mathbf{C}^{n-p}$. We may also consider τ_{n-p} as an $(n-p, n-p)$-form on $\mathbf{C}^n$. Since τ_{n-p} and π are dual (with $\tau_{n-p} \wedge \pi = \tau_n$), for each $\beta \in \bigwedge^{k,k}\mathbf{C}^n$ (with $p + k = n$), the equations $\beta \wedge \pi = \lambda\tau_n$ and $\sigma^*(\beta) = \lambda\tau_{n-p}$ define the same number λ. Therefore the following definition of a weakly positive (k, k)-form is equivalent to the original definition.

DEFINITION 1.14. (b′) Suppose $\beta \in \bigwedge^{k,k}\mathbf{C}^n$. If the $2(n-p)$-form $\sigma^*(\beta)$ on $\mathbf{C}^k$ is positive for each complex linear map $\sigma : \mathbf{C}^k \to \mathbf{C}^n$ then β is said to be (*weakly*) *positive*. In particular, if $k = 1$, then $\beta = \sum \beta_{ij}\,(i/2) dz_i \wedge d\bar{z}_j$ is *positive* if and only if

$$(1.25) \qquad \sum \beta_{ij}\sigma_i\bar{\sigma}_j \geqq 0 \quad \text{for each } \sigma \in \mathbf{C}^n.$$

(Take $\sigma : \mathbf{C} \to \mathbf{C}^n$ to be $\sigma(w) = w \cdot \sigma$ for each vector $\sigma \in \mathbf{C}^n$, and note that $\sigma^*(\beta) = (\sum \beta_{ij}\sigma_i\bar{\sigma}_j)(i/2)\, dw \wedge d\bar{w}$.)

REMARK. Note that the definitions of strongly positive and weakly positive forms are both independent of the hermitian structure on $\mathbf{C}^n$, and hence these notions extend (pointwise) to forms on a complex manifold.

1.7. *Positive currents.* Now we can formulate the important notion of a positive current.

DEFINITION 1.18. A current T of bidimension p, p defined on a complex manifold is said to be *weakly positive* (or *positive*) if $T(\varphi) \geqq 0$ for each test form φ which is strongly positive.

Note that if T is weakly positive, then T is real.

Alternately if T is a current of bidimension p, p defined on an open subset Ω of $\mathbf{C}^n$ then T is *weakly positive* (or *positive*) if

$$T \wedge \pi = \mu_\pi \tau_n,$$

where μ_π is a positive measure on Ω, for each $\pi = (i/2)\,\pi_1 \wedge \bar{\pi}_1 \wedge \cdots \wedge (i/2)\pi_p \wedge \bar{\pi}_p \in G_C(p, n)$.

In addition, if $p = 1$, and $T = \sum T_{ij}(i/2)\, dz_i \wedge d\bar{z}_j$ then T is *positive* if $\sum T_{ij}\sigma\bar{\sigma}_j$ is a positive measure on Ω, for each vector $\sigma \in \mathbf{C}^n$.

REMARK. The two notions (of a weakly positive (p, p)-current on an open subset Ω of $\mathbf{C}^n$) given above agree. This can be seen as follows. Obviously, if $T(\varphi) \geqq 0$ for all strongly positive test forms φ then each μ_π is a positive measure (take $\varphi = \psi \cdot \pi$ with $\psi \in C_0^\infty(\Omega)$ and $\psi \geqq 0$). Conversely, if each μ_π is a positive measure, one must show that $T(\varphi) \geqq 0$, if φ is strongly positive. Since φ is the limit of $\varphi + \varepsilon\omega^p$ with ω the standard Kähler form on $\mathbf{C}^n$, we may replace φ by $\varphi + \varepsilon\omega^p$, and hence assume that at each point z, the (p, p)-form $\varphi(z)$ actually lies in the interior of the cone $\mathrm{SP}^{p,p}$ (see Corollary 1.17). In addition, by using a partition of unity we may assume that φ has very small support with $\varphi(z)$ remaining in an arbitrarily small relatively compact subcone of $\mathrm{SP}^{p,p}$ about the point $\varphi(z_0)$. Now φ can be expressed in terms of a constant basis $\{\varphi_j\}$ for $\bigwedge_{\mathbf{R}}^{p,p}\mathbf{C}^n$ with each $\varphi_j \in \mathrm{SP}^{p,p}$ and such that $\varphi(z) = \sum \lambda_j(z)\,\varphi_j$, and where the coefficients $\lambda_j(z)$ are > 0 for all z. The inequality $T(\varphi) \geqq 0$ now follows easily.

EXAMPLE 1. A holomorphic p-chain $T = \sum n_j[V_j]$ is a positive current if and only if each $n_j \geqq 0$.

EXAMPLE 2. A current T of the form $T = i\partial\bar{\partial}\varphi$ is positive if and only if φ is plurisubharmonic. (The reader should consult Lelong **[32]** or Vladimirov **[53]** for a thorough discussion of plurisubharmonic functions.)

DEFINITION 1.19. A current T of bidimension p, p defined on a complex manifold is said to be *strongly positive* if $T(\varphi) \geqq 0$ for each test form φ which is weakly positive.

Both the cones $\mathrm{WP}^{p,p}$ and $\mathrm{SP}^{p,p}$ (as well as how they relate) are poorly understood, except for $p = 1$ (or $p = n - 1$). However, strongly positive currents are natural with regard to Wirtinger's inequality (see the next sections), while weakly positive currents seem to enjoy many of the other properties of strongly positive currents (for example, see Theorem 1.22 or, for that matter, the next lemma).

LEMMA 1.20. *A weakly positive current T of bidimension p, p has measure coefficients.*

PROOF. The conclusion is local so we may assume that T is defined on an open subset Ω of $\mathbf{C}^n$. We can choose a basis $\mathscr{B}$ for $\bigwedge_{\mathbf{R}}^{p,p}\mathbf{C}^n$ with each element of $\mathscr{B}$ belonging to $\mathrm{SP}^{p,p}$ by Lemma 1.16. Express T as $T = \sum T_j\varphi_j$ where $\mathscr{B}^* = \{\varphi_j\}$ is the basis for $\bigwedge_{\mathbf{R}}^{k,k}\mathbf{C}^n$ dual to $\mathscr{B} = \{\beta_j\}$. Since $T_j(\psi) = T(\psi\beta_j) \geqq 0$ for $\psi \in C_0^\infty(\Omega)$, $\psi \geqq 0$, each coefficient T_j must be a positive measure. Therefore, expressing T in terms of an arbitrary basis, the coefficients must be measures.

Suppose T is a current with measure coefficients of dimension r defined on an open subset Ω of $\mathbf{R}^n$. If A is a subset of Ω, let χ_A denote the characteristic function of A. For each compact subset K of Ω let,

(1.26) $$M_K(T) = \sup_{\|\varphi\|^* \leqq 1} |\chi_K T(\varphi)|.$$

(The product of χ_K and T is defined since T has measure coefficients. Here $\|\varphi\|^* = \sup\|\varphi(z)\|^*$ over all $z \in \Omega$ where the comass norm $\|\varphi(z)\|^*$ was defined in §1.1.) $M_K(T)$ is called the *r-mass* or *r-volume* of T on K. Note that if $T = [X]$ with X any locally closed oriented r-dimensional submanifold of Ω, then $M_K(T)$ is precisely the induced r-volume of $X \cap K$. The measure which assigns the number $M_K(T)$ to each compact set $K \subset \Omega$ is called the *mass* or *volume measure* and denoted $\|T\|$, so that $M_K(T) = \|T\|(K)$. If T is of the special form $T = [X]$ with X a submanifold as mentioned above, we sometimes use the more classical notation σ_X for the volume measure $\|T\|$. An application of the Daniell integral construction shows that there exists a $\|T\|$ measurable function, $\mathbf{T}$, with values in the mass unit ball in $\bigwedge_r \boldsymbol{R}^m$ (i.e., $\|\mathbf{T}(x)\| = 1$ for $\|T\|$-a.e. points x), such that

(1.27) $$T(\varphi) = \int_{\operatorname{supp} T} \langle \mathbf{T}(x), \varphi(x)\rangle \|T\|,$$

for all test r-forms φ on Ω (see Federer [**10**, 4.1.5 and 4.1.7]).

Using this representation (1.27) we can reformulate the notions of weakly and strongly positive given in Definitions 1.18 and 1.19, in the following equivalent definitions.

DEFINITION 1.21. Suppose T is a current with measure coefficients of dimension $2p$ defined on a complex manifold X. If (for some choice of a hermitian structure on X):

(1.28) (a) $\mathbf{T}(z) \in \mathrm{WP}^{p,p}$ for $\|T\|$-a.e. z then T is said to be *weakly positive.*

(1.29) (b) $\mathbf{T}(z) \in \mathrm{SP}^{p,p}$ for $\|T\|$-a.e. z then T is said to be *strongly positive.*

REMARK. Obviously, these definitions are local and independent of the coordinate system or the hermitian metric.

We conclude by mentioning (without proof) an interesting connection between positive currents and holomorphic chains due to Siu [**47**] (see Harvey and Polking [**24**] for an alternate proof). As before χ_V denotes the characteristic function of V.

THEOREM 1.22. *Suppose T is a (weakly) positive current of bidimension p, p on a complex manifold X with $dT = 0$. If V is a pure p-dimensional subvariety of X with irreducible components $\{V_j\}$ then $\chi_V \cdot T = \sum c_j[V_j]$ with each $c_j \geqq 0$.*

Of course (using Lemma 1.9), the thing one must prove is that $d(\chi_V T) = 0$. Or going one step further (using the support Theorem 1.7(a)) one need only consider the case where V is a submanifold, and hence the case where V is a complex p-plane in $\boldsymbol{C}^n$.

1.8. *Positive currents and the generalized Plateau problem.* Now we can state and prove the current version of Wirtinger's inequality (Corollary 1.2). As before χ_K denotes the characteristic function of K.

THEOREM 1.23. *Suppose T is a current with measure coefficients of dimension $2p$ defined on a complex hermitian manifold X (with associated $(1, 1)$-form ω) and K is a compact subset of X. Then*

(1.30) $$(\chi_K T)(\omega^p/p!) \leqq M_K(T),$$

and equality holds if and only if $\chi_K T$ is strongly positive. In particular, if V is a pure p-dimensional subvariety then

$$\sigma_V = \omega^p/p! \wedge [V]. \tag{1.31}$$

PROOF. Since $\|\omega^p/p!\|^* = 1$ the inequality portion of (1.30) follows from the definition of $M_K(T)$ in (1.26). To prove the equality portion of (1.30) it suffices to consider T to have compact support (replace T by $\chi_K T$). Using the representation (1.27),

$$T(\omega^p/p!) = \frac{1}{p!}\int \langle \mathbf{T}(z), \omega^p(z)\rangle \|T\|,$$

and applying Wirtinger's inequality (Theorem 1.1) to the integrand $\langle \mathbf{T}(z), \omega^p(z)/p!\rangle$ pointwise (a.e. with respect to $\|T\|$) we see that the equality portion of Theorem 1.1 yields the equality portion of (1.30) above (see Definition 1.21(b)). Note that the equality portion of (1.30) implies that for a strongly positive current T, the trace measure $\sigma_T = T \wedge \omega^p/p!$ dominates each coefficient of T.

Theorem 1.23 has important applications to the (homology) Plateau problem (see §4.5) on a Kähler manifold. Here we consider $\boldsymbol{C}^n = \boldsymbol{R}^{2n}$.

Suppose B is a d-closed current of dimension $p - 1$ with compact support in $\boldsymbol{R}^m$ (the interesting geometric case is when $B = [M]$ is given by integration over a compact oriented submanifold M of $\boldsymbol{R}^m$). The *Plateau problem* (over $\boldsymbol{R}$ instead of $\boldsymbol{Z}$) is to find a compactly supported current T of dimension p with $dT = B$ which has minimal mass; that is, such that $M(T) \leqq M(S)$, if S is any other compactly supported p-dimensional current with $dS = B$. Or, said differently (let $S = T + R$),

$$M(T) \leqq M(T + R) \tag{1.32}$$

if R is any compactly supported, d-closed, p-dimensional current. Since we are considering $\boldsymbol{R}^m$, the condition $dR = 0$ is equivalent to $R = dR'$ for some compactly supported current R' of dimension $p + 1$. Consequently, (1.32) may be reformulated as

$$M(T) \leqq M(T + dR') \tag{1.33}$$

for all compactly supported $(p + 1)$-dimensional currents R'. Note that the boundary B has disappeared in (1.32) and (1.33) (of course $B = dT$). Now if we replace $\boldsymbol{R}^m$ by an open subset Ω of $\boldsymbol{R}^m$ the conditions (1.32) and (1.33) are not necessarily equivalent (of course (1.32) implies (1.33)). If (1.32) is satisfied for all R supported in Ω then the completely supported current T is said to be *absolutely volume minimizing in Ω (over $\boldsymbol{R}$)*. If (1.33) is satisfied for all R' supported in Ω then T is said to be *homologically volume minimizing in Ω (over $\boldsymbol{R}$)*.

We wish to consider d-closed (not necessarily compactly supported) currents T with the property that each cut-off $\chi_K T$ solves the Plateau problem with boundary $d(\chi_K T)$. More precisely we have the following definition.

DEFINITION 1.24. Suppose T is a d-closed current of dimension p with measure coefficients defined on an open subset Ω of $\boldsymbol{R}^m$. If, for each compact subset K of Ω,

$$M_K(T) \leqq M(\chi_K T + R) \tag{1.34}$$

for all compactly supported, d-closed p-dimensional currents R on Ω, then T is said to be *absolutely volume minimizing in* Ω *(over* $\boldsymbol{R}$*).*

If T is a d-closed locally rectifiable current on Ω (see the appendix to §1), and (1.34) holds for all (compactly supported) d-closed rectifiable p-dimensional currents R on Ω then T is said to be *absolutely volume minimizing in* Ω *(over* $\boldsymbol{Z}$*).*

If T is locally rectifiable and absolutely volume minimizing (over $\boldsymbol{R}$) then of course T is absolutely volume minimizing (over $\boldsymbol{Z}$). Let ω denote the standard Kähler form on $\boldsymbol{C}^n$.

COROLLARY 1.25. *Suppose T is a strongly positive, d-closed current of bidimension p, p defined on an open subset Ω of $\boldsymbol{C}^n$. In particular, T may be of the form $T = [V]$ with V a pure p-dimensional subvariety of Ω. Then T is absolutely volume minimizing in Ω (over $\boldsymbol{R}$).*

PROOF. Suppose K is a compact subset of Ω and R is a d-closed current of dimension $2p$ with compact support in Ω. Then, by Theorem 1.23, $M_K(T) = \chi_K T(\omega^p/p!)$. Now $\omega^p/p! = d\eta/p!$ where $\eta = (1/4)d^c|z|^2 \wedge \omega^{p-1}$. Therefore $\chi_K T(\omega^p/p!) = (\chi_K T + R)\cdot(\omega^p/p!)$. Now applying the inequality (1.30) to $\chi_K T + R$ we have $M_K(T) \leqq M(\chi_K T + R)$ which is the desired inequality.

Next we relate

$$\bar{M}_K(T) = (\chi_K T)(\omega^p/p!) \tag{1.35}$$

(the *Kähler mass of T on K*) to $M_K(T)$ (the mass of T on K) in the case where T is only weakly positive (of dimension p, p).

LEMMA 1.26. *There exists a constant c, $0 < c < 1$, such that*

$$cM_K(T) \leqq \bar{M}_K(T) \leqq M_K(T) \tag{1.36}$$

for all weakly positive currents T.

PROOF. $\bar{M}_K(T) = (\chi_K T)(\omega^p/p!) \leqq M_K(T)\,\|\omega^p/p!\|^*$, which equals $M_K(T)$, because $\|\omega^p/p!\|^* = 1$ by Wirtinger's inequality.

Let c denote the infimum of $\omega^p/p!$ on the compact set $\mathrm{WP}^{k,k} \cap \{\zeta \in \bigwedge_{\boldsymbol{R}}^{k,k}\boldsymbol{C}^n : \|\zeta\| = 1\}$, where we identify $\bigwedge_{p,p}^{\boldsymbol{R}}\boldsymbol{C}^n$ with $\bigwedge_{\boldsymbol{R}}^{k,k}\boldsymbol{C}^n$ $(p + k = n)$. (Actually we have used this identification several times already.) Then using the representation (1.27)

$$\bar{M}_K(T) = \int_K \langle \mathbf{T}, \omega^p/p! \rangle \,\|T\| \geqq c\int_K \|T\| = cM_K(T).$$

The constant c cannot be zero, because $\omega^p/p!$ belongs to the interior of the cone $\mathrm{SP}^{p,p}$ by Corollary 1.17 (in fact, in the proof of Corollary 1.17 we showed that $\zeta \in \mathrm{WP}^{k,k}$ and $\zeta \wedge \omega^p/p! = 0$ implies $\zeta = 0$).

Using Lemma 1.26 we can give an application of Lemma 1.9. (Recall Lemma 1.9 says that if a d-closed locally flat current T of dimension $2p$ is supported in a pure p-dimensional subvariety V then $T = \sum c_j[V_j]$ with $c_j \in \boldsymbol{R}$.)

THEOREM 1.27. *Suppose V is an irreducible p-dimensional complex subvariety of a complex manifold X. Then $\{\lambda[V] : \lambda > 0\}$ is an extreme ray in the cone of weakly positive d-closed currents on X.*

PROOF (HARVEY-SHIFFMAN [**21**, THEOREM 3.8]). For a weakly positive current T

of bidimension p, p on an open subset Ω of $\boldsymbol{C}^n$, supp T agrees with the support of the positive measure $T \wedge \omega^p/p!$ by Lemma 1.26.

Suppose $[V] = T + S$ with T and S weakly positive d-closed currents on X. Since T is weakly positive it can be decomposed as $T = \sum T_j$ where T_j is weakly positive of bidimension j, j. Similarly we have $S = \sum S_j$. Note that in a coordinate chart $\Omega \subset \boldsymbol{C}^n$ for X

$$\text{(1.37)} \qquad \operatorname{supp} T_j = \operatorname{supp}(T_j \wedge \omega^j) \subset \operatorname{supp}(T_j + S_j) \wedge \omega^j = \operatorname{supp}(T + S)_j.$$

Consequently, if either T or S has a nonzero term of bidimension j, j then $T + S = [V]$ must have a nonzero term of bidimension j, j. This proves that T and S are of bidimension p, p.

Also (1.37) implies that supp $T \subset V$ and hence, by Lemma 1.9, $T = c[V]$. Since T is weakly positive the constant c must be positive. Similarly S is a positive multiple of $[V]$.

REMARK. It is natural to conjecture that, for X Stein, all extreme rays of the cone of strongly positive d-closed currents are of the form $\{\lambda[V] : \lambda > 0\}$ with V an irreducible subvariety of X. See Phelps [**37**] for implications of this conjecture.

1.9. *The density, Lelong number, and multiplicity*. First, suppose T is a current of dimension p defined on an open subset Ω of $\boldsymbol{R}^m$. Let $c(p)$ denote the volume of the unit ball in $\boldsymbol{R}^p$. If T has measure coefficients then the *density* $\Theta(T, x)$ *of* T *at* x is defined to be

$$\text{(1.38)} \qquad \lim_{r \to 0^+} M_{B(x,r)}(T)/c(p)r^p,$$

whenever this limit exists. Also let $\Theta(T, x, r)$ denote $M_{B(x,r)}(T)/c(p)r^p$. Here $B(x, r)$ denotes the closed ball of radius r about x.

The density is a measure of the mass (volume) of T near x compared to the volume of a p-plane through x (see Federer [**10**]).

Second, if T is a weakly positive current of bidimension p, p defined on an open subset Ω of $\boldsymbol{C}^n$ then in (1.38) the mass M can be replaced by the Kähler mass $\bar{M}$ (cf. (1.35)). Thus we obtain the concept of the *Lelong number* $n(T, z)$ *of* T *at* z which is defined to be

$$\text{(1.39)} \qquad \lim_{r \to 0^+} \chi_{B(z,r)}\, T(\omega^p/p!)/c(2p)r^{2p},$$

whenever this limit exists. Also let $n(T, z, r)$ denote $(\chi_{B(z,r)} T)(\omega^p/p!)/c(2p)r^{2p}$ (see Lelong [**32**]).

REMARK. By (1.36), if T is strongly positive then the density and Lelong number agree, while for T weakly positive $c\Theta(T, z) \leqq n(T, z) \leqq \Theta(T, z)$.

Third, if $T = \sum n_j[V_j]$ is a positive holomorphic p-chain on an open subset Ω of $\boldsymbol{C}^n$, we can, in many equivalent ways, define the multiplicity of T at z. (The multiplicity is additive so we need only consider the case where $T = [V]$.) The *multiplicity of* $T = [V]$ *at* z can be defined as:

(1) the multiplicity of the local ring ${}_z\mathcal{O}/{}_z\mathcal{I}_V$,

(2) the minimal sheeting number of V near z over a complex p-plane through z,

(3) the degree of the tangent cone $C(V, z)$ to V at z considered as a subvariety of $\boldsymbol{P}^{n-1}$ (there are various ways of computing the degree of C in $\boldsymbol{P}^{n-1}$),

(4) the density (or Lelong number) of $T = [V]$ at z.

Now, suppose T is a d-closed current with measure coefficients defined on $|x| < R$ in $\mathbf{R}^m$. A result of Federer's theory of slicing [10] is that for almost all r, $0 < r < R$, $[\partial B(0, r)] \wedge T$ is meaningful and defines a current with measure coefficients (supported on $\partial B(0, r)$), in the following sense. Let $f(x) \equiv |x|$, and choose an approximate identity $\{\varphi_\varepsilon\}$ for δ_r in $\mathbf{R}$ (with $\varphi_\varepsilon(x) = (1/\varepsilon)\ \varphi((x-r)/\varepsilon)$ where $\varphi \in C_0^\infty((-1, 1))$, $\varphi \geqq 0$). Let $[\partial B(0, r)]_\varepsilon$ denote the smoothing $f^*(\varphi_\varepsilon)$ of $[\partial B(0, r)]$. Then for almost all r, $0 < r < R$, $\lim_{\varepsilon\to 0} [\partial B(0, r)]_\varepsilon \wedge T$ exists (in the locally flat topology). We denote this limit (when it exists) by $[\partial B(0, r)] \wedge T$.

Using this result we can prove that for d-closed weakly positive currents the Lelong number exists at each point. Let $\omega = \frac{1}{4}\, dd^c|z|^2$ denote the standard Kähler form on $\mathbf{C}^n$, and let $\tilde{\omega}$ denote the positive current $\frac{1}{4} dd^c \log|z|^2$.

THEOREM 1.28. *Suppose T is a weakly positive d-closed current of bidimension p, p defined on $\{z: |z| < R\} \subset \mathbf{C}^n$. If $[\partial B(0, r)] \wedge T$ exists then:*

$$(1) \qquad \bar{M}_{B(0,r)}(T) = (1/p!)([\partial B(0, r)] \wedge T)\left(\tfrac{1}{4}\, d^c|z|^2 \wedge \omega^{p-1}\right),$$

$$(2) \qquad \begin{aligned} n(T, 0, r) &= \bar{M}_{B(0,r)}(T)/c(2p)r^{2p} \\ &= (1/c(2p)p!)\,([\partial B(0, r)] \wedge T)\left(\tfrac{1}{4}\, d^c \log|z|^2 \wedge \tilde{\omega}^{p-1}\right). \end{aligned}$$

In addition, if $0 < s < r < R$ then

$$(3) \qquad n(T, 0, r) - n(T, 0, s) = (1/c(2p)p!)\,(\chi(r, s)T)\,(\tilde{\omega}^p)$$

where $\chi(r, s) = \chi_{B(0,r)} - \chi_{B(0,s)}$ is the characteristic function of the annulus $B(0, r) - B(0, s)$.

REMARK 1. Suppose $T = [V]$ with V a complex submanifold, and that $\partial B(0, r)$ is transverse to V. Let $M_r = \partial B(0, r) \cap V$ denote the oriented boundary of $B(0, r) \cap V$. Let $\nu(z)$ denote the (unique) outward pointing normal to M_r at z, and let $d\sigma$ denote the volume form on M_r. Then a short computation shows that the boundary integral in (1) can be replaced by (note that $M = \bar{M}$ since $T = [V]$)

$$(1') \qquad M_{B(0,r)}(T) = \int_{M_r} \langle \nu(z), z\rangle\, d\sigma,$$

where $\langle\ ,\ \rangle$ denotes the standard inner product on $\mathbf{C}^n = \mathbf{R}^{2n}$.

REMARK 2. One can easily check that the current $[\partial B(0, r)] \wedge d^c|z|^2$ is positive. (In fact, the approximations $f^*(\varphi_\varepsilon) \wedge d^c|z|^2$ are all positive.) Consequently, the measures

(1) $(1/p!)[\partial B(0, r)] \wedge \frac{1}{4}\, d^c|z|^2 \wedge \omega^{p-1} \wedge T$,
(2) $(1/p!)[\partial B(0, r)] \wedge \frac{1}{4}\, d^c \log|z|^2 \wedge \tilde{\omega}^{p-1} \wedge T$, and
(3) $(1/p!)\, \chi(r, s)\, \tilde{\omega}^p \wedge T$,

occurring on the right-hand side above, are all positive measures. In particular, in (3), $n(T, 0, r) - n(T, 0, s) \geqq 0$ if $r \geqq s$, so that we have the following corollary.

COROLLARY 1.29. *If T is a weakly positive d-closed current defined on an open subset Ω of $\mathbf{C}^n$ then, for each point $z \in \Omega$, $n(T, z, r)$ is monotonic decreasing as r approaches zero, and hence $n(T, z) = \lim_{r\to 0} n(T, z, r)$ exists.*

PROOF OF THEOREM 1.28. The right-hand side of

$$(1.40) \qquad \frac{1}{p!}\chi_{B(r)}\, \omega^p - \frac{1}{p!}[\partial B(r)] \wedge \tfrac{1}{4}\, d^c|z|^2 \wedge \omega^{p-1} = d\left(\frac{1}{p!}\chi_{B(r)} \tfrac{1}{4} d^c|z|^2 \wedge \omega^{p-1}\right)$$

pairs with T to yield zero, and hence, by pairing the left-hand side with T we obtain (1) (to establish this rigorously—if T is not smooth—approximate $\chi_{B(r)} = f^*(\chi_{(-\infty,r)})$ by $f^*(\varphi_\varepsilon^* \chi_{(-\infty,r)})$ where $f(x)$ is as above and $\{\varphi_\varepsilon\}$ is an approximate identity).

To prove (2) first note that on $\partial B(r)$, $|z|^2$ equals the constant r^2, and hence $d|z|^2$ restricts to 0 on $\partial B(r)$. Therefore $[\partial B(r)] \wedge d|z|^2 = 0$. Now since $dd^c \log|z|^2 = dd^c|z|^2/|z|^2 - d|z|^2 \wedge d^c|z|^2/|z|^4$, we obtain that

$$[\partial B(r)] \wedge dd^c \log|z|^2 = [\partial B(r)] \wedge dd^c|z|^2/|z|^2.$$

Consequently,

$$r^{-2p}[\partial B(r)] \wedge d^c|z|^2 \wedge \omega^{p-1} = [\partial B(r)] \wedge d^c \log|z|^2 \wedge \bar{\omega}^{p-1},$$

which enables us to deduce (2) from (1).

To prove (3) consider the formula

$$\frac{1}{p!}[\partial B(r)] \wedge \tfrac{1}{4} d^c \log|z|^2 \wedge \bar{\omega}^{p-1} - \frac{1}{p!}[\partial B(s)] \wedge \tfrac{1}{4} d^c \log|z|^2 \wedge \bar{\omega}^{p-1}$$

$$= \chi(r, s) \frac{1}{p!} \bar{\omega}^p - d\left(\frac{1}{p!} \chi(r, s) \tfrac{1}{4} d^c \log|z|^2 \wedge \bar{\omega}^{p-1}\right).$$

Pairing both sides with T and using (2) twice on the left we obtain (3) (again one must smooth to complete the argument).

We conclude this subsection by discussing the sets $E_c(T) \equiv \{z \in \Omega: n(T, z) \geqq c\}$ for $c > 0$.

PROPOSITION 1.30. *Suppose T is a weakly positive d-closed current defined on a subset Ω of $\mathbf{C}^n$. Then the sets $E_c(T)$ ($c > 0$), where the Lelong number is $\geqq c$, are closed subsets of Ω with locally finite $\mathscr{H}_{2p}$ measure. In particular, if T is locally rectifiable, positive and d-closed then* supp T *has locally finite $\mathscr{H}_{2p}$ measure.*

PROOF. The sets $E_c(T)$ are closed if and only if $n(T, z)$ is upper semicontinuous in z. That is, $n(T, a) \geqq \overline{\lim}_{z\to a}\, n(T, z)$ for each $a \in \Omega$. To prove this we may assume $a = 0$. Then since $B(z, r) \subset B(0, r + |z|)$,

$$n(T, z, r) \leqq (1 + |z|/r)^{2p}\, n(T, 0, r + |z|).$$

By monotonicity $n(T, z) \leqq n(T, z, r)$, so that $n(T, z) \leqq (1 + |z|/r)^{2p}\, n(T, 0, r + |z|)$. Now letting $r^2 = |z|$ and taking $\overline{\lim}_{z\to 0}$ of both sides we obtain the desired inequality. The fact that $E_c(T)$ has locally finite $\mathscr{H}_{2p}$ measure can be proved via a standard covering argument, such as given in Bishop [**3**, p. 290] (cf. Federer [**10**, 2.10.6, 2.10.19(3)]). Finally, if T is locally rectifiable then $E_1(T) =$ supp T by Federer [**10**, 4.1.28(5)].

REMARK. Bombieri [**5**] proved much more about $E_c(T)$ ($c > 0$) if T is of bidimension $n - 1, n - 1$ (bidegree 1,1) and as in Proposition 1.30. Namely, he proved that each set $E_c(T)$ ($c > 0$) is contained in a proper subvariety. Bombieri's results lead naturally to the conjecture that, for a weakly positive current T of bidimension p, p which is d-closed, each set $E_c(T)$ ($c > 0$) is a subvariety of dimension $\leqq p$ (Harvey and King [**18**, p. 52]). Recently Siu [**47**] proved this conjecture using Bombieri's results.

1.10. *The tangent cone.* Suppose T is of dimension p in $\boldsymbol{R}^m$ and defined near the origin. Then, for each $r \neq 0$ small, the map $1/r\colon B(0, r) \to B(0, 1)$ allows us to expand $T|_{B(0,r)}$ to the current $(1/r)_*(T)$ on $B(0,1)$. If $\lim_{r\to 0^+} (1/r)_*(T)$ exists on $B(0,1)$ then it is called *the tangent cone to T at* 0. Note that

$$M_{B(0,1)}((1/r)_*(T)) = (1/r^p)\, M_{B(0,r)}(T).$$

In particular, if $\Theta(T, 0, r)$ is bounded for $r \leqq r_0$, then the family $\{(1/r)_*(T)\}$ has bounded volume.

THEOREM 1.31. *Suppose* $T = \sum n_j[V_j]$ *is a holomorphic p-chain defined near the origin in* $\mathbf{C}^n$. *Then there exists a holomorphic p-chain* $C(T, 0)$ *whose support is a homogeneous subvariety of* $\mathbf{C}^n$ *such that*

$$\lim_{r\to 0^+} (1/r)_*(T) = C(T, 0)$$

in the locally flat topology on $B(0,1)$.

PROOF. See Federer [**10**, 4.3.18, 4.3.19] for the general case. Here we present a different proof in the hypersurface case utilizing the Poincaré-Lelong equation. We may assume that T is positive and of the form $T = [V]$ with V a hypersurface defined by $f = 0$ near the origin. Express f as a series $f = \sum_{j\ge m}^{\infty} f_j$ in homogeneous polynomials f_j of degree j. Let $g = \sum_{j\ge m+1} f_j$, so that $f = f_m + g$. Then with $\varphi = (1/\pi) \log |f|$ we have $T = i\partial\bar\partial\varphi$. Therefore,

$$\left(\frac{1}{r}\right)_*(T) = i\partial\bar\partial\left(\frac{1}{r}\right)_*(\varphi) = i\partial\bar\partial\varphi(rz) = \frac{i}{\pi}\,\partial\bar\partial \log\left|r^m(f_m(z) + r^{-m}g(rz))\right|$$
$$= \frac{i}{\pi}\,\partial\bar\partial \log\left|f_m(z) + r^{-m}g(rz)\right|.$$

Now one can show, since $\log|f_m(z) + r^{-m}g(rz)|$ converges pointwise to $\log|f_m(z)|$ and is locally bounded above, that $\log|f_m(z) + r^{-m}g(rz)|$ converges in L^1_{loc} to $\log|f_m(z)|$. Consequently, $\lim_{r\to 0}(1/r)_*(T) = (i/\pi)\, \partial\bar\partial \log|f_m|$. Therefore, by defining $C(T, 0)$ to be the divisor of f_m the proof is complete.

REMARK. The proof presented above easily generalizes if V is a complete intersection near 0. In fact, using King's generalization of the Poincaré-Lelong formula [**27**] this type of proof can be completed in general.

CONJECTURE 1.32. *If T is a strongly positive d-closed current defined near the origin then* $\lim_{r\to 0^+} (1/r)_*(T)$ *exists.*

This conjecture is unsolved even in the most tractable case where $T = i\partial\bar\partial\varphi$ with φ plurisubharmonic.

1.11. *The Poisson-Jensen equation.* Our approach is a little different from the standard development in that first we consider an underlying differential equation which can be verified locally.

Let σ_r denote the normalized volume measure $d\sigma_r$ on $\partial B(0,r)$ considered as a current of dimension 0, so that $\sigma_r(\varphi) = \int \varphi\, d\sigma_r$ denotes the mean of φ over the sphere $\partial B(0, r)$. Let χ_r denote the characteristic function of the ball $B(0, r)$.

THEOREM 1.33 (THE JENSEN EQUATION).

$$(1.41) \qquad \frac{c_n}{\pi}\, i\partial\bar\partial(\chi_r \log r/|z|(i\partial\bar\partial \log |z|^2)^{n-1}) = \sigma_r - [0], \qquad c_n = (n-1)!/(2\pi)^{n-1}.$$

PROOF. The left-hand side is supported in $\partial B(0, r) \cup \{0\}$. This is because the hermitian matrix $\tilde{\omega} = i\partial\bar{\partial} \log|z|^2$ has rank $n - 1$ and hence $\tilde{\omega}^n = 0$ on $\boldsymbol{C}^n - \{0\}$. We define μ_r and μ_0, supported in $\partial B(0, r)$ and $\{0\}$ respectively, by

$$\mu_r - \mu_0 \equiv \frac{c_n}{2\pi} dd^c(\chi_r \log r/|z|\, \tilde{\omega}^{n-1}).$$

It suffices to prove $\mu_r = \sigma_r$ since $\lim_{r\to 0} \chi_r \log r/|z|\, \tilde{\omega}^{n-1} = 0$ in $L^1(\boldsymbol{C}^n)$, which implies that $\mu_0 = \lim_{r\to 0} \mu_r$; and of course, $[0] = \lim_{r\to 0} \sigma_r$.

First we calculate $dd^c(\chi_r \log r/|z|)$ near $\partial B(0, r)$. $d^c(\chi_r \log r/|z|) = \log(r/|z|)d^c\chi_r - \chi_r d^c \log|z|$, and, since $d^c\chi_r$ has measure coefficients and $\log r/|z|$ vanishes on $\operatorname{supp} d^c\chi_r = \partial B(0, r)$, the first term $\log(r/|z|)\, d^c\chi_r$ drops out. Therefore.

$$dd^c(\chi_r \log r/|z|) = -d\chi_r \wedge d^c \log|z| - \chi_r dd^c \log|z|.$$

Consequently (since $\tilde{\omega}^n = 0$ on $\boldsymbol{C}^n - \{0\}$),

$$\frac{c_n}{2\pi} dd^c\left(\chi_r \log \frac{r}{|z|}\, \tilde{\omega}^{n-1}\right) = \frac{c_n}{2\pi} [\partial B(0, r)] \wedge d^c \log|z| \wedge \tilde{\omega}^{n-1} \quad \text{on } \boldsymbol{C}^n - \{0\}.$$

This current is easily seen to be positive (with mass one) and invariant under rotations, and hence it must agree with σ_r. (Alternately, one can directly calculate that this current equals

$$[\partial B(0, r)] \wedge \frac{k_n}{|z|^{2n}} \sum_{j=1}^{n} (x_j\, \widehat{dx}_j - y_j\, \widehat{dy}_j)$$

which is the standard expression for σ_r.)

If $\zeta \in B(0, r)$ is fixed then let $h_\zeta\colon B(0, r) \to B(0, r)$ denote a biholomorphism taking ζ to 0. The Poisson kernel P_ζ satisfies $P_\zeta \sigma_r = (h_\zeta)^*(\sigma_r)$.

COROLLARY 1.34 (THE POISSON-JENSEN EQUATION).

$$\frac{c_n}{\pi} i\partial\bar{\partial}(\chi_r \log r/|h_\zeta(z)|\, (i\partial\bar{\partial} \log|h_\zeta(z)|^2)^{n-1}) = P_\zeta\, \sigma_r - [\zeta]. \tag{1.42}$$

PROOF. Apply $(h_\zeta)_*$ to both sides of (1.41).

THEOREM 1.35 (THE CLASSICAL JENSEN FORMULA). *Suppose φ is a plurisubharmonic function on a neighborhood of $\overline{B(0, r)}$, and let T denote the positive current $(c_n/\pi)\, i\partial\bar{\partial}\varphi$. If $\varphi(0) \neq -\infty$ then*

$$\int_{B(0,r)} \log \frac{r}{|z|}\, \tilde{\omega}^{n-1} \wedge T = \int \varphi(z)\, d\sigma_r - \varphi(0). \tag{1.43}$$

In particular, note the important special case $\varphi = \log|f|$, with f holomorphic on a neighborhood of $\overline{B(0, r)}$, where T is the divisor of f.

PROOF. If φ is smooth then (1.43) is immediately obtained by pairing φ with both sides of (1.41) and switching the $i\partial\bar{\partial}$ to φ on the left-hand side. If φ is not smooth then approximate φ by smooth plurisubharmonic functions in the standard way.

The Jensen equation (1.41) has many applications. In order to briefly illustrate its usefulness in the differentiated form given in (1.41) we give a short derivation of the first main theorem in Nevanlinna theory for $n = 1$.

Let $\tilde{\omega}$ denote the standard Kähler form on $\boldsymbol{P}^1$ (i.e., $\tilde{\omega} = (i/2\pi)\, \partial\bar{\partial} \log(1 + |w|^2)$

on a chart $C \subset P^1$). Choose a point $a \in P^1$, and then with a fixed coordinate chart $C \subset P^1$ with $a = 0 \in C$ let $\varphi_a \in L^1(P^1)$ be defined by

$$\varphi_a = \tfrac{1}{2}\log(1 + |w|^2) - \tfrac{1}{2}\log|w|^2 \quad \text{on } C.$$

Note $\varphi_a \in C^\infty(P^1 - \{0\})$. Then

$$(1.44) \qquad (i/\pi)\,\partial\bar\partial\varphi_a = \bar\omega - [a] \quad \text{on } P^1$$

extends the Poincaré-Lelong equation.

Now suppose $f\colon C \to P^1$. Then one can show that $f^*([a])$ is meaningful (in the sense that $\lim f^*(\psi_\varepsilon)$ exists, if ψ_ε is an approximate identity at $a \in P^1$) and equals the divisor $D_f(a)$ of $f(z) - a$. Consequently, pulling back (1.44) by f we obtain the current equation $(i/\pi)\,\partial\bar\partial\varphi_a f = f^*(\bar\omega) - D_f(a)$ on C. Now pairing $\varphi_a \circ f$ with the Jensen equation (in differentiated form) (1.41) one immediately obtains the F. M. T. ("in integrated form")

$$(1.45) \qquad \int_{B(0,r)} \log\frac{r}{|z|} f^*(\bar\omega) - \int_{B(0,r)} \log\frac{r}{|z|} D_f(a) = \frac{1}{2\pi}\int_0^{2\pi} \varphi_a(f(re^{i\theta}))\,d\theta - \varphi_a(f(a)).$$

REMARK. Actually one should view the Jensen equation (1.41) as a limiting case of the following equation. Suppose X is an oriented Riemannian manifold. Let Δ denote the induced Laplacian on X. Suppose $\Omega \subset\subset X$ with $\partial\Omega$ smooth and $\zeta \in \Omega$. Let $G_\zeta(z)$ denote the Green's function for Ω and $P_\zeta(z)$ the Poisson kernel for Ω, with singularity at ζ. We define $G_\zeta(z) = 0$ for $z \notin \Omega$. Let $d\sigma$ denote surface measure on $\partial\Omega$. The following differential equation is basic:

$$(1.46) \qquad \Delta G_\zeta = P_\zeta d\sigma - \delta_\zeta.$$

Pairing with a test function φ we see that (1.46) is equivalent to

$$(1.47) \qquad \int_\Omega G_\zeta(z)\Delta\varphi(z) = \int_{\partial\Omega} \varphi(z)P_\zeta(z)\,d\sigma(z) - \varphi(\zeta).$$

(This equation can easily be extended to the case where φ is subharmonic.) The classical equation (1.47) is referred to in several different ways. For example, it is called the Poisson-Jenson equation, the Riesz decomposition theorem, or the third Green's identity. It is sometimes useful to consider the form (1.46). (Then G_ζ can be viewed as a kind of Laplacian homology between the two positive masses, $P_\zeta d\sigma$, supported on $\partial\Omega$, and δ_ζ, supported at ζ.)

If the manifold X is complex with a Kähler form ω one can easily check that

$$(1.48) \qquad \Delta f = *\, i\partial\bar\partial(f \wedge \omega^{n-1}/(n-1)!).$$

Now it is convenient to identify currents of degree zero with currents of top degree $2n$ via the $*$ operator. Then the equation (1.46) becomes:

THEOREM 1.36 (POISSON-JENSEN EQUATION ON A KÄHLER MANIFOLD).

$$(1.49) \qquad zi\partial\bar\partial(G_\zeta\, \omega^{n-1}/(n-1)!) = P_\zeta d\sigma - [\zeta].$$

Note that if we apply (1.49) to the degenerate Kähler metric $\omega = i\partial\bar\partial \log|z|^2$ on C^n we obtain the Jensen equation (1.41).

Pairing the Poisson-Jensen equation with a test function we obtain the Poisson-Jensen formula

$$\int_{\Omega} G_\zeta(z)\mu(\zeta) = \int_{\partial\Omega} \varphi(z)P_\zeta(z)\, d\sigma(z) - \varphi(\zeta) \tag{1.50}$$

where $\mu = (1/(n-1)!)\omega^{n-1}2i\partial\bar\partial\varphi$.

The above differential equation (1.49) has so many important applications to several complex variables that it deserves recognition similar to the Poincaré-Lelong equation. Unfortunately space prevents us from pursuing the point further.

Appendix. Rectifiable currents. In this appendix we briefly discuss three (equivalent) definitions of the important concept of a locally rectifiable current (a *rectifiable current* is just a locally rectifiable current with compact support). These currents, which were introduced by Federer and Fleming **[12]**, are of fundamental geometric importance. Suppose Ω is an open subset of $\boldsymbol{R}^m$, and that X is an oriented, locally closed submanifold of Ω which has locally finite volume on Ω. The basic examples of locally rectifiable currents on Ω are the natural geometic currents, that is, the currents T of the form $T = [X]$ (integration of forms over X) with X as above. The locally finite volume insures a sup norm estimate: $|T(\varphi)| \leqq C_K\|\varphi\|^*$ for all test forms φ with supp $\varphi \subset K$, on each compact subset K of Ω. This estimate implies that T, when expressed in the form $T = \sum T_I dx^I$, has measure coefficients T_I. (The mass norm $M_K(T)$ is comparable with the combined total variations of the measures T_I on K.) The group of locally rectifiable currents on $\Omega \subset \boldsymbol{R}^m$ of dimension p (degree k with $p + k = m$) will be denoted $\mathscr{R}_p^{\text{loc}}(\Omega)$ or $\mathscr{R}^k_{\text{loc}}(\Omega)$. The locally rectifiable currents of dimension zero are particularly easy to describe. $\mathscr{R}_0^{\text{loc}}(\Omega)$ consists of all currents T of the form $T = \sum n_j[a_j]$ where $\{a_j\}$ is a discrete subset of Ω and each n_j $(j = 1, \cdots)$ is an integer. Here $[a_j] = \delta_{a_j}$.

In order to formulate the first definition of $\mathscr{R}_p^{\text{loc}}(\Omega)$, let σ_p denote the standard oriented p-simplex in $\boldsymbol{R}^p$. Thus $[\sigma_p]$ is a p-dimensional current on $\boldsymbol{R}^p$ with supp$[\sigma_p] = \sigma_p$ compact. A *Lipschitzian p-chain T on Ω* is a finite sum $T = \sum n_j f^j_*[\sigma_p]$ where each f^j is a Lipschitzian map of $\boldsymbol{R}^p$ into Ω and each n_j is an integer. We refer the reader to Federer **[10**, 4.1.14] for the precise definition of the push-forward $f_*([\sigma_p])$ with f Lipschitzian; however, if f is continuously differentiable then the push-forward $f_*([\sigma_p])$ is defined by $(f_*([\sigma_p]), \varphi) = \int_{\sigma_p} f^*(\varphi)$ as usual. (Each rectifiable arc α in Ω is a Lipschitzian 1-chain on Ω—parametrize α by arc length.) Recall from (1.26) the *mass seminorms*

$$M_K(T) = \sup\{|\chi_K T(\varphi)| : \|\varphi\|^* \leqq 1\}$$

defined for each compact subset K of Ω. These seminorms $\{M_K\}$ determine a topology on the space of currents with measure coefficients which we will refer to as the *mass topology*.

DEFINITION A.1. A current T of dimension p on Ω with measure coefficients is said to be *locally rectifiable* ($T \in \mathscr{R}_p^{\text{loc}}(\Omega)$) if T can be approximated by Lipschitzian p-chains on Ω in the mass topology.

REMARK. One can modify this definition by replacing Lipschitzian p-chains by continuously differentiable p-chains, since it follows from Rademacher's theorem and Whitney's extension theorem that Lipschitzian p-chains can be approximated in the mass topology by continuously differentiable p-chains (see Federer [**10**, 3.1.6, 3.1.14, 3.1.15, 4.1.14]).

Note that since convergence in the mass topology implies convergence in the topology on $F_p^{\mathrm{loc}}(\Omega)$, this definition of $\mathscr{R}_p^{\mathrm{loc}}(\Omega)$ implies that $\mathscr{R}_p^{\mathrm{loc}}(\Omega) \subset F_p^{\mathrm{loc}}(\Omega)$.

Now we give a second definition of $\mathscr{R}_p^{\mathrm{loc}}(\Omega)$. A Borel set $B \subset \Omega$ is said to be *locally* $(\mathscr{H}_p, p)$ *rectifiable* if for each compact subset K of Ω and each $\varepsilon > 0$ there exists a Lipschitzian map f: $\boldsymbol{R}^p \to \boldsymbol{R}^m$ and a bounded set $E \subset \boldsymbol{R}^p$ such that $f(E)$ approximates B on K in the sense that $\mathscr{H}_p(K \cap (B - f(E))) < \varepsilon$ (see Federer [**10**, 3.2.14, 3.2.18]). If B is locally $(\mathscr{H}_p, p)$ rectifiable then there exists an approximate tangent space to B at x for $\mathscr{H}_p$-almost all $x \in B$ (see Federer [**10**, 3.3]).

In order to associate a current with a locally $(\mathscr{H}_p, p)$ rectifiable set B, one must prescribe an orientation and a multiplicity η on B. More precisely, consider a pair (B, η) where B is locally$(\mathscr{H}_p, p)$ rectifiable and η is a p-vector field on B with $\eta(x)$ an integer multiple of either of the two decomposable p-vectors representing the (nonoriented) approximate tangent space to B at x (for $\mathscr{H}_p$-almost all $x \in B$). If in addition, η is $\mathscr{H}_{p|B}$ locally integrable then we will refer to the pair (B, η) as an *oriented p-rectifold.* (An oriented p-rectifold is an infinite sum of "oriented p rectifiable sets" as defined in §1 of Lawson and Simons [**30**].) Using the second definition of $\mathscr{R}_p^{\mathrm{loc}}(\Omega)$ we see that the basic geometric currents $[X]$ mentioned above are obvious examples.

DEFINITION A.2. A current T on Ω is said to be *locally rectifiable* of dimension p on Ω if for some oriented p-rectifold (B, η), T can be expressed in the form

$$T(\varphi) = \int_B \langle \eta, \varphi \rangle \, d\mathscr{H}_p$$

for all test p-forms φ on Ω.

For the third definition recall the definition of the density function $\Theta(T, x)$ given in (1.38) of §1.9 and the (volume) measure $\|T\|$ associated with a current T having measure coefficients.

DEFINITION A.3. A current T of dimension p on Ω with measure coefficients is said to be *locally rectifiable* if T is locally flat and $\Theta(T, x)$ exists and is an integer for $\|T\|$-a.e. $x \in \Omega$.

For the proof that the first two definitions of $\mathscr{R}_p^{\mathrm{loc}}(\Omega)$ given above are equivalent, see (1) and (4) of Theorem 4.1.28 in Federer [**10**]. (Our first definition is stated slightly differently from Federer's definition in [**10**, 4.1.24]. Their equivalence follows easily from the fact that if $T \in \mathscr{R}_k^{\mathrm{loc}}(\Omega)$ and if E is a Borel subset of Ω then $\chi_E \cdot T \in \mathscr{R}_p^{\mathrm{loc}}(\Omega)$, where χ_E denotes the characteristic function of E.)

See Theorem 1.4 in Harvey and Shiffman [**21**] for the proof that the third definition is equivalent to the first two. For an application of this third definition see Remark 3 following Lemma 2.2 in the next section.

2. A characterization of holomorphic chains. First we describe the (trivial) classical version of our characterization theorem. Suppose that X is a connected, oriented $2p$-dimensional real submanifold of an open subset Ω of $\boldsymbol{C}^n$. Let $T_z(X)$ denote the *oriented* tangent space to X at z, and let $-T_z(X)$ denote the tangent space with the reversed orientation. Integration of forms over X defines a current $[X]$ (i.e., $[X](\varphi) = \int_X \varphi$ for all test forms φ on Ω) on Ω. Consider a current $T = [X]$ of this form. Then T is a holomorphic p-chain (i.e., either X or its reverse $-X$ is a complex submanifold) if and only if $\pm T_z(X)$ is a complex linear subspace.

Our characterization theorem is simply a generalization of this result where X is allowed to have corners and a certain amount of scarring.

2.1. *Properties of holomorphic chains.* We briefly summarize some of the properties of holomorphic p-chains (cf. §1.3). If V is a pure p-dimensional subvariety with irreducible components $\{V_j\}$ and we prescribe integer multiplicities n_j to each V_j then $T = \sum n_j[V_j]$ defines a current, called a *holomorphic p-chain*, where $[V_j]$ denotes the current corresponding to integration (of $2p$-test forms) over Reg V_j (the manifold points of V). Moreover we proved:

(2.1) If $T = \sum n_j[V_j]$ is a holomorphic p-chain then T is a current with measure coefficients. In fact more is true.

(2.2) If $T = \sum n_j[V_j]$ is a holomorphic p-chain then T is a locally rectifiable current of dimension $2p$. (See the appendix to §1 for a brief discussion of rectifiable currents.)

Suppose the holomorphic chain T is of the form $\sum n_j[V_j]$ with each $n_j \neq 0$. For each manifold point z of V with $z \in V_j$, let $\mathbf{T}(z)$ denote $T_z(V)$ if $n_j > 0$ and $-T_z(V)$ if $n_j < 0$. $\mathbf{T}(z)$ represents the (oriented) tangent space to T at z. Since the singular locus of V has locally finite Hausdorff $2p-2$ measure we see that:

(2.3) If $T = \sum n_j[V_j]$ is a holomorphic p-chain then for $\mathscr{H}_{2p}$-a.e. points $z \in$ supp $T = V$ either $\mathbf{T}(z)$ or $-\mathbf{T}(z)$ is complex linear.

It is useful to reinterpret this as follows. Consider a real oriented $2p$-plane ζ as a decomposable $2p$-vector in the grassmann algebra. Then (considering ζ as a real $2p$-plane) $\pm\zeta$ is complex linear if and only if (considering ζ as a $2p$-vector) ζ is of bidegree p,p. Recall that a current T is said to be of *bidimension* p,p if $T(\varphi) = 0$ for all test forms φ of bidegree r, s unless $r = s = p$. Now (2.3) is essentially equivalent to

(2.3′) If $T = \sum n_j[V_j]$ is a holomorphic p-chain then T is of bidimension p, p.

In Lemma 1.8 we proved that a holomorphic p-chain has no boundary.

(2.4) If T is a holomorphic p-chain then $dT = 0$.

2.2. *The structure theorem.* The following result characterizes holomorphic p-chains. Since the characterization is local it immediately generalizes with Ω replaced by any complex manifold:

THEOREM 2.1. *Suppose T is a locally rectifiable current of bidimension p, p with $dT = 0$ on an open subset Ω of $\mathbf{C}^n$. Assume that $\mathscr{H}_{2p+1}(\text{supp } T) = 0$ (the support hypothesis). Then $T = \sum n_j[V_j]$ is a holomorphic p-chain on Ω.*

Before further amplification of Theorem 2.1 we mention an alternate (more geometric) way of stating the hypothesis that T is of bidimension p,p.

LEMMA 2.2. *Suppose T is a locally rectifiable current of dimension $2p$ on an open subset Ω of $\mathbf{C}^n$; i.e., T is of the form $T(\varphi) = \int_B\langle\varphi, \eta\rangle \, d\mathscr{H}_{2p}$, where (B, η) is an oriented $2p$-rectifold. Then:*

(a) *T is of bidimension p,p if and only if $\pm\eta(z)$ represents a complex subspace of $\mathbf{C}^n$ for $\mathscr{H}_{2p}$-a.e. $z \in B$.*

(b) *T is positive (of bidimension p, p) if and only if $\eta(z)$ represents a complex subspace of $\mathbf{C}_n$ for $\mathscr{H}_{2p}$-a.e. $z \in B$.*

PROOF. The decomposable $2p$-vector $\eta(z)$ is of bidegree p,p if and only if $\pm\eta(z)$ represents a complex subspace of $\mathbf{C}^n$, so part (a) follows from the representation for T. Similarly, $\vec{\eta}(z)$ is positive if and only if $\vec{\eta}(z)$ represents a complex subspace of $\boldsymbol{C}^n$ so (b) follows.

The support hypothesis $\mathscr{H}_{2p+1}(\text{supp } T) = 0$ in Theorem 2.1 is probably not necessary, even though d-closed locally rectifiable currents with large support do exist. For example, let $\{a_j\}$ be a dense sequence in $\boldsymbol{R}^m$ and, for each point a_j, choose a circle S_j of radius 2^{-j} containing a_j. Then $S = \sum[S_j]$ is a locally rectifiable current of dimension one on $\boldsymbol{R}^m$ with $dS = 0$, but supp $S = \boldsymbol{R}^m$.

Each of the following five remarks allows us to replace the support hypothesis with an alternate hypothesis. (In addition, see Theorem A.4 in the appendix.) They are crucial for applications. However, they are stated as remarks rather than corollaries since the support hypothesis is probably unnecessary.

REMARK 1. If T is positive and d-closed then the support hypothesis is automatic by Proposition 1.29. In this case the holomorphic chain $T = \sum n_j[V_j]$ obtained in the conclusion must, of course, have positive multiplicities (i. e., $n_j \geqq 0$ for j). This important positive case of Theorem 2.1 was first proved by King [**26**].

In §1.4 we have already given a proof of Theorem 2.1 in the case where T is positive, d-closed, locally rectifiable, and of bidimension $n-1, n-1$. The construction of the holomorphic function satisfying $(i/\pi)\,\partial\bar{\partial}\log|f| = T$, which was given in the proof of Theorem 1.13, remains valid as written for such T except the pole set P is empty. (Since supp T has locally finite Hausdorff $(2n-2)$-dimensional measure the Riemann removable singularity theorem is still applicable.)

REMARK 2. If T has compact support in $\boldsymbol{C}^n$ then the support hypothesis may be dropped and the same proof will work (choose $B \times \Delta$ in Step 2 below so that supp $T \subset B \times \Delta$).

REMARK 3. Instead of assuming T is locally rectifiable and $\mathscr{H}_{2p+1}(\text{supp } T) = 0$, if one assumes that T has measure coefficients and the $2p$-density $\Theta_{2p}(T, z)$ exists and is an integer for $\mathscr{H}_{2p}$-a.e. points in supp T, then Theorem 2.1 remains valid. In fact, these hypotheses imply that T is locally rectifiable and that $\mathscr{H}_{2p}(\text{supp } T)$ is locally finite. This can be seen as follows. Since T has measure coefficients and $dT = 0$, T is locally normal. Therefore T is locally flat. Let E denote the set of points $z \in \text{supp } T$ such that $\Theta_{2p}(T, z)$ is not a positive integer. Since T is locally flat and $\mathscr{H}_{2p}(E) = 0$ it follows from 4.2.14 in Federer [**10**] that $\|T\|(E) = 0$. Therefore, using Definition A.3, $T \in \mathscr{R}_{2p}^{\text{loc}}(\Omega)$. Finally, since $\Theta(T, z) \geqq 1$ for $\mathscr{H}_{2p}$-a.e. points in supp T, just as in Proposition 1.30, supp T must have locally finite $\mathscr{H}_{2p}$ measure.

REMARK 4. If T is of bidimension $n-1, n-1$ or equivalently bidegree 1, 1 (the hypersurface case) then the support hypothesis can be considerably weaker. Namely, if Ω is connected one need only assume that supp T is a proper subset of Ω. The idea of the proof of this result is to replace the E.E. Levi theorem (used in Step 5 of the proof of Theorem 2.1 presented below) by a more delicate theorem of Rothstein [**44**]. This stronger result for the hypersurface case then yields a stronger result for the general case. Namely, it suffices to assume $\mathscr{H}_{2p+2}(\text{supp } T) = 0$. See §2.4 of Harvey and Shiffman [**21**] for the details.

REMARK 5. Suppose T is locally rectifiable m-dimensional current on an oriented Riemannian manifold. If T is stationary then $\mathscr{H}_m$ is locally finite on supp T. See §4.6 for an application, and see p. 580 in [**21**] for a discussion of the proof (cf. Proposition 1.30). Using a result of Allard [**2**], this is proved in much the same way as Corollary 1.29 (see Harvey and Shiffman [**21**]). §4.6 on applications includes a definition of stationary and an application of this remark.

The proof of Theorem 2.1 is divided into six steps below. This proof can be briefly described as follows. The general case can be reduced to the hypersurface case (where T is of bidegree 1,1 or bidimension $n - 1, n - 1$). Next, consider this case, and solve $i\partial\bar{\partial}\varphi = T$ as explicitly as possible. Then one shows that φ is of the form $(1/\pi) \log |R|^2$ where R is a meromorphic function. Finally one concludes from the Poincaré-Lelong formula that T is a holomorphic $(n - 1)$-chain.

2.3. *Step* 1: *Setting up the geometry*. Let $G(p, n)$ denote the grassmannian of complex p-planes π in $\boldsymbol{C}^n$ and let $\pi^\perp$ denote the orthogonal complement of π. We also find it convenient to let $\pi: \boldsymbol{C}^n \to \boldsymbol{C}^p$ denote orthogonal projection onto the p-plane $\pi = \boldsymbol{C}^p$ and hope this causes no confusion.

First we discuss the hypersurface case $p = n - 1$. That is, suppose T is of bidimension $n - 1, n - 1$ or equivalently bidegree 1,1. Note that in this case the support hypothesis says $\mathscr{H}_{2n-1}(\operatorname{supp} T) = 0$. The objective is to find a hyperplane π in $\boldsymbol{C}^n$ so that supp T can be realized as an analytic cover over $\pi = \boldsymbol{C}^{n-1}$. Suppose $z \in \operatorname{supp} T$ is arbitrary (and we might as well take $z = 0$). We will apply the next lemma, which is due to Shiffman [**46**], to the set $A = \operatorname{supp} T$.

LEMMA 2.3. *Suppose A is a closed subset of $\boldsymbol{C}^n$ with $\mathscr{H}_{2n-1}(A) = 0$. For almost all hyperplanes π in $\boldsymbol{C}^n$, there exists an open disk Δ about $0 \in \boldsymbol{C}$ and a neighborhood Δ' of $0 \in \boldsymbol{C}^{n-1} = \pi$ such that $\pi: A \cap (\Delta' \times \Delta) \to \Delta'$ is proper, or equivalently $A \cap (\Delta' \times \partial\Delta) = \emptyset$.*

PROOF. Obviously π restricted to $A \cap (\Delta' \times \Delta)$ is proper if and only if $A \cap (\Delta' \times \partial\Delta) = \emptyset$. We will choose Δ' and Δ so that $A \cap (\Delta' \times \partial\Delta) = \emptyset$. Let σ: $\boldsymbol{C}^n - \{0\} \to \boldsymbol{P}^{n-1}$ denote the natural quotient map. Remove a small ball $B(0, \varepsilon)$ from $\boldsymbol{C}^n$. Then σ: $\boldsymbol{C}^n - \overline{B(0, \varepsilon)} \to \boldsymbol{P}^{n-1}$ is Lipschitz of order one. Therefore, by a result of Federer [**9**],

$$\int^*_{\boldsymbol{P}^{n-1}} \mathscr{H}_1(\sigma^{-1}(\pi^\perp) \cap A - B(0, \varepsilon))\, d\pi^\perp \leqq C_\varepsilon \mathscr{H}_{2n-1}(A)$$

for some constant C_ε. Taking $\varepsilon = 1/j, j = 1, 2, \cdots$, one finds that for almost all $\pi^\perp \in \boldsymbol{P}^{n-1}$, $\mathscr{H}_1(\pi^\perp \cap A) = 0$. For such a line $\pi^\perp$ we may choose a disk Δ about 0 in $\pi^\perp \cong \boldsymbol{C}$ such that $(\{0\} \times \partial\Delta) \cap A = \emptyset$. (Replacing σ by the function $\sigma(w) = |w|^2$ on $\pi^\perp = \boldsymbol{C}$ one obtains an integral inequality similar to the one above.) Since A is closed we may now fatten up $\{0\} \times \partial\Delta$ to $\Delta' \times \partial\Delta$ so that $(\Delta' \times \partial\Delta) \cap A$ remains empty, where Δ' is a neighborhood of 0 in $\boldsymbol{C}^{n-1}$.

To handle the general case of higher codimension $p < n - 1$ we need, in addition to the analogue of Lemma 2.3, a result insuring that the supports of the slices $\langle T, \pi, z' \rangle$ fill up all of supp T.

LEMMA 2.4. *Suppose T is as in Theorem* 2.1 *and* $0 \in \operatorname{supp} T$. *Then for almost all p-planes π in $\boldsymbol{C}^n$*:

(a) *there exists an open disk Δ'' about $0 \in \boldsymbol{C}^{n-p} = \pi^\perp$ and a neighborhood Δ' of $0 \in \pi = \boldsymbol{C}^p$ such that*

$$(\operatorname{supp} T) \cap (\Delta' \times \partial\Delta'') = \emptyset$$

(*i. e.*, π: (supp T) $\cap$ $(\Delta' \times \Delta'') \to \Delta'$ *is proper*),

(b) *and replacing T by T restricted to $\Delta' \times \Delta''$*

$$\operatorname{supp} T = \overline{\bigcup_{z' \in E(\pi)} \operatorname{supp}\langle T, \pi, z' \rangle},$$

where $E(\pi)$ denotes the set of points $z' \in \Delta'$ such that the slice $\langle T, \pi, z'\rangle$ exists.

PROOF. Since $\mathscr{H}_{2p+1}(\operatorname{supp} T) = 0$, for almost all p-planes $\pi \in G(p, n)$ in $\boldsymbol{C}^n$, $\mathscr{H}_1(\pi^\perp \cap \operatorname{supp} T) = 0$ (see Shiffman **[46]** for the proof). Now part (a) follows exactly as in the proof of the hypersurface case $p = n - 1$ in Lemma 2.3.

The basic fact relating T to the slices of T is as follows. Let τ_p denote the volume form on $\boldsymbol{C}^p$. For each p-plane π, identify π with $\boldsymbol{C}^p$ and consider the orthogonal projection (also denoted π), $\pi: \boldsymbol{C}^n \to \boldsymbol{C}^p$, so that $\pi^*\tau_p$ is a (p, p)-form on $\boldsymbol{C}^n$. Since T is locally flat

$$T(\varphi\pi^*\tau_p) = \int_{\boldsymbol{C}^p} \langle T, \pi, z'\rangle(\varphi)\, d\mathscr{H}_{2p} \tag{2.5}$$

for all test functions φ.

Now we prove part (b) using (2.5). It suffices to show that, for each point $z_0 \in \operatorname{supp} T$ fixed, we have:

$$z_0 \in \overline{\bigcup_{z' \in E(\pi)} \operatorname{supp}\langle T, \pi, z'\rangle} \quad \text{for almost all } \pi \in G(p, n) \tag{2.6}$$

(apply this fact to a countable dense subset of supp T). Consequently, we consider the point $0 \in \operatorname{supp} T$ and prove (2.6) for $z_0 = 0$. Again by another countability argument (take $\varepsilon = 1/j$) it suffices to show that for each fixed ball $B(0, \varepsilon)$ almost all $\pi \in G(p, n)$ have the property that $B(0, \varepsilon)$ contains a point z with $z' \in E(\pi)$ and $z \in \operatorname{supp}\langle T, \pi, z'\rangle$.

Now choose $\varphi \in C_0^\infty(B(0, \varepsilon))$ such that $T_{IJ}(\varphi) \neq 0$ for some I, J, where $T = \sum T_{IJ} dz^I \wedge d\bar{z}^J$. If, for a fixed π, $P(\pi) \equiv T(\varphi\pi^*\alpha) \neq 0$ then by (2.5) there must be a subset of $E(\pi)$ of positive measure such that for each z' in this subset $(\operatorname{supp}\langle T, \pi, z'\rangle) \cap B(0, \varepsilon) \neq \varnothing$. One can easily see that $P(\pi) = T(\varphi\pi^*\alpha)$ is of the form $\sum T_{IJ}(\varphi)P_{IJ}(\pi)$ so that P is a polynomial function of $\bigwedge^{2p}\boldsymbol{C}^n$ and hence a rational function on $G(p, n)$. By the proof of Lemma 1.19 each T_{IJ} is a linear combination of the measures $T \wedge \pi^*\tau_n$ as π varies over $G(p, n)$. Since $\varphi T \neq 0$, $P(\pi)$ is not identically zero. Consequently, the zero set of P is a proper subvariety of $G(p, n)$ and hence the set of π with $P(\pi) \neq 0$ is of full measure.

2.4. *Step* 2: *Solving* $i\partial\bar{\partial}\varphi = T$.

LEMMA 2.5. *Suppose $T \in \mathscr{D}'^{1,1}(B \times \boldsymbol{C})$ is d-closed and π restricted to* supp T *is proper where $\pi(z) \equiv z'$, $z = (z', z_n)$. Then*

$$\varphi \equiv [(2/\pi)\, \delta_0(z')\log|z_n|] * T_{nn}$$

satisfies $i\partial\bar{\partial}\varphi = T$ on $B \times \boldsymbol{C}$.

In subsection 3.6 we will denote this convolution by $(2/\pi)\delta_0(z')\log|z_n|\, d\lambda(z') \# T$, where $d\lambda(z')$ is the volume form in $\boldsymbol{C}^{n-1}$.

PROOF. It is a basic fact (see the Poincaré-Lelong formula) that

$$\frac{\partial^2}{\partial w \partial \bar{w}} \frac{2}{\pi} \log|w| = \delta_0(w) \quad \text{on } \boldsymbol{C}.$$

Consequently

$$\frac{\partial^2 E}{\partial z_r \partial \bar{z}_r} = \delta_0(z) \quad \text{with } E = \frac{2}{\pi}\, \delta_0(z') \log|z_n|.$$

Let $T = \sum T_{ij}\,(i/2)\,dz_i \wedge d\bar{z}_j$. Then $\partial^2\varphi/\partial z_n \partial\bar{z}_n = T_{nn}$ as desired. In general,

$$\frac{\partial\varphi}{\partial\bar{z}_j} = \frac{\partial}{\partial\bar{z}_j}(E*T_{nn}) = E*\frac{\partial T_{nn}}{\partial\bar{z}_j} = E*\frac{\partial T_{nj}}{\partial\bar{z}_n} = \frac{\partial}{\partial\bar{z}_n}(E*T_{nj}) = \frac{\partial E}{\partial\bar{z}_n}*T_{nj},$$

with the crucial third equality following because $dT = 0$. A similar calculation is valid with $\partial/\partial\bar{z}_j$ replaced by $\partial/\partial z_j$. Therefore $\partial^2\varphi/\partial z_i\partial\bar{z}_j = (\partial^2 E/\partial z_n\partial\bar{z}_n)*T_{ij} = \delta_0 * T_{ij} = T_{ij}$.

2.5. *Step* 3: *Analysis of the solution to* $i\partial\bar{\partial}\varphi = T$. The hypothesis that T is locally rectifiable is only used to prove the following real variable result of Federer **[10]**. For a point $a \in \mathbf{C}^n$ let δ_a or $[a]$ denote the delta function at a.

LEMMA 2.6. *If T is locally rectifiable and $dT = 0$ with $\pi: B \times \Delta \to B$ as above then there exists a set $F \subset B$ of full measure such that for each $z' \in F$ fixed the slice $\langle T, \pi, z'\rangle$ exists, and is of the form (a holomorphic* 0*-chain)*

$$\sum_{j=1}^{P}[(z', r_j^+(z'))] - \sum_{j=1}^{Q}[(z', r_j^-(z'))]$$

with each $r_j^{\pm} \in \Delta$.

We will not need all of the following additional information.

LEMMA 2.7. *Assuming the notation of Lemma* 2.6, *let*

$$R(z'; z_n) = \prod_{j=1}^{P}(z_n - r_j^+(z')) \Big/ \prod_{j=1}^{Q}(z_n - r_j^-(z')).$$

(1) *Then for $z' \in F$ fixed: $\varphi_{z'}(z_n) = \log|R(z'; z_n)|$; and hence $i\partial\bar{\partial}\varphi_{z'} = \sum_1^P[r_j^+(z')] - \sum_1^Q[r_j^-(z')]$ on $\mathbf{C}$. (Also note $\varphi = \log|R|$ a.e. on $B \times \mathbf{C}$.)*

(2) *Let $C_m(z') = \sum_{j=1}^P(r_j^+(z'))^m - \sum_{j=1}^Q(r_j^-(z'))^m$, $m = 0, 1, 2, \cdots$, denote the power functions in the roots $r_j^{\pm}$. Then for $w \notin \Delta$ and $z' \in F$ both fixed*:

(a) $\log R(z'; w) = (P - Q)\log w + \sum_1^\infty m^{-1} C_m(z')w^{-m}$,

(b) $R(z'; w) = w^{P-Q}\exp\left(\sum_1^\infty m^{-1} C_m(z')w^{-m}\right)$,

(c) $(\partial R/\partial w)(z'; w)/R(z', w) = (P - Q)/w + \sum_{m=1}^\infty C_m(z')w^{-m-1}$.

(3) *The power functions $C_m(z')$ are equal to $\pi_*(z_n^m T)$, and hence each C_m is holomorphic on B.*

(4) *$R(z'; w)$ is holomorphic on $B \times (\mathbf{C} - \bar{\Delta})$; the convergence in* 2(b) *being uniform on compact subsets.*

(5) *If, in addition, T is positive then there are no negative roots (i.e., $Q = 0$) and hence formula* 2(b) *in combination with* (4) *shows that $R(z'; w) = \sum_0^P R_j(z')w^j$ with each $R_j \in \mathcal{O}(B)$.*

REMARK. Note that 2(b) expresses the coefficients $R_j(z')$ of the Laurent expansion $R(z'; w) = \sum_{-\infty}^{P-Q} R_j(z')w^j$ as polynomials in the C_j's independent of T. These polynomials are called the Newton polynomials, and if $R(z'; w) = \sum_0^P R_j(z')w^j$ is a polynomial in w the Newton polynomials express the elementary symmetric functions R_j (of the roots r_j^+) in terms of the power functions C_m (of the roots r_j^+).

PROOF. (1) If T is smooth then

$$\varphi(z', z_n) = \left[\frac{2}{\pi}\,\delta_0(z')\log|z_n|\right]*T_{nn}$$

$$= \lim_{\varepsilon \to 0} \frac{2}{\pi} \int \varphi_\varepsilon(z' - w') \log |z_n - w_n| \, T_{nn}(w)$$
$$= \frac{2}{\pi} \int \log |z_n - w_n| \, dw_n \lim_{\varepsilon \to 0} \int \varphi_\varepsilon(z' - w') \, T_{nn}(w) \, dw'$$
$$= \langle T, \pi, z' \rangle \left(\frac{2}{\pi} \log |z_n - w_n| \right),$$

since $\langle T, \pi, z' \rangle = \lim_{\varepsilon \to 0} \varphi_\varepsilon(z') T_{nn} \, d\lambda(z)$. For general T approximate by smooth T.

(2) Let $\operatorname{Log}(1 - u) = \sum_1^\infty u^m/m$ denote the principal branch of the logarithm.

(a) Then

$$\log R(z', w) = \log w^{P-Q} + \operatorname{Log} \pi \left(1 - \frac{r_j^+}{w}\right) - \operatorname{Log} \pi \left(1 - \frac{r_j^-}{w}\right)$$
$$= (P - Q) \log w + \sum_1^P \sum_1^\infty \frac{1}{m} \left(\frac{r_j^+}{w} \right)^m - \sum_1^Q \sum_1^\infty \frac{1}{m} \left(\frac{r_j^-}{w} \right)^m$$
$$= (P - Q) \log w + \sum_1^\infty \frac{1}{m} C_m(z') w^{-m}.$$

Part (b) follows by exponentiating part (a).

Part (c) follows by applying $\partial/\partial w$ to part (a).

(3) Since T is d-closed and of bidegree k, k both $\bar\partial T$ and $\partial T = 0$. Therefore ϕT is $\bar\partial$-closed for any holomorphic function ϕ; and therefore $\pi_*(\phi T)$ is also $\bar\partial$-closed, since π_* and $\bar\partial$ commute.

$$[\pi_*(\phi T)](z') = \lim_{\varepsilon \to 0} \langle \pi_*(\phi T), \varphi_\varepsilon(z' - w') d\lambda(w') \rangle$$
$$= \lim_{\varepsilon \to 0} \langle \phi T, \varphi_\varepsilon(z' - w') d\lambda(w') \rangle$$
$$= \lim_{\varepsilon \to 0} \langle \varphi_\varepsilon(z' - w') d\lambda(w') T(w), \phi \rangle = \langle T, \pi, z' \rangle (\phi).$$

Applying this to $\phi(w) = w_n^m$ we obtain (3).

(4) The proof of 2(a) now shows that the convergence of $\sum_1^\infty m^{-1} c_m(z') w^{-m}$ is uniform on compact subsets of $B \times \boldsymbol{C}$; and hence (4) follows from 2(a) and 2(b).

(5) If T is positive then each slice $\langle T, \pi, z' \rangle$ is the limit of positive currents $\varphi_\varepsilon(w' - z') T(w) d\lambda(w')$ and hence is positive. Consequently no r_j^- can occur. This completes the proof of the central Lemma 2.7.

The two facts in Lemma 2.7 that we will use are:

(A) For each $z' \in F$ fixed, $\varphi_{z'}(z_n) = \log |R(z'; z_n)|$ on $\boldsymbol{C}$, with $R(z'; z_n)$ rational in z_n.

(B) $R(z', z_n)$ is holomorphic and never vanishing on $B \times (\boldsymbol{C} - \bar\Delta)$.

REMARK. If T is positive the proof of Theorem 2.1 in the hypersurface case is completed as follows. By Lemma 2.7(5), $R(z'; z_n)$ is a polynomial in z_n with coefficients that are holomorphic in z' on B, and $\varphi = \log |R|$. Therefore $T = i\partial\bar\partial\varphi = i\partial\bar\partial \log |R|$ which by the Poincaré-Lelong formula proves that T is a positive holomorphic $(n - 1)$-chain. For a current T that is not necessarily positive we need the next step in order to complete the hypersurface case.

2.6 *Step* 4: *The E. E. Levi theorem.* The facts (A) and (B) in Step 3 above in combination with the classical theorem of E. E. Levi (stated as Corollary 2.9 below) on meromorphic continuation immediately imply that $R(z'; z_n)$ (which is a rational function of z_n) has coefficients in $\mathcal{O}(B)$. The object of this subsection is to prove Corollary 2.9 (the E. E. Levi theorem). The first result is essentially a one-complex variable result which goes back to Hadamard.

THEOREM 2.8. *Suppose A is any integral domain. The formal power series*

$$F(w) = \sum_{j=-\infty}^{m} F_j w^j, \quad \textit{with each } F_j \in A,$$

represents a rational function of the form $P(w)/Q(w)$ with P, $Q \in A[w]$ and with $\deg Q \leqq q$ *if and only if*

$$\det a^K = 0 \tag{2.7}$$

for each increasing multi-index $K = (k_1, \cdots, k_{q+1})$ of positive integers, where a^K denotes the matrix whose ith row is $F^{-k_i} \equiv (F_{-k_i}, F_{-k_i-1}, \cdots, F_{-k_i-q})$.

PROOF. The problem is to find a polynomial $Q(w) = \sum_{j=0}^{q} b_j w^j$ such that $F(w) \cdot Q(w)$ is a polynomial $P(w)$. The coefficient of w^k in the expansion of $F(w)Q(w)$ is $F^k \cdot B = \sum_{j=0}^{q} F_{k-j} \cdot b_j$ where F^k denotes $(F_k, F_{k-1}, \cdots, F_{k-q})$ and B denotes $(b_0, \cdots, b_q)$. Consequently, the problem of finding Q with FQ a polynomial is equivalent to finding a (nontrivial) solution to the infinite system of equations

$$F^k \cdot B = 0, \qquad k = -1, -2, \cdots. \tag{2.8}$$

This system can be solved with $B \in A^{q+1}$, $B \neq 0$, if and only if it can be solved with $B \in \bar{A}^{q+1}$ where $\bar{A}$ denotes the quotient field of A. Considering (2.8) as a system of equations over the field $\bar{A}$ we see that (2.8) can be solved if and only if the vectors $F^{-1}, F^{-2}, \cdots$ span at most a hypersurface in $\bar{A}^{q+1}$; that is, if and only if any $q + 1$ of these vectors, say $F^{-k_1}, \cdots, F^{-k_{q+1}}$, are linearly dependent. Or equivalently that the matrix with the rows equal to the vectors $F^{-k_1}, \cdots, F^{-k_{q+1}}$, should have zero determinant for each multi-index $K = (k_1, \cdots, k_{q+1})$.

Let $A(\rho, r)$ denote the annulus about the origin of inner radius ρ and outer radius r in the plane $\boldsymbol{C}$. Let $\Delta(r)$ denote the disk of radius r. Suppose $F \in \mathcal{O}(U \times A(p, r))$, where U is an open subset of $\boldsymbol{C}^{n-1}$. Then F has a Laurent expansion

$$F(z', z_n) = \sum_{j=-\infty}^{\infty} F_j(z') z_n^j$$

on $U \times A(p, r)$ with each $F_j \in \mathcal{O}(U)$. Let $F^+(z', z_n) = \sum_{j=0}^{\infty} F_j(z') z_n^j$ and $F^-(z', z_n) = \sum_{-\infty}^{-1} F_j(z') z_n^j$. Then $F^+ \in \mathcal{O}(U \times \Delta(r))$ and $F^- \in \mathcal{O}(U \times (\boldsymbol{C} - \overline{\Delta(p)}))$. Therefore F has a meromorphic extension across $U \times \overline{\Delta(p)}$. Consequently we need only prove part (b) of the following corollary.

COROLLARY 2.9 (E. E. LEVI). *Let $F \in \mathcal{O}(U \times A(\rho, r))$, where U is an open connected subset of $\boldsymbol{C}^{n-1}$.*

(a) *If for each fixed $z' \in E \subset U$, the holomorphic function $F_{z'}(z_n) \equiv F(z', z_n)$ has a meromorphic extension to $\Delta(r)$, and E is not contained in the countable union of complex hypersurfaces in U, then F has a meromorphic extension to $U \times \Delta(r)$.*

(b) *Moreover, if in addition $F = F^-$ then the extension has the form $P(z', z_n)/Q(z', z_n)$ with P, $Q \in \mathcal{O}(U)[z_n]$ polynomials in z_n whose coefficients belong to $\mathcal{O}(U)$.*

PROOF OF (b). Let E_q denote the set of points $z' \in E$ such that the meromorphic extension of $F_{z'}(z_n)$ to $\Delta(r)$ has at most q poles (counting multiplicities). Since $E = \bigcup_{q=0}^{\infty} E_q$ is not contained in the countable union of hypersurfaces we may choose q sufficiently large so that E_q is not contained in a complex hypersurface in U. First applying Theorem 2.8 with $z' \in E_q$ fixed (so the ring A is $\boldsymbol{C}$) to $F_{z'}$ we conclude that $\det a^K(z') = 0$ for each increasing multi-index $K = (k_1, \cdots, k_{q+1})$. Consequently, each function $\det a^K(z')$ vanishes on E_q. But since E_q is not contained in a hypersurface and $\det a^K(z')$ is holomorphic on U, $\det a^K(z')$ must vanish identically. Now Theorem 2.8 applies to F (the ring A is now taken to be $\mathcal{O}(U)$) to yield the desired result.

2.7. *Step* 5: *Concluding that T is a holomorphic chain in the hypersurface case.* Since R is rational in z_n with $\mathcal{O}(B)$ coefficients, the Poincaré-Lelong formula implies $(2/\pi)i\partial\bar{\partial}\log|R|$ is a holomorphic $(n-1)$-chain; however, $\varphi = \log|R|$ so that $T = i\partial\bar{\partial}\varphi = (2/\pi)i\partial\bar{\partial}\log|R|$ is a holomorphic chain which completes the proof of Theorem 2.1 in the hypersurface case $p = 1$.

2.8. *Step* 6: *Reduction to the hypersurface case.* Because of Lemma 1.9 it suffices to prove that T is supported in a p-dimensional subvariety. We shall prove this as follows. Suppose $0 \in \operatorname{supp} T$ and choose a projection $\pi: \boldsymbol{C}^n \to \boldsymbol{C}^p$ satisfying all of Lemma 2.4. We adopt the notation of that lemma. In particular $(\operatorname{supp} T) \cap (\Delta' \times \partial\Delta'') = \emptyset$. Let T denote T restricted to $\Delta' \times \Delta''$ in the rest of the proof. Suppose $\phi \in (\boldsymbol{C}^{n-p})^*$ is a linear functional on $\pi^\perp$. Define $\rho^\phi: \boldsymbol{C}^n \to \boldsymbol{C}^{p+1}$ by $\rho^\phi(z) = (z', \phi(z''))$. This map ρ^ϕ will enable us to reduce the proof of the hypersurface case in $\boldsymbol{C}^{p+1}$. The push-forward $(\rho^\phi)_*(T)$ is a current of bidimension p, p in $\Delta' \times \boldsymbol{C} \subset \boldsymbol{C}^{p+1}$, which satisfies all the hypotheses of Theorem 2.1. Therefore $W^\phi = \operatorname{supp}(\rho^\phi)_*(T)$ is a hypersurface in $\Delta' \times \boldsymbol{C}$. Let V^ϕ denote the corresponding hypersurface $(\rho^\phi)^{-1}(W^\phi)$ in $\boldsymbol{C}^n$. The proof will be complete if we can find a basis $\mathscr{B}$ for $(\boldsymbol{C}^{n-p})^*$ such that $\operatorname{supp} T \subset V^\phi$ for each $\phi \in \mathscr{B}$, because then we have $\operatorname{supp} T \subset \bigcap_{\phi \in \mathscr{B}} V^\phi$ and this intersection must be a complex subvariety of $\Delta' \times \Delta''$ of dimension p.

It remains to show that we can find $\phi \in (\boldsymbol{C}^{n-p})^*$ so that $\rho^\phi(\operatorname{supp} T) \subset \operatorname{supp} \rho^\phi_*(T)$. Choose a countable dense subset D of $E(\pi) = \{z' \in \Delta' : \langle T, \pi, z'\rangle \text{ exists}\}$. For each point $z' \in D$ the current $\langle T, \pi, z'\rangle$ is a holomorphic 0-chain (i.e., a finite linear combination of delta functions).

Consequently, except for a set of measure zero in $(\boldsymbol{C}^{n-p})^*$, each ϕ separates the points of $\operatorname{supp}\langle T, \pi, z'\rangle$. Since D is countable we can choose a basis $\mathscr{B}$ for $(\boldsymbol{C}^{n-p})^*$ consisting of linear functionals ϕ with the property that for each point $z' \in D \subset E(\pi)$, ϕ separates the points of $\operatorname{supp}\langle T, \pi, z'\rangle$. Now observe that since ϕ separates the points of $\operatorname{supp}\langle T, \pi, z'\rangle$ we have $\rho^\phi(\operatorname{supp}\langle T, \pi, z'\rangle) = \operatorname{supp} \rho^\phi_*(\langle T, \pi, z'\rangle)$. However $\rho^\phi_*(\langle T, \pi, z'\rangle)$ equals $\langle \rho^\phi_* T, \pi^\phi, z'\rangle$ where $\pi^\phi: \boldsymbol{C}^{p+1} \to \boldsymbol{C}^p$ is defined by $\pi^\phi(z', w) = z'$ (note $\pi^\phi \circ \rho^\phi = \pi$). Of course, $\operatorname{supp}\langle \rho^\phi_* T, \pi^\phi, z'\rangle \subset \operatorname{supp} \rho^\phi_*(T)$ so we have proved that

$$\rho^\phi(\operatorname{supp}\langle T, \pi, z'\rangle) \subset \operatorname{supp} \rho^\phi_*(T) \quad \text{for all } z' \in D.$$

The part (b) of Lemma 2.4 now implies that $\rho^\phi(\operatorname{supp} T) \subset \operatorname{supp} \rho^\phi_*(T)$ as desired.

3. A characterization of boundaries of holomorphic chains. The question that we discuss in this section is which odd-dimensional compact submanifolds of $\boldsymbol{C}^n$ are boundaries of complex submanifolds of $\boldsymbol{C}^n$. The main result, Theorem 3.3 below, is taken from Harvey and Lawson [23]; however, the proof presented here differs in several ways from the proof given in [23]. Just as in the characterization of holomorphic chains given in §2, the hypersurface case is the basic case to which the general result is reduced (the "hypersurface case" refers to the situation where the complex submanifold of $\mathbf{C}^n$ has complex dimension $n - 1$ and the boundary has real dimension $2n - 3$).

In §3.12 we briefly discuss a generalization from $\mathbf{C}^n = \mathbf{P}^n - \mathbf{P}^{n-1}$ $(q = 1)$ to $\mathbf{P}^n - \mathbf{P}^{n-q}$ $(q \geqq 1)$ (see Harvey and Lawson [24a]).

3.1. *Maximal complexity.* Although an odd-dimensional real C^1 submanifold M of real dimension m in $\mathbf{C}^n$ cannot be a complex submanifold, it can be "partially complex" in the following sense. The tangent space $T_z(M)$ to M at z can be considered a real vector subspace of the tangent space $T_z(\mathbf{C}^n)$ to $\mathbf{C}^n$ at z. Let J denote the real linear operator from $T_z(\mathbf{C}^n)$ into itself defined by multiplication by $\sqrt{-1}$. Let $H_z(M)$ denote the set of tangent vectors $v \in T_z(M)$ such that Jv also belongs to $T_z(M)$. Obviously, $H_z(M)$ is the largest complex vector subspace of $T_z(\mathbf{C}^n)$ which is also contained in the tangent space $T_z(M)$ to M. The complex linear space $H_z(M)$ is a measure of the "partial complex structure" in M at z. Of course, if the real dimension m of M is even then it may happen that $H_z(M) = T_z(M)$ at each point z, in which case M is a complex submanifold. On the other hand, if $m = 2p - 1$ is odd, the largest $H_z(M) \subset T_z(M)$ can be is of complex dimension $p - 1$ (or real codimension 1 in $T_z(M)$).

DEFINITION 3.1(a). Suppose M is an oriented C^1 real submanifold of $\mathbf{C}^n$ of real dimension $2p - 1$. If $H_z(M) = \{v \in T_z(v)\colon Jv \in T_z(M)\}$ has complex dimension $p - 1$ at each point $z \in M$ then M is called *maximally complex.*

Now suppose the pair (V, M) is a C^1 submanifold with boundary of some open subset of $\mathbf{C}^n$, with M denoting the boundary of V. If V is a complex p-dimensional submanifold then the boundary M inherits as much of the complex structure of V as is possible. In particular, if $z \in M$, then $H_z(M)$ has complex dimension $p - 1$. More precisely, let v denote the real normal vector to (V, M) at z. Then $Jv \in T_z(M)$ and $H_z(M)$ is the orthogonal complement of Jv in $T_z(M)$. The orthogonal complement to Jv in $T_z(M)$ must be complex linear since it can also be described as the orthogonal complement to the complex line spanned by v in the complex linear space $T_z(V)$. In summary, if M bounds a complex manifold V then M is maximally complex.

Of course, maximal complexity only imposes a condition on M if the real dimension of M is greater than one (i.e., every real curve M in $\mathbf{C}^n$ is automatically maximally complex). Assume as before that (V, M) is a C^1 submanifold with boundary in $\mathbf{C}^n$, with M denoting the boundary of V, but suppose that V is of real dimension two. If (V, M) is compact then Stokes' theorem is applicable:

$$\int_V d\omega = \int_M \omega \quad \text{for all smooth one-forms } \omega \text{ on } \mathbf{C}^n.$$

In particular, if $\omega = \sum_{j=1}^n \omega_j(z)dz_j$ is a holomorphic 1-form on $\mathbf{C}^n$ (i.e., ω is of bidegree 1, 0 and each function $\omega_j(z)$ is entire—or equivalently $\bar\partial\omega = 0$) then $\int_M \omega = \int_V \partial\omega$ since $d = \partial + \bar\partial$. If V is a complex manifold then $\int_V \partial\omega = 0$ since $\partial\omega$ is of bidegree 2, 0.

DEFINITION 3.1(b). If M is an oriented compact one-dimensional submanifold of

$\mathbf{C}^n$ and $\int_M \omega = 0$ for all holomorphic one-forms ω on $\mathbf{C}^n$ then M is said to satisfy the *moment condition.*

In summary, if (V, M) is a compact C^1 submanifold with boundary of $\mathbf{C}^n$ and V is a complex submanifold of complex dimension one then the real curve M which bounds V satisfies the moment condition.

REMARK. The moment condition (for M of dimension 1) generalizes to M of arbitrary dimension $2p - 1$. However, for $p > 1$ maximal complexity implies the moment condition (see §3 of **[23]** for the proof of this fact and several related results).

The main result, which is stated in the next subsection, provides a converse to the fact that if M bounds a complex manifold V then M is maximally complex (alternatively if M is one-dimensional then M satisfies the moment condition). The reader may wish to proceed directly to the next subsection as the remaining material of this subsection is not essential.

Suppose as before that M is a real m-dimensional submanifold of a complex n-dimensional manifold X. Then $(T_z(X), J)$ is a complex vector space, and we may define $H_z(M)$ to be $T_z(M) \cap JT_z(M)$. The real codimension r of $H_z(M)$ in $T_z(M)$ is of particular significance. We saw above that an odd-dimensional real submanifold M is maximally complex if and only if $r = 1$, and, of course, an even-dimensional real submanifold M is a complex submanifold if and only if $r = 0$. We may consider the tangent space $T_z(M)$ to M at z as a decomposable element ζ of degree m in the grassmann algebra $\bigwedge^* T_z(X)$. If we complexify $\bigwedge^* T_z(X)$ then the complexification $\bigwedge^m T_z(X) \otimes_{\mathbf{R}} \mathbf{C}$ has the direct sum decomposition $\sum_{p+q=m} \wedge^{p,q} T_z(X)$ into vectors of bidegree p, q. The decomposition of the m-vector ζ into $\sum_{p+q=m} \zeta^{p,q}$ relates to the amount of complex structure in $\zeta \cong T_z(M)$. Namely, $\zeta = \sum_{|p-q| \leq r} \zeta^{p,q}$, where r is the real codimension of $H_z(M)$ in $T_z(M)$; and conversely if $\zeta = \sum_{|p-q| \leq r} \zeta^{p,q}$ with r minimal then r is the real codimension of $H_z(M)$ in $T_z(M)$.

A basic elementary fact is that if there exists a complex subspace E of $T_z(X)$ which contains $T_z(M)$ then r must be less than or equal to the real codimension of $T_z(M)$ in E. To prove this note that J sends a basis $e_1, \cdots, e_r$ for the complement of $H_z(M)$ in $T_z(M)$ into a basis $Je_1, \cdots, Je_r$ for a subspace complementary to $T_z(M)$ in E.

As suggested by Griffiths, "Hodge radius" seems appropriate terminology for the codimension of $H_z(M)$ in $T_z(M)$. If M has the property that the Hodge radius at each point z is the same then M is called a Cauchy-Riemann, or CR, submanifold (of Hodge radius r). Using this terminology a maximally complex submanifold is the same thing as a CR submanifold of Hodge radius one, while a CR submanifold M with Hodge radius equal to the dimension of M is referred to as a totally real submanifold.

3.2. *The main theorem.* As mentioned above the main result presented here is a converse to the fact that if M bounds a complex submanifold V of $\mathbf{C}^n$ then M is maximally complex. Before stating the precise theorem we mention some of the boundary behavior that must be taken into account.

EXAMPLE. Even if M is a real-analytic submanifold and V is a complex submanifold, the pair (V, M) may not be a C^1 submanifold with boundary. Let $\Phi : \mathbf{C}^{n+1} \to \mathbf{C}^{n+2}$ be defined by $\Phi(t, z) = (t^2, t^3 - t, z)$ with $z = (z_1, \cdots, z_n)$. Then

Φ is an immersion with points of the form $(1, 0, z)$ covered twice (by $(1, z)$ and $(-1, z)$), while all other points are covered once. Let B denote the open unit ball in $\mathbf{C}^{n+1}$ centered about $(-1, 0, \cdots, 0) \equiv a$ and of radius 2 and denote $(1, 0, \cdots, 0) \in \mathbf{C}^{n+1}$ by b. Then $\Phi(a) = \Phi(b) = (1, 0, \cdots, 0) \in \mathbf{C}^{n+2}$ while Φ is one-to-one on $\bar{B} - \{a, b\}$ (i.e., Φ folds $\bar{B}$ over bringing the center a of B and the boundary point b together). Let $M = \Phi(\partial B)$ and $V = \Phi(B - \{a\})$. Then M is a real-analytic submanifold, V is a complex submanifold of $\mathbf{C}^{n+2} - M$, $\bar{V} = V \cup M$ is compact, V has finite volume, and $d[V] = [M]$. However, near $(1, 0, \cdots, 0) \in M$, the pair (V, M) is not a manifold with boundary. This example is not locally irreducible near $(1, 0, \cdots, 0)$, and hence not even a topological manifold with boundary near $(1, 0, \cdots, 0)$. However, examples with (V, M) a topological manifold with boundary but (V, M) still not a C^k submanifold also exist (see §10 of Harvey and Lawson [**23**]).

In the main result of this section it is important to allow the proposed boundary M to have a small scar set S.

DEFINITION 3.2. Suppose M is a compact subset of a riemannian manifold X. Suppose that there is a closed subset S (the *scar set*) of M of Hausdorff $2p - 1$ measure zero such that $M - S$ is an oriented $2p - 1$ dimensional submanifold of $X - S$ of class C^r. Furthermore, assume that $M - S$ has finite volume and that the current $[M]$ in X given by integration over $M - S$ has no boundary, i.e., $d[M] = 0$. Then M will be called a *scarred $2p - 1$ cycle* (*of class C^r*).

If M is a scarred $2p - 1$ cycle in a complex manifold X then M is said to be *maximally complex* if the submanifold $M - S$ of $X - S$ is maximally complex.

Now we state the main result of this section.

THEOREM 3.3. *Suppose M is a scarred $2p - 1$ cycle of C^1 in a Stein manifold X.* (*Actually it suffices to assume that $M - S$ is an oriented immersed submanifold of $X - S$ instead of an embedded submanifold.*)

Suppose that M is maximally complex, or if $p = 1$, suppose M satisfies the moment condition. Then there exists a unique holomorphic p-chain T in $X - M$ with supp $T \subset\subset X$, *and with finite mass, such that $dT = [M]$ in X. Furthermore, there is a compact subset A of M with Hausdorff $2p - 1$ measure zero such that each point of $M - A$, near which M is of class C^k, $1 \leqq k \leqq \infty$, has a neighborhood in which* (supp T) $\cup$ *M is a regular C^k submanifold with boundary.*

In particular, if M is connected, then there exists a unique precompact irreducible complex p-dimensional subvariety of $X - M$ such that $d[V] = \pm[M]$ with boundary regularity as above.

REMARK. The uniqueness holds in greater generality, and is a nice application of the structure Theorem 2.1. Consequently, see Theorem 4.6 in §4.4 for an alternate proof of uniqueness.

We divide the proof into nine steps which, however, are not of equal difficulty.

3.3. *Step* 1: *Replacing X Stein by $X = \mathbf{C}^n$.* The Stein manifold X may be properly embedded in $\mathbf{C}^N$ for N large. Assuming the holomorphic p-chain T with boundary M has been constructed in $\mathbf{C}^N$, we must show that T is in fact supported in X. However, each entire function f vanishing on X vanishes on the boundary $M \subset X$ and hence must vanish on supp T by the maximum principle.

This proves that T is supported in X.

3.4. *Step* 2: *Setting up the geometry.* There are two different objectives of this section. Notation is established which will be used constantly throughout the remainder of the proof. In addition, two important lemmas are proved. For example, Lemma 3.4 below is our basic geometric lemma.

Occasionally it is convenient to denote by π both a complex p-plane in $\boldsymbol{C}^n$ and orthogonal projection onto the complex p-plane. Let $\pi^\perp$ denote the $(n-p)$-plane orthogonal to the p-plane π (i.e., $\pi^\perp$ is the kernel of the projection π). Let $G(p, n)$ denote the grassmannian of complex p-planes in $\boldsymbol{C}^n$.

Relative to a fixed projection $\pi \in G(p, n)$ we employ the following notations. Choose coordinates $z = (z', z'')$ with $z' = (z_1, \cdots, z_p)$, $z'' = (z_{p+1}, \cdots, z_n)$ so that $\pi(z) = z'$. Let $\pi|_M$ denote π restricted to M and denote $\pi(M)$ by $\mathfrak{M}$. We will be concerned with a notion of "good" points in $\boldsymbol{C}^n$ with respect to the projection π. Let G_π (or sometimes G) denote the set of points $z \in \boldsymbol{C}^n$ such that each point of $\pi^{-1}(\pi(z)) \cap M$ is a regular point of $\pi|_M$. Let K denote the set of critical points of $\pi|_M$, so that $\pi(G)$ is the complement of $\pi(K)$ in $\boldsymbol{C}^p$.

LEMMA 3.4. *Suppose $\pi \in G(p, n)$ is a fixed projection.*

(a) *Then $\pi(K)$ has Hausdorff $2p-1$ measure zero, and hence $\pi(G)$ is an open, dense, connected subset of $\boldsymbol{C}^p$. Moreover, for each point $z' \in \mathfrak{M} \cap \pi(G)$, there are only finitely many points of $M \cap \pi^{-1}(z')$, and hence there exists a neighborhood Δ' of z' such that $M \cap \pi^{-1}(\Delta')$ consists of r connected manifolds $M_1, \cdots, M_r$ with $\pi\colon M_j \to \Delta'$ an embedding, $j = 1, \cdots, r$.*

(b) *Let* Sing $\mathfrak{M}$ *denote the union of the critical values $\pi(K)$ and the nonmanifold points of $\mathfrak{M}$. Then* Sing $\mathfrak{M}$ *is nowhere dense in $\mathfrak{M}$ and, most important, $\boldsymbol{C}^p -$* Sing $\mathfrak{M}$ *is a connected (open) set. In fact, if ρ denotes central projection of $\boldsymbol{C}^p$ onto a sphere S^{2p-1} about any fixed point of $\boldsymbol{C}^p - \mathfrak{M}$, then $\rho($*Sing $\mathfrak{M})$ *is a compact nowhere dense subset of S^{2p-1}.*

Compare part (b) with Stolzenberg [**50**].

REMARK. Some additional notation regarding part (a) will be useful in later sections. Namely let $\mathfrak{M}_j$ denote $\pi(M_j)$ with the induced orientation, and let $f_j \in C^1(\mathfrak{M}_j)$ denote the function (vector-valued) whose graph is precisely M_j $(j = 1, \cdots, r)$.

PROOF. The fact that the set $\pi(K)$, of critical values for $\pi|_M$, has Hausdorff $2p-1$ measure zero is a generalization of the classical "Sard's theorem." See Federer [**10**, 3.4.1—3.4.3 or 3.2.3] for two different proofs. The regular points of $\pi|_M$ contained in $M \cap \pi^{-1}(z')$ are all isolated points of $M - S$. Since $M \cap \pi^{-1}(z')$ is closed and M is compact this implies that, for $z' \in \pi(G)$, $M \cap \pi^{-1}(z')$ must be finite. The remainder of part (a) now follows.

Suppose $z' \in \mathfrak{M} - \pi(K)$ and that Δ' is an arbitrary neighborhood of z' satisfying part (a). We will show that Δ' contains manifold points of $\mathfrak{M}$. This implies that Sing $\mathfrak{M}$ is nowhere dense.

The union $\bigcup_{j=1}^r \mathfrak{M}_j$ equals $\mathfrak{M} \cap \Delta'$. Associated to each point $\zeta' \in \Delta'$ is the number $k(\zeta')$ of the sets $\mathfrak{M}_1, \cdots, \mathfrak{M}_r$ to which ζ' belongs. Let k denote the minimum of $k(\zeta')$ over all the points of $\Delta' \cap \mathfrak{M}$. Choose a point $\zeta' \in \Delta'$ for which the minimum is attained. Say $\zeta' \in \mathfrak{M}_1 \cap \cdots \cap \mathfrak{M}_k$ but $\zeta' \notin \mathfrak{M}_j$ for $j = k+1, \cdots, n$. Now choose a neighborhood $\Delta'(\zeta')$ of ζ' contained in Δ' with $\Delta'(\zeta') \cap \mathfrak{M}_j$ empty for $j = k+1$, $\cdots, n$. Consequently, each point of $\Delta'(\zeta') \cap \mathfrak{M}$ can only be contained in $\mathfrak{M}_1, \cdots, \mathfrak{M}_k$. Since k was chosen to be minimal for Δ', each point of $\Delta'(\zeta') \cap \mathfrak{M}$ must belong to all of the sets $\mathfrak{M}_1, \cdots, \mathfrak{M}_k$. That is, $\mathfrak{M} \cap \Delta'(\zeta') = \mathfrak{M}_j \cap \Delta'(\zeta')$ for $j = 1, \cdots, k$ as desired.

By Sard's theorem the regular values $\Theta \subset S^{2p-1}$ of $\rho \circ \pi|_M$ form an open dense set. Moreover, each regular point of $\rho \circ \pi|_M$ is automatically a regular point of $\pi|_M$, and hence $(\rho \circ \pi)^{-1}(\Theta)$ is contained in $M \cap G$. The fact, proven above, that (Sing $\mathfrak{M}$) $\cap\ \pi(G)$ is nowhere dense in $\mathfrak{M} \cap \pi(G)$ implies that $\pi^{-1}(\text{Sing } \mathfrak{M}) \cap M \cap G$ is nowhere dense in $M \cap G$. Therefore $\pi^{-1}(\text{Sing } \mathfrak{M}) \cap (\rho \circ \pi)^{-1}(\Theta)$ is nowhere dense in $(\rho \circ \pi)^{-1}(\Theta)$, and hence

$$\rho(\text{Sing } \mathfrak{M}) \cap \Theta = (\rho \circ \pi)\,[\pi^{-1}(\text{Sing } \mathfrak{M}) \cap (\rho \circ \pi)^{-1}(\Theta)]$$

is nowhere dense in Θ. Since Θ is nowhere dense in S^{2p-1} the lemma follows.

In order to compare the holomorphic chains T_π which we will obtain for each projection π we need the following connectedness result. For a fixed point $z \in \boldsymbol{C}^n - S$, let $\mathscr{G}(z)$ denote the set of projections $\pi \in G(p, n)$ such that z belongs to G_π (i. e., $\pi \in \mathscr{G}(z)$ if each point of $M \cap \pi^{-1}(\pi(z))$ is a regular point of $\pi|_M$).

LEMMA 3.5. *Suppose $z \in \boldsymbol{C}^n - S$ is fixed. Then the subset $\mathscr{G}(z)$ of $G(p, n)$ is open, dense, and connected.*

PROOF. Suppose $\pi \in \mathscr{G}(z)$. By Lemma 3.4 there exists an open neighborhood $\Delta'(z')$ of $z' = \pi(z)$ in $\boldsymbol{C}^p$ such that $\pi^{-1}(\Delta') \cap M$ consists of r disjoint connected sets M_j with $\pi|_{M_j}$ an embedding into Δ' for $j = 1, \cdots, r$. Now if π' is sufficiently close to π we can choose a neighborhood $\Delta'(\pi'(z))$ of $\pi'(z)$ such that, above $\Delta'(\pi'(z))$, M splits into r pieces exactly as for π and Δ'. This proves that $\mathscr{G}(z)$ is open.

To prove that $\mathscr{G}(z)$ is dense we recall the general fact proved in Shiffman **[46]**. If $Y \subset \boldsymbol{C}^n - \{0\}$ has Hausdorff $2p$ measure zero, then $\{\pi \in G(p, n)\colon \pi^\perp \cap Y \neq \varnothing\}$ has measure zero in $G(p, n)$. Applied to $Y = M - \{z\}$ (with z assumed to be the origin) we have that $A = \{\pi \in G(p, n)\colon \pi^\perp \cap (M - \{z\}) \neq \varnothing\}$ has measure zero. Also, if $z = 0 \in M$, then $B = \{\pi \in G(p,n)\colon \text{the plane } \pi^\perp \text{ is not transverse to } T_z(M)\}$ has measure zero. Since $G(p, n) - (A \cup B)$ is contained in $\mathscr{G}(z)$, $\mathscr{G}(z)$ must be dense.

To prove that $\mathscr{G}(z)$ is connected we first consider the case $p = n - 1$, where $\boldsymbol{P}^{n-1} = \{\pi^\perp\colon \pi \in G(p, n)\}$ consists of all the complex lines $\pi^\perp$ in $\boldsymbol{C}^n$ through the origin. Assume that $z = 0$, and consider the natural quotient map $\sigma\colon \boldsymbol{C}^n - \{0\} \to \boldsymbol{P}^{n-1}$. Again consider the set $B = \{\pi^\perp \in \boldsymbol{P}^{n-1}\colon$ the line $\pi^\perp$ is contained in the complex span of $T_0(M)$ in $T_0(\boldsymbol{C}^n)\}$. Since $B \cong \boldsymbol{P}^{n-2}$, it has, in particular, Hausdorff $2n - 3$ measure zero. Next consider σ restricted to $M - \{0\}$. By the general version of Sard's theorem mentioned above, the critical values A of σ form a set of Hausdorff $2n - 3$ measure zero. Finally note that $\mathscr{G}(z) \cong \boldsymbol{P}^{n-1} - (A \cup B)$ since $\pi^\perp \in \boldsymbol{P}^{n-1} - B$ has the property that $\pi \in \mathscr{G}(z)$ if and only if $\pi^\perp$ is a regular value for $\sigma|_{M-\{0\}}$. Since $A \cup B$ has Hausdorff $2n - 3$ measure zero it cannot disconnect the space $\boldsymbol{P}^{n-1}$ of real dimension $2n - 2$. For the general case $p \leqq n - 1$, we wish to show that any two p-planes π_1 and $\pi_2 \in \mathscr{G}(z)$ can be connected in $\mathscr{G}(z)$. First we assume that π_1 and π_2 are "simply related." That is, assume that the span of $\pi_1 \cup \pi_2$ is of dimension $p+1$. Let $\boldsymbol{C}^{p+1}$ denote this span and let $\rho\colon \boldsymbol{C}^n \to \boldsymbol{C}^{p-1}$ denote the corresponding projection. Let $\sigma\colon \boldsymbol{C}^{p+1} - \{0\} \to \boldsymbol{P}^p$ denote the natural projection. Then just as in the hypersurface case, let A denote the critical values of $\sigma \circ \rho|_{M-\{0\}}$ and let B denote the set of lines in $\boldsymbol{C}^{p+1}$ which are contained in the complex span of $\rho(T_0(M))$. Then $\{\pi \in \mathscr{G}(z)\colon \pi \subset \boldsymbol{C}^{p+1}\} \cong \boldsymbol{P}^p - (A \cup B)$ is connected as in the hypersurface case. Now if π_1 and π_2 belong to $\mathscr{G}(z)$ but are not simply related, first choose connected neighborhoods $\mathscr{N}(\pi_1)$ and $\mathscr{N}(\pi_2)$ of π_1 and π_2

respectively in $\mathscr{G}(z)$. Second, choose a finite number of p-planes $\pi_1, \cdots, \pi_r$ (and relabel so that the original π_2 is denoted π_r) such that π_j and π_{j+1} are simply related ($j = 1, \cdots, r - 1$). Proceeding by induction on r, and using the fact that $\mathscr{G}(z)$ is open and dense, we can find another sequence $\pi'_1, \cdots, \pi'_r$ with π'_j and π'_{j+1} simply related, $j = 1, \cdots, r - 1$, and, in addition, such that $\pi'_1 \in \mathscr{N}(\pi_1)$, $\pi'_r \in \mathscr{N}(\pi_r)$ and each $\pi'_j \in \mathscr{G}(z)$. Consequently $\mathscr{G}(z)$ must be connected.

3.5. *Step* 3: *Analyzing the main hypothesis.* The hypothesis that M is maximally complex has various equivalent formulations. See Harvey and Lawson **[23]** for a more complete discussion. The next result is local. The superscript $S^{0,1}$ on a current S of degree 1 denotes the component of S of bidegree (0, 1).

LEMMA 3.6. *Suppose $\mathfrak{M}$ is an oriented real hypersurface of class C^1 embedded in $\boldsymbol{C}^p$. Suppose $f \in C^1(\mathfrak{M})$ (vector-valued) and let M denote the graph of f over $\mathfrak{M}$. The following conditions are equivalent.*

(i) *M is maximally complex.*

(ii) *The differential f_* restricted to $H_{z'}(\mathfrak{M})$ is complex linear.*

(iii) *$\bar\partial_{\mathfrak{M}} f = 0$ on $\mathfrak{M}$, where $\bar\partial_{\mathfrak{M}}$ denotes the induced Cauchy-Riemann operator on $\mathfrak{M} \subset \boldsymbol{C}^p$ (acting on components of f).*

(iv) $\bar\partial(f[\mathfrak{M}]^{0,1}) = 0$.

REMARK. Sometimes it is also useful to replace (i) by the condition

(i′) $[M]$ has the bidimension decomposition $[M] = M_{p,p-1} + M_{p-1,p}$.

(i) is trivially equivalent to (i′).

PROOF. The equivalence of (i) and (ii) is a direct consequence of the implicit function theorem and the following elementary fact. A real linear map $L: \boldsymbol{C}^r \to \boldsymbol{C}^s$ is complex linear if and only if its graph in $\boldsymbol{C}^r \times \boldsymbol{C}^s$ is a complex linear subspace. Applying this fact to $L = f_*|_{H_z(\mathfrak{M})}$ completes the proof.

Before proving that (ii) and (iii) are equivalent we define $\bar\partial_{\mathfrak{M}}$ more precisely. Consider multiplication by $\sqrt{-1}$ as a real-linear transformation J of $H_z(\mathfrak{M})$. Then J extends to the complexification $H_z(\mathfrak{M}) \otimes_{\boldsymbol{R}} \boldsymbol{C}$ with eigenvalues $\pm i$. Let $H_z(\mathfrak{M}) \otimes_{\boldsymbol{R}} \boldsymbol{C} = H_z^{1,0}(\mathfrak{M}) \oplus H_z^{0,1}(\mathfrak{M})$ denote the decomposition into the eigenspaces for i and $-i$ respectively. Then there is a natural $\boldsymbol{C}$-linear isomorphism from $(H_z(\mathfrak{M}), J)$ onto $(H_z^{1,0}(\mathfrak{M}), i)$ given by sending X into $Z = \frac{1}{2}(X - iJX)$. Also note that complex conjugation is naturally defined on $H_z(\mathfrak{M}) \otimes_{\boldsymbol{R}} \boldsymbol{C}$ and we have $H_z^{0,1}(\mathfrak{M}) = \overline{H_z^{1,0}(\mathfrak{M})}$.

Suppose now that $f: \mathfrak{M} \to \boldsymbol{C}$ is a function of class C^1. Then the differential f_* is complex-linear at z if and only if $f_*(JX) = if_*(X)$ for all $X \in H_z(\mathfrak{M})$. This is equivalent to the condition that $(JX - iX)(f) = 0$. Therefore, f satisfies condition (ii) (f_* is complex-linear) on $\mathfrak{M}$ if and only if $\bar{Z}(f) = 0$ for all (class C^1) vector fields $Z \in \Gamma(H^{1,0}(\mathfrak{M}))$. The operator $\bar\partial_{\mathfrak{M}}: \mathscr{E}'^0(\mathfrak{M}) \to \mathscr{E}'^1(\mathfrak{M})$ can be defined as follows. Choose a complementary line field l to $H(\mathfrak{M})$ in $T(\mathfrak{M})$. The splitting $T(\mathfrak{M}) = H(\mathfrak{M}) \oplus l$ determines a splitting $T^*(\mathfrak{M}) = H^*(\mathfrak{M}) \oplus l^*$ of the cotangent bundle of $\mathfrak{M}$. (The spaces $H_z(\mathfrak{M})$ and $l_z^* = H_z^{\perp}(\mathfrak{M})$ are canonical. The other two depend on the choice.) Let $z_0 \in \mathfrak{M}$ be fixed and choose local vector fields $X_1, \cdots, X_{p-1} \in \Gamma(H)$, $V \in \Gamma(l)$, such that $X_1, JX_1, \cdots, X_{p-1}, JX_{p-1}, V$ are linearly independent at each point. Let $X_1^*, JX_1^*, \cdots, X_{p-1}^*, JX_{p-1}^*, V^*$ be the dual covector fields. Then we define

$$\bar\partial_{\mathfrak{M}}(f) = \sum_{k=0}^{p-1} \bar{Z}_k(f)\,\bar{Z}_k^*$$

where $\bar{Z}_k = \frac{1}{2}(X_k + iJX_k)$ and $\bar{Z}_k^* = (X_k^* - iJX_k^*)$. Note that the definition of $\bar{\partial}_{\mathfrak{M}}$ is independent of the choice of local vector fields, but does depend on the splitting of $T(\mathfrak{M})$. As noted above, $f_*|_{H_*(\mathfrak{M})}$ is complex linear if and only if $\bar{Z}(f) = 0$ for all $Z \in \Gamma(H^{1,0}(\mathfrak{M}))$, and hence (ii) is equivalent to (iii).

If we extend the local vector fields used in defining $\bar{\partial}_{\mathfrak{M}}$ above to a neighborhood in $\boldsymbol{C}^p$ (and extend f to a neighborhood), then we have that $\bar{\partial} f = \sum \bar{Z}_k(f)\bar{Z}_k^* + \bar{W}(f)\bar{W}^*$ where $\bar{W} = \frac{1}{2}(V + iJV)$ and $\bar{W}^* = (V^* - iJV^*)$. If furthermore $\mathfrak{M}$ is given implicitly by $\rho(z) = 0$ then along $\mathfrak{M}$ we may assume that $W^* = \partial\rho = \frac{1}{2}(d\rho + id^c\rho)$. Now let $i: \mathfrak{M} \to \boldsymbol{C}^p$ be the inclusion map. In order to prove that (iii) and (iv) are equivalent it suffices to prove that $\bar{\partial}(f[\mathfrak{M}]^{0,1}) = i_*(\bar{\partial}_{\mathfrak{M}} f)^{0,2}$, since $i_*(\bar{\partial}_{\mathfrak{M}} f)^{0,2} = 0$ if and only if $\bar{\partial}_{\mathfrak{M}} f = 0$. Let $\chi(t)$ denote the characteristic function of $(-\infty, 0)$. Then $[\mathfrak{M}] = d(\chi(\rho)) = \delta_0(\rho)d\rho$ so that $[\mathfrak{M}]^{0,1} = \mu\bar{\partial}\rho = \mu\bar{W}^*$ where μ is Hausdorff $2p-1$ measure restricted to $\mathfrak{M}$. Therefore $\bar{W}^* \wedge [\mathfrak{M}]^{0,1} = 0$. Since $\bar{\partial} = \sum \bar{Z}_k(\ )\cdot\bar{Z}_k^* + \bar{W}(\ )\bar{W}^*$, we have $\bar{\partial}(f[\mathfrak{M}]^{0,1}) = \sum \bar{Z}_k(f)\bar{Z}_k^* \wedge [\mathfrak{M}]^{0,1} = i_*(\bar{\partial}_{\mathfrak{M}}(f))^{0,2}$, which proves the desired identity.

DEFINITION 3.7. If any of the (equivalent) conditions of Lemma 3.6 hold for $f \in C^1(\mathfrak{M})$ then f is called a CR function on $\mathfrak{M}$ (Cauchy-Riemann function).

The fact that (i) implies (iv) in the above lemma has a simpler proof which admits a useful global generalization.

LEMMA 3.8. *If M is maximally complex and ϕ is holomorphic in a neighborhood of M then $\pi_*(\phi[M])^{0,1}$ is $\bar{\partial}$-closed.*

PROOF. As remarked above, since M is maximally complex, $[M] = M_{p,p-1} + M_{p-1,p}$. Since $d[M] = 0$, $\bar{\partial} M_{p,p-1}^{k,k+1} = 0$ where $p + k = n$. Therefore

$$\bar{\partial}[\pi_*(\phi[M])]^{0,1} = \bar{\partial}[\pi_*(\phi M_{p,p-1}^{k,k+1})] = \pi_*(\phi\bar{\partial} M_{p,p-1}^{k,k+1}) = 0,$$

which proves the lemma.

Taking $\phi = z_j$ $(j = p+1, \cdots, n)$, $\pi_*(\phi[M]) = f_j[\mathfrak{M}]$ and hence part (iv) of Lemma 3.6 is a special case of Lemma 3.8.

3.6. *Step* 4: *Solving* $\bar{\partial}$. Throughout this subsection assume that $\pi : \boldsymbol{C}^n \to \boldsymbol{C}^p$ is a fixed projection and let $\mathfrak{M} = \pi(M)$. Also let $U_0, U_1, \cdots$ denote the connected components of $\boldsymbol{C}^p - \mathfrak{M}$ with U_0 the unbounded component. As before let $z = (z', z'')$ with $\pi(z) = z'$.

Our next major objective is to construct, for each component U_j of $\boldsymbol{C}^p - \mathfrak{M}$, the desired holomorphic p-chain T on $\pi^{-1}(U_j)$. This is done by constructing a function $R_j^\phi(z'; w)$ on $U_j \times \boldsymbol{C}$ which is rational in $w \in \boldsymbol{C}$ for each linear function $\phi : \boldsymbol{C}^{n-p} \to \boldsymbol{C}$.

REMARK. If we are in the hypersurface case $p = n - 1$ then $z'' = z_n$ and we need only consider the one linear function $\phi(z'') = z_n$. As noted above, the reader may wish to reflect on this simpler but central case $\phi(z'') = z_n$ in the following development. The holomorphic p-chain T will then be defined (on $\pi^{-1}(U_j)$) as the simultaneous zeros of $\{R_j^\phi(z'; w)\}_\phi$ minus the simultaneous poles of $\{R_j^\phi(z', w)\}_\phi$.

Now we consider ϕ fixed. The construction of $R^\phi(z'; w)$ is closely related to the proof of the structure Theorem 2.1. In fact, more information than was logically necessary was included in the proof of Theorem 2.1 to give the reader more insight into the construction of R^ϕ.

We begin this construction of R^ϕ by defining the coefficients of the Laurent expansion of $(\partial R^\phi/\partial w)/R^\phi$ about infinity in w. Because of the following motivational remark it is natural to consider the equation:

$$\bar\partial C_m^\phi = [\pi_*(\phi^m[M])]^{0,1} \quad \text{on } \mathbf{C}^p \ (m = 0, 1, \cdots), \tag{3.1}$$

in the unknown distribution C_m^ϕ. The right-hand side is a compactly supported current of bidegree (0, 1) on $\mathbf{C}^p$. By Lemma 3.8, and the hypothesis that M is maximally complex if $p > 1$, the right-hand side is also $\bar\partial$-closed. Our objective after the next remark is to solve this equation for C_m^ϕ.

MOTIVATIONAL REMARK. If we assume that the holomorphic p-chain T with boundary $[M]$ is available, then (working backwards) let R^ϕ denote the rational function defining the divisor $\rho_*(T)$ in $\mathbf{C}^{p+1}$ where $\rho(z) \equiv (z', \phi(z''))$. Looking at the proof of Theorem 2.1 (see part 2(c) of Lemma 2.7) one easily checks that the coefficients C_m^ϕ of the Laurent expansion of $(\partial R^\phi/\partial w)/R^\phi$ about $w = \infty$ are given by $C_m^\phi = \pi_*(\phi^m T)$; and now an easy calculation shows that $\bar\partial\pi_*(\phi^m T) = \pi_*(\phi^m[M])^{0,1}$, since $dT = [M]$. This proves that (assuming T given) equation (3.1) must be satisfied by the Laurent coefficients of $(\partial R^\phi/\partial w)/R^\phi$. Summarizing, we have shown that if Theorem 3.1 is valid, then defining the C_m^ϕ as solutions to (3.1) is not just a guess, but a necessity.

The solution to (3.1) is obtained in analogy with Lemma 2.5 in Step 2 of Theorem 2.1. First we introduce some notation. Let

$$K = \sum_{j=1}^p K_j \frac{\partial}{\partial \bar z_j}, \qquad K_j \in \mathscr{D}'^0(\mathbf{C}^p)$$

be a vector field of type (0, 1) with distribution coefficients, and define the $\bar\partial$-divergence of K to be

$$\bar\partial iv(K) = \sum_{j=1}^p \frac{\partial K_j}{\partial \bar z_j}.$$

Suppose we are given a current $u \in \mathscr{E}'^{r,s}(\mathbf{C}^p)$ and express $u = \sum u_{IJ} dz^I \wedge d\bar z^J$, where $u_{IJ} \in \mathscr{E}'^0(\mathbf{C}^p)$ and where the sum is taken over all I, J with $|I| = r$ and $|J| = s$. Then for $s \geqq 1$ we define a current $K \# u \in \mathscr{D}'^{r,s-1}(\mathbf{C}^p)$ by the formula

$$K \# u = \sum_{|I|=r;|J|=s} \sum_{j\in J} (K_j * u_{IJ})\sigma_{IJ}(j) dz^I \wedge d\bar z^{J-\{j\}}$$

where $*$ denotes convolution and where $\sigma_{IJ}(j) = (-1)^{|I|+k+1}$ for $j = j_k$ in the multi-index $J = \{j_1, \cdots, j_s\}$. We will call $K \# u$ a *convolution contraction* of K and u. The main significance of this operation is in the following relation.

LEMMA 3.9 (THE HOMOTOPY FORMULA). *For each vector field* $K = \sum K_j \partial/\partial \bar z_j$ *of type* (0, 1) *with coefficients* $K_j \in \mathscr{D}'^0(\mathbf{C}^p)$, *and for each* $u \in \mathscr{E}'^{(r,s)}(\mathbf{C}^p)$, *we have*

$$\bar\partial(K \# u) + K \# (\bar\partial u) = \bar\partial iv(K) * u. \tag{3.2}$$

The proof is a straightforward calculation which is omitted.

COROLLARY 3.10. *Let K be a* (0, 1)*-vector field with distribution coefficients in* $\mathbf{C}^p$ *such that* $\bar\partial iv(K) = \delta_0$. *Then for* $u \in \mathscr{E}'^{r,s}(\mathbf{C}^p)$, $s \geqq 1$, *with* $\bar\partial u = 0$, *we have* $\bar\partial(K \# u) = u$.

Recall that from Serre duality **[45]** and the fact that $H^{p-1}(\mathbf{C}^p; \Omega^p) = 0$, we know that the cohomology group $H^1_*(\mathbf{C}^p; \mathcal{O}) = [H^{p-1}(\mathbf{C}^p; \Omega^p)]' = 0$ for $p > 1$. However, a more constructive proof will follow from Corollary 3.10 by using the following vector field:

$$K^C(z) = \frac{\delta_0(z_2, \cdots, z_p)}{\pi z_1} \frac{\partial}{\partial \bar{z}_1}. \tag{3.3}$$

This vector field will be called the *Cauchy kernel* on the slice $\tau^{-1}(0)$ where $\tau(z) = (z_2, \cdots, z_p)$, and will sometimes be denoted by $K^C_\tau(z)$. For certain questions it is often useful to construct solutions using the following *Bochner-Martinelli kernel.*

$$K^{BM}(z) = \frac{(p-1)!}{\pi^p |z|^{2p}} \sum_{j=1}^{p} \bar{z}_j \frac{\partial}{\partial \bar{z}_j}. \tag{3.4}$$

The fundamental property of these kernels for our purposes is the following.

LEMMA 3.11. *Suppose u is a compactly supported current in* $\mathbf{C}^p$ *of bidegree* (0, 1).

(i) ($p > 1$.) *If u is* $\bar\partial$*-closed then there exists a* (*unique*) *compactly supported distribution* φ *satisfying* $\bar\partial\varphi = u$ *on* $\mathbf{C}^p$.

(ii) ($p = 1$.) *If* $u(z^m dz) = 0$ *for* $m = 0, 1, \cdots$ *then there exists a* (*unique*) *compactly supported distribution* φ *satisfying* $\bar\partial\varphi = u$ *on* $\mathbf{C}$.

Furthermore, φ *is given explicitly by either*

(a) (*the Bochner-Martinelli kernel*) $\varphi = K^{BM} \# u$, *or*

(b) (*the Cauchy kernel on a slice* τ) $\varphi = K^C \# u$.

In particular, (a) *implies that if u has measure coefficients then* φ *is locally integrable.*

PROOF. It is a standard fact that $\bar\partial iv(K^C) = \bar\partial iv(K^{BM}) = \delta_0$. Hence the equations follow directly from Corollary 3.10. It remains to prove the compactness of the supports. For this we consider the Cauchy kernel. Note that for fixed z with $|z_2|^2 + \cdots + |z_n|^2$ sufficiently large, the supports of $K^C(\zeta - z)$ and $u(\zeta)$ are disjoint, and hence $(K^C \# u)(z) = 0$. Since $K^C \# u$ is holomorphic in $\mathbf{C}^n - \text{supp}(u)$, it follows that $K^C \# u$ vanishes identically in the unbounded component of $\mathbf{C}^n - \text{supp}(u)$. (In part (ii) expand $(\zeta - z)^{-1}$ about $\zeta = \infty$.)

Of course there can be at most one compactly supported solution to $\bar\partial F = u$ since the difference of any two such solutions is a compactly supported holomorphic function. Let $F = K^C \# u$ denote this solution. Then $K^{BM} \# u = K^{BM} \# (\bar\partial F) = \bar\partial iv\, K^{BM} * F = (\delta_0) \# F = F$ by the homotopy formula in Lemma 3.9. This completes the proof.

REMARK. Note that the above proof actually shows that for any K with $\bar\partial iv(K) = \delta_0$, we have $K \# u = K^C \# u$ under the stated conditions on u.

Applying the lemma to $u = \pi_*(\phi^m[M])^{0,1}$ we obtain a unique compactly supported solution to equation (3.1) which will henceforth be denoted C^ϕ_m.

REMARK. If $p = 1$, then the fact that M satisfies the moment condition implies that $[\pi_*(\phi^m[M])^{0,1}](z^r dz) = 0$ for each $r = 0, 1, \cdots$ so that, again equation (3.1) has a unique solution C^ϕ_m with compact support (this time part (ii) of Lemma 3.11 must be employed). Now that (3.1) can be solved with compact support, the remainder of the proof of Theorem 3.3 will make no distinction between M being maximally complex ($p > 1$) and M satisfying the moment condition ($p = 1$).

Note that C_0^ϕ, the (unique) solution to $dC_0 = [\mathfrak{M}] = \pi_*([M])$ (with compact support), is integer-valued. In fact, $C_0 = \sum_{j=0}^\infty m_j[U_j]$ where each $m_j \in \boldsymbol{Z}$ and $m_0 = 0$.

Again, the proof of Theorem 2.1 presented in §2 provides some motivation for the next step.

Choose a large ball $B(0, R)$ containing M. For each $w \in \boldsymbol{C} - \overline{\phi(B(0, R))}$, choose an open simply connected subset of $\boldsymbol{C} - \{0\}$ which contains both $w + \phi(B(0, R))$ and the disk of radius 1 centered at 1. Let Log denote the branch of the logarithm on this open set which is zero at $\zeta = 1$. Consider the equation:

$$\bar\partial \Phi_w = [\pi_*(\mathrm{Log}(w - \phi)[M])]^{0,1} \quad \text{on } \boldsymbol{C}^p. \tag{3.5}$$

Just as for (3.1), there exists a unique solution Φ_w with compact support, which we take as the definition of Φ_w.

The $\bar\partial$ equations (3.1) and (3.5) are related as follows. The expansion $\mathrm{Log}(w - \phi) = \mathrm{Log}\, w + \mathrm{Log}(1 - \phi/w) = \mathrm{Log}\, w - \sum_1^\infty m^{-1} \phi^m w^{-m}$ converges uniformly (modulo $2\pi i\boldsymbol{Z}$) for values of ϕ taken on M and $w \in \boldsymbol{C} - \overline{\phi(B(0, R))}$. Therefore using (3.1) and the uniqueness of the solution to (3.5) we find

$$\Phi_w = C_0 \log w - \sum_1^\infty \frac{1}{m} C_m^\phi w^{-m}. \tag{3.6}$$

It is natural to make the following definition (see Lemma 2.5).

DEFINITION 3.12. For $(z', w) \in (\boldsymbol{C}^p - \mathfrak{M}) \times (\boldsymbol{C} - \overline{\phi(B(0, R))})$, let $R^\phi(z'; w) \equiv \exp \Phi_w(z')$.

LEMMA 3.13. *$R^\phi(z'; w)$ is a single-valued holomorphic function on $(\boldsymbol{C}^p - \mathfrak{M}) \times (\boldsymbol{C} - \overline{\phi(B(0, R))})$ with the following properties.*

(a) *$R^\phi \equiv 1$ on $U_0 \times (\boldsymbol{C} - \overline{\phi(B(0, R))})$.*

(b) *$R^\phi(z'; w) = \sum_{m=-\infty}^{C_0} A_m^\phi(z') w^m$ with uniform convergence on compact subsets of $(\boldsymbol{C}^p - \mathfrak{M}) \times (\boldsymbol{C} - \overline{\phi(B(0, R))})$.*

(c) *Each coefficient A_m^ϕ in* (b) *is given by a polynomial function of a finite number of the C_m^ϕ's (independent of ϕ, etc.).*

PROOF. Since $C_0(z')$ is integer-valued on $\boldsymbol{C}^p - \mathfrak{M}$, it follows from (3.6) that R^ϕ is well defined in $(\boldsymbol{C}^p - \mathfrak{M}) \times (\boldsymbol{C} - \overline{\phi(B(0, R))})$, $R^\phi = \exp \Phi_w$ must be holomorphic in $(z'; w)$.

Since Φ_w is compactly supported in $\boldsymbol{C}^p$ and $\bar\partial\Phi_w = 0$ on U_0, $\Phi_w \equiv 0$ on U_0. Therefore $R^\phi \equiv 1$ on U_0.

The series (3.6) for $\Phi_w(z')$ (modulo $2\pi i\boldsymbol{Z}$) converges normally on $(\boldsymbol{C}^p - \mathfrak{M}) \times (\boldsymbol{C} - \overline{\phi(B(0, R))})$ and the coefficients are C_m^ϕ/m. Therefore, the series for $R^\phi = \exp \Phi_w$ must also converge normally, and its coefficients are polynomials in the C_m^ϕ's.

So far, we only have complete information about R^ϕ above U_0; namely $R^\phi \equiv 1$ on $\pi^{-1}(U_0) \times (\boldsymbol{C} - \overline{\phi(B(0, R))})$. Our next objective is to analyze the boundary and jump behavior of R^ϕ on $\mathfrak{M}$ so we can determine more about R_j^ϕ for $j \neq 0$.

3.7. *Step* 5: *Boundary and jump behavior.* Because of the independent interest of the boundary value theory and for the sake of completeness this subsection contains a thorough discussion which is independent of the previous subsections.

Suppose Ω is an open subset of C^n ($n > 1$) with a connected and smooth boundary $d\Omega$. Bochner [4] and Martinelli [35] independently provided rigorous proofs of what is called Hartogs' theorem. Namely, that each function holomorphic in some neighborhood of $d\Omega$ has a holomorphic extension to a neighborhood of $\bar{\Omega}$. Bochner showed further that a smooth function defined only on $d\Omega$, for which $\bar{\partial}_{d\Omega} f = 0$, extends to a function holomorphic on Ω and smooth on $\bar{\Omega}$.

Serre [45] found an elegant proof of Hartogs' theorem which utilized his duality to reduce the proof to statements about solving the $\bar{\partial}$ equation. Later Ehrenpreis [8] provided a proof of Hartogs' theorem using the idea of solving $\bar{\partial}$ with compact supports. Hörmander [25] continued the development of the Serre-Ehrenpreis approach to Hartogs' theorem, making use of the fact that $H^1_*(C^n, \mathcal{O}) = 0$. He also used this fact to give a proof of Bochner's theorem in the following explicit form: If $d\Omega$ is C^4 and the function f is $C^4(d\Omega)$, then the extension is C^1 on $\bar{\Omega}$. (He notes that the differentiability assumptions can be reduced by two units.)

Finally, Fichera [13] and Weinstock [54], [55] have investigated generalizations to several variables of the standard one-variable hypothesis that $\int_{d\Omega} f(z)z^n\,dz = 0$ for all $n \in Z^+$. Weinstock assumes f is continuous and proves the holomorphic extension is continuous on $\bar{\Omega}$.

We present a development which essentially contains all of the results on boundary values of holomorphic functions which are referenced above. Bochner's theorem can be stated in various forms by trading off the regularity of f for regularity of $d\Omega$ in the hypothesis; we choose to emphasize the case where it is assumed that $d\Omega$ is of class C^1 and $f \in C^1(d\Omega)$ and then prove that: If f satisfies $\bar{\partial}_{d\Omega} f = 0$ on $d\Omega$ there exists a function $F \in C^1(\bar{\Omega}) \cap \mathcal{O}(\Omega)$ with $F|_{d\Omega} = f$. The case where $f \in L^2(d\Omega)$ and $d\Omega \in C^\infty$ is discussed in Rossi [43]; and the case where $d\Omega$ real-analytic and f is a hyperfunction on $d\Omega$ is discussed in Polking and Wells [39].

THEOREM 3.14 (BOCHNER ET AL.). *Suppose Ω is an open, relatively compact subset of a noncompact, connected complex manifold X of dimension n. Further suppose $d\Omega$ is a connected C^k submanifold* ($1 \leqq k \leqq \omega$).

(A) *For $n \geqq 2$ the following conditions on the function $f \in C^p(d\Omega)$, $0 \leqq p \leqq k$, are equivalent*:

(1) $\bar{\partial}_{d\Omega} f = 0$.

(2) *$f[d\Omega]^{0,1}$ is $\bar{\partial}$-closed, that is, $\int_{d\Omega} f\bar{\partial}\alpha = 0$ for all $\alpha \in \mathcal{E}^{n,n-2}(X)$.*

(3) *f is a CR function, i.e., the differential of f when restricted to the complex subspaces of $T(d\Omega)$ is C-linear* (*for* $p \geqq 1$).

(4) *The graph of f is maximally complex* (*for* $p \geqq 1$).

(B) *For $n \geqq 1$ the following conditions on a function $f \in C^p(d\Omega)$, $0 \leqq p \leqq k$, are equivalent*:

(1) *There exists $F \in \mathcal{E}'^0(X)$ with $\bar{\partial}F = f[d\Omega]^{0,1}$.*

(2) *$\int_{d\Omega} f\omega = 0$ for all $\bar{\partial}$-closed $(n, n-1)$-forms ω on X.*

(3) *There exists $\tilde{F} \in C^p(\bar{\Omega}) \cap \mathcal{O}(\Omega)$ with $\tilde{F}|_{d\Omega} = f$.*

If $H^1_*(X, \mathcal{O}) = 0$ (for example, if X is Stein and $n \geqq 2$), then the conditions in (A) *imply the conditions in* (B).

The distribution in (B1) *is the unique $F \in \mathcal{E}'^0(X)$ with $\bar{\partial}F = f[d\Omega]^{0,1}$, and it corresponds to the classical extension $\tilde{F}$ in* (B3). *In the special case that $X = C^n$, F is given explicitly by any of the following formulae*:

(i) $$F(z) = K^C \# (f[d\Omega]^{0,1})(z) = \frac{1}{2\pi i}\int_{d\Omega\cap\{\zeta'=z'\}} \frac{f(\zeta_1, z')d\zeta_1}{\zeta_1 - z_1}$$

where K^C is the Cauchy kernel (3.3).

(ii) $$F(z) = K^{BM} \# (f[d\Omega]^{0,1})(z) = \int_{d\Omega} f k_z^{BM}$$

where K^{BM} is the Bochner-Martinelli kernel (3.4) *and where k_z^{BM} is the corresponding $(n, n-1)$-form given by*

$$k_z^{BM}(\zeta) = \frac{(-1)^{n(n-1)/2}(n-1)!}{(2\pi i)^n |\zeta - z|^{2n}} \sum_{j=1}^n (-1)^{j+1}(\bar\zeta_j - \bar z_j)d\zeta_1 \wedge \cdots \wedge d\zeta_n \wedge d\bar\zeta_1 \wedge \cdots \wedge \widehat{d\bar\zeta_j} \wedge \cdots \wedge d\bar\zeta_n.$$

(iii) *If $d\Omega$ is C^2 or better, then*

$$F(z) = \int_{d\Omega} f d^c(G_z\omega^{n-1}) \quad (\textit{Poisson's equation})$$

where $G_z(\zeta)$ is the Green's function for Ω, $\omega(\zeta) = \sum_{j=1}^n (i/2)d\zeta_j \wedge d\bar\zeta_j$, and $d^c(G_z\omega^{n-1}) \wedge [d\Omega]$ is harmonic measure.

REMARKS. (1) The fact that on a Stein manifold X each CR function on $\partial\Omega$ has a holomorphic extension to Ω (i.e., (A3) implies (B3)) is perhaps the central portion of the above theorem.

(2) The differentiability conditions required for the statement that (A3) implies (B3) above are an improvement on Theorem 2.3.2′ **[25]**.

(3) The fact that the moment condition (B2) for $f \in C(d\Omega)$ implies that there is an extension (B3) as stated above generalizes a result of Weinstock **[34]** in two ways. First $\boldsymbol{C}^n$ is replaced by an arbitrary connected noncompact complex manifold X. Second $d\Omega$ is only required to be C^1.

(4) Compare the portion of the above theorem which concludes that for $f \in C^1(d\Omega)$ the extension F is C^1 on $\bar\Omega$ with the fact that the solution to the Dirichlet problem for $\Delta = \{z \in C: |z| < 1\}$, with general boundary values f, need not be C^1 on $\bar\Delta$ even for $f \in C^1(d\Delta)$.

(5) There are many examples of complex manifolds X for which $H^1_*(X, \mathcal{O}) = 0$. In particular, X can be any submanifold of $\boldsymbol{P}^n - \boldsymbol{P}^r$ with $r \geq 1$. (*Hint*: use Mayer-Vietoris for $H^j_*(-, \mathcal{O})$.)

PROOF. (A) The proof that all the conditions in (A) are equivalent is precisely Lemma 3.6. (Note that all these conditions (A1)—(A4) are local conditions on f—and as such are still equivalent.)

(B) First we prove that (B1) and (B2) are equivalent. The statement (B1) says that $f[d\Omega]^{0,1}$ is in the range of $\bar\partial: \mathcal{E}'^{0,0}(X) \to \mathcal{E}'^{0,1}(X)$. The statement (B2) says that $f[d\Omega]^{0,1}$ is perpendicular to the kernel of the adjoint $\bar\partial: \mathcal{E}^{n,n-1}(X) \to \mathcal{E}^{n,n}(X)$. Since by the Hahn-Banach theorem the closure of the range of the first operator equals the perpendicular to the kernel of the second operator (the adjoint), it suffices to prove that the first operator has closed range. Because of the closed range theorem for Frechét spaces it is equivalent to show that the second operator $\bar\partial: \mathcal{E}^{n,n-1}(X) \to \mathcal{E}^{n,n}(X)$ has closed range. This range is closed since it equals all of $\mathcal{E}^{n,n}(X)$ for any noncompact complex manifold X (see Malgrange **[34]** and choose a real-analytic hermitian metric).

To prove that (B3) implies (B1) choose a C^1 function ρ defined in an open set U containing $d\Omega$ with $d\rho$ never vanishing. $d\Omega = \{z \in U\colon \rho(z) = 0\}$, and $U \cap \Omega = \{z \in U\colon \rho(z) < 0\}$. Then define $\Omega_\varepsilon = \{z \in \Omega\colon \text{if } z \in U \text{ then } \rho(z) < -\varepsilon\}$, so that as $\varepsilon \leqq 0$ approaches 0, the C^1 submanifolds $d\Omega_\varepsilon$ approach $d\Omega$ as C^1 submanifolds.

Then $F[d\Omega_\varepsilon]^{0,1}$ converges weakly to $f[d\Omega]^{0,1}$ as currents because by hypothesis $F \in C(\bar{\Omega})$ and $F|_{d\Omega} = f$. Let χ_ε denote the characteristic function of Ω_ε. Then $\chi_\varepsilon F$ converges in L^1 to F and hence $\bar{\partial}(\chi_\varepsilon F)$ converges weakly as currents to $\bar{\partial}F$. But $\bar{\partial}(\chi_\varepsilon F) = F[\partial\Omega_\varepsilon]^{0,1}$, and hence $\bar{\partial}F$ must equal $f[d\Omega]^{0,1}$.

To complete the equivalences in (B), it remains to prove that (B1) implies (B3). Since $d\Omega$ is connected and X is connected, a standard argument shows that $X - d\Omega$ can have at most two components. By assumption $X - \bar{\Omega}$ and Ω are nonempty. Hence $X - d\Omega$ must have precisely the following two components: Ω which is relatively compact, and $X - \bar{\Omega}$ which is not relatively compact. Now assume that F is a compactly supported distribution on X satisfying $\bar{\partial}F = f[d\Omega]^{0,1}$. Then F must be holomorphic on $X - d\Omega$. Therefore, by the uniqueness of analytic continuation, F must vanish on $X - \bar{\Omega}$, i.e., supp $F \subset \bar{\Omega}$. This proves that (B1) is equivalent to:

(B1′) *There exists a unique* $F \in \mathscr{E}'^0(X)$ *with* $\bar{\partial}F = f[d\Omega]^{0,1}$, *and* supp $F \subset \bar{\Omega}$.

Finally, to complete the proof of (B) we will show via a series of lemmas that this condition (B1′) implies that $F|_\Omega$ has a C^p extension $\tilde{F}$ to $\bar{\Omega}$ with boundary values f if the given boundary function f belongs to $C^p(d\Omega)$. This is a local problem and will be treated as such in the proof.

THEOREM 3.15. *Suppose X is a complex manifold, and that Ω is an open subset of X with $d\Omega \in C^k$. Suppose F is a distribution on X with* supp $F \subset \bar{\Omega}$ *and* $\bar{\partial}F = f[d\Omega]^{0,1}$ *where $f \in C^p(d\Omega)$ with $0 \leqq p \leqq k$ and $1 \leqq k$. Then $F|_\Omega$ has an extension $\tilde{F} \in C^p(\bar{\Omega}) \cap \mathscr{O}(\Omega)$ with $\tilde{F}|_{d\Omega} = f$.*

REMARK. Locally there is a necessary and sufficient condition for the existence of a solution to $\bar{\partial}F = f[d\Omega]^{0,1}$ with support in $\bar{\Omega}$ for all $\bar{\partial}$-closed right-hand sides. See for example Theorem 10.2 in Harvey and Lawson [**23**].

The next lemma reduces the case where $f \in C^p(d\Omega)$ (general p) to the case where $f \in C^0(d\Omega)$. Let B denote an open ball about the origin in $\mathbf{C}^n$.

LEMMA 3.16. *Suppose Ω is an open subset of B with $d\Omega$ of class C^1. Suppose F is a distribution on B satisfying $\bar{\partial}F = f[d\Omega]^{0,1}$ where $f \in C^1(d\Omega)$. Then $G_j \equiv \partial F/\partial z_j - f\,\partial\chi_\Omega/\partial z_j$ is locally integrable on B and $\bar{\partial}G_j = (\partial f/\partial z_j)\,[d\Omega]^{0,1}$ on B (where $\partial f/\partial z_j$ is a well-defined continuous function on $d\Omega$ because f uniquely determines a continuous 1-jet $\tilde{f}$ on $d\Omega$ with the property $\bar{\partial}\tilde{f} = 0$, since $\bar{\partial}_b f = 0$ on $d\Omega$).*

PROOF. Choose a function $u \in C^1(B)$ with the 1-jet of u on $d\Omega \cap B$ equal to the 1-jet determined by f on $d\Omega \cap B$. Then $u|_{d\Omega} = f$, $\partial u/\partial z_j|_{d\Omega} = \partial f/\partial z_j$, and the coefficients of $\bar{\partial}u$ vanish on $d\Omega \cap B$. Applying $\partial/\partial z_j$ to $\bar{\partial}F = f[d\Omega]^{0,1}$ we have

$$\begin{aligned}\bar{\partial}\left(\frac{\partial F}{\partial z_j}\right) &= \frac{\partial}{\partial z_j}[u\bar{\partial}\chi_\Omega] = \frac{\partial u}{\partial z_j}\bar{\partial}\chi_\Omega + u\bar{\partial}\left(\frac{\partial \chi_\Omega}{\partial z_j}\right) \\ &= \frac{\partial f}{\partial z_j}[d\Omega]^{0,1} + u\bar{\partial}\left(\frac{\partial \chi_\Omega}{\partial z_j}\right).\end{aligned}$$

Hence,

$$\begin{aligned}\bar\partial G_j &= \bar\partial\left(\frac{\partial}{\partial z_j}F\right) - \bar\partial\left(f\,\frac{\partial\chi_\Omega}{\partial z_j}\right)\\ &= \frac{\partial f}{\partial z_j}[d\Omega]^{0,1} + u\bar\partial\left(\frac{\partial\chi_\Omega}{\partial z_j}\right) - \bar\partial\left(u\,\frac{\partial\chi_\Omega}{\partial z_j}\right)\\ &= \frac{\partial f}{\partial z_j}[d\Omega]^{0,1} - (\bar\partial u)\,\frac{\partial\chi_\Omega}{\partial z_j} = \frac{\partial f}{\partial z_j}[d\Omega]^{0,1}.\end{aligned}$$

The term $(\bar\partial u)\,\partial\chi_\Omega/\partial z_j$ vanishes since $\partial\chi_\Omega/\partial z_j$ is a measure and $\bar\partial u$ has continuous coefficients which vanish on $\operatorname{supp}(\partial\chi_\Omega/\partial z_j)$. This completes the proof of Lemma 3.16.

Now the only case remaining for the proof of Theorem 3.15 is the case $f \in C(d\Omega)$ (i.e., $p = 0$). It suffices to prove the following local result in $\mathbf{C}^n$.

LEMMA 3.17. *Suppose Ω is an open subset of B with $d\Omega$ of class C^1 and $0 \in d\Omega$. Suppose F is a distribution on B with* supp $F \subset \bar\Omega \cap B$ *and $\bar\partial F = f[d\Omega]^{0,1}$ on B where $f \in C(d\Omega)$. Then $F|_\Omega$ extends to a continuous function $\tilde F$ on $\bar\Omega \cap B$ with boundary values f on $d\Omega$.*

PROOF. Choose $\phi \in C_0^\infty(B)$ with $\phi \equiv 1$ near 0. Then $\bar\partial(\phi F) = \phi f[d\Omega]_{0,1} + F\bar\partial\phi$ on $\mathbf{C}^n$ and hence

$$\phi(z)F(z) = \int_{d\Omega}\phi(\zeta)f(\zeta)k_z^{BM}(\zeta) + [k^{BM} \# F\bar\partial\phi](z). \tag{3.7}$$

Consider the term $G(z) = \int_{d\Omega}\phi(\zeta)f(\zeta)k_z^{BM}(\zeta)$. Since the last term in (3.7) is real-analytic near 0, it suffices to prove the following three facts about the Bochner-Martinelli transform:

(1) $G|_{B-\bar\Omega}$ has a continuous extension G^- to $B - \Omega$ near 0.
(2) $G|_\Omega$ has a continuous extension G^+ to $\bar\Omega$ near 0.
(3) The jump $G^+(z) - G^-(z) = f(z)$ for all $z \in d\Omega$ near 0.

Note that by hypothesis $\phi F = 0$ on $B - \bar\Omega$ and hence statement (1) follows from the fact that the last term of (3.7) is real-analytic near 0. Therefore, statements (2) and (3) are a consequence of part A of the next theorem. This theorem generalizes results of Plemelj [**38**] from the Cauchy transform in $\mathbf{C}$ to the Bochner-Martinelli transform in $\mathbf{C}^n$.

THEOREM 3.18. *Suppose M is an oriented C^1 closed submanifold of an open set U in $\mathbf{C}^n$ and assume $U - M$ has two components U^+ and U^- with $dU^+ = M$. Given $f \in C_0(M)$, let $F(z) \equiv \int_M f(\zeta)k_z^{BM}(\zeta)$ denote the Bochner-Martinelli transform of f.*

(A) *If $F|_{U^-}$ has a continuous extension F^- to $U^- \cup M$ then $F|_{U^+}$ also has a continuous extension F^+ to $U^+ \cup M$ and the jump formula $F^+(z) - F^-(z) = f(z)$ is valid on the boundary M.*

(B) *If we assume more differentiability on f, for example assume $f \in C_0^1(M)$, then $F^\pm|_{U^\pm}$ has a continuous extension to $U^\pm \cup M$. Moreover, on M these extensions satisfy the Plemelj formulas:*

$$F^\pm(z) = \pm\tfrac{1}{2}f(z) + P.V.\int_M f(\zeta)k_z^{BM}(\zeta) \tag{i}$$

or equivalently

$$F^+ - F^- = f \quad \text{and} \quad F^+ + F^- = 2\,P.V.\int_M f(\zeta)k_z^{BM}(\zeta) \tag{ii}$$

where P.V. denotes the Cauchy principal value. In addition, for any given compact set $K \subset M$, there exists a constant C such that:

$$|F^{\pm}|_{\infty,M} \leqq C\|f\|_{C^1(M)} \tag{3.8}$$

for all $f \in C^1(M)$ *with* $\operatorname{supp} f \subset K$.

We refer the reader to §5 of Harvey and Lawson **[23]** for the proof of part (A) and to Appendix B in **[23]** for the proof of part (B).

The proof that the three conditions in part (B) of Theorem 3.14 are all equivalent is now complete because of part (A) of Theorem 3.18. Finally, we complete the proof of Theorem 3.14.

If $H^1_*(X, \mathcal{O}) = 0$ then for each $u \in \mathscr{E}'^{0,1}(X)$ with $\bar\partial u = 0$ there exists $v \in \mathscr{E}'^{0,0}(X)$ with $\bar\partial v = u$. Therefore, if $H^1_*(X, \mathcal{O}) = 0$ then the condition (A2), that $\bar\partial(f[d\Omega]^{0,1}) = 0$, implies (B1), that $\bar\partial F = f[d\Omega]^{0,1}$ can be solved with $F \in \mathscr{E}'^{0,0}(X)$.

If X is Stein and $n \geqq 2$, then $H^1_*(X; \mathcal{O}) = H^{n-1}(X; \Omega^n)' = 0$ by Serre duality. When $X = \boldsymbol{C}^n$, one can use the methods of §3.4, Lemma 3.11 to give an explicit construction of the solution F. This yields formulas (i) and (ii) in Theorem 3.14. Formula (iii) in Theorem 3.14 is valid for any harmonic function F on Ω which is continuous on $\bar\Omega$ with boundary values f. The assumption $d\Omega \in C^2$ implies that, for $z \in \Omega$, $G_z(\zeta) \in C^1(\bar\Omega - \{z\})$ so that d^cG_z makes sense on $d\Omega$.

REMARK. If $d\Omega \in C^2$ so that harmonic measure on $d\Omega$ is given by $d^c(G_z(\zeta)\omega^{n-1}(\zeta)) \wedge [d\Omega]$ then a direct proof can be given that: If $\bar\partial F = f[d\Omega]^{0,1}$ with $f \in C(d\Omega)$ then representation (iii) is valid; and hence $F \in C(\bar\Omega)$ by the usual boundary regularity for the Dirichlet problem (see Weinstock **[54]** and **[55]**).

The E. E. Levi theorem on meromorphic continuation (Corollary 2.9) must be reformulated in the context of boundaries. First, we need the following lemma on boundary zeros of holomorphic functions.

LEMMA 3.19. *Suppose $\Omega \subset\subset \boldsymbol{C}^n$ is an open set with boundary $d\Omega$. If g is holomorphic on Ω and continuous on $\bar\Omega$ then the boundary zero set $\{z \in d\Omega: g(z) = 0\}$ has no interior in $d\Omega$ (in fact it has harmonic measure zero). If, in addition near a given point $\zeta \in d\Omega$ the boundary is of class C^2 then the set $\{z \in d\Omega: g(z) = 0\}$ has $(2n-1)$-dimensional measure zero near ζ.*

SKETCH OF PROOF. Let $G_z(\zeta)$ denote the Green's function for Ω with singularity at $z \in \Omega$, and let μ_z denote harmonic measure at z. Define $G_z(\zeta) \equiv 0$ for $\zeta \notin \Omega$. Let ν denote the $(2n-2)$-dimensional volume measure of the divisor of g on Ω. Then one can verify:

$$\int_\Omega G_z(\zeta)\nu(\zeta) = \frac{1}{2\pi}\int_{d\Omega} \log|g(\zeta)|\,d\mu_z(\zeta) - \frac{1}{2\pi}\log|g(z)| \tag{3.9}$$

if $z \in \Omega$ and $g(z) \neq 0$, as in §1.11.

In particular, one shows that the upper semicontinuous function $\log|g(\zeta)|$ is integrable with respect to $d\mu_z$ (using the fact that $G_z \geqq 0$), and hence that $\mu_z(\{z \in d\Omega: g(z) = 0\}) = 0$. Now μ_z assigns positive measure to open subsets of $d\Omega$, and if $d\Omega$ is class C^2 then μ_z and $d\sigma = \mathscr{H}^{2n-1}|_{d\Omega}$ are mutually absolutely continuous, with $\mu_z = P_z d\sigma$ where P_z is the Poisson kernel.

Note. It is a well-known conjecture that μ_z and $\mathscr{H}^{2n-1}|_{d\Omega}$ are mutually absolutely continuous for $d\Omega$ of class C^1. This conjecture implies that Lemma 3.19 remains valid for $d\Omega$ of class C^1.[1]

Again let $A(\rho, r)$ denote the annulus about the origin of inner radius ρ and outer radius r in the complex plane $\boldsymbol{C}$.

COROLLARY 3.20. *Let* $F \in \mathcal{O}(\Omega \times A(\rho, r)) \cap C(\bar{\Omega} \times A(\rho, r))$, *where* Ω *is an open connected subset of* $\boldsymbol{C}^{n-1}$. *If for each fixed* $z' \in E \subset \partial\Omega$, *the holomorphic function* $F_{z'}(z_n) \equiv F(z'; z_n)$ *has a meromorphic extension to* $\Delta(r)$ *and E contains an open subset of* $\partial\Omega$ *then F has a meromorphic extension to* $\Omega \times \Delta(r)$. *Moreover, if* $F_{z'}$ *has a rational extension to* $\boldsymbol{C}$ *for each* $z' \in E$ *then F has an extension* $P(z'; z_n)/Q(z'; z_n)$ *to* $\Omega \times \Delta(r)$ *with* $P, Q \in \mathcal{O}(\Omega)[z_n]$ *polynomials in* z_n *whose coefficients belong to* $\mathcal{O}(\Omega)$.

With the aid of Lemma 3.19 (and the Baire category theorem) the proof of Corollary 3.20 proceeds exactly as the proof of Corollary 2.9 from the basic Theorem 2.8. Consequently, we omit the proof.

Now applying the above facts about boundary and jump behavior to the special situation set up in §3.6 (Step 4: Solving $\bar{\partial}$), we can proceed with the proof of Theorem 3.3.

3.8. *Step* 6: *Rationality of* $R^\phi(z'; w)$. Throughout this subsection we adopt the notation of §3.6. In particular, $\pi: \boldsymbol{C}^n \to \boldsymbol{C}^p$ denotes a fixed projection, $\mathfrak{M} = \pi(M)$, $U_0, U_1, \cdots$ denote the connected components of $\boldsymbol{C}^p - \mathfrak{M}$ with U_0 the unbounded component, and ϕ is a fixed linear function on $\boldsymbol{C}^{n-p} = \pi^{-1}(0)$.

Recall that solving the $\bar{\partial}$ equations (3.1) for the C^ϕ_m, and (3.5) for Φ_w, enabled us to define $R^\phi(z'; w)$ for large w (Definition 3.12), and then in Lemma 3.13 we collected together various facts about $R^\phi(z'; w)$. (Also recall the functions A^ϕ_m occurring in the Laurent expansion of $R^\phi(z'; w)$.) Using the jump relations of the last section we continue investigating $R^\phi(z'; w)$. As in §3.4, let Sing $\mathfrak{M}$ denote the union of the set of critical values of $\pi|_M$ and the set of nonmanifold points of $\mathfrak{M}$.

LEMMA 3.21. *In addition to properties* (a), (b), *and* (c) *of Lemma* 3.13, R^ϕ *satisfies the following.*

(d) *The restriction of* A^ϕ_m *to* U_j *extends to a continuous function on* $\bar{U}_j$ – Sing $\mathfrak{M}$ (*let* $\bar{A}^{\phi,j}_m = \bar{A}^j_m$ *denote the extension*).

(e) *The restriction of* $R^\phi(z'; w)$ *to* $U_j \times (\boldsymbol{C} - \phi(\overline{B(0, R)}))$ *extends to a continuous function on* $(\bar{U}_j$ – Sing $\mathfrak{M}) \times (\boldsymbol{C} - \phi(\overline{B(0, R)}))$ (*let* $R^\phi_j(z; w)$ *denote this extension*).

(f) *For each* $z' \in U_j$ – Sing $\mathfrak{M}$ *fixed*, $R^\phi_j(z'; w)$ *is holomorphic in* w, *with Laurent expansion*

$$R^\phi_j(z'; w) = \sum_{-\infty}^{C_0} \bar{A}^j_m(z')w^m \tag{3.10}$$

on $\boldsymbol{C} - \phi(\overline{B(0, R)})$.

(g) *Suppose that* $\mathfrak{M}_0$ *is a component of the manifold* $\mathfrak{M}$ – Sing $\mathfrak{M}$ *with* $\mathfrak{M}_0$ *contained in* $\bar{U}_i \cap \bar{U}_j$, *and that* $\pi^{-1}(\mathfrak{M}_0) \cap M$ *has r components* $M_1, \cdots, M_r$ *with* M_s *the graph of* $f_s \in C^1(\mathfrak{M}_0)$, $s = 1, \cdots, r$. *Then*

---

[1]This conjecture (and more) has recently been proved by B. Dahlberg, *A note on sets of harmonic measure zero* (to appear). Consequently, $\partial\Omega$ need only be of class C^1 in the final conclusion of Lemma 3.19 concerning boundary regularity.

(3.11) $$R_i^\phi(z'; w) = \prod_{s=1}^{r} (w - \phi(f_s(z')))^{\pm 1} R_j^\phi(z'; w) \quad \text{on } \mathfrak{M}_0 \times (\boldsymbol{C} - \phi(\overline{B(0,R)})),$$

where the + sign occurs if $\pi(M_s)$ is the oriented boundary of U_i and the − sign occurs if $\pi(M_s)$ is the oriented boundary of U_j (near $\mathfrak{M}_0 = \pi(M_s)$).

(h) *For each $z' \in \bar{U}_j - \operatorname{Sing} \mathfrak{M}$ fixed, $R_j^\phi(z'; w)$ is rational in w and the coefficients are holomorphic on U_j and continuous on $\bar{U}_j - \operatorname{Sing} \mathfrak{M}$.*

PROOF. For fixed $w \in \boldsymbol{C} - \phi(\overline{B(0, R)})$, Φ_w satisfies

(3.1′) $$\bar{\partial}\Phi_w = \sum_{s=1}^{r} \log(w - \phi(f_s(z')))[\mathfrak{M}]^{0,1}$$

on the open set $U_i \cup \mathfrak{M}_0 \cup U_j$. Similarly, each $C_m^\phi(z')$, $m = 0, 1, \cdots$, satisfies

(3.5′) $$\bar{\partial}C_m^\phi = \phi(f(z'))^m[\mathfrak{M}]^{0,1} \quad \text{on } U_i \cup \mathfrak{M}_0 \cup U_j.$$

Since each f_s is C^1 on $\mathfrak{M}_0$, Theorem 3.18B implies that the functions Φ_w and C_m^ϕ, $m = 0, 1, \cdots$, when restricted to U_j, all have continuous extensions to $\bar{U}_j - \operatorname{Sing} \mathfrak{M}$. Consequently, for each fixed $w \in \boldsymbol{C} - \phi(\overline{B(0, R)})$ we have a continuous extension of $R^\phi(z'; w) = \exp \Phi_w(z')$ to $\bar{U}_j - \operatorname{Sing} \mathfrak{M}$. Because of the formula for the A_m's in terms of C_m's in part (c) of Lemma 3.13 this proves the first part (d) above.

It follows from the estimate (3.8) in Theorem 3.18 that the function $\Phi_w(z')$ is bounded on compact subsets on $(\bar{U}_j - \operatorname{Sing} \mathfrak{M}) \times (\boldsymbol{C} - \phi(\overline{B(0, R)}))$. Consequently, $R_j^\phi(z'; w)$ is also bounded on compact subsets of $(\bar{U}_j - \operatorname{Sing} \mathfrak{M}) \times (\boldsymbol{C} - \phi(\overline{B(0, R)}))$.

Therefore, if K is a compact subset of $U_j - \operatorname{Sing} \mathfrak{M}$ then the family $\{R_j^\phi(z'; w)\}_{z' \in K - \mathfrak{M}}$ of holomorphic functions on $\boldsymbol{C} - \phi(\overline{B(0, R)})$ is a normal family. Since for each w fixed, $R_j^\phi(z'; w)$ is continuous on $\bar{U}_j - \operatorname{Sing} \mathfrak{M}$, precompactness of normal families implies that, as z' approaches $z_0' \in \bar{U}_j - \operatorname{Sing} \mathfrak{M}$, $R_j^\phi(z_0'; w) = \lim_{z' \to z_0} R_j^\phi(z'; w)$ with normal convergence on $\boldsymbol{C} - \phi(\overline{B(0, R)})$. One easily deduces that $R_j^\phi(z'; w)$ is jointly continuous on $(\bar{U}_j - \operatorname{Sing} \mathfrak{M}) \times (\boldsymbol{C} - \phi(\overline{B(0, R)}))$ which proves part (e). Also, the normal convergence in w implies that for each $z' \in \bar{U}_j - \operatorname{Sing} \mathfrak{M}$ fixed R^ϕ is holomorphic in w on $\boldsymbol{C} - \phi(\overline{B(0, R)})$. Computing the Laurent expansion in w (with $z_0' \in \bar{U}_j - \operatorname{Sing} \mathfrak{M}$ fixed) we obtain,

$$R_j^\phi(z_0'; w) = \sum_{-\infty}^{\infty} \left[\frac{1}{2\pi i} \int_{|\zeta| = R + \varepsilon} R_j^\phi(z_0'; \zeta)\, \zeta^{-m-1} \right] w^m.$$

However,

$$\int_{|\zeta| = R+\varepsilon} R_j^\phi(z_0'; \zeta)\, \zeta^{-m-1} = \lim_{z' \to z_0'} \int_{|\zeta| = R+\varepsilon} R_j^\phi(z'; \zeta)\, \zeta^{-m-1}$$
$$= \lim_{z' \to z_0'} \bar{A}_m^j(z') = \bar{A}_m^j(z')$$

which completes the proof of part (f).

Equation (3.11) is immediately proved by applying the jump formula in Theorem 3.18(B) to Φ_w.

Finally to prove that $R_j^\phi(z'; w)$ is rational in w proceed by induction from the unbounded component U_0 (where $R_0^\phi \equiv 1$ is rational). Since $\boldsymbol{C}^p$ − Sing $\mathfrak{M}$ is connected every component of $\boldsymbol{C}^p - \mathfrak{M}$ is accessible via a finite number of boundaries $\mathfrak{M}_0$ in $\mathfrak{M}$ − Sing $\mathfrak{M}$. Consequently (with the notation of part (g)), we may assume that $R_j^\phi(z'; w)$ is rational. Then equation (3.11) implies that $R_i^\phi(z'; w)$ is rational in w for all $z' \in \mathfrak{M}_0$. The version of the E. E. Levi theorem given in Corollary 3.20 immediately implies that $R_i^\phi(z, w)$ is rational in w (with coefficients holomorphic in U_i and continuous on $\bar{U}_i$ − Sing $\mathfrak{M}$).

3.9. *Step 7: Construction of the holomorphic p-chain T on G in the hypersurface case.* In this subsection we assume that $p = n - 1$, i.e., we are in the hypersurface case. We adapt the previous notation but only consider $\phi(z) = z_{p+1} = z_n$. The objective is to show that the holomorphic p-chain T, defined by

$$T = (i/\pi)\, \partial\bar{\partial} \log|R| \quad \text{on } (\boldsymbol{C}^p - \mathfrak{M}) \times \boldsymbol{C}, \tag{3.12}$$

extends to G (we will also denote the extension by T) in such a way that T is a holomorphic p-chain on $G - M$ and $dT = [M]$ on G with appropriate boundary regularity (i.e., we shall prove Theorem 3.3 on G). Let T_i denote T restricted to $U_i \times \boldsymbol{C}$, consistent with $T_i = (i/\pi)\, \partial\bar{\partial} \log|R_i|$ on $U_i \times \boldsymbol{C}$. Since T is a holomorphic p-chain, the support of T is a pure p-dimensional subvariety V of $(\boldsymbol{C}^p - \mathfrak{M}) \times \boldsymbol{C}$. Let V_i denote supp $T_i = V \cap (U_i \times \boldsymbol{C})$.

LEMMA 3.22. *T extends to a holomorphic p-chain in $G - M$.*

PROOF. For each point $\zeta \in (\boldsymbol{C}^{p+1} - M) \cap G$ we will exhibit a neighborhood B of ζ and a holomorphic p-chain T_B on B such that T_B agrees with T on $B - \pi^{-1}(\mathfrak{M})$.

First, we consider the (easier) case where $\pi(\zeta) \notin$ Sing $\mathfrak{M}$. The neighborhood B will be chosen to be of the form $B = \Delta' \times \Delta$ with Δ' a ball about ζ' in $\boldsymbol{C}^p$ and Δ a disk about ζ_{p+1} in $\boldsymbol{C}$. Choose Δ' so that $\mathfrak{M} \cap \Delta'$ is a manifold and $\Delta' - \mathfrak{M}$ consists of precisely two components $U_i \cap \Delta'$ and $U_j \cap \Delta'$. Moreover, if $M \cap \pi^{-1}(\Delta')$ has components $M_1, \cdots, M_r$, let $f_s \in C^1(\mathfrak{M} \cap \Delta')$ denote the function whose graph is M_s ($s = 1, \cdots, r$).

Since the coefficients of R_i (or R_j) are continuous on $\bar{U}_i \cap \Delta'$ (or $\bar{U}_j \cap \Delta'$) the points of $\pi^{-1}(\zeta')$ which belong to $\bar{V}_i = \overline{\text{supp } T_i}$ (or $\bar{V}_j = \overline{\text{supp } T_j}$) form a finite set corresponding to the zeros and poles of $R_i(\zeta'; w)$ (or $R_j(\zeta'; w)$ respectively). Therefore we can choose a small disk Δ about ζ_{p+1} with $\{\zeta'\} \times \bar{\Delta}$ containing no points of $[\bar{V}_i \cup \bar{V}_j]$ (other than ζ) and disjoint from M. Now choose a small ball Δ' about ζ' so that $\Delta' \times \partial\Delta$ remains disjoint from $\bar{V}_i \cup \bar{V}_j$ and $\Delta' \times \bar{\Delta}$ remains disjoint from M. Let B denote $\Delta' \times \Delta$. It follows that π restricted to $B \cap \overline{\text{supp } T}$ is proper.

The holomorphic chain T_B on B will be defined by $T_B = (i/\pi)\, \partial\bar{\partial} \log|r|$ on B where $r(z'; w)$ is a rational function in w with coefficients holomorphic on Δ'. The function $r(z'; w)$ will be chosen so that $T_B = T$ on $B - \pi^{-1}(\mathfrak{M})$. Let T' denote T restricted to $B - \pi^{-1}(\mathfrak{M})$. Namely, for $z' \in \Delta' - \mathfrak{M}$, $r(z'; w)$ is defined as follows. First, let

$$c_m \equiv \pi_*(w^m T'), \quad m \geq 0. \tag{3.13}$$

Since π restricted to ((supp T) $\cap B$) $- \pi^{-1}(\mathfrak{M})$ is a proper map, each c_m is a well-defined holomorphic function on $\Delta' - \mathfrak{M}$. Define a_m to be the mth Newton polynomial in the c's for $-\infty < m \leqq c_0$, and define $r(z'; w)$ by $\sum_{-\infty}^{c_0} a_m(z')w^m$. It is obvious from the discussion in §2 that

$$T = (i/\pi)\, \partial\bar{\partial} \log|r| \quad \text{on } B - \pi^{-1}(\mathfrak{M}). \tag{3.14}$$

It remains to show that the coefficients of $r(z'; w)$ extend holomorphically from $\Delta' - \mathfrak{M}$ to all of Δ'. It suffices to show that each $c_m(z')$ extends holomorphically from $\Delta' - \mathfrak{M}$ to Δ'. Equations (3.13) and (3.14) imply that

$$c_m(z') = \frac{1}{2\pi i} \int_{\partial\Delta} w^m \frac{\partial R_i}{\partial w}(z'; w)/R_i(z'; w)\, dw \tag{3.15}$$

on $U_i \cap \Delta'$ with an analogous formula for $c_m(z')$ on $U_j \cap \Delta'$.

Let $c_m^i(z')$ denote the continuous extension of $c_m(z')|_{U_i}$ to $\bar{U}_i \cap \Delta'$ and let c_m^j denote the continuous extension of $c_m(z')|_{U_j}$ to $\bar{U}_j \cap \Delta'$. Then using (3.15) (for i and j) we obtain that, for $z' \in \bar{U}_i \cap \bar{U}_j \cap \Delta'$,

$$c_m^i(z') - c_m^j(z') = \pm \frac{1}{2\pi i} \int_{\partial\Delta} w^m \sum_s \pm(w - f_s(z'))^{-1}\, dw, \tag{3.16}$$

since the jump relation

$$R_i(z'; w) = \sum_s (w - f_s(z'))^{\pm 1} R_j(z'; w) \quad \text{on } \mathfrak{M} \cap \Delta'$$

implies that

$$\frac{\partial R_i}{\partial w} \Big/ R_i - \frac{\partial R_j}{\partial w} \Big/ R_j = \sum_s \pm (w - f_s(z'))^{-1} \quad \text{on } \mathfrak{M} \cap \Delta'.$$

(The choice of $\pm$ is determined by whether $\Delta' \cap \pi(M_s)$ is the oriented boundary of $\Delta' \cap U_i$ or $\Delta' \cap U_j$.)

Since $\sum_{s=1}^r \pm(w - f(z'))^{-1}$ is holomorphic on Δ for $z' \in \mathfrak{M} \cap \Delta'$ fixed, the right-hand side of (3.16) vanishes. Therefore, c_m on $\Delta' - \mathfrak{M}$ extends continuously to Δ'. A standard removable singularities result now implies that c_m extends holomorphically to Δ' (see, for example, **[24]**). This completes the proof if $\pi(\zeta) \notin \operatorname{Sing} \mathfrak{M}$.

The (harder) case where ζ is an arbitrary point of $G - M$ is quite similar to the case where $\pi(\zeta) \notin \operatorname{Sing} \mathfrak{M}$ treated above once we have the jump information given in (3.18) below. Suppose $\zeta \in G - M$ is given, and choose a neighborhood Δ' of $\zeta' = \pi(\zeta)$ as in part (a) of Lemma 3.4. Then $M \cap \pi^{-1}(\Delta')$ consists of r connected manifolds $M_1, \cdots, M_r$ such that π is an embedding of each M_s into Δ'. Let $\mathfrak{M}_s$ denote $\pi(M_s)$ including orientation, and let $f_s \in C^1(\mathfrak{M}_s)$ denote the (graphing) function for M_s over $\mathfrak{M}_s$ $(s = 1, \cdots, r)$.

Now in general, suppose $\tilde{G}$ satisfies

$$\bar{\partial}\tilde{G} = \sum_{s=1}^{r} g_s[\mathfrak{M}_s]^{0,1} \quad \text{on } \Delta' \tag{3.17}$$

where each $g_s \in C^1(\mathfrak{M}_s)$. Let $G(z')$ denote the restriction of $\tilde{G}$ to $\Delta' - \mathfrak{M}$, which is holomorphic. A number α will be referred to as a cluster value of G at ζ' if there exists a sequence $\{\zeta_k'\}$ in $\Delta' - \mathfrak{M}$ such that $\alpha = \lim_{k\to 0} G(\zeta_k')$. We need the following fact. If α, β are any two cluster values of G at ζ' then

$$(3.18) \qquad \alpha - \beta = \sum_{s=1}^{r} \varepsilon_s g_s(\zeta')$$

where each ε_s equals 0, +1, or −1. In particular, the set of cluster values for G at ζ' is finite.

In order to prove (3.18) assume that $\alpha = \lim G(\zeta'_k)$ and $\beta = \lim G(\eta'_k)$. Solve $\bar{\partial} G_s = g_s[\mathfrak{M}_s]$ on Δ' ($s = 1, \cdots, r$). Then $\tilde{G} = \sum G_s + H$ where H is holomorphic on Δ'. Passing to subsequences if necessary we may assume that, for each s, the sequence $\{\zeta'_k\}$ lies in one of the components of $\Delta' - \mathfrak{M}_s$ and similarly for $\{\eta'_k\}$. Since Theorem 3.18(B) is applicable to each G_s, $\lim G_s(\zeta'_k) - \lim G_s(\eta'_k)$ must be either zero or $\pm\, g_s(\zeta')$, which proves (3.18) above.

Now referring back to the proof in the easier case $\pi(\zeta) \notin \operatorname{Sing} \mathfrak{M}$, first applying (3.18) to the functions $C_m(z')$, $m = 0, 1, \cdots$, we find that the cluster values of each C_m at ζ' must be finite. Therefore, the cluster values of the coefficients of $R(z'; w)$ at ζ' form a finite set, and hence as before we have proved $\overline{\operatorname{supp} T} \cap \pi^{-1}(\zeta')$ is finite. The proof now proceeds exactly as in the case $\pi(\zeta) \notin \operatorname{Sing} \mathfrak{M}$, except that we have the following replacement for (3.16). Suppose α and β are any two cluster values for c_m at ζ'. Then, for an appropriate choice of $\varepsilon_s \in \{0, 1, -1\}$,

$$(3.16') \qquad \alpha - \beta = \frac{1}{2\pi i} \int_{\partial\Delta} w^m \sum_{s=1}^{r} \varepsilon_s (w - f_s(\zeta'))^{-1} \, dw.$$

As before, the right-hand side vanishes and hence the functions c_m used to define $r(z'; w)$ are continous on Δ' (which in turn implies they are holomorphic on Δ').

COROLLARY 3.23. *The holomorphic p-chain T on $G - M$ obtained in the previous lemma has locally finite volume near $M \cap G$ and hence extends as a locally rectifiable $2p$-current on G. Moreover, $dT = \sum n_j [M_j]$ where the M_j are the components of $M \cap G$ and each n_j is an integer.*

PROOF. The fact that T has locally finite volume across M is proved exactly as Lemma 1.5. Consequently T extends to a locally rectifiable current on G (which we again denote by T). In particular T, and hence dT, are locally flat. Now dT is of dimension $2p - 1$ and supported in $M \cap G$. The support Theorem 1.7(b) says that a locally flat current of real dimension r supported in an r-dimensional submanifold M is of the form $h[M]$ with h a locally integrable function on M. Since $d(dT) = 0$, in our case h must be locally constant. It also follows that, since T is rectifiable, h must be integer-valued; however we will not use this last fact. This proves the corollary.

The remaining objective of this section is to establish boundary regularity on an open dense subset of $M \cap G$. In this process we show each n_j in Corollary 3.23 equals +1 so that $dT = [M]$.

LEMMA 3.24. *There exists a closed nowhere dense[2] subset A of M (take A to contain $M - G$) such that each point $\zeta \in M - A$ has a neighborhood B with the following properties. The subvariety $W = \operatorname{supp} T$ of $B - M$ is a submanifold and either*

Case (i): *W is connected and the pair ($\pm W$, $M \cap B$) is a C^1 submanifold with boundary, or*

[2]The set A is of Hausdorff $2p - 1$ measure zero. See the note and footnote following Lemma 3.19.

Case (ii): *$\overline{W} \equiv W \cup (M \cap B)$ is a connected complex submanifold of B containing $M \cap B$ as a real C^1 hypersurface dividing $\overline{W}$ into two components W_i and W_j over $\Delta' \cap U_i$ and $\Delta' \cap U_j$ respectively.*

In either case, if the labeling is chosen so that $\pi(M \cap B)$ is the oriented boundary of $U_i \cap \Delta'$ then $T|_B$ is of the form $(m+1)[W_i] + m[W_j]$ so that $dT = [M]$ on B. (In case (i), *$m = -1$ or $m = 0$.)*

PROOF. As before let $V_j = \text{supp } T_j$. Consider the polynomial $p_j(z'; w)$ in w which vanishes precisely on V_j and to order 1 at each manifold point. That is, $p_j(z'; w)$ consists of all the irredundant factors occurring in both the numerator and the denominator of $R_j(z'; w)$. Consequently the coefficients of $p_j(z'; w)$ are holomorphic on U_j and continuous on $\overline{U}_j - \text{Sing } \mathfrak{M}$. Let δ_j denote the discriminant of p_j and let $D_j = \{z' \in \overline{U}_j - \text{Sing } \mathfrak{M}: \delta_j(z') = 0\}$ denote the discriminant locus in $\overline{U}_j - \text{Sing } \mathfrak{M}$.

Let $\mathscr{A}$ denote the subset of $\mathfrak{M}$ consisting of all points z' which belong to at least one of the discriminant loci D_j or belong to Sing $\mathfrak{M}$. We take $A = M \cap \pi^{-1}(\mathscr{A})$. Note that Lemma 3.19 implies that A is nowhere dense in M.

Now suppose $\zeta \in M - A$. As in the proof of Lemma 3.22, $\overline{\text{supp } T} \cap \pi^{-1}(\pi(\zeta))$ is finite and hence we can choose a disk Δ about $\zeta_{p+1} = f(\zeta')$ and a ball Δ' about ζ' so that the neighborhood $B = \Delta' \times \Delta$ of ζ has the property that $(\Delta' \times \Delta) \cap M$ equals the graph of f over $\Delta' \cap \mathfrak{M}$.

Since $\Delta' \cap \overline{U}_i$ is simply connected and disjoint from the discriminant locus D_i, $\overline{\text{supp } T_i} \cap B$ is the graph of a single-valued function F_i holomorphic on $\Delta' \cap U_i$ and continuous on $\Delta' \cap \overline{U}_i$. Similarly we can find F_j defined on $\Delta' \cap \overline{U}_j$. Let W_i denote the graph of F_i over $\Delta' \cap U_i$ and W_j denote the graph of F_j over $\Delta' \cap U_j$. Then the complex submanifold $W = W_i \cup W_j$ of $B - M$ is equal to the support of T on $B - M$ (a priori either W_i, W_j or both may be empty).

Although either W_i or W_j may be empty (a priori both may be empty), we first consider the case where $W_i \neq \varnothing$ and $W_j \neq \varnothing$. Suppose $F^i(z') \neq f(z')$ at some point $z' \in \Delta' \cap \mathfrak{M}$. Then $(z', F^i(z')) \in \overline{\text{supp } T_i - M}$. Therefore, by Lemma 3.22, $\overline{\text{supp } T_i} \cup \overline{\text{supp } T_j}$ must be a subvariety near $(z', F^i(z'))$. Consequently, F_i and F_j must agree on a neighborhood of z' in $\mathfrak{M}$. This implies that $\overline{W} \cap B$ is a complex submanifold of B. To complete the proof in case (ii) it remains to show that $M \subset \overline{W}$ and $dT = [M]$ on B. Let T' denote T restricted to $B - M$. Then T' must be of the form $T' = a[W_i] + b[W_j]$ for some integers a and b. We will show that $a - b = 1$, so that T' is of the form $T' = (m+1)[W_i] + m[W_j]$. Since $dT' = 0$ on $B - M$ by Lemma 3.22, this implies, first, that $M \subset \overline{W}$, and, second, that $dT = [M]$ on B.

The proof that $a - b = 1$ proceeds exactly as in the proof of Lemma 3.22. Defining $c_0(z')$ to be $\pi_*(T')$ and using formula (3.16) with $m = 0$ we obtain the jump

$$c_0^i(z') - c_0^j(z') = 1 \quad \text{for each } z' \in \mathfrak{M} \cap \Delta', \tag{3.19}$$

since the right-hand side of (3.16) equals $(1/2\pi i)\int_{\partial\Delta}(w - f(z')^{-1})$ which equals one since $f(z') \in \Delta$. Of course, $c_0(z') = \pi_*(T') = a[U_i \cap \Delta'] + b[U_j \cap \Delta']$, so the jump in (3.19) also equals $a - b$ which completes the proof.

The easier case (i) has been proved at the same time since this is just the case $b = 0$.

3.10. *Step* 8: *Varying the projection π in the hypersurface case.* The proof of Theorem 3.3 (in the hypersurface case $p = n - 1$) is easily completed with the aid of the following lemma; the only problem then remaining is the scar set S.

LEMMA 3.25. *Suppose π_1 and $\pi_2 \in \boldsymbol{P}^{n-1}$ are two arbitrary projections. The currents T_{π_1} and T_{π_2} constructed in the last section agree on the intersection $G_{\pi_1} \cap G_{\pi_2}$, and hence we obtain a (unique) current T on $\boldsymbol{C}^n - S$ which satisfies all of Theorem 3.3 on $\boldsymbol{C}^n - S$.*

PROOF. Suppose z belongs to both C_{π_1} and G_{π_2}. Consider the collection $\mathscr{G}(z)$ of projections $\pi \in \boldsymbol{P}^{p-1}$ such that $z \in G_\pi$. Recall that by Lemma 3.5, $\mathscr{G}(z)$ is an open connected (and nonempty for each point $z \in \boldsymbol{C}^n - S$) subset of $\boldsymbol{P}^{n-1}$. Therefore, it suffices to prove that for $\pi \in \mathscr{G}(z)$ the currents $T_{\pi'}$ on $G_{\pi'}$ all agree near z if π' is sufficiently close to π.

More precisely, we will show that for each projection $\pi \in \boldsymbol{P}^{p-1}$ there exists a neighborhood $N(z)$ of z in $\boldsymbol{C}^n$ and a neighborhood $\mathscr{N}(\pi)$ in $\mathscr{G}(z)$ such that for each $\pi' \in \mathscr{N}(\pi)$ we have $N(z) \subset G_{\pi'}$ and $T_{\pi'} = T_\pi$ on $N(z)$. Recall that, by Lemma 3.4, $\pi(G_\pi)$ is connected. Therefore, we can choose an open connected subset U' of $\pi(G_\pi)$ with $\pi(z) \in U'$ and with U' intersecting the unbounded component U_0^π of $\boldsymbol{C}^p - \mathfrak{M}_\pi$. In addition choose U' relatively compact in $\pi(G_\pi)$. Then there exists a neighborhood $\mathscr{N}(\pi)$ in $\boldsymbol{P}^{n-1}$ such that $\pi^{-1}(U')$ remains in $G_{\pi'}$ for all $\pi' \in \mathscr{N}(\pi)$. Let $N(z) = \pi^{-1}(U')$. Then we have two currents $T_\pi|_{N(z)}$ and $T_{\pi'}|_{N(z)}$ which satisfy Theorem 3.3 on $N(z)$. Since the mapping $\pi\colon (\operatorname{supp} T_{\pi'}) \cap N(z) \to U'$ is proper, we obtain a second set of power functions C'_m on U' by letting $C'_m \equiv \pi_*(w^m T_{\pi'})$ on U'. Both $C_m \equiv \pi_*(w^m T_\pi)$ and C'_m satisfy the same equation $\bar\partial(\ ) = \pi_*(w^m[M])^{0,1}$ on U'. Therefore the difference $C_m - C'_m$ is holomorphic on U'. Now by choosing U' to include points in $\boldsymbol{C}^p - \pi(B(0, R))$ where $M \subset B(0, R)$, we can insure that C'_m (as well as C_m) vanishes on $U' - \pi(B(0, R))$. Hence $C'_m \equiv C_m$ on U'. Since the functions C'_m uniquely determine $T_{\pi'}$ on $N(z)$ this proves that T_π and $T_{\pi'}$ agree on $N(z)$, which completes the proof of Lemma 3.25 in the hypersurface case (i.e., this proves Theorem 3.3 on $\boldsymbol{C}^n - S$).

REMARK. A careful reading of the proof that we have presented of Theorem 3.3 (in the hypersurface case) on $\boldsymbol{C}^n - S$ will show that $M - S$ can be allowed to be an immersed, instead of an embedded, submanifold.

The remainder of this section is devoted to completing the proof of Theorem 3.3, taking into account the scar set S.

It suffices to prove that T has finite mass across S, because then T can be trivially extended across S to a current (also denoted T) on all of $\boldsymbol{C}^n$, which is flat. That is, in case we can show that T has finite mass, $dT - [M]$ is a flat current of dimension $2p - 1$ supported in S. Since by hypothesis $\mathscr{H}_{2p-1}(S) = 0$, the support Theorem 1.7(a) then implies that $dT - [M] = 0$ as desired.

In order to prove that T has locally finite mass across S we first consider the case where T is positive on $\boldsymbol{C}^n - S$. By Wirtinger's equality the mass of T can be (locally) estimated above (see the proof of Lemma 1.3) by adding together the mass of the functions $C_0^\pi = \pi_*(T_\pi)$ over a suitable choice of coordinate projections π. However, since C_0^π satisfies $dC_0^\pi = \pi_*([M])$ where $\pi_*([M])$ has finite mass and C_0^π has compact support, standard estimates show that C_0^π must have finite mass.

If T is not positive the proof proceeds as follows. On $\boldsymbol{C}^n - M$, $T = \sum n_j[v_j]$ is a holomorphic chain. Let T^+ denote $\sum_{n_j>0} n_j[v_j]$ and T^- denote $-\sum_{n_j<0} n_j[v_j]$ so that both T^+ and T^- are positive and $T = T^+ - T^-$. Since T has locally finite mass in $C^n - S$ the same is true for T^+ and T^-. Consequently, both T^+ and T^- are locally flat currents on $C^n - S$ which are supported in $M - S$. Let $M^{\pm}$ denote $dT^{\pm}$ on $C^n - S$. The submanifold $M - S$ is the disjoint union of connected oriented submanifolds M_j with $[M] = \sum[M_j]$. It follows from the support Theorem 1.7(b) that $M^{\pm}$ is of the form $\sum n_j^{\pm}[M_j]$ on $C^n - S$, where each $n_j^{\pm} \in Z$. Consider a point $z \in M_j$ of boundary regularity. Then, by Lemma 3.24, T is of the form $(m + 1) \cdot [W_i] + m[W_j]$ near z, where $m \in Z$. Of course, it is impossible for both the integer m to be negative and $m + 1$ to be positive. Therefore, in a neighborhood of z either both m and $m + 1$ are negative (in which case $T^+ \equiv 0$ and $n_j^+ \equiv 0$) or both m and $m + 1$ are positive (in which case $T^- \equiv 0$ and $n_j^- \equiv 0$). This proves that supp M^+ and supp M^- are disjoint in $C^n - S$. Consequently, the mass of $[M]$ is the sum of the masses of M^+ and M^- on $C^n - S$. Therefore, M^+ and M^- have finite mass and hence define rectifiable currents on C^n.

Now the proof that T^+ (and similarly T^-) has finite mass across S proceeds as in the case where T is positive (see above). However, we no longer know that $dC_0^+ = \pi_*(M^+)$ except on $C^p - \pi(S)$, where $C_0^+ = \pi_*(T^+)$. It suffices to show that $\pi_*(M^+)$ is d-closed (a priori $d\pi_*(M^+) = \pi_*(dM^+)$ may have some support in $\pi(S)$). This is because one can then solve $d\chi = \pi_*(M^+)$ on C^p where χ has compact support and finite mass, which by uniqueness implies $C_0^+ = \chi$ has finite mass.

Let D_n denote the open set where $C_0^+ \equiv n$. Then $C_0^+ = \sum_{n=1}^{\infty} n[D_n]$. Let χ_N denote the minimum of C_0^+ and N, or $\chi_N = \sum_{n=1}^{N} n[D_n] + \sum_{n=N+1}^{\infty} [D_n]$. Then χ_N has mass less than N times the measure of supp C_0^+, which is finite since supp C_0^+ is compact. Consequently, $d\chi_N$ is a d-closed flat current in C^p. The idea is to show that $\pi_*(M^+) = \lim d\chi_N$ and hence that $\pi_*(M^+)$ is d-closed.

Let S' denote $\pi(S) \cup \pi(K)$ where K denotes the set of critical points of $\pi|_{M-S}$. By Lemma 3.4(a), $\mathscr{H}_{2p-1}(S') = 0$. Locally on $C^p - S'$ (say on Δ') $\pi_*(M^+)$ is of the form $\sum_{j=1}^{k} [\mathfrak{M}_j]$ where each $\mathfrak{M}_j$ is an oriented $(2p - 1)$-dimensional submanifold (see Lemma 3.4(a)). By choosing Δ' small we can insure that $\Delta' - \mathfrak{M}_j$ consists of exactly two connected open sets. Let Ω_j denote the component whose oriented boundary is $\mathfrak{M}_j$ in Δ'. On Δ', C_0^+ is of the form $\sum_1^k [\Omega_j] + m$, for some integer m. For $N \leqq m$, $\chi_N \equiv N$ on Δ' and hence $d\chi_N = 0$. For $N \geqq m + k$, $\chi_N = C_0^+$ on Δ' and hence $d\chi_N = \pi_*(M^+)$. Consequently, on Δ', $\lim_{N\to\infty} d\chi_N = \pi_*(M^+)$ since equality holds for $N \geqq m + k$.

Next we show that the mass of $d\chi_N$ is bounded above by the mass of $\pi_*(M^+)$ independent of N. By the compactness Theorem 4.2.17 [10] this implies that $\{d\chi_N\}$ must have a convergent subsequence in the flat topology. Say $R = \lim d\chi_{N_i}$. By the above, $R = \pi_*(M^+)$ on $C^p - S'$; however since $R - \pi_*(M^+)$ is flat and $\mathscr{H}_{2p-1}(S') = 0$ we must have $R = \pi_*(M^+)$ on all of C^p by the support Theorem 1.7(a). Since R is d-closed this would prove that $\pi_*(M^+)$ is d-closed.

To prove that $d\chi_N$ has smaller mass than $\pi_*(M^+)$ only requires a local calculation on $C^p - S'$. Suppose $z \in C^p - S'$ is fixed. It suffices to prove that the density of $d\chi_N$ at z is less than or equal to the density of $\pi_*(M^+)$ at z. Let k denote the density of $\pi_*(M^+)$ at z. We can then choose a neighborhood Δ' of z and $\mathfrak{M}_1, \cdots, \mathfrak{M}_k$ embedded submanifolds of Δ' as in the above. Since $C_0^+ = \sum_{j=1}^{k} [\Omega_j] + m$ on Δ', the values of C_0^+ must lie between m and $m + k$.

Now let D_n' denote the open subset of $C^p - \pi_*(M^+)$ where $C_0^+ \geqq n$. Then

$C_0^+ = \sum_{n=1}^{\infty} [D'_n]$ and, on Δ', $C_0^+ = \sum_{n=m}^{m+k} [D'_n]$. Therefore on Δ' (for $N \leqq m + k$) $\chi_N = \sum_{n=m}^{N} [D'_n]$. By the Gauss-Green Theorem 4.5.6 **[10]** the density of $d[D'_n]$ at z is either 0 or 1. Therefore the density of $d\chi_N$ at z is $\leqq N - m \leqq k$ as desired. This proves that T^+ has finite mass across S completing the proof of Theorem 3.3 in the hypersurface case.

3.11. *Step* 9: *Reduction to the hypersurface case.* In this section we complete the proof of Theorem 3.3 in higher codimension where M is of real dimension $2p - 1$ in $\mathbf{C}^n$. By choosing linear functions ϕ carefully (employing the geometric analogue of the algebraic theorem on the primitive element) the proof is reduced to the hypersurface case. We adopt the notation used in §3.8 (rationality of $R^\phi(z'; w)$). Suppose $\pi: \mathbf{C}^n \to \mathbf{C}^p$ is a fixed projection, and that $\phi \in (\mathbf{C}^{n-p})^*$ is a linear function. We define $\rho^\phi: \mathbf{C}^n \to \mathbf{C}^{p+1}$ by $\rho^\phi(z) = (z', \phi(z''))$. Let K denote the set of critical points of $\rho^\phi|_{M-S}$ union the set S. By the general version of Sard's theorem $\rho^\phi(K)$ has $(2p - 1)$-dimensional Hausdorff measure zero. Let $S^\phi = \rho^\phi(K)$ and $M^\phi = \rho^\phi(M)$. Then $M^\phi - S^\phi$ is an immersed submanifold, and $\rho^\phi_*([M]) = [M^\phi]$ so that $M^\phi - S^\phi$ has finite volume. Consequently, Theorem 3.3 in the hypersurface case is applicable, and we obtain a (unique) current T^ϕ satisfying all the conditions of Theorem 3.3 with $dT^\phi = [M^\phi]$. Let $\pi^\phi(z'; w) = z'$ and note that $\pi = \pi^\phi \circ \rho^\phi$. We construct T on $(\mathbf{C}^p - \text{Sing } \mathfrak{M}) \times \mathbf{C}^{n-p}$, starting with the unbounded component U_0 and defining $T \equiv 0$ on $U_0 \times \mathbf{C}^{n-p}$. Suppose U is an open subset of $\mathbf{C}^p$. We will say that a current T *satisfies Theorem* 3.3 *above* U (with respect to M) if T is defined on $U \times \mathbf{C}^{n-p}$, satisfies all the conditions in Theorem 3.3 and $\overline{\text{supp } T}$ is compact in $\mathbf{C}^n$.

LEMMA 3.26. *Suppose T is a current which "satisfies Theorem* 3.3 *above U". If U is connected and has nonempty intersection with both the unbounded component U_0 and another component, say U_i, then T has a (unique) extension to a current which satisfies Theorem* 3.3 *above $U \cup U_i$.*

PROOF. First note that if T satisfies Theorem 3.3 above U, and U is connected with $U \cap U_0 \neq \varnothing$, then:

$$\rho^\phi_*(T) = T^\phi \quad \text{on } U \times \mathbf{C} \text{ for each } \phi \in (\mathbf{C}^{n-p})^*. \tag{3.20}$$

This is because both $\rho^\phi_*(T)$ and T^ϕ satisfy Theorem 3.3 above U (with respect to M^ϕ), and we have uniqueness in the hypersurface case above U since U is connected with $U \cap U_0 \neq \varnothing$. (For this last fact—uniqueness—consider that the power functions C_m, restricted to U, are uniquely determined by the facts that each has bounded support and satisfies the $\bar{\partial}$ equation (3.1) on U.)

Choose $\Delta' \subset U \cap U_i$, and let V denote the subvariety $\text{supp}(T|_{\Delta'})$ of $\pi^{-1}(\Delta')$. Since $\pi: V \to \Delta'$ is proper, and V is of pure dimension p, there exists a proper subvariety D of Δ' such that $\pi: V - \pi^{-1}(D) \to \Delta' - D$ is an m-sheeting covering map (cf. Narasimhan **[36]** or Gunning and Rossi **[14]**). Now we choose the geometric analogue of a primitive element. Namely choose a linear function $\phi \in (\mathbf{C}^{r-p})^*$ which separates the points of $V \cap \pi^{-1}(z')$ for some $z' \in \Delta' - D$. Then (by enlarging D if necessary) the proper map $\rho^\phi: V \to \Delta' \times \mathbf{C}$, when restricted to $V - \pi^{-1}(D)$, is an embedding into $(\Delta' - D) \times \mathbf{C}$. Moreover, by 3.20, $\pi(V)$ is just V^ϕ (=supp T^ϕ) on $\Delta' \times \mathbf{C}$. Let $z_{p+1} = \phi_{p+1}(z'')$ denote this choice of ϕ. There exists a finite family $\mathfrak{F} \subset (C^{n-p})^*$ of linear functionals, which includes ϕ_{p+1}, as well as a basis for $(C^{n-p})^*$, and has the property that the hypersurfaces

$$(\rho^\phi)^{-1}((\operatorname{supp} T^\phi) \cap (\Delta' \times C))$$

in $\Delta' \times C^{n-p}$ intersected over $\phi \in \mathfrak{F}$ yield V. (Note that, in general, $\mathfrak{F}$ cannot be chosen as a basis for $(C^{n-p})^*$, but must be larger.) Now we define the subvariety V_i to consist of all p-dimensional components of the intersection $\bigcap_{\phi \in \mathfrak{F}} (\rho^\phi)^{-1}$ (supp T_i^ϕ) of hypersurfaces in $U_i \times C^{n-p}$. Then $V_i \cap \pi^{-1}(\Delta') = V$. (However, a single component of V_i may, when intersected with $\Delta' \times C^{n-p}$, yield several components of V.)

It remains to define T by attaching multiplicities to each component of V_i. Since V_i is bounded, the map $\rho(z) \equiv (z', \phi_{p+1}(z''))$ restricted to V_i is proper. Therefore $\rho\colon V_i \to V_i^{\phi_{p+1}}$ has generic sheeting number above each component of $V_i^{\phi_{p+1}}$. Since each component of $V_i^{\phi_{p+1}}$ intersects $\Delta' \times C$ and above $V_i^{\phi_{p+1}} \cap (\Delta' \times C)$ the generic sheeting number is one,

$$\rho\colon V_i \to V_i^{\phi_{p+1}} \tag{3.21}$$

must be single-sheeted outside of a subvariety. Since ρ is generically single-sheeted, the multiplicities for $T_i^{\phi_{p+1}}$ supported on $V_i^{\phi_{p+1}}$ uniquely determine multiplicities for a current T_i supported on V_i with the property that $\rho_*^{\phi_{p+1}}(T_i) = T_i^{\phi_{p+1}}$. This T_i provides the desired extension of T over U to a current (also denoted T) over $U \cup U_i$.

LEMMA 3.27. *Suppose T is a current which satisfies Theorem* 3.3 *above U, where U is connected, has nonempty intersection with the unbounded component U_0, and contains U_i. Then there exists a nowhere dense subset A_i of ∂U_i, with each point $\zeta' \in \partial U_i - A_i$ a manifold point of $\mathfrak{M}$, such that, for some neighborhood Δ' of ζ', T has a (unique) extension to a current which satisfies Theorem* 3.3 *over $U \cup \Delta'$. In addition, the C^1 boundary regularity occurs at each point of M above Δ'.*

PROOF. Let V_i denote supp T_i. Choose a primitive $\phi_{p+1} \in (C^{n-p})^*$ for V_i and a family $\mathfrak{F}$ containing ϕ_{p+1}, as in the proof of Lemma 3.26, so that V_i is given as the intersection of the hypersurfaces $(\rho^\phi)^{-1}(V_i^\phi)$ over all $\phi \in \mathfrak{F}$, and V_i is generically single-sheeted over the hypersurface $V_i^{\phi_{p+1}}$.

We will consider only points $\zeta' \in \partial U_i$ which are manifold points of $\mathfrak{M}$ (i.e., we will choose A_i to contain Sing $\mathfrak{M}$). The fibre $\pi^{-1}(\zeta') \cap \overline{\operatorname{supp} T_i}$ must be finite, since each linear functional $\phi \in \mathfrak{F}$ can take on at most a finite number of values on the points of $\pi^{-1}(\zeta) \cap \overline{(\operatorname{supp} T_i)}$. That is, the values of ϕ must be zeros or poles of $R_i^\phi(\zeta'; w)$.

Now, by choosing a new family $\mathfrak{F}$ if necessary, we can assume ϕ_{p+1} and all the other ϕ belonging to $\mathfrak{F}$ separate the points of $\pi^{-1}(\zeta') \cap [\overline{(\operatorname{supp} T_i)} \cup M]$.

First suppose $\zeta \in \pi^{-1}(\zeta') \cap [\overline{(\operatorname{supp} T_i)} - M]$. Since $\rho^\phi(\zeta) = (\zeta', \phi(\zeta)) \notin M^\phi$ for each $\phi \in \mathfrak{F}$, the intersection $\bigcap_{\phi \in \mathfrak{F}} (\rho^\phi)^{-1}$ (supp T^ϕ) defines a subvariety of C^n near ζ. Let V denote the union of all the p-dimensional components near ζ. Note that in the intersection of a neighborhood of ζ with $\pi^{-1}(U_i)$, V and V_i agree. As in the proof of Lemma 3.26, the fact that V is generically single-sheeted over supp $T^{\phi_{p+1}}$ can be used to assign the correct multiplicities to V so that V becomes the holomorphic p-chain defined near ζ which agrees with T_i on the intersection of neighborhood of ζ with $\pi^{-1}(U_i)$.

Next suppose that $\zeta \in M$. Define A_i so that if $\zeta' = \pi(\zeta) \notin A_i$ then $\rho^\phi(\zeta)$ is a point of C^1 boundary regularity for each $\phi \in \mathfrak{F}$. That is, in a neighborhood of $\rho^\phi(\zeta)$ we have C^1 boundary regularity for the manifolds supp T^ϕ, M^ϕ. (Note M^ϕ is a manifold at $\rho^\phi(\zeta)$ since ϕ separates points of $M \cap \pi^{-1}(\zeta)$.) As before, we define the support of T near ζ to be the intersection of the supports of the T^ϕ, and then use the primitive element ϕ_{p+1} in assigning the multiplicities near ζ. This completes the proof of Lemma 3.27.

COROLLARY 3.28. *For each projection $\pi \in G(p, n)$ there exists a current T satisfying Theorem* 3.3 *on G_π.*

PROOF. Lemmas 3.26 and 3.27 enable us to define a current T satisfying Theorem 3.3 on $(\mathbf{C}^p - A) \times \mathbf{C}^{n-p}$ where A is a closed subset of $\mathfrak{M}$ with Hausdorff $2p - 1$ measure zero. The same argument used in the proof of Lemma 3.27 to show that T extends past a point $\zeta \notin M$ with $\pi(\zeta) \notin A$ also shows that T extends as a holomorphic p-chain past any point $\zeta \notin M$, $\zeta \in G$. Thus we obtain a holomorphic p-chain T on $G - M$. Corollary 3.28 now follows as in §3.9 in the hypersurface case (see, in particular, the proof of Corollary 3.23).

To complete the proof of Theorem 3.3 one argues that for two different projections π_1 and π_2 the corresponding currents T_{π_1} and T_{π_2} agree on $G_{\pi_1} \cap G_{\pi_2}$ exactly as in the hypersurface case. Also the scar set S is handled just as in the hypersurface case.

Now suppose M is connected (i.e., $M - S$ is connected). To complete the proof of Theorem 3.3. we must show that T is of the form $T = [V]$ with V irreducible. Suppose V is any component of the pure p-dimensional subvariety supp $T - M$ of $\mathbf{C}^n - M$. As in Corollary 3.23, we must have $d[V] = m[M]$ for some integer m. One concludes $m = \pm 1$ as follows. Choose a point $\zeta' \in \partial U_0$ which is a manifold point of $\mathfrak{M}$. Then by Lemma 3.24 there exists a neighborhood Δ' of ζ' such that ((supp $T) - M$, M) is a C^1 submanifold with boundary on $\pi^{-1}(\Delta')$, and supp $T - M$ is the graph of r holomorphic functions on $\Delta' \cap U_1$ (i.e., we must be in case (i)). Since m cannot be zero, one can now easily conclude that $m = \pm 1$. This completes the proof of Theorem 3.3.

3.12. *A generalization to boundaries in $\mathbf{P}^n - \mathbf{P}^{n-q}$.* In this section we replace $\mathbf{C}^n = \mathbf{P}^n - \mathbf{P}^{n-1}$ by a somewhat larger subset of $\mathbf{P}^n$, namely $\mathbf{P}^n - \mathbf{P}^{n-q}$ (with the codimension q arbitrary) and obtain a result generalizing Theorem 3.3; this result is closely related to the results of Rossi [**42**].

THEOREM 3.29. *Suppose M is a compact, oriented submanifold of real dimension* $2p - 1$ *and of class C^1 in a complex submanifold X of $\mathbf{P}^n - \mathbf{P}^{n-q}$ (or more generally, M is allowed to have a scar set S exactly as in Theorem* 3.3). *If M is maximally complex and p is strictly larger than the codimension q then the entire conclusion of Theorem* 3.3 *remains valid for $M \subset X$.*

The basic idea of the proof is to pull M back to $\tilde{M} \subset \mathbf{C}^{n+1} - \mathbf{C}^{n-q+1}$ using homogeneous coordinates. Now, even though $\tilde{M}$ is a *noncompact* cone, the problem of filling in M with a complex subvariety can be solved. See Harvey and Lawson [**24**a, Part II] for further discussion and the proof.

4. Some applications. In this section we present several applications of the characterization of holomorphic chains given in §2. The applications in §§4.1, 4.2, and 4.5 use the simpler characterization of positive holomorphic chains, while §§4.3, 4.4, and 4.6 definitely require the (nonpositive) general case. In fact, the work of Lawson and Simons [**30**] on stable currents in $\boldsymbol{P}^n$ provided motivation for the work of Harvey and Shiffman [**21**]. In the last section the main result of §3 is used to deduce a generalization of the CR extension theorem of Bochner. Except for the closely related §§4.5 and 4.6 the parts of this section are independent of one another, and the interested reader may just as well turn to the part of his choice.

4.1. *The Remmert-Stein-Shiffman theorem.* The result presented here is due to Shiffman [**46**]. The important special case, where the exceptional set A is a complex subvariety (of dimension $\leqq p-1$), is due to Remmert and Stein [**40**]. Deducing the theorem from the structure theorem is presented in King [**26**].

THEOREM 4.1. *Suppose A is a closed subset of a complex manifold X and $\mathscr{H}_{2p-1}(A)=0$. If Y is a pure p-dimensional subvariety of $X-A$ then $\bar{Y}$ is a pure p-dimensional subvariety of X.*

PROOF. First one must show that Y has locally finite volume across A. Since this is done in much the same way as Lemma 1.3 or Lemma 1.5 (utilizing the Shiffman Lemma 2.3) the proof is omitted. Consequently, $T=[Y]$ defines a locally rectifiable current of bidimension p, p on all of X, which is a holomorphic p-chain on $X-A$. Therefore dT is supported in A. However, T and hence dT are locally flat so that, by the support Theorem 1.7(a), $dT=0$. Consequently, the structure Theorem 2.1 (Remark 1) is applicable and T is a holomorphic p-chain on X. In particular, supp $T=\bar{Y}$ is a pure p-dimensional subvariety of X.

REMARK 1. Theorem 4.1 can be generalized as follows. Suppose X and the exceptional set A are exactly as in the above theorem. If T is a weakly positive d-closed current of bidimension p, p defined on $X-A$ then there exists a weakly positive d-closed current $\bar{T}$ of bidimension p, p defined on all of X with $\bar{T}$ equal to T on $X-A$. See Harvey [**15**]. The difficulty here is that supp T may be large, in which case there are no projections as in Lemma 2.3 which are proper on supp T.

4.2. *A compactness theorem.* The structure Theorem 2.1 enables us to apply a compactness theorem of Federer to holomorphic p-chains (see Harvey and Shiffman [**21**]).

THEOREM 4.2. *Suppose $\{T_j\}$ is a sequence of holomorphic p-chains, on an open subset Ω of $\boldsymbol{C}^n$, with uniformly bounded mass on each compact subset of Ω. Then there exists a subsequence of $\{T_j\}$ that converges in the locally flat topology to a holomorphic p-chain T on Ω.*

PROOF. First note that we can assume without loss of generality that each T_j is positive. Otherwise, let $T_j=T_j^+ + T_j^-$, where T_j^+ is the positive part of T_j and T_j^- is the negative part, and consider the sequences $\{T_j^+\}$ and $\{T_j^-\}$. Each sequence has locally finite volume since $M_K(T_j)=M_K(T_j^+)+M_K(T_j^-)$ for all compact subsets K of Ω.

By Federer's compactness Theorem 4.2.17(2) in [**10**] there exists a subsequence of $\{T_j\}$ which converges in the locally flat topology to a locally rectifiable current T. The properties of being d-closed and positive of bidimension p, p are preserved under weak limits on test forms and hence T is a positive current satisfying all the hypotheses of Theorem 2.1 (Remark 1). Therefore T is a holomorphic p-chain on Ω.

Next we wish to deduce a corollary. If A_n ($n = 1, 2, \cdots$), and A are closed subsets of Ω we say that A_n converges to A if the following two conditions are satisfied:

(4.1) For each point $X \in A$, there exist $X_n \in A_n$ ($n = 1, \cdots$) such that $X_n \to X$.

(4.2) For each compact subset K of $\Omega - A$, $A_n \cap K = \emptyset$ for n large.

This notion of convergence can also be described in terms of the Hausdorff metric (Stolzenberg [**51**, pp. 22–23]).

Part (b) of the next result is due to Bishop [**3**] (see also [**51**]).

COROLLARY 4.3. (a) *Suppose $\{T_j\}$ is a sequence of holomophic p-chains with uniformly bounded mass on each compact subset of Ω. If $\{T_j\}$ converges weakly (as currents) to a current T on Ω then T is a holomorphic p-chain.*

(b) *Suppose $\{V_j\}$ is a sequence of pure p-dimensional analytic subvarieties of Ω with uniformly bounded volume on each compact subset of Ω. If $\{V_j\}$ converges to a closed subset A of Ω then A is a pure p-dimensional subvariety.*

PROOF. (a) By the above theorem $\{T_j\}$ has a subsequence which converges to a holomorphic p-chain in the locally flat topology and hence weakly as currents.

(b) By Theorem 4.2, we may assume, after passing to a subsequence, that $\{[V_j]\}$ converges to a holomorpihic p-chain T on Ω.

To complete the proof we shall show that supp $T = A$.

The condition (4.2) implies that supp $T \subset A$. Suppose $z \in A$; then by the condition (4.1) we can choose $z_j \in V_j$ with $z_j \to z$. The monotonicity theorem (Corollary 1.28) implies that $n([V_j], z_j, r) \geqq 1$ if $B(z_j, r) \subset\subset \Omega$. Since $[V_j]$ converges to T weakly, one easily concludes that $n(T, z, r) \geqq 1$ for $B(z, r) \subset\subset \Omega$. Therefore $z \in \text{supp } T$.

4.3. *Bishop's extension theorem.* In this application we prove a result due to Bishop [**3**] which requires the general (nonpositive) case of Theorem 2.1. However, we must use the first half of Bishop's original proof in order to verify the support hypothesis. This is unsatisfactory, and illustrates one of the advantages that would be gained if the support hypothesis could be removed from Theorem 2.1.

THEOREM 4.4. *Suppose A is a complex subvariety of a complex manifold X and that V is a pure p-dimensional subvariety of $X - A$. If V has locally finite volume (across A) in Ω then $\bar{V}$ is a pure p-dimensional subvariety of X.*

REMARK. Of course, for the volume to be defined we need to impose a hermitian metric on X. However, one can easily check that the property of having locally finite volume across A is independent of the hermitian metric on X.

PROOF. There are several obvious reductions. Suppose that the theorem is true when A is nonsingular. To prove the theorem in general, first extend V across the regular points of A and then extend across the singular points of A by induction on the dimension of A. Thus we may assume that A is nonsingular. Since the conclusion is local, by making a biholomorphic coordinate change we may assume that A is a complex linear subspace of $\boldsymbol{C}^n$ with $0 \in \bar{V} \cap A$; and it suffices to prove that $\bar{V}$ is a subvariety in a neighborhood Ω of 0. By enlarging A if necessary, we may further assume that A has dimension $l \geqq p$. After making a complex linear change of coordinates, we may assume that A is defined by $z_{l+1} = \cdots = z_m = 0$, and that $\{z \in V: z_1 = \cdots = z_p = 0\}$ is zero-dimensional (see Narasimhan [**19**, p. 124]). Therefore we may choose a product neighborhood $\Delta = \Delta' \times \Delta''$ about the origin (with $\Delta' \subset \boldsymbol{C}^l, \Delta'' \subset \boldsymbol{C}^{n-l}$) such that $(\Delta' \times \partial\Delta'') \cap V = \emptyset$. Then the orthogonal projection $\pi: \Delta \to \Delta'$ onto $\boldsymbol{C}^l$ when restricted to $\bar{V} \cap \Delta$ is proper.

Now that we have simplified the geometry (except for the support hypothesis) the theorem is an easy corollary of the structure theorem.

Since $V \cap \Delta$ has finite volume, $T = [V \cap \Delta]$ defines a locally rectifiable current of bidimension p, p on Δ. Since $\pi: \Delta \cap (\text{supp } T) \to \Delta'$ is proper, $\pi_*(T)$ is also a well-defined locally rectifiable current of bidimension p, p. Therefore, $T - \pi_*(T)$ has the same properties. Moreover, since dT is locally flat and supported in $\Delta' \times \{0\}$ (with π equal to the identity on $\Delta' \times \{0\}$), a theorem of Federer [**10**, 4.1.15] says that $\pi_*(dT) = dT$. Consequently $T - \pi_*(T)$ is d-closed.

In order to apply the structure Theorem 2.1 to $T - \pi_*(T)$ and conclude that $T - \pi_*(T)$ is a holomorphic p-chain (and hence supp$(T - \pi_*(T))$ is a pure p-dimensional subvariety which extends V) we must verify the support hypothesis (of course $\mathscr{H}_{2p+1}(V \cap (\Delta - A)) = 0$). However, Bishop [**3**] has shown that the hypothesis of locally finite volume for V across A implies that $\mathscr{H}_{2p}(\bar{V} \cap A) = 0$, which completes the proof.

Remark 1. Notice that in the special case, where A and V have the same dimension p, the support hypothesis $\mathscr{H}_{2p+1}(\bar{V}) = 0$ is automatic. This special case is due to Stoll [**49**].

Remark 2. In the special case where A and V have the same dimension p, Theorem 4.4 has been generalized (Harvey and Polking [**24**]), by replacing $T = [V]$ on $X - A$ by any weakly positive (of dimension p, p) d-closed current T on $X - A$ which has locally finite mass across A and then concluding that T has an extension across A which is also weakly positive (of dimension p, p) and d-closed on X (cf. Siu [**47**]).

4.4. *Uniqueness in the Plateau problem.* In this application we use the (general) nonpositive version of the structure Theorem 2.1 to examine the question of uniqueness in the Plateau problem for special boundaries. The Plateau problem (we consider) is as follows: Given a rectifiable current B (the prescribed boundary) of dimension $r - 1$ on $\boldsymbol{R}^m$, find a rectifiable current T of dimension r on $\boldsymbol{R}^m$ with $dT = B$ which has minimal mass (i.e., $M(T) \leqq M(S)$) for all rectifiable currents S satisfying $dS = B$. Solutions are not always unique. For example, let I denote the positively oriented unit interval [0, 1] in $\boldsymbol{R}$ and consider the boundary $B \equiv d[I] \times d[I]$. Both $T = [I] \times d[I]$ and $T = d[I] \times [I]$ are solutions.

The portion of the structure Theorem 2.1 that we shall use can be stated as a lemma.

Lemma 4.5. *Suppose S is a locally rectifiable, d-closed current of bidimension p, p defined on $\boldsymbol{C}^n$. If $p > 0$ and S has compact support then $S = 0$.*

Proof. By Theorem 2.1 (Remark 2), S is a holomorphic p-chain on $\boldsymbol{C}^n$ and, in particular, supp S is a pure p-dimensional subvariety. However, all compact subvarieties of $\boldsymbol{C}^n$ are of dimension zero, so that $S = 0$.

REMARK. The alternate proof of uniqueness in Theorem 3.3 which was mentioned earlier is now immediate. Because, if T and S are two holomorphic p-chains on $\mathbf{C}^n - M$ satisfying Theorem 3.3 (in particular $dT = dS = [M]$), then $T - S$ is compactly supported, d-closed, locally rectifiable, and of bidimension p, p. Therefore $T = S$ by Lemma 4.5.

Recall that a locally rectifiable current with compact support is the same thing as a rectifiable current.

THEOREM 4.6 (UNIQUENESS). *Suppose B is a rectifiable current of dimension $2p - 1$ on $\mathbf{C}^n$. Moreover, assume that T is a rectifiable current on $\mathbf{C}^n$ which is positive of bidimension p, p and has boundary B (so that, by Corollary 1.24, T is a solution to the Plateau problem with boundary B). If S is any other rectifiable current, satisfying $dS = B$ and of minimal mass, then $S = T$. In particular, if S is positive (of bidimension p, p) and $dS = B$ then $S = T$.*

PROOF. By (Wirtinger's inequality) Theorem 1.22,

$$M(T) = T(\omega^p/p!) = S(\omega^p/p!) \leqq M(S),$$

since $d(T - S) = 0$. If S is of minimal mass then by the equality portion of Wirtinger's inequality S must be of bidimension p, p. Therefore $T - S$ is a rectifiable d-closed current of bidimension p, p on $\mathbf{C}^n$ (with $p > 1$), and hence, by Lemma 4.5, $T - S = 0$.

The last part follows since S positive implies that S is of minimal mass.

REMARK. Lemma 4.5 and Theorem 4.6 immediately generalize to the case of an arbitrary Kähler manifold (instead of $\mathbf{C}^n$) which has no compact pure p-dimensional subvarieties. For example, $\mathbf{P}^n - Y$, where Y is a subvariety of pure codimension q, can have no subvarieties of dimension $\geqq q$. Therefore, the proof of uniqueness for Theorem 3.3 (given in the remark above after Lemma 4.5) generalizes to the case where the boundary M is contained in $\mathbf{P}^n - Y$, if M is of real dimension $2p - 1$ with $p \geqq$ codimension Y.

4.5. *Holomorphic chains on compact Kähler manifolds*. The purpose of this section is to show how the structure Theorem 2.1 and Wirtinger's inequality can be used to reformulate the Hodge conjecture in terms of a (homology) Plateau problem. These results are basically taken from Harvey and Knapp [**20**], but incorporate the discussion in Lawson [**28**].

First, we survey some real variables results and formulate the (homology) Plateau problem (see Federer [**10**]). Suppose X is a compact Riemannian manifold. Let $N_p(X)$ denote the space of normal currents on X of dimension p. Let $\hat{R}_p(X)$ denote the space of p-dimensional currents T on X such that both T and dT are rectifiable. Then Federer and Fleming [**12**] have shown that:

$$H_p(X, \mathbf{R}) \cong H_p(N_*(X)), \tag{4.1}$$

$$H_p(X, \mathbf{Z}) \cong H_p(\hat{R}_*(X)). \tag{4.2}$$

(It will be convenient to consider these isomorphisms as equality throughout this section.) If $\alpha \in H_p(X, \mathbf{Z}) (= H_p(\hat{R}_*(X)))$ then we will let:

(4.3) $\|\alpha\|_{\mathbf{Z}}$ denote the infimum of $M(T)$ over all $T \in \alpha$.

Similarly, if $\beta \in H_p(X, \mathbf{R})$ $(=H_p(N_*(X)))$ then we will let:

$$\|\beta\|_{\mathbf{R}} \text{ denote the infimum of } M(T) \text{ over all } T \in \beta. \tag{4.4}$$

Given a class $\alpha \in H_p(X, \mathbf{Z})$ the (*homology*) *Plateau problem* (*over* $\mathbf{Z}$) is:

$$\text{find a current } T \in \alpha \text{ with } \|\alpha\|_{\mathbf{Z}} = M(T). \tag{4.5}$$

Similarly, given a class $\beta \in H_p(X, \mathbf{R})$ the (*homology*) *Plateau problem* (*over* $\mathbf{R}$) is:

$$\text{find a current } T \in \beta \text{ with } \|\beta\|_{\mathbf{R}} = M(T). \tag{4.6}$$

It follows from Federer [**10**] that (4.5) and (4.6) always have a solution.

Let $i\colon H_p(X, \mathbf{Z}) \to H_p(X, \mathbf{R})$ denote the natural inclusion. Then, in terms of the identifications (4.1) and (4.2) above, $\alpha \subset i(\alpha)$ so that

$$\|i(\alpha)\|_{\mathbf{R}} \leqq \|\alpha\|_{\mathbf{Z}} \quad \text{for } \alpha \in H_p(X, \mathbf{Z}). \tag{4.7}$$

Since, obviously $\|m\beta\|_{\mathbf{R}} = m\|\beta\|_{\mathbf{R}}$ and $\|m\alpha\|_{\mathbf{Z}} \leqq m\|\alpha\|_{\mathbf{Z}}$ for each $m \in \mathbf{Z}^+$, the inequality (4.7) implies that

$$\|i(\alpha)\|_{\mathbf{R}} \leqq (1/m)\,\|m\alpha\|_{\mathbf{Z}} \leqq \|\alpha\|_{\mathbf{Z}} \quad \text{for } \alpha \in H_p(X, \mathbf{Z}). \tag{4.8}$$

These inequalities can be strict for all m; however, Federer has shown that

$$\|i(\alpha)\|_{\mathbf{R}} = \lim(1/m)\,\|m\alpha\|_{\mathbf{Z}} \quad \text{for each } \alpha \in H_p(X, \mathbf{Z}) \tag{4.9}$$

(see Federer [**11**]). Following Lawson [**28**], we will say that a class $\alpha \in H_p(X, \mathbf{Z})$ is *stable* if $\|i(m\alpha)\|_{\mathbf{R}} = \|m\alpha\|_{\mathbf{Z}}$ for some $m \in \mathbf{Z}^+$. The question of when $\|i(\alpha)\|_{\mathbf{R}} = \|\alpha\|_{\mathbf{Z}}$ seems very important.

Throughout the remainder of this section we assume that X is a compact Kähler manifold with Kähler form ω. Using Wirtinger's inequality we have the analogue of Corollary 1.24 on Kähler manifolds.

THEOREM 4.7. *Suppose the class $\beta \in H_{2p}(X, \mathbf{R})$ contains a strongly positive current T of bidimension p, p. Then*

(1) (*existence*) $\|\beta\|_{\mathbf{R}} = M(T)$.

(2) (*uniqueness*) *If $S \in \beta$ is any other mass minimum* (*that is,* $\|\beta\|_{\mathbf{R}} = M(S)$) *then S is also a strongly positive current of bidimension p, p.*

PROOF. Suppose $S \in \beta$. Then $S - T = dR$ so that $S(\omega^p/p!) = T(\omega^p/p!)$. Therefore,

$$M(T) = T(\omega^p/p!) = S(\omega^p/p!) \leqq M(S) \tag{4.10}$$

where the last inequality is a consequence of Wirtinger's inequality (Theorem 1.22). This proves (1). To prove (2) note that equality in (4.10) implies that S is strongly positive, again by Theorem 1.22.

Using the structure Theorem 2.1 and Federer's solution to the (homology) Plateau problem (over $\mathbf{Z}$) we can now prove the following result.

LEMMA 4.8. *Suppose that $\alpha \in H_{2p}(X, \mathbf{Z})$ is an integral class and that the real class $\beta = i(\alpha)$ contains a strongly positive current T of bidimension p, p. If $\|\alpha\|_{\mathbf{Z}} = \|i(\alpha)\|_{\mathbf{R}}$ then α contains a positive holomorphic p-chain $T = \sum n_j[V_j]$.*

PROOF. Suppose $T \in \alpha$ is a solution to (4.5) (i.e., a rectifiable current $T \in \alpha$ of least mass). By part (b) of Theorem 4.7, T is positive (of bidimension p, p). Consequently by the structure Theorem 2.1 (Remark 1) T must be a positive holomorphic p-chain.

These observations led to the following conjecture (Harvey and Knapp [**20**]). A (k, k)-form $\zeta \in \bigwedge^{k,k}\boldsymbol{C}^n$ is said to be (strongly) positive *definite* if ζ belongs to the interior of the cone $\mathrm{SP}^{p,p}$ (i.e., small perturbations of ζ remain strongly positive). A form ϕ on X is said to be (strongly) positive *definite* if $\varphi(z)$ is (strongly) positive definite at each point $z \in X$.

If $p + k = n$, the dimension of X, then a current of bidimension $2p$ can be considered a current of bidegree $2k$ via the isomorphism $\bigwedge_{p,p}\boldsymbol{C}^n = \bigwedge^{k,k}\boldsymbol{C}^n$. Therefore,

$$H_{2p}(X, \boldsymbol{R}) = H^{2k}(X, \boldsymbol{R}) \quad \text{for } p + k = n. \tag{4.11}$$

We implicitly assume this identification in the following.

CONJECTURE I. *Suppose X is a projective algebraic manifold. If a class $\alpha \in H_{2p}(X, \boldsymbol{Z})$ has the property that $i(\alpha)$ contains a strongly positive definite form ϕ then there exists an integer $m \in \boldsymbol{Z}^+$ such that $\|i(m\alpha)\|_{\boldsymbol{R}} = \|m\alpha\|_{\boldsymbol{Z}}$ (i.e., α is stable).*

A current $T = \sum r_j[V_j]$ with each $r_j \in \boldsymbol{Q}$ rational and $\{V_j\}$ the components of a pure p-dimensional subvariety will be referred to as a *rational holomorphic p-chain*.

CONJECTURE I′. *Suppose X is a projective algebraic manifold. If a class $\alpha \in H_{2p}(X, \boldsymbol{Z})$ has the property that $i(\alpha)$ contains a strongly positive definite form ϕ then $i(\alpha)$ contains a positive rational holomorphic p-chain $T = \sum r_j[V_j]$.*

REMARK 1. Conjectures I and I′ are equivalent. Suppose $\|i(m\alpha)\|_{\boldsymbol{R}} = \|m\alpha\|_{\boldsymbol{Z}}$. Then, by Lemma 4.8, $m\alpha$ contains a positive holomorphic p-chain S. Therefore $i(\alpha)$ contains the positive rational holomorphic p-chain $T = (1/m)\, S$. Conversely, suppose $i(\alpha)$ contains a positive rational holomorphic p-chain T. Then T equals $(1/m)S$ for some $m \in \boldsymbol{Z}^+$ and some positive holomorphic p-chain S. Since $S \in m\alpha \subset i(m\alpha)$ is positive, Theorem 4.7(1) implies that $\|i(m\alpha)\|_{\boldsymbol{R}} = \|m\alpha\|_{\boldsymbol{Z}}$.

REMARK 2. Under the same hypotheses as in Conjectures I and I′ one might conjecture that $\|\alpha\|_{\boldsymbol{Z}} = \|i(\alpha)\|_{\boldsymbol{R}}$. However, an example provided by David Mumford shows this to be false even in the simplest case $p = 1$ (see Harvey and Knapp [**20**]).

REMARK 3. The assumption that ϕ is definite cannot be dropped from the above conjectures, as shown by the example of Lawson [**28**, Theorem 6.7] (X is a torus of complex dimension $2p$ with $p \geqq 2$).

Recall the Hodge decomposition

$$H^r(X, \boldsymbol{C}) = \sum_{p+q=r} H^{p,q}(X), \tag{4.12}$$

where $H^{p,q}(X)$ denotes the space of harmonic (p, q)-forms on X.

HODGE CONJECTURE. *Suppose X is a projective algebraic manifold. If a class $\alpha \in H_{2p}(X, \boldsymbol{Z})$ has the property that the harmonic representative ϕ of $i(\alpha)$ is of bidegree k, k $(p+k = n)$ then $i(\alpha)$ contains a rational holomorphic p-chain $T = \sum r_j[V_j]$.*

Conjecture I-I′ implies the Hodge conjecture. To see this suppose that X is embedded in $P^N(C)$ for some N, and that ω is the standard Kähler form on $P^N(C)$ restricted to X. Let $h_p \in H_{2p}(X, Z)$ denote the homology class containing a p-dimensional linear section $L_p \subset X$ of X in $P^N(C)$. (That is, $L_p = X \cap P^r(C)$ for some appropriate r.) Then ω^k is the harmonic representative of $i(h_p)$ where $p+k=n$, the dimension of X, and $[L_p] \in h_p$. If ϕ is a representative of a class $i(\alpha)$ of bidegree k, k with $\alpha \in H_{2p}(X, Z)$ $(p+k=n)$ then for some $q \in Z^+$ sufficiently large $\phi + q\omega^k$ is strongly positive definite, because ω^k belongs to the interior of the cone of strongly positive (k, k)-forms (see Corollary 1.16). Consequently, if Conjecture I-I′ were true for $\phi + q\omega^k$ then $\alpha + qh_p$ would contain a positive rational holomorphic p-chain T. In this case $i(\alpha)$ would contain the rational holomorphic p-chain $T - q[L_p]$.

REMARK 4. This proves that Conjecture I-I′ implies the Hodge conjecture. In fact, this proves that the following weaker Conjecture II implies the Hodge conjecture.

CONJECTURE II. *Suppose X is a projective algebraic manifold with $h_p \in H_{2p}(X, Z)$ as above. For each class $\alpha \in H_{2p}(X, Z)$ such that the harmonic representative ϕ of $i(\alpha)$ is of bidegree k, k $(p+k=n)$ Conjecture* I-I′ *is valid for the class $\alpha + qh_p$ for q sufficiently large.*

REMARK 5. Conjecture II is actually equivalent to the Hodge conjecture. To prove this suppose that the Hodge conjecture is true and that α satisfies the hypotheses of II. Then there exists a rational holomorphic p-chain T contained in $i(\alpha)$. Replacing α by a high integer multiple of α we may assume that $T = \sum n_j[V_j]$ is a holomorphic p-chain. Let $-[W]$ denote a typical negative term in T with W irreducible.

There exist k hypersurfaces $H_1, \cdots, H_k$ such that $\bar{W} = \bigcap_1^k H_j \cap X$ is of pure dimension p and contains W. (This can be established locally at a point $0 \in W \subset X$ with 0 the origin in $C^N \subset P^N$ by simultaneously expressing W and X as analytic covers over C^p and C^n, respectively, with $C^p \subset C^n$, and then utilizing the $n-p$ polynomials in the variables $z_{p+1}, \cdots, z_n$ with coefficients depending on $z_1, \cdots, z_p$ which vanish on W near 0, to define the hypersurfaces $H_1, \cdots, H_k$. The $H_1, \cdots, H_k$ extend to P^N with the desired properties.)

Consequently $[\bar{W}] = [W] + [U]$ and, since $\bar{W}$ is a complete intersection, $[\bar{W}] \sim r[L_p]$ for some integer r, where L_p denotes a linear section of X. Therefore $-[W] \sim [U] - r[L_p]$.

Let J denote the set of indices j with $n_j < 0$. Applying this argument to each component V_j with $j \in J$, we obtain that, for some integer q,

$$T + \sum_{j \in J} |n_j| \, [U_j] - q[L_p] \in \alpha. \tag{4.13}$$

Therefore $\alpha + qh_p$ contains the positive holomorphic p-chain $T + \sum_{j \in J} |n_j| \, [U_j]$, which completes the proof that the Hodge conjecture implies Conjecture II.

REMARK 6. The Conjecture I-I′ is true for the hypersurface case $p = n-1$. (In particular, since Conjecture I-I′ implies the Hodge conjecture this proves, as is well known, the Hodge conjecture for $p = n-1$.) The case $p = n-1$ is deduced as follows. Suppose that $\alpha \in H_{2n-2}(X, Z)$ has the property that $i(\alpha) \in H^2(X, R)$ contains a strongly positive definite form ϕ of bidegree 1, 1. A short argument shows that there exists a line bundle E on X with first Chern form ϕ. By the Kodaira embedding theorem there exists an integer m such that the natural mapping $f\colon X \to$

$P(H^0(X, (L^{-1})^m))$ is an embedding in projective space. Since $m\phi$ and $f^*([H])$ both belong to $i(m\alpha)$, where H is a hyperplane section of $f(X)$, this proves that $i(m\alpha)$ contains the positive holomorphic $(n-1)$-chain $f^*([H])$.

4.6. *Stable and stationary currents on $P^n(C)$.* In this subsection we indicate how the structure Theorem 2.1 can be used to characterize the stationary and stable currents on complex projective space. Parts (a) and (b) of Theorem 4.9 below are due to King and Lawson-Simons [**30**], respectively.

First, we give the definitions of stationary and stable. Let X denote a Riemannian manifold. Let V be a compactly supported vector field on X, and consider the flow $\{h_t\}$ corresponding to V. This flow $\{h_t\}$ is a 1-parameter local group of diffeomorphisms on X (defined for $|t| < \varepsilon$) with $h_t(x) = x$ for $x \notin \operatorname{supp} V$. For each locally rectifiable current T of dimension p on X let

(4.14) $$(J_{T,V}(t) = M((h_t)_*(T)$$

the mass at time t. The first derivative of this function of t evaluated at $t = 0$ is called the first variation of T with respect to V (see [**10**, 5.1.7, 5.4.1] and [**30**]) and is denoted $\delta^{(1)}(T, V)$. Similarly, the second derivative evaluated at $t = 0$ is called the second variation of T with respect to V and denoted $\delta^{(2)}(T, V)$. A locally rectifiable current T on X is said to be

(4.15) *stationary* if $\delta^{(1)}(T, V) = 0$ for all compactly supported vector fields V on X, or

(4.16) *stable* if $J_{T,V}(t)$ has a local minimum at $t = 0$, for all compactly supported vector fields V or X.

Of course, if T is stable then T is stationary.

Federer [**10**, 5.4.19] proved part (a) of the following proposition.

PROPOSITION 4.9. *Suppose X is a compact Kähler manifold.*

(a) *If T is a positive holomorphic p-chain on X then T is of minimal mass in its homology class $\beta \in H_{2p}(X, \mathbf{R})$ (i.e., $\|\beta\|_{\mathbf{R}} = M(T)$).*

(b) *If T is a holomorphic p-chain on X then T is stable.*

PROOF. Part (a) is just a special case of Theorem 4.7(a). To prove part (b), first note that T can be divided into a positive part T^+, and a negative part $-T^-$, such that $T = T^+ - T^-$ and $A = (\operatorname{supp} T^+) \cap (\operatorname{supp} T^-)$ has $\mathscr{H}_{2p}$ measure zero (and hence $\|T^+\|$ and $\|T^-\|$ measure zero). Now suppose $\{h_t\}$ is a 1-parameter family of diffeomorphisms on X. Since A has $\|T^+\|$ and $\|T^-\|$ measure zero, $M((h_t)_*(T)) = M((h_t)_*(T^+)) + M((h_t)_*(T^-))$. By part (a) and [**10**, 5.1.7], $M(T^\pm) \leqq M((h_t)_*(T^\pm))$. Therefore,

$$\begin{aligned} M(T) &= M(T^+) + M(T^-) \\ &\leqq M((h_t)_*(T^+)) + M((h_t)_*(T^-)) = M((h_t)_*(T)), \end{aligned}$$

and hence $J_{T,V}(t)$ has a minimum at $t = 0$.

The next result is a converse to Proposition 4.9 where $X = P^n(C)$ is taken to be projective space.

THEOREM 4.10. *Suppose T is a rectifiable current on $P^n(C)$ of dimension $2p$ and $dT = 0$.*

(a) *T is of minimal mass with respect to its homology class $\alpha \in H_{2p}(\boldsymbol{P}^n, \boldsymbol{Z})$ if and only if either T or $-T$ is a positive holomorphic p-chain.*

(b) *T is stable if and only if T is a holomorphic p-chain.*

PROOF. (a) The homology class $\alpha \in H_{2p}(\boldsymbol{P}^n, \boldsymbol{Z})$must contain $m[\boldsymbol{P}^p]$ where $\boldsymbol{P}^p$ is a linear subspace and $m \in \boldsymbol{Z}$. We may assume $m > 0$. Then we have

$$M(m[\boldsymbol{P}^p]) = m = \|\alpha\|_{\boldsymbol{Z}} = \|i(\alpha)\|_{\boldsymbol{R}},$$

and hence $M(T) = \|i(\alpha)\|_{\boldsymbol{R}}$ so that by Theorems 4.7 and 2.1 T is a positive holomorphic p-chain.

(b) Suppose T is stable. Then $\delta^{(2)}(T, V) \geqq 0$ for each vector field V. Lawson and Simons conclude that the tangent $2p$-planes $\mathbf{T}(z)$ are of bidimension p, p for $\|T\|$-a.e. points z. Therefore by the structure Theorem 2.1 (Remark 5), T must be a holomorphic p-chain on $\boldsymbol{P}^n(\boldsymbol{C})$.

REMARK. Note that for part (a) we used the positive case of the structure Theorem 2.1, while for part (b) the general (nonpositive) case was necessary.

4.7. *A generalization of Bochner's extension theorem.* In this subsection we generalize the Bochner extension Theorem 3.17, for open subsets $\Omega \subset\subset \boldsymbol{C}^n$ with smooth boundary $\partial\Omega$, to submanifolds-with-boundary $(\Omega\ ,\ \partial\Omega)$ in $\boldsymbol{C}^n$. There are many possible variations as the proof will indicate. However, we present, perhaps, the simplest statement (see Harvey and Lawson [**23**, §12]). Recall the various equivalent notions of a CR function given in Lemma 3.6.

THEOREM 4.11. *Suppose $(\Omega, \partial\Omega)$ is a compact C^1 submanifold-with-boundary in $\boldsymbol{C}^n$, where Ω is a complex p-dimensional submanifold of $\boldsymbol{C}^n - \partial\Omega$. Assume $p > 1$ and that $\partial\Omega$ is connected. For each CR function $f \in C^1(\partial\Omega)$ there exists a function $F \in C^1(\bar{\Omega}) \cap \mathcal{O}(\Omega)$ with $F|_{\partial\Omega} = f$.*

REMARK. If Ω is allowed to have a finite number of isolated singularities then the theorem is still valid if the conclusion is modified so that the extension F is only claimed to be weakly holomorphic on Ω. The proof is the same.

PROOF. Let M denote the oriented graph of f over $\partial\Omega$ in $\boldsymbol{C}^{n+1}$. (The orientation of $\partial\Omega$ is induced by the orientation of Ω.) By Lemma 3.6, M is maximally complex, since f is a CR function. By Theorem 3.3, there exists a holomorphic chain $T = \pm[V]$ on $\boldsymbol{C}^{n+1} - M$, where V is an irreducible p-dimensional complex subvariety of $\boldsymbol{C}^{n-1} - M$, with $dT = [M]$. Let $\pi\colon \boldsymbol{C}^{n+1} \to \boldsymbol{C}^n$ denote orthogonal projection. Then, pushing forward $dT = [M]$ to $\boldsymbol{C}^n$ with π_* we obtain $d\pi_*(T) = [\partial\Omega]$. Since $S = [\Omega]$ is the *unique* holomorphic chain in $\boldsymbol{C}^n - \partial\Omega$ with $dS = [\partial\Omega]$ (by the uniqueness part of Theorem 3.3), $\pi_*(T)$ must agree with $[\Omega]$. Consequently, $T = [V]$ and V is (generically) single-sheeted over Ω. Since Ω is smooth, V must be single-sheeted over Ω, and hence we can define $F \in \mathcal{O}(\Omega)$ so that the graph of F over Ω equals V. Since M is the topological boundary of V, F has a continuous extension to $\bar{\Omega}$, which we also denote as F, and $F|_{\partial\Omega} = f$.

To complete the proof we must show that $F \in C^1(\bar{\Omega})$. The argument is local. Suppose $\zeta \in \partial\Omega$ and choose coordinates $z = (z_1, \cdots, z_n)$ for $\boldsymbol{C}^n$ so that $\boldsymbol{C}^p \times \{0\}$ is the tangent space to $\bar{\Omega}$ at ζ. Let $\rho : \boldsymbol{C}^n \to \boldsymbol{C}^p$ be orthogonal projection onto this tangent space ($\rho(z) = (z_1, \cdots, z_p)$). There exists a neighborhood B of ζ in $\boldsymbol{C}^n$ such that the submanifold $\rho(\partial\Omega)$ of $\rho(B)$ splits $\rho(B)$ into exactly two components U^+ and U^-. We can assume that, on $\rho(B)$, $\rho(\partial\Omega)$ is the oriented boundary of U^+. Also, there exists $G \in \mathcal{O}(U^+) \cap C^1(\bar{U}^+)$ such that $\Omega \cap B$ is the graph of G over U^+ and $\partial\Omega \cap B$ is the graph of $g \equiv G|_{\partial U^+}$ over $\partial U^+ \cap \rho(B)$. Now $\rho_* \circ \pi_*(z_{n+1}[V]) = \rho_*(F[\Omega]) = (F \circ G)[U^+]$ on $\rho(B)$, and hence

$$\bar{\partial}((F \circ G)[U^+]) = \rho_* \circ \pi_*(z_{n+1}\bar{\partial}[V]) = (f \circ g)[\rho(\partial\Omega)]^{0,1}$$

on $\rho(B)$. It follows from Theorem 3.18 (since $f \circ g \in C^1$) that $F \circ G \in C^1(\bar{U}^+ \cap \rho(B))$, and hence that $F \in C^1(B \cap \bar{\Omega})$.

Using Theorem 3.29, a Bochner type extention theorem for "meromorphic CR function" is obtained in [**24**a]. This extends Theorem 4.11 above.

References

1. H. Alexander, *A note on polynomial hulls*, Proc. Amer. Math. Soc. **33** (1972), 389–391. MR **45** #3757.

2. W. K. Allard, *On the first variation of a varifold*, Ann. of Math. (2) **95** (1972), 417–491. MR **46** #6136.

3. E. Bishop, *Conditions for the analyticity of certain sets*, Michigan Math. J. **11** (1964), 289–304. MR **29** #6057.

4. S. Bochner, *Analytic and meromorphic continuation by means of Green's formula*, Ann. of Math. (2) **44** (1943), 652–673. MR **5**, 116.

5. E. Bombieri, *Algebraic values of meromorphic maps*, Invent. Math. **10** (1970), 267–287. MR **46** #5328.

6. ———, *Addendum to algebraic values of meromorphic maps*, Invent. Math. **11** (1970), 163–166.

7. R. Draper, *Intersection theory in analytic geometry*, Math. Ann. **180** (1969), 175–204. MR **40** #403.

8. L. Ehrenpries, *A new proof and an extension of Hartog's theorem*, Bull. Amer. Math. Soc. **67** (1961), 507–509. MR **24** #A1511.

9. H. Federer, *Some theorems on integral currents*, Trans. Amer. Math. Soc. **117** (1965), 43–67. MR **29** #5984.

10. ———, *Geometric measure theory*. Die Grundlehren der math. Wissenschaften, Band 153, Springer-Verlag, New York, 1969. MR **41** #1976.

11. ———, *Real flat chains, cochains and variational problems* (to appear).

12. H. Federer and W. H. Fleming, *Normal and integral currents*, Ann. of Math. (2) **72** (1960), 458–520. MR **23** #A588.

13. G. Fichera, *Caratterizzazione della traccia, sulla frontiera di un campo, di una funzione analitica di piu variabili complesse*, Atti Accad. Naz. Liencei Rend. Cl. Sci. Fis. Mat. Nat. (8) **22** (1957), 706–715. MR **20** #121.

14. R. C. Gunning and H. Rossi, *Analytic functions of several complex variables*, Prentice-Hall, Englewood Cliffs, N. J., 1965. MR **31** #4927.

15. R. Harvey, *Removable singularities for positive currents*, Amer. J. Math. **96** (1974), 67–78.

16. ———, *Removable singularities of cohomology classes in several complex variables*, Amer. J. Math. **96** (1974), 498–504.

17. ——— *Three structure theorems in several complex variables*, Bull. Amer. Math. Soc. **80** (1974), 633–641. MR **50** #7574.

18. R. Harvey and J. King, *On the structure of positive currents*, Invent. Math. **15** (1972), 47–52. MR **45** #5409.

19. R. Harvey and B. Lawson, *Boundaries of complex analytic varieties*, Bull. Amer. Math. Soc. **80** (1974), 180–183.

20. Reese Harvey and A. W. Knapp, *Positive (p, p) forms, Wirtinger's inequality, and currents*, Value distribution theory, Part A (Proc. Tulane Univ. Program on Value-Distribution Theory in Complex Analysis and Related Topics in Differential Geometry, 1972–1973), pp. 43–62. Dekker, New York, 1974.

21. R. Harvey and B. Shiffman, *A characterization of holomorphic chains*, Ann. of Math. (2) **99** (1974), 553–587. MR **50** #7572.

22. R. Harvey and B. Lawson, *Extending minimal varieties*, Invent. Math. **28** (1975), 209–226.

23. ———, *On boundaries of complex analytic varieties*. I, Ann. of Math. **102** (1975), 233–290.

24. R. Harvey and J. Polking, *Extending analytic objects*, Comm. Pure Appl. Math. **28** (1975), 701–727.

24a. R. Harvey and B. Lawson, *On boundaries of complex analytic varieties*. II (to appear).

25. L. Hörmander, *An introduction to complex analysis in several variables*, Van Nostrand, Princeton, N. J., 1966. MR **34** #2933.

26. J. King, *The currents defined by analytic varieties*, Acta. Math. **127** (1971), 185–220.

27. J. R. King, *A residue formula for complex subvarieties*, Proc. Carolina Conf. on Holomorphic Mappings and Minimal Surfaces (Chapel Hill, N. C., 1970), Dept. of Math., Univ. of North Carolina, Chapel Hill, N. C., 1970, pp. 43–56. MR **42** #7942.

28. H. Blaine Lawson, Jr., *The stable homology of a flat torus*, Math. Scand. **36** (1975), 49–73.

29. ———, *Minimal varieties in real and complex geometry*, University of Montreal Press, 1973.

30. H. Blaine Lawson, Jr. and J. Simons, *On stable currents and their application to global problems in real and complex geometry*, Ann. of Math. (2) **98** (1973), 427–450. MR **48** #2881.

31. P. Lelong, *Intégration sur un ensemble analytique complexe*, Bull. Soc. Math. France **85** (1957), 239–262. MR **20** #2465.

32. ———, *Fonctions plurisousharmoniques et formes différentielles positives*, Gordon and Breach, New York; distributed by Dunod Éditeur, Paris, 1968. MR **39** #4436.

32a. ———, *Fonctions entiéres (n variables) et fonctions plurisousharmoniques d'ordre fini dans* C^n, J. Analyse Math. **12** (1964), 365–407. MR **29** #3668.

33. H. Lewy, *On the local character of the solutions of an atypical linear differential equation in three variables and a related theorem for regular functions of two complex variables*, Ann. of Math. (2) **64** (1956), 514–522. MR **18**, 473.

34. B. Malgrange, *Faisceaux sur des variétiés analytiques-réelles*, Bull. Soc. Math. France **85** (1957), 231–237. MR **20** #1340.

35. E. Martinelli, *Sopra una dimonstrazione di R. Fueter per un teorema di Hartogs*, Comment. Math. Helv. **15** (1943), 340–349. MR **6**, 61.

36. R. Narasimhan, *Introduction to the theory of analytic spaces*, Lecture Notes in Math., No. 25, Springer-Verlag, Berlin-New York, 1966. MR **36** #428.

37. R. Phelps, *Lectures on Choquet's theorem*, Van Nostrand, Princeton, N. J., 1966. MR **33** #1690.

38. J. Plemelj, *Ein Erganzungssatz zur Cauchyschen Integraldarstellung analytischer Funktionen, Randwerte betreffend*, Monatsh. Math. und Phys. **19** (1908), 205–210.

39. J. Polking and R. Wells, *Boundary values of Dolbeault cohomology classes and a generalized Bochner-Hartogs theorem*, Abh. Math. Sem. Univ. Hamburg (to appear).

40. R. Remmert and K. Stein, *Über die wesentilchen Singularitäten analytischer Mengen*, Math. Ann. **126** (1953), 263–306. MR **15**, 615.

41. H. Rossi, *Attaching analytic spaces to an analytic space along a pseudoconcave boundary*, Proc. Conf. Complex Analysis (Minneapolis, 1964), pp. 242–256. Springer, Berlin, 1965. MR **31** #381.

42. ———, *Continuation of subvarieties of projective varieties*, Amer. J. Math. **91** (1969), 565–575. MR **39** #5830.

43. ———, *A generalization of a theorem of Hans Lewy*, Proc. Amer. Math. Soc. **19** (1968), 436–440. MR **36** #5379.

44. W. Rothstein, *Ein neuer Beweis des Hartogsschen Hauptsatzes und seine Ausdehnung auf meromorphe Funktionen*, Math. Z. **53** (1950), 84–95. MR **12**, 252.

45. J.-P. Serre, *Un théorème de dualité*, Comment. Math. Helv. **29** (1955), 9–26. MR **16**, 736.

46. B. Shiffman, *On the removal of singularities of analytic sets*, Michigan Math. J. **15** (1968), 111–120. MR **37** #464.

47. Y.-T. Siu, *Analyticity of sets associated to Lelong numbers and the extension of closed positive currents*, Invent. Math. **27** (1974), 53–156. MR **50** #5003.

48. H. Skoda, *Sous-ensembles analytiques d'ordre fini ou infini dans* $\boldsymbol{C}^n$, Bull. Soc. Math. France **100** (1972), 353–408. MR **50** #5004.

49. W. Stoll, *Über die Fortsetzbarkeit analytischer Mengen endlichen Oberflächaninhaltes*, Arch. Math. **9** (1958), 167–175. MR **21** #729.

50. G. Stolzenberg, *Uniform approximation on smooth curves*, Acta Math. **115** (1966), 185–198. MR **33** #307.

51. ———, *Volumes, limits and extensions of analytic varieties*, Lecture Notes in Math., No. 19, Springer-Verlag, Berlin-New York, 1966. MR **34** #6156.

52. P. Thie, *The Lelong number of a point of a complex analytic set*, Math. Ann. **172** (1967), 269–312. MR **35** #5661.

53. V. S. Vladimirov, *Methods of the theory of functions of many complex variables*, Izdat. "Nauka", Moscow, 1964; English transl., M. I. T. Press, Cambridge, Mass., 1966. MR **30** #2163; **34** #1551.

54. B. Weinstock, *Continuous boundary values of analytic functions of several complex variables*, Proc. Amer. Math. Soc. **21** (1969), 463–466. MR **38** #6106.

55. ———, *An approximation theorem for* $\bar{\partial}$*-closed forms of type* $(n, n-1)$, Proc. Amer. Math. Soc. **26** (1970), 625–628. MR **42** #547.

56. J. Wermer, *The hull of a curve in* $\boldsymbol{C}^n$, Ann. of Math. (2) **68** (1958), 550–561. MR **20** #6536.

57. ———, *Banach algebras and several complex variables*, Markham Publishing Co., Chicago, Ill., 1971. MR **46** #672.

58. G. M. Henkin, *Solutions with estimates of the H. Lewy and Poincaré-Lelong equations. Construction of functions of the Nevanlinna class with prescribed zeros in strictly pseudoconvex domains*, Dokl. Akad. Nauk SSSR, **224**, no. 4, 1975, pp. 3–13. (Russian)

59. H. Skoda, *Valeurs au bord pour les solutions de l'operateur d'' et caracterisation des zeros des fonctions de la classe de Nevanlinna* (preprint).

RICE UNIVERSITY

AUTHOR INDEX

I or II indicates in which part of this volume the pages occur.

Roman numbers refer to pages on which a complete reference to a work by the author is given.

Boldface numbers indicate the first page of the articles in each part of this volume.